Guest Editor
Christian Liebsch
Institute of Orthopaedic Research and Biomechanics
Ulm University Medical Centre
Ulm
Germany

Editorial Office
MDPI AG
Grosspeteranlage 5
4052 Basel, Switzerland

This is a reprint of the Special Issue, published open access by the journal *Bioengineering* (ISSN 2306-5354), freely accessible at: https://www.mdpi.com/journal/bioengineering/special_issues/EY4RX9989T.

For citation purposes, cite each article independently as indicated on the article page online and as indicated below:

Lastname, A.A.; Lastname, B.B. Article Title. *Journal Name* **Year**, *Volume Number*, Page Range.

ISBN 978-3-7258-7729-4 (Hbk)
ISBN 978-3-7258-7730-0 (PDF)
https://doi.org/10.3390/books978-3-7258-7730-0

Contents

About the Editor

Christian Liebsch

Christian Liebsch is a postdoctoral researcher at the Institute of Orthopaedic Research and Biomechanics at the Ulm University Medical Centre in Germany and is qualified as a university lecturer in the field of experimental orthopaedics and traumatology. His main research interest comprises experimental studies on the biomechanics of the spine with a specific focus on the effects of degenerative changes, traumatic injuries, novel implants, and the rib cage. He has published more than 45 original works and book chapters, has held more than 50 presentations at national and international congresses, is an active member of the German and European Spine Societies as well as the German and European Societies of Biomechanics, has been awarded, for example, with the Eurospine Full Paper Award, and has served as a reviewer for more than 200 submissions to peer-reviewed journals.

bioengineering

MDPI

Editorial

Recent Findings and Developments in Spine Biomechanics

Christian Liebsch

Institute of Orthopaedic Research and Biomechanics, Ulm University Medical Centre, 89081 Ulm, Germany; christian.liebsch@uni-ulm.de

1. Introduction

As the central musculoskeletal element of the human body, the spine simultaneously enables trunk movement, upright posture, and load transfer from the upper to the lower body. Consequently, the spine must withstand a variety of forces and moments and exhibit unique material properties and kinematics. However, the fundamental importance of the spine for human biomechanics is also accompanied by multiple musculoskeletal spinal pathologies, with spine-related pain being one of the main causes of disability worldwide. A more detailed knowledge of spinal biomechanics is therefore essential regarding the prevention and treatment of musculoskeletal spinal diseases. To contribute to this, the Special Issue "Spine Biomechanics" aimed to collate new findings and developments in biomechanical research of the spine, comprising in vivo (clinical) trials, in vitro studies, and numerical modelling studies on the spine, the design and validation of novel research methodologies for spinal biomechanics, and investigations of novel technologies and devices for the orthopaedic and traumatological treatment of the spine, as well as studies on the effects of influencing factors on spinal biomechanics, such as ageing, degeneration, and trauma.

2. In Vivo (Clinical) Trials

In recent years, evolving technologies regarding in vivo measurements of spinal kinematics have resulted in an increasing number of studies investigating intervertebral motion in both healthy subjects and specific patient groups. These data are not only valuable for the understanding of normal spine function; they can also serve as input and validation data for numerical simulations and experimental models and may support clinical diagnostics as well as the assessment of surgical outcomes. In their scoping review, Breen et al. identified six categories of research questions being addressed by in vivo studies analysing lumbar spinal motion, comprising 'normal biomechanical mechanisms', 'direct kinematic measurement', 'dynamic radiography', 'pathological and injury mechanisms', 'spinal stabilisation', and 'clinical markers', thus proposing a novel taxonomy for in vivo measurements of spinal kinematics [1]. Concerning clinical markers, the prospective cross-sectional study of Hoehl et al. provided a unique dataset comprising both structural and functional changes in participants suffering from chronic or intermittent low back pain as well as participants without any back pain [2]. In this study, Hoehl et al. found that chronic and intermittent low back pain were both associated with lower rates of high physical activity as well as with more prevalent morphological and functional impairments compared to painlessness, respectively, indicating the requirement for a more accurate differentiation between specific types of low back pain and their potential causes and therapeutic consequences. Regarding acute low back pain, the retrospective clinical trial of Löchel et al. was furthermore able to show that patients with lumbar disc herniation at the

L4–L5 level exhibit significantly different spinopelvic anatomy in terms of the clinical parameters of pelvic incidence, pelvic tilt, and relative lumbar lordosis, as well as the absolute difference between pelvic incidence and lumbar lordosis compared to patients with lumbar disc herniation at the L5–S1 level [3]. The findings of this study indicate that spinopelvic anatomy represents a relevant factor for the segmental level where lumbar disc herniations occur, potentially affecting the biomechanical load distribution in the lower lumbar spine, which should be further investigated by future in silico studies. In a longitudinal pilot study, Rossi et al. furthermore found indications for relationships between asymmetry in the neurodynamic function of the brachial plexus, determined by the Upper Limb Neural Tension Test, and upper thoracic spine kinematics, quantified by a novel motion analysis technique [4]. In particular, the authors detected altered motion patterns in lateral bending and proposed the use of manual therapy to potentially address both the neurodynamic and biomechanical impairments—a suggestion which should be substantiated in any follow-up studies. Furthermore, in their systematic literature review, Yoseph et al. concluded that about half of all pregnant women are predisposed to low back pain due to increased lumbar lordosis, ligamentous laxity, altered gait mechanics, and muscular deconditioning [5]. Moreover, they determined associations between these pregnancy-related biomechanical changes and sacroiliac joint laxity, anterior pelvic tilt, and multiparity, as well as potential long-term risks of degenerative disc disease and spondylolisthesis from previous studies, highlighting the clinical relevance of spinal biomechanics in maternal spinal health.

3. In Vitro Studies

Experimental studies investigating spinal biomechanics are essential to directly address clinical questions related to the primary and long-term stability of spinal implants and to quantitatively assess the potential effects of pathologies and surgical approaches on spinal flexibility and kinematics. In vitro studies generally have the advantage of being able to evaluate novel treatment options under standardised and thus reproducible loading conditions, which are essential for the objective assessment of their clinical applicability. One common issue in this context is the testing of implants and implant combinations in cases of reduced bone quality. The study of Jacob et al. investigated whether a novel screw augmentation technique using two additional subcortical screws would be able to prevent cage subsidence when performing Transforaminal Lumbar Interbody Fusion [6]. Compared to the control group, who underwent surgery using the conventional cage placement technique, higher failure loads as well as a higher number of load cycles until reaching cage subsidence were found, indicating the advantages of the novel technique regarding long-term stability. However, different failure modes also implied potential complications related to the novel approach, requiring further investigations. Another clinical challenge in the context of reduced bone quality is the treatment of osteoporotic vertebral body fractures. In cases of severe wedge-shaped morphologies of the fractured vertebral body, causing kyphotic spinal deformity, the so-called kyphoplasty procedure is usually performed to straighten the affected vertebra. The findings of Riesenbeck et al. revealed that two-compartment kyphoplasty was not significantly superior in height reconstruction and load-bearing capacity compared to one-compartment kyphoplasty and showed that higher cement volume correlated with the occurrence of adjacent vertebral fractures [7]. Moreover, Riesenbeck et al. determined a significant reduction in post-traumatic segmental instability when applying kyphoplasty; however, this did not lead to complete restoration of native spinal stability [8]. Another clinically relevant problem in spine surgery, often associated with reduced bone quality, is pedicle screw loosening. In addition to these studies investigating the potential effects of bone quality and pedicle morphology on the anchorage capacity of pedicle screws, screw malpositioning is also widely discussed in

https://doi.org/10.3390/bioengineering13040482

the literature as a potential influencing factor for pedicle screw loosening. In their study, however, Schleifenbaum et al. were able to prove that minor deviations in screw placement had no major effect on the fixation strength of pedicle screws [9]. As the main clinically relevant consequence of pedicle screw malpositioning is the potential injury caused to adjacent structures, particularly the spinal cord, Kurz et al. investigated whether screw malpositioning can be prevented during the insertion process by directly measuring the screw insertion resistance [10]. However, using standardised testing conditions, Kurz et al. did not find any evidence that torque-based mechanical resistance is a reliable and consistent real-time indicator of pedicle screw malposition. Another important aspect in clinical spinal biomechanics is the kinematic behaviour of motion-preserving implants, especially in the cervical spine. To better understand the complex motion characteristics of disc prostheses under in vivo-like movements, Ansaripour et al. developed a novel testing method to simulate combined rotational–translational motions [11]. Validating their test setup, Ansaripour et al. could show that simultaneous rotation–translation motion provoked subluxation of ball-and-socket prostheses at lower motion extents compared to isolated motion types.

4. In Silico Investigations

Compared to in vitro studies, which can also provide input and validation data for numerical models of the spine, finite element analyses offer insights into load distributions, stress concentrations, and local strains in both spinal structures and implants, offering risk estimations regarding potential tissue and implant failure following surgical treatment and allowing direct comparisons between surgical techniques under the exact same boundary conditions. In the study of Dhar et al., this principle was used to compare the traditional bilateral pedicle screw–rod system with a novel reverse transdiscal screw system, which can be applied for the minimally invasive surgical treatment of lumbar degenerative disc disease [12]. Findings of this study revealed that the use of the reverse transdiscal screws resulted in lower segmental range of motion, lower screw and cage stresses, and higher anterior and posterior shear load resistance compared to the use of the pedicle screw–rod system. In a similar study, Lodde et al. investigated the effectiveness of different fixation techniques for the treatment of unilateral non-displaced fragility fracture of the pelvis [13]. Lodde et al. could show that both bilateral iliosacral and transsacral screws produced less dislocation and local stress concentrations compared to unilateral iliosacral screw fixation, with bilateral iliosacral screws providing higher rotational stability compared to transsacral screws. Apart from the evaluation of implant performance, finite element models of the spine can be used to investigate alterations in the load distribution in cases of spinal deformity and degeneration. Wang et al. used finite element analysis to determine stresses in the discs adjacent to spinal fixation following scoliosis treatment [14]. Deriving parametrised finite element models from biplanar pre- and post-operative X-ray images of an adolescent idiopathic scoliosis patient, Wang et al. determined a decrease in maximum endplate and disc stress in the cranial adjacent segment as well as an increase in stress in the caudal adjacent segment following spinal fixation, indicating a high risk for adjacent segment disease below scoliosis instrumentation. Using another parametric finite element model, Ardatov et al. investigated the impact of progressing disc degeneration on lumbar spinal biomechanics by gradually adjusting disc height, nucleus pulposus volume, and tissue stiffness to simulate altered dehydration and fibrosis [15]. Applying displacement-controlled compressive loading, Ardatov et al. could show that anulus fibrosus stresses largely increased while nucleus pulposus stresses as well as intradiscal pressure values decreased, indicating significant load redistribution during the process of disc degeneration,

https://doi.org/10.3390/bioengineering13040482

including a shift in primary compressive load absorption from the nucleus pulposus to the anulus fibrosus.

5. Design and Validation of Novel Methods

New and further developments of methods to investigate spinal biomechanics are essential to answer unacknowledged or complex research questions related to the spine. Advancements and innovations in the field of engineering and programming enable novel, more detailed, and highly specific designs of experimental and numerical models. Remus et al. developed a muscle-driven numerical model of the human torso within the open-source modelling framework ArtiSynth, combining forward dynamic simulation with finite element modelling, enabling them to study interactions between active and passive spinal structures and spinal implants as well as orthoses under physiologically realistic loading conditions [16]. Using this model, investigations of loads inside spinal instrumentation, intervertebral discs, and surrounding soft tissue can be performed under physiological loading conditions, allowing studies on clinical phenomena such as adjacent segment degeneration as well as facilitating predictions of treatment outcomes. Apart from basic and translational research, the design and validation of novel diagnostic methods are of major importance for their potential clinical use. Kniepert et al. introduced Dynamic Surface Topography as a novel technique to analyse spine kinematics in patients with back pain under dynamic conditions in an easy-to-use, radiation-free setup [17]. Validating this novel method in a prospective observational study, Kniepert et al. observed more movement in patients with more pain and concluded that Dynamic Surface Topography can provide valuable therapeutic information for an individual patient. In another study, Konradi et al. showed that Surface Topography proves extensive inter-individual variability as well as high intra-individual consistency of spinal motion during gait [18]. To enable replication and validation of their findings, Konradi et al. made all tools and visualisations freely available in repositories. Finally, van Santbrink et al. presented a novel, artificial intelligence-assisted technique for the image recognition of cervical vertebrae in dynamic X-ray recordings in order to investigate qualitative motion patterns in the cervical spine [19]. Using a U-Net architecture, van Santbrink et al. could prove the feasibility of implementing deep learning models to detect qualitative cervical motion patterns and thus found promising approaches for the use of this technique in clinical research.

6. Conclusions

Research on spinal biomechanics has been constantly growing due to progressive advancements in methodology and technology, and will further diversify due to the necessity to address upcoming topics related to the ageing society and novel diagnostic and therapeutic strategies in orthopaedics and traumatology. While in vivo studies have been shown to allow for increasingly precise measurements of three-dimensional spinal motion and the detection of specific motion patterns influenced by spinal pathology, injury, and pain, in vitro studies enable quantifications of spinal flexibility, kinematics, and intra-articular pressure under standardised loading conditions, which are essential for implant testing and biomechanical evaluations of novel surgical treatment strategies, as well as the generation of input and validation data for numerical models of the spine. The number of finite element analyses for the investigation of spinal biomechanics is constantly increasing, as relevant information regarding stress and strain distributions inside anatomical structures and implants that cannot be directly quantified from in vivo and in vitro studies is discovered. Future in silico studies will presumably use more complex musculoskeletal models, comprising inverse and forward dynamics in addition to finite element modelling, in order to investigate spinal biomechanics under more physiologically realistic loading conditions.

https://doi.org/10.3390/bioengineering13040482

Conflicts of Interest: The author declares no conflict of interest.

References

1. Breen, A.; Breen, A.; Branney, J.; du Rose, A.; Nematimoez, M. Advances in the Measurement and Interpretation of Intervertebral Motion in the Lumbar Spine: A Scoping Review. *Bioengineering* **2026**, *13*, 239. [CrossRef] [PubMed]
2. Hoehl, B.U.; Taheri, N.; Schonnagel, L.; Becker, L.A.; Modl, L.; Reitmaier, S.; Pumberger, M.; Schmidt, H. Comprehensive Analysis of Chronic Low Back Pain: Morphological and Functional Impairments, Physical Activity Patterns, and Epidemiology in a German Population-Based Cross-Sectional Study. *Bioengineering* **2025**, *12*, 878. [CrossRef] [PubMed]
3. Löchel, J.; Hanisch, M.; Burger, J.; Labbus, K.; Zahn, R. Association of Spinopelvic Anatomy with the Level of Lumbar Disc Herniation. *Bioengineering* **2025**, *12*, 993. [CrossRef] [PubMed]
4. Rossi, M.; Signorini, M.; Baram, A.; De Robertis, M.; Capo, G.; Riva, M.; Fornari, M.; Pessina, F.; Brembilla, C. Upper Limb Neural Tension Test and Spinal Biomechanics: Insights from a Longitudinal Pilot Study. *Bioengineering* **2025**, *12*, 487. [CrossRef] [PubMed]
5. Yoseph, E.T.; Taiwo, R.; Kiapour, A.; Touponse, G.; Massaad, E.; Theologitis, M.; Wu, J.Y.; Williamson, T.; Zygourakis, C.C. Pregnancy-Related Spinal Biomechanics: A Review of Low Back Pain and Degenerative Spine Disease. *Bioengineering* **2025**, *12*, 858. [CrossRef] [PubMed]
6. Jacob, A.; Feist, A.; Zderic, I.; Gueorguiev, B.; Caspar, J.; Wirtz, C.R.; Richards, G.; Loibl, M.; Haschtmann, D.; Fekete, T.F. Augmenting Screw Technique to Prevent TLIF Cage Subsidence: A Biomechanical In Vitro Study. *Bioengineering* **2025**, *12*, 337. [CrossRef] [PubMed]
7. Riesenbeck, O.; Czarnowski, N.; Raschke, M.J.; Oeckenpohler, S.; Hartensuer, R. Biomechanical Comparisons between One- and Two-Compartment Devices for Reconstructing Vertebrae by Kyphoplasty. *Bioengineering* **2024**, *11*, 795. [CrossRef] [PubMed]
8. Riesenbeck, O.; Czarnowski, N.; Raschke, M.J.; Oeckenpohler, S.; Hartensuer, R. Primary Stability of Kyphoplasty in Incomplete Vertebral Body Burst Fractures in Osteoporosis: A Biomechanical Investigation. *Bioengineering* **2024**, *11*, 798. [CrossRef] [PubMed]
9. Schleifenbaum, S.; Metzner, F.; Schultze, J.; Kurz, S.; Heyde, C.E.; Pieroh, P. Does Malpositioning of Pedicle Screws Affect Biomechanical Stability in a Novel Quasistatic Test Setup? *Bioengineering* **2025**, *12*, 781. [CrossRef] [PubMed]
10. Kurz, S.; Fischer, B.; Schultze, J.; Metzner, F.; Wendler, T.; Heyde, C.E.; Schleifenbaum, S. Does Resistance Indicate Malposition? A Standardized Comparison of Pedicle Screw Placement. *Bioengineering* **2025**, *12*, 1254. [CrossRef] [PubMed]
11. Ansaripour, H.; Ferguson, S.J.; Flohr, M. Evaluation of Load on Cervical Disc Prosthesis by Imposing Complex Motion: Multiplanar Motion and Combined Rotational-Translational Motion. *Bioengineering* **2024**, *11*, 857. [CrossRef] [PubMed]
12. Dhar, U.K.; Aghayev, K.; Sultan, H.; Rajendran, S.; Tsai, C.T.; Vrionis, F.D. Finite Element Analysis of Biomechanical Assessment: Traditional Bilateral Pedicle Screw System vs. Novel Reverse Transdiscal Screw System for Lumbar Degenerative Disc Disease. *Bioengineering* **2025**, *12*, 671. [CrossRef] [PubMed]
13. Lodde, M.F.; Klimek, M.; Herbst, E.; Peez, C.; Riesenbeck, O.; Raschke, M.J.; Rosslenbroich, S. Bilateral Iliosacral and Transsacral Screws Are Biomechanically Favorable and Reduce the Risk for Fracture Progression in Fragility Fractures of the Pelvis-A Finite Element Analysis. *Bioengineering* **2025**, *12*, 27. [CrossRef] [PubMed]
14. Wang, T.H.; Chou, P.H.; Chen, C.S. Using Two X-Ray Images to Create a Parameterized Scoliotic Spine Model and Analyze Disk Stress Adjacent to Spinal Fixation-A Finite Element Analysis. *Bioengineering* **2025**, *12*, 1212. [CrossRef] [PubMed]
15. Ardatov, O.; Fernandes, S.R.; Kilikevicius, A.; Alekna, V. Parametric Finite Element Evaluation of Load Redistribution Under Progressive Lumbar Disc Degeneration. *Bioengineering* **2026**, *13*, 234. [CrossRef] [PubMed]
16. Remus, R.; Lipphaus, A.; Ritter, M.; Neumann, M.; Bender, B. A Muscle-Driven Spine Model for Predictive Simulations in the Design of Spinal Implants and Lumbar Orthoses. *Bioengineering* **2025**, *12*, 263. [CrossRef] [PubMed]
17. Kniepert, J.; Ronsch, H.; Betz, U.; Konradi, J.; Huthwelker, J.; Wolf, C.; Westphal, R.; Drees, P. Dynamic Surface Topography for Thoracic and Lumbar Pain Patients-Applicability and First Results. *Bioengineering* **2025**, *12*, 289. [CrossRef] [PubMed]
18. Konradi, J.; Betz, U.; Huthwelker, J.; Wolf, C.; Schmidtmann, I.; Westphal, R.; Cerpa, M.; Lenke, L.G.; Drees, P. Using Surface Topography to Visualize Spinal Motion During Gait-Examples of Possible Applications and All Tools for Open Science. *Bioengineering* **2025**, *12*, 348. [CrossRef] [PubMed]
19. van Santbrink, E.; Schuermans, V.; Cerfonteijn, E.; Breeuwer, M.; Smeets, A.; van Santbrink, H.; Boselie, T. AI-Assisted Image Recognition of Cervical Spine Vertebrae in Dynamic X-Ray Recordings. *Bioengineering* **2025**, *12*, 679. [CrossRef] [PubMed]

https://doi.org/10.3390/bioengineering13040482

bioengineering

MDPI

Review

Pregnancy-Related Spinal Biomechanics: A Review of Low Back Pain and Degenerative Spine Disease

Ezra T. Yoseph [1,*], Rukayat Taiwo [1], Ali Kiapour [2], Gavin Touponse [1], Elie Massaad [3], Marinos Theologitis [1], Janet Y. Wu [1], Theresa Williamson [2] and Corinna C. Zygourakis [1]

1 Department of Neurosurgery, Stanford University Medical Center, Stanford, CA 94305, USA
2 Department of Neurosurgery, Massachusetts General Hospital, Harvard Medical School, Boston, MA 02115, USA
3 Department of Neurosurgery, Perelman School of Medicine, University of Pennsylvania, Philadelphia, PA 19104, USA
* Correspondence: ezyoseph@stanford.edu

Abstract

Pregnancy induces substantial anatomical, hormonal, and biomechanical changes in the spine and pelvis to accommodate fetal growth and maintain postural adaptation. This narrative review synthesizes peer-reviewed evidence regarding pregnancy-related spinal biomechanics, with a particular focus on low back pain, spinopelvic alignment, sacroiliac joint dysfunction, and potential contributions to degenerative spinal conditions. A systematic search of PubMed, Embase, and Google Scholar was conducted using Boolean operators and relevant terms, yielding 1050 unique records, with 53 peer-reviewed articles ultimately cited. The review reveals that increased lumbar lordosis, ligamentous laxity, altered gait mechanics, and muscular deconditioning elevate mechanical load on the lumbar spine, predisposing up to 56% of pregnant individuals to low back pain. These changes are often associated with sacroiliac joint laxity, anterior pelvic tilt, and multiparity. Long-term risks may include degenerative disc disease and spondylolisthesis. Conservative interventions such as pelvic floor muscle training, prenatal exercise, and surface topography monitoring offer symptom relief and support early rehabilitation, although standardized protocols and longitudinal outcome data remain limited. Pregnancy-related spinal changes are multifactorial and clinically relevant; an interdisciplinary approach involving spinal biomechanics, physical therapy, and obstetric care is critical for optimizing maternal musculoskeletal health.

Keywords: pregnancy biomechanics; low back pain; LBP; degenerative spondylolisthesis; pelvic tilt; lumbar lordosis; spinopelvic parameters

1. Introduction

Women undergo significant anatomical and physiological changes during pregnancy to meet the extensive demands required for fetal development and childbirth. These changes include musculoskeletal adaptations, such as accommodating increased body mass index (BMI) and pelvic adjustments during delivery. Past studies have examined the association between gravidity or parity and biomechanical alterations to the spine during pregnancy, hypothesizing that these changes may be associated with future back pain or degenerative spine disorders. Understanding these biomechanical spinal changes is crucial, as up to 56% of women experience low back pain during pregnancy, significantly impacting quality of life [1,2].

While the biomechanical changes in pregnancy have been individually described, their integration into a broader understanding of long-term spinal degeneration, including conditions like spondylolisthesis and disc degeneration, remains poorly defined. Few studies synthesize how hormonal, postural, and neuromuscular factors intersect to influence the onset or exacerbation of spinal pathology in postpartum populations. Interest in pregnancy-related spinal biomechanics has evolved considerably since the 1980s. Earlier investigations largely focused on pain prevalence and obstetric ergonomics, while more recent research has incorporated gait analysis, spinal alignment modeling, and hormonal profiling. This temporal shift reflects advances in biomechanics methodology and growing awareness of pregnancy as a complex state of mechanical adaptation. By recognizing this progression, our review contextualizes the current literature within the broader scientific evolution of the field.

This narrative review consolidates the current literature on pregnancy-induced spinal adaptations and their potential contribution to spinal degeneration. We examine alterations in spinopelvic parameters, discuss mechanisms underlying pregnancy-related LBP, and evaluate conservative interventions aimed at prevention and symptom mitigation. By bridging biomechanics with clinical implications, our aim is to support a more comprehensive understanding of pregnancy's impact on spinal health and inform multidisciplinary approaches to prenatal musculoskeletal care.

2. Methods

2.1. Study Design

This review employed a narrative review methodology to synthesize the literature on spinal biomechanics and low back pain related to pregnancy. A narrative review was chosen for its suitability in integrating diverse types of clinical, biomechanical, and anatomical evidence, particularly when heterogeneous methodologies or outcome measures across studies preclude formal meta-analysis. This approach enabled a comprehensive understanding of anatomical adaptations, spinopelvic parameters, sacroiliac joint pain, and related conservative interventions. Literature searches were conducted across three databases: PubMed, Embase, and Google Scholar, without date restrictions, with the last search conducted in March 2025. Boolean operators ("AND", "OR") were used to combine search terms including the following: "pregnancy biomechanics," "spinopelvic parameters," "sacroiliac joint pain," "lumbar lordosis," "low back pain," and "degenerative spondylolisthesis." No restrictions were placed on publication year to incorporate foundational studies in spine biomechanics and pregnancy-related musculoskeletal research. As such, landmark studies from the 1980s–1990s were included due to their enduring relevance to modern spine biomechanical understanding. This narrative review was conducted in accordance with the JBI Critical Appraisal Checklist for Narrative, Expert Opinion and Text. The checklist was used to guide assessment of source credibility, analytical rigor, and peer-supported interpretation in line with best practices for narrative evidence synthesis.

2.2. Search Strategy

A total of 1050 records were identified after duplicate removal across databases. Initial screening by title and abstract yielded 200 articles for full-text assessment (Figure 1). Studies were included if they were peer-reviewed human studies published in English; focused on spinal, spinopelvic, or sacroiliac adaptations during or after pregnancy; evaluated musculoskeletal, biomechanical, diagnostic, or non-surgical therapeutic approaches; and included clinical relevance or implications for spinal care. Exclusion criteria included animal studies, surgical-only outcomes, reviews lacking original data, and non-English or unpublished sources such as books, theses, and conference abstracts. Narrative synthesis was used to organize findings into key domains: anatomical and physiological adaptations, biomechanical compensation, low back and sacroiliac joint pain etiologies, degenerative spine risk, and conservative management strategies. Given the descriptive nature of this review, no formal risk-of-bias or quality assessment was performed. However, only peer-reviewed, full-length articles were included to enhance consistency and reliability. Reference management and duplicate removal were conducted using Zotero (Version 6).

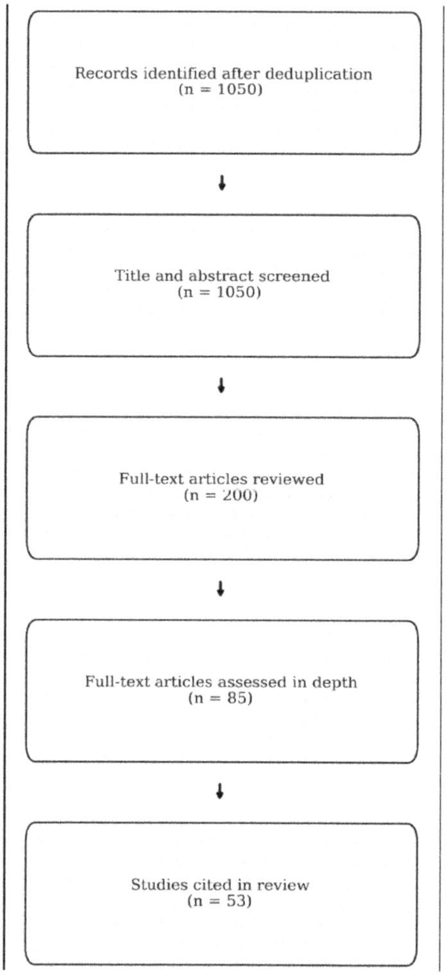

Figure 1. Methods flowchart. Diagram illustrating the identification, screening, and inclusion process of articles reviewed in this narrative synthesis.

2.3. Study Selection

The final manuscript includes 53 cited studies. These citations were selected from a larger pool of approximately 85 full-text articles that were reviewed for relevance and thematic contribution. Not all full-text studies were directly cited in the manuscript, as only those providing central or representative data were retained.

3. Results

3.1. Anatomic and Physiologic Changes During Pregnancy

The long-lasting impact of pregnancy on the spine and pelvis can vary widely among individuals, with several factors contributing to this heterogeneity, including a woman's lifestyle before and during pregnancy, the size and position of the fetus, or even pre-existing spinal and musculoskeletal conditions. Research on the long-lasting effects of pregnancy on the spine has investigated how vaginal delivery and parity can impact the spinopelvic parameters. One study of 320 women who underwent whole-spine radiographs showed that pelvic incidence and sacral slope were significantly greater in women who delivered vaginally compared to those who delivered by C-section, while both of them significantly increased with the number of vaginal deliveries [3]. However, vertebral body dimensions remain unaffected, as evidenced by a population-based middle-aged birth cohort study by Oura et al., revealing no association between gravidity or parity and vertebral cross-sectional area and height ratio [4].

Pregnancy induces a range of physiological and anatomical adaptations designed to accommodate fetal growth while maintaining maternal function. Among the earliest and most pronounced changes is an increase in joint laxity, mediated largely by hormonal influences such as relaxin and estrogen. Calguneri et al. (1982) demonstrated measurable increases in ligamentous flexibility in pregnant women, particularly during the second and third trimesters, highlighting the body's preparation for parturition through altered connective tissue integrity [5]. This generalized increase in joint laxity, while adaptive, may compromise the mechanical stability of load-bearing joints, including the spine and pelvis.

These hormonal changes intersect with biomechanical demands as maternal weight and abdominal girth increase. Hartmann and Bung (1999) emphasize the importance of addressing these changes through physical conditioning, noting that insufficient neuromuscular control and core strength may exacerbate postural imbalances and spinal strain [6]. Similarly, Artal and O'Toole (2003) underscore the need for structured guidelines surrounding peripartum exercise to prevent injury and maintain spinal health [7]. Their recommendations suggest that proper musculoskeletal conditioning may mitigate the risk of pain and dysfunction in late pregnancy and the postpartum period.

The hormonal driver relaxin plays a critical role in modifying ligamentous properties. Dehghan et al. (2014) explored its systemic effects, reporting that increased relaxin levels can reduce the tensile strength of ligaments, further increasing the susceptibility to pelvic instability and sacroiliac dysfunction [8]. Taken together, these studies indicate that pregnancy-related changes in laxity, muscle function, and posture must be understood not as isolated phenomena, but as an integrated response with direct biomechanical implications for spinal load distribution and compensatory mechanisms.

As pregnancy progresses, there is also a notable increase in body mass, along with a significant expansion of the abdominal girth, with weight gain ranging between 5 and 18 kg, depending on the woman's BMI before pregnancy [9,10]. Several studies have examined how increases in body mass impact spinopelvic parameters, with Pauk and Swinarska reporting that maternal BMI was significantly associated with increased thoracic kyphosis in late pregnancy ($R = 0.50$, $p < 0.05$) [11]. Contrastingly, a different cross-sectional study demonstrated that there was a poor correlation between different BMI categories and spinopelvic values or lumbar lordosis types. When they categorized non-pregnant patients according to the presence and absence of obesity, the spinopelvic parameters increased in the obese group but did not reach statistical significance. Similarly, in order to investigate how fat distribution influences spinal anatomy, the researchers made a dichotomous comparison between patients with normal and elevated abdominal circumference, but they did not find a statistically significant correlation between abdominal obesity and frequency of lumbar lordosis [12]. This disparity ultimately suggests that pregnancy involves additional hormonal and musculoskeletal adaptations, which further influence spinal alignment and spinopelvic balance. Ultimately, these findings underscore the need to view pregnancy-related biomechanical changes not as isolated structural phenomena, but as part of a hormonally mediated continuum that informs both spinal vulnerability and the rationale for targeted clinical interventions.

3.2. Biomechanical Changes During Pregnancy

Pregnancy induces dynamic and multifaceted biomechanical changes in the spine and pelvis to accommodate fetal growth, hormonal fluctuations, and shifting weight distribution. These adaptations are essential for maintaining postural stability and upright gait but can increase mechanical stress on the spine and contribute to musculoskeletal discomfort (Table 1). Developing a comprehensive understanding of the biomechanical changes in the spine during pregnancy may allow for understanding pathologies, including chronic low back pain, which pose a significant burden to patients in their daily lives. In describing the evolution of the spine, Whitcome et al. highlight several adaptations of the Hominin axial skeletons that support bipedalism, including elongated lumbar regions in both the number and length of the vertebrae as well as the marked posterior concavity of wedged lumbar vertebrae called lordosis [13]. This lordotic curvature functions to stabilize the torso over the pelvis and counteracts anterior displacement of the center of mass (Figure 2). During pregnancy, this forward shift in mass requires compensatory spinal realignment to prevent postural decompensation and mechanical overload.

Table 1. Key biomechanical changes during pregnancy.

Biomechanical Change	Mechanism	Clinical Significance	References
Increased Lumbar Lordosis	Forward shift in the center of mass and compensatory hyperextension of the lower spine	May elevate facet joint loads; can predispose to low back pain and spondylolisthesis	Whitcome et al., 2007 [13]; Yoo et al., 2015 [14]
Altered Pelvic Parameters (↑ Sacral Slope, ↑ Pelvic Tilt)	Hormonal relaxation of ligaments and change in weight distribution lead to greater pelvic incidence and sacral slope	Excessive pelvic tilt alters lumbopelvic biomechanics; may contribute to postpartum spinal issues and degenerative changes over time	Yamada et al., 2021 [3]; Wang et al., 2016 [15]
Changes in Gait and Balance	Abdomen enlargement shifts center of gravity anteriorly, altering stride length, cadence, and stance width	Altered balance can increase mechanical load on lower spine and pelvis; may cause postural instability and higher fall risk	Foti et al., 2000 [16]; Wu et al., 2004 [17]
Reduced Lumbopelvic Stabilization	Abdominal expansion reduces stability of transverse abdominis, multifidus, and pelvic floor	Reduced spinal support increases stress on lumbar structures, potentially aggravating low back pain	Fast et al. 1987 [1]; Borg-Stein et al., 2005 [18]; Fast et al. 1992 [19]
Pelvic Girdle and SI Joint Laxity	Elevated relaxin and progesterone levels lead to ligamentous laxity in the pubic symphysis and SI joints	Instability in the pelvic ring can co-occur with low back pain and is unique to pregnancy (not observed with non-pregnant weight gain)	Wu et al., 2004 [17]; Mens et al., 2009 [20]
Anterior Loading from the Growing Uterus	Progressive uterine and fetal enlargement places an anterior pull on the lower spine	Increases shear forces across the lumbar region; can amplify lordotic posture and contribute to degenerative disc or facet changes	Ritchie JR et al., 2003 [21]

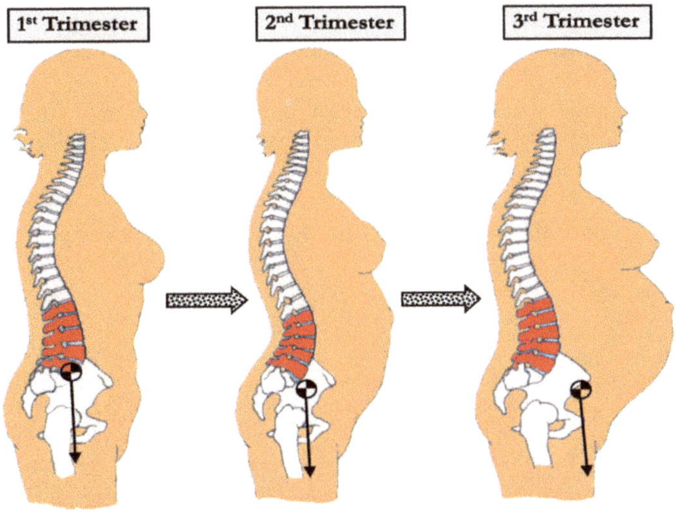

Figure 2. Changes in center of mass shift during pregnancy progression. An increase in abdominal mass during pregnancy shifts the center of mass forward, leading to mechanical strain.

As pregnancy progresses, increasing body mass and anterior abdominal expansion further accentuate spinal curvature. One adaptive mechanism is the exaggeration of lumbar lordosis to maintain spinal alignment above the pelvis and preserve upright posture (Figure 3). Studies comparing pregnant women during the 2nd and 3rd trimesters with non-pregnant women found increased curvature of both the thoracic and lumbar spine [14]. Thoracic kyphosis increased from 10.7 degrees in the 2nd trimester to 11.5 degrees in the 3rd trimester. Similarly, lumbar lordosis increased from 9.0 degrees in the 2nd trimester to 10.0 degrees in the 3rd trimester, while non-pregnant women had an average lumbar lordosis of 7.3 degrees. Another study using 3D measurements with surface topography found a 7.4-degree change in thoracic kyphosis angle and an 8.4-degree change in lumbar lordosis angle in a uniparous pregnant woman from the 17th to the 37th week of pregnancy [22,23]. Whitcome et al. revealed that women at full term extend their lower back by nearly 60%, from a mean angle of $32 \pm 12°$ in early pregnancy to $50 \pm 12°$ at term [13]. These spinal adjustments, while protective, may increase load on the facet joints and intervertebral discs, potentially contributing to the development of degenerative pathology (Figures 2 and 4).

In addition to spinal curvature changes, pregnancy introduces biomechanical challenges to postural stability. As gestation progresses, the anterior shift in the center of mass requires neuromuscular compensation to maintain balance. This adaptation is reflected in center of pressure (CoP) displacement during static stance, with studies showing a consistent anterior shift in CoP and increased sway area in pregnant individuals compared to non-pregnant controls, changes indicative of reduced postural control [9]. These balance alterations may be associated with the development or exacerbation of low back pain, though individual variability in compensatory strategies appears substantial. Dynamic assessments show that postural stability declines further in late pregnancy, particularly during gait initiation and transitions. In a motion analysis study, McCrory et al. demonstrated that women in their third trimester exhibited poorer dynamic balance scores and delayed neuromuscular responses during walking compared to non-pregnant controls, suggesting increased instability during daily activities [24]. Gait analysis studies have produced mixed findings; while some report no significant changes in walking velocity or stride length, others have observed reduced velocity, wider stride, and altered joint kinematics, including increased hip flexion and reduced ankle plantarflexion [16,17,25]. These discrepancies are likely due to individual differences in trunk control and neuromuscular adaptation, as supported by Krkeljas, who found that reduced trunk stiffness and compensatory gait changes were common responses to gestational biomechanical demands [26]. These findings illustrate that pregnancy significantly affects both static and dynamic balance mechanisms, contributing to the multifactorial etiology of low back pain and underscoring the need for individualized assessment and preventive intervention. In summary, these postural and neuromuscular adjustments though protective in nature appear to impose a cumulative mechanical load that may underlie the increased prevalence of musculoskeletal discomfort and spinal pathology in later pregnancy stages or postpartum.

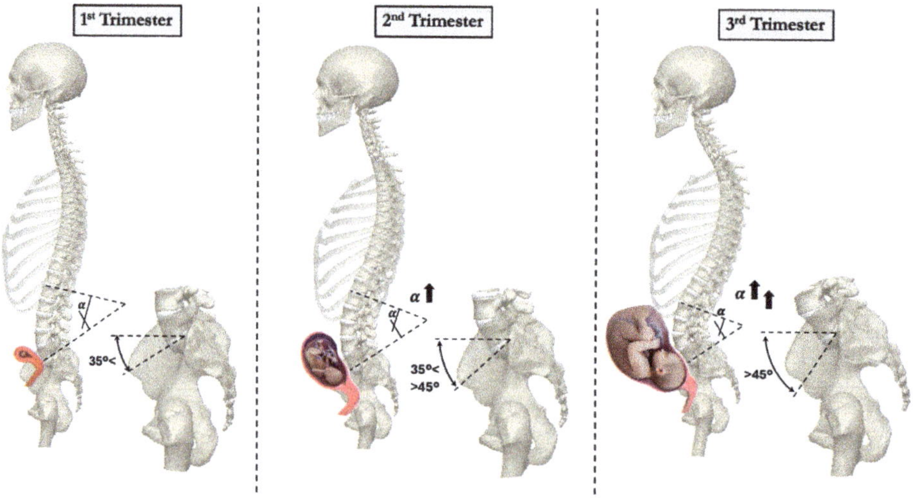

Figure 3. Increased lumbar lordosis and sacral slope during pregnancy progression. Changes in spinopelvic parameters, such as lumbar lordosis and sacral slope, occur throughout pregnancy. These adaptations maintain balance and mobility but may also contribute to musculoskeletal discomfort and long-term spinal degeneration.

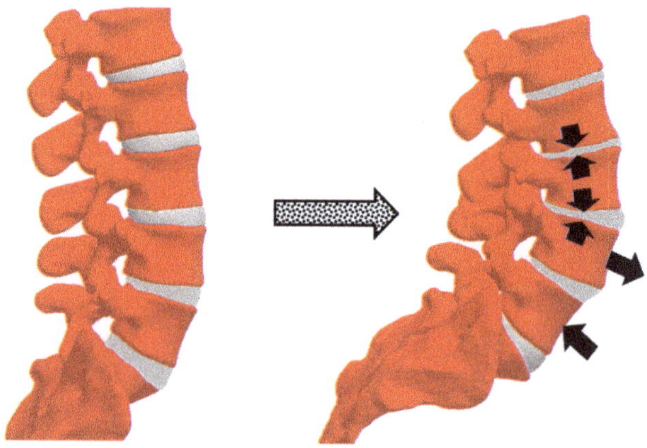

Figure 4. Bulging disc secondary to increased load and lumbar lordosis during pregnancy. Compensatory changes in spinopelvic parameters may predispose individuals to degenerative conditions such as intervertebral disc degeneration, leading to chronic back pain.

3.3. Etiology of Pregnancy-Related Back Pain

Lower back pain typically begins in the second trimester and is the most common musculoskeletal complaint in pregnancy, affecting up to 56% of pregnant individuals [9]. Low back pain during pregnancy can be up to four times worse than in non-pregnant women, significantly impacting day-to-day functioning and quality of life [22,23]. However, these symptoms are not isolated to the pregnancy period and can translate to chronic postpartum disability. The prevalence of postpartum low back pain is reported to be as high as 5–43% at six months postpartum and 20% at three years after delivery [14,27]. Furthermore, 15% of women with chronic low back pain report initial onset during pregnancy. Pregnancy-related lower back pain is defined by pain in the lumbar region, is dull in character, and is often elicited by forward flexion. It is also characterized by the restriction of spine movement in the lumbar region with worsening pain on palpation of erector spinae muscles [28].As expected, lower back pain in pregnancy can significantly impact quality of life and may have socioeconomic consequences due to work absenteeism [29]. It is important to distinguish between lumbar-origin and sacroiliac joint-related pain, the latter of which is also highly prevalent during pregnancy and may present similarly but require distinct management strategies [30].

The pathophysiological causes of lower back pain in pregnancy are multifactorial and include mechanical strain, pelvic ligamentous laxity, and vascular compression, which will each be discussed in further detail. Risk factors for lower back pain include pregestational history of back pain, back pain during prior pregnancies, and advanced maternal age. The enlarging uterus during pregnancy also leads to weakening of abdominal muscles and reduced muscle tone, which places additional strain on compensatory lumbar musculature [21]. This weakening may be due to mechanical stretch, neuromuscular inhibition, and impaired recruitment of core stabilizers like the transversus abdominis and multifidus [31]. It is estimated that about half of weight gain during pregnancy is gained in the abdomen [29]. The enlarging abdomen necessitates postural compensations such as increased lumbar lordosis and forward pelvic tilt. In addition to the weakening of lumbopelvic stabilizing musculature, these biomechanical alterations lead to posterior chain loading on the lower back [18].

Sacroiliac (SI) joint pain is a distinct and highly prevalent contributor to pregnancy-related low back pain, often misattributed to lumbar spine pathology. It is typically localized just below the posterior superior iliac spine and may radiate into the buttocks or thighs, differing from the more central presentation of lumbar-origin pain. SI joint dysfunction is driven largely by hormonal ligamentous laxity and pelvic asymmetry, particularly during the later trimesters. Accurate clinical differentiation is important, as the underlying mechanisms and optimal treatment strategies differ from those used for lumbar spine conditions. Joint laxity increases during pregnancy to allow for the passage of the fetus. This has been attributed to the pregnancy hormone relaxin, with estrogen and progesterone as contributing factors [21,32]. The increase in joint laxity of the pubis symphysis has been implicated in lower back pain (Figure 5) [20].

Despite the well-documented biomechanical changes caused by the gravid uterus, the relationship between pregnancy-related weight gain and low back pain remains unclear in the literature. A retrospective study by Matsuda et al. examining the relationship between pregnancy-related weight gain and low back pain found that women reporting moderate-to-severe back pain at four months after delivery had 10.4 kg gestational weight gain compared to 9.0 kg in women without back pain [27]. Contrastingly, Boarg-stein et al. found no relationship between pregnancy-related low back pain and weight of the mother, weight gain during pregnancy, or weight of the baby [18].

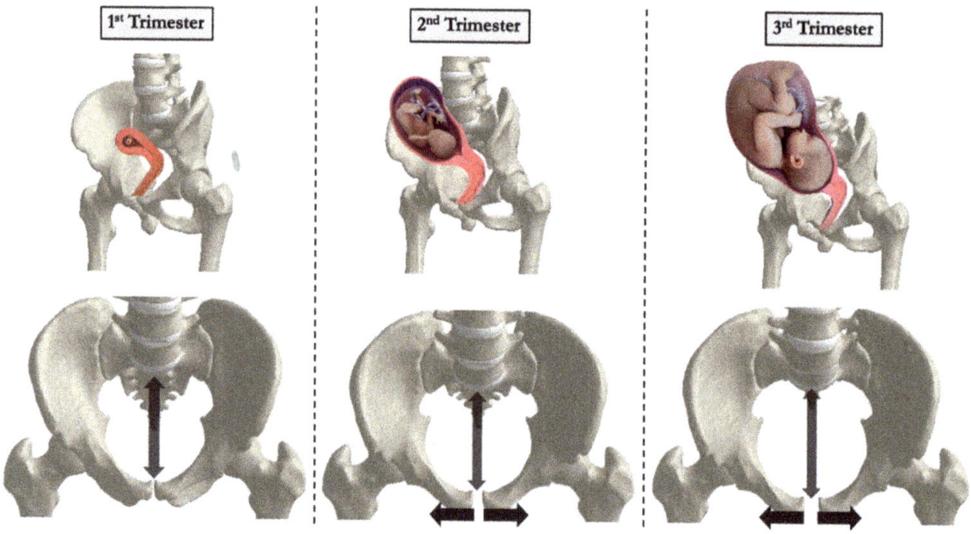

Figure 5. Widening of the pelvis and pubic symphysis during pregnancy. An increase in hormone-mediated joint laxity during pregnancy allows for the widening of the pubic symphysis to accommodate for childbirth. These changes have been implicated in lower back pain.

The emerging literature suggests that vascular factors may also contribute to the pathogenesis of pregnancy-related low back pain. Fast et al. reported that nocturnal back pain was higher in pregnant women who exhibited decreased basal oxygen saturation and spent longer in the supine position. This finding supports the theory that compression of the inferior vena cava by the gravid uterus leads to venous stasis, reduced cardiac output, and congestion in the lumbar venous plexus. Compromised metabolic supply and edema in the lumbosacral neural microcirculation may result in pain, which in some patients is mitigated by adequate collateral circulation [19]. These findings underscore the importance of interventions such as compression stockings and regular ambulation to prevent pregnancy-related back pain (Table 2). Compression stockings enhance blood circulation, reducing venous stasis and associated discomfort, while regular movement prevents prolonged pressure on the inferior vena cava, which mitigates venous congestion [33]. These multifactorial contributors, ranging from anatomical and vascular to hormonal, likely interact in complex ways rather than act in isolation. For instance, hormonal laxity may potentiate musculoskeletal strain from rapid abdominal growth, while vascular compression could amplify nociceptive signaling, together exacerbating the subjective experience of pain. Building upon these findings, the following Sections 3.4 and 3.5 explore a range of intervention strategies, including neuromuscular training, postural stabilization, and pain management, that address the biomechanical disruptions identified above.

Table 2. Evidence-based interventions for pregnancy-related low back pain.

Intervention	Mechanism	Evidence	References
Prenatal Exercise (PFMT, aquatics, stability ball, etc.)	Strengthens the deep core (transverse abdominis, multifidus) and pelvic floor (levator ani complex) while improving posture and offsetting excessive lumbar lordosis. Specific routines like aquatics and PFMT ("Kegels") focus on lumbopelvic stability	Numerous RCTs and reviews link pelvic floor and core strengthening to reduced low back and pelvic pain, improved function, and better postural stability. Prenatal yoga/Pilates programs (with appropriate modifications) may also alleviate discomfort and reduce stress	Borg-Stein et al., 2005 [18]; Nascimento et al., 2012 [34]; Mørkved & Bø, 2014 [35]; Woodley et al., 2020 [36]; Wang XQ et al., 2012 [37]
Maternity Support Belts and Sacroiliac (SI) belts	Provides external support to the lower abdomen and lumbopelvic region. SI belts, a subtype of maternity belts, are specifically designed to stabilize the sacroiliac joints and control pelvic ring laxity. These supports reduce spinal load and improve functional mobility in patients with SI joint pain	Some trials indicate short-term pain relief and reduced postpartum pelvic pain, especially in cases of sacroiliac joint dysfunction. Evidence is variable, but SI belts are often included in conservative management	Casagrande et al., 2015 [29]; Mens et al., 2009 [20]
Acupuncture and Related Modalities	May modulate pain pathways and reduce pelvic girdle instability; believed to stimulate endorphin release and increase local blood flow	Cochrane reviews suggest beneficial effects for pelvic girdle and low back pain in pregnancy	Pennick & Liddle, 2013 [38]
Physical Therapy and Spinal Manipulation	Manual therapy techniques can address restricted spinal segments and muscular imbalances; guided exercises improve muscle activation	Mild to moderate relief in some pregnant populations; safety often considered good if performed by experienced clinicians	Ritchie, 2003 [21]; Borg-Stein et al., 2005 [18], Mens et al., 2009 [20]
Postural Re-education and Ergonomics	Teaches pregnant individuals to distribute weight more evenly, maintain neutral spine alignment, and use proper lifting mechanics	Anecdotally effective and often recommended, though high-quality RCT data may be sparse; widely included in comprehensive prenatal programs	Conder et al., 2019 [9]; Branco et al., 2014 [10]
Compression Stockings and Regular Ambulation	Reduces venous stasis in the inferior vena cava region (especially in 3rd trimester with prolonged standing) and improves blood flow to pelvic/lumbar tissues	Shown to alleviate nocturnal back pain associated with decreased basal oxygen saturation and supine positioning; may help reduce edema-related pain	Fast et al., 1992 [19]; Szkwara et al., 2019 [33]
Pharmacological Management	Limited safe options in pregnancy (e.g., acetaminophen, possibly muscle relaxants like cyclobenzaprine); opioids and NSAIDs often restricted by trimester-specific risks	Generally recommended only if pain is debilitating and after nonpharmacologic strategies; usage minimized for fetal safety	Black & Hill, 2003 [39]; Rathmell et al., 1997 [40]
Surface Topography and AI-Based Monitoring	Non-invasive imaging and predictive algorithms to track spinal curvature changes, posture, and potential LBP onset	Emerging data support early intervention if major postural alterations are detected; may reduce postpartum persistence of pain by guiding exercise	Michoński et al., 2016 [23], Betsch et al., 2015 [41]

3.4. Degenerative Spine Disease

The relationship between parity, gravidity, and spinal degeneration remains debated, with studies reporting both protective and deleterious associations depending on age, number of pregnancies, and anatomical focus. A 1996 study by Sanderson and Fraser found that the incidence of degenerative spondylolisthesis was significantly higher in multiparous women compared to nulliparous women and men, suggesting that repeated pregnancies may contribute to spinal instability through mechanical or hormonal mechanisms [42]. More recently, Cholewicki et al. identified parity and hysterectomy as independent predictors of degenerative spondylolisthesis in older women, even after adjusting for age and BMI [43]. Proposed contributing factors include increased joint laxity from hormonal shifts, cumulative axial loading of the spine during pregnancy, and weakening of abdominal musculature due to mechanical stretch and neuromuscular inhibition. These findings suggest that repeated pregnancies may disrupt spinal stability not only via direct mechanical strain but also through cumulative hormonal exposures that induce ligamentous laxity. Over time, this laxity, which is compounded by core muscle inhibition, may increase shear forces across lumbosacral joints, potentiating degenerative changes, particularly in predisposed individuals.

Conversely, a study by Çevik et al. reported that in women aged 20–40, increasing parity was associated with a lower Pfirrmann grade of disc degeneration, postulating that elevated estrogen levels during pregnancy may exert protective effects on intervertebral discs through receptor-mediated mechanisms [44]. This supports a broader view that hormonal shifts during pregnancy can play dual roles: protective in some anatomical contexts (e.g., intervertebral discs via estrogen receptors) yet deleterious in others (e.g., facet joint strain due to ligamentous laxity). Such differential effects may help explain contradictory findings in parity-degeneration associations across studies. A 2024 investigation by Güngör et al. found that higher parity (particularly five or more pregnancies) was associated with accelerated spinal degeneration, likely due to cumulative mechanical strain and altered hormonal regulation [45]. These mechanical and hormonal influences, especially increased lumbar lordosis, anterior mass loading, and ligamentous laxity, may contribute to intervertebral disc bulging or degeneration, particularly in the context of altered spinopelvic alignment (Figure 4). In this biomechanical context, the relative magnitude and persistence of lumbar lordosis may be key. While lordosis increases are often adaptive during pregnancy, exaggerated or poorly compensated curvature may elevate posterior joint compression and alter facet loading, promoting localized degenerative changes.

Changes in spinopelvic parameters are observed in patients with degenerative spondylolisthesis (DS). In a study of patients with lumbar spinal canal stenosis, both male and female patients who additionally had degenerative spondylolisthesis had significantly increased pelvic incidence (PI), sacral slope (SS), L4 slope, L5 slope, thoracic kyphosis (TK), and lumbar lordosis (LL). These parameters were also higher in two-level versus single-level DS patients, though comparison to non-DS controls was not made directly. Differentiating between single-level and two-level lumbar degenerative spondylolisthesis, these spinopelvic parameters were significantly larger in patients with two-level disease [15]. In another study on the pathogenesis of lumbar degenerative spondylolisthesis, weight was found to be a highly correlated factor, as the average body mass index (BMI) and percent of patients with BMI > 25 were significantly increased in the degenerative spondylolisthesis cohort [43,46]. Taken together with compensatory biomechanical changes during pregnancy, a theory for the pathogenesis of degenerative lumbar spondylolisthesis is that increased PI and LL are predisposing factors for the disease, and adding an overweight body constitution forces the body to compensate by decreasing PT (entering a "compensated balance" state), which then facilitates the development of degenerative disease (an "unbalanced" state) [46,47]. Parity has also been shown to be associated with significantly greater PI-LL and thoracic kyphosis, the first of which has been shown in studies detailed above to be correlated with lumbar degenerative spondylolisthesis. However, in this study, parity itself was not found to be directly correlated with lumbar disc/spine degeneration [48]. These theories suggest that the compensatory changes in spinopelvic parameters during pregnancy may be a predisposing factor to spine degeneration. Overall, the collective evidence indicates that spine degeneration in pregnancy is not solely dependent on parity or BMI, but rather emerges from a confluence of anatomical, hormonal, and postural changes. These interact dynamically, and their cumulative effect may render the spine increasingly susceptible to degenerative spondylolisthesis in a subset of individuals.

3.5. Treatment of Pregnancy-Related Lower Back Pain

Several studies have been conducted to understand how the risk of back pain associated with pregnancy can be minimized. For conservative management of pain, there is evidence suggesting maternity support belts and spinal manipulation are helpful with pain relief. Although chiropractic manipulative therapy (CMT) is gaining popularity for pregnancy-related back pain, evidence remains limited, and current guidelines recommend it as an adjunct rather than a primary treatment [49]. It is important to distinguish between interventions aimed at preventing pregnancy-related back pain and those intended to treat it once symptoms are present. Preventive strategies tend to emphasize biomechanical conditioning and postural stabilization, while treatment strategies often target symptom management and functional mobility. The strongest evidence suggests prenatal exercise and acupuncture are best for preventing pregnancy-related back pain. Prenatal exercise interventions, including stability ball training and aquatic-based exercise programs, activate the transverse abdominis and multifidus to enhance lumbopelvic stability and reduce pregnancy-related low back pain [29,48]. Beyond muscular strengthening, prenatal exercise likely confers benefits through neuromuscular re-education, fostering more efficient motor patterns during pregnancy-related biomechanical shifts. In addition, prenatal exercise is thought to prevent pregnancy-related weight gain within the recommended range, which may help mitigate lower back pain. Acupuncture has been found to be particularly effective at treating pelvic girdle pain, explaining its efficacy in back pain prevention.

Beyond standard prenatal fitness guidelines discussed above, targeted exercise interventions can also help bolster lumbopelvic stability and help mitigate pregnancy-related low back pain [34,38]. Pelvic floor muscle training (PFMT) including "Kegels" has been shown to have measurable benefits for pelvic stability along with reducing urinary incontinence and pelvic discomfort [35,36]. A meta-analysis of 11 randomized controlled trials involving 2347 pregnant women found that exercises, including pelvic floor strengthening and aerobics, reduced the risk of pregnancy-related LBP by 9% [50]. Deep core activation exercises act on the transverse abdominis to counteract trunk instability and have been shown to lower the risk of persistent low back symptoms [37,51]. For patients with sacroiliac joint dysfunction, pelvic stabilization exercises, targeted physical therapy, and sacroiliac belts have demonstrated effectiveness in reducing SI-related pain and improving functional mobility during pregnancy [52]. Prenatal yoga offers low-impact flexibility, balance, and postural exercises, with systematic reviews pointing to enhanced overall musculoskeletal function and comfort in pregnant populations. Yoga is thought to reduce pain during pregnancy by stimulating pressure receptors, which enhance vagal activity, lower cortisol and substance P, increase serotonin, and ultimately decrease stress, blood pressure, and the risk of pregnancy complications [53]. Yoga's effects, including hormonal, postural, and autonomic regulation, position it as a uniquely integrative modality, aligning physical and psychological benefits during pregnancy. Altogether, these targeted programs provide a comprehensive approach for reinforcing core and pelvic floor musculature, potentially diminishing biomechanical stresses that accompany pregnancy.

Pharmacological management of lower back pain during pregnancy is challenging due to health considerations of the developing fetus. Pain-relieving medications such as non-steroidal anti-inflammatory drugs (NSAIDs) are contraindicated late in pregnancy due to their teratogenic effects, while acetaminophen and the muscle relaxant cyclobenzaprine are usually considered safe to use throughout pregnancy [39,40]. Epidural steroid injection is typically only indicated when there is evidence of lumbar nerve compression and spine surgery is not typically indicated for pain during pregnancy unless the pain renders the patient incapacitated or with a neurologic deficit [28]. Surface topography studies, which are non-invasive imaging techniques using optical scanning systems to assess spinal curvature, represent a viable alternative to radiation-based imaging and can help monitor spine curvature and postural changes throughout pregnancy [23,41]. These techniques may be especially valuable in detecting early maladaptive compensations in spinal curvature or pelvic tilt that precede the onset of pain, thus enabling preemptive intervention. These minimally invasive methods could become routine in prenatal care, potentially preventing lower back pain and biomechanical changes that might lead to degenerative spine disease later in life.

4. Future Directions

Future studies examining back pain during pregnancy should focus on recruiting a more representative patient sampling, as many of the present studies lack heterogeneous patient populations. Understanding how the spine is affected during pregnancy for a more diverse patient population will make the results gleaned from these studies more generalizable. Additionally, there is a need for more longitudinal studies that focus on the long-term impact of pregnancy-related spinal changes. Most of the existing literature emphasizes acute pain and treatments during pregnancy, with fewer studies investigating chronic pain and its management in the postpartum period.

Exploring the use of non-invasive methods like surface topography to monitor spine curvature and postural changes throughout pregnancy could yield valuable insights. These technologies might offer a preventive strategy for both pregnancy-related back pain and degenerative spine disorders later in life. Another exciting avenue is the use of artificial intelligence (AI)-driven predictive models to create large datasets, including patient demographics, pre-pregnancy spinal imaging, gait analysis, and physiological changes, to forecast musculoskeletal complications such as low back pain and degenerative spinal disease. Using AI modeling can help stratify risk factors and personalize preventive strategies, such as targeted prenatal exercise programs or early postural interventions.

5. Limitations

Several studies investigating the spine during pregnancy are limited in generalizability due to the homogeneity of their sampled populations. Historical and ongoing medical mistrust may contribute to underrepresentation of marginalized groups in pregnancy-related research, limiting the applicability of findings to diverse patient cohorts. Additionally, many of the studies cited in this review are retrospective or observational in nature, inherently limiting the ability to draw causal inferences. Finally, biomechanical factors during pregnancy are difficult to isolate due to the influence of co-occurring variables such as hormonal fluctuations, psychosocial stressors, and varying levels of physical activity, all of which may contribute to pregnancy-related back pain.

6. Conclusions

Ethical considerations are particularly important in this field, as the safety of both mother and fetus must be prioritized. Imaging techniques such as CT and X-rays raise concerns about fetal harm from radiation exposure, making non-invasive alternatives preferable. Another important consideration is the informed consent process. Pregnant individuals may face additional stress or anxiety due to the nature of their condition, and it is essential that researchers ensure that participants fully understand the risks, benefits, and purposes of the studies in which they are participating. The autonomy of pregnant participants must be respected, particularly in vulnerable or high-risk pregnancies where added pressures may complicate decision-making.

Understanding the biomechanical changes that occur in the spine during pregnancy is crucial for improving maternal health and preventing long-term musculoskeletal issues for half of our world's population. Pregnancy induces significant anatomical and physiological adaptations, particularly in the spine and pelvis, which can lead to discomfort and increase the risk of conditions such as low back pain and degenerative spine disorders. This review emphasizes the importance of addressing these changes through early, non-invasive monitoring (e.g., surface topography) and preventive strategies such as targeted prenatal exercise. However, the existing literature is limited by homogeneous sampling and a lack of long-term follow-up, highlighting the need for more inclusive and longitudinal research. Advancing this understanding will enable earlier, more personalized interventions to support spinal health throughout pregnancy and beyond.

Funding: This research received no external funding.

Acknowledgments: We thank the reviewers for their valuable feedback which helped improve the manuscript.

Conflicts of Interest: The authors declare no conflict of interest.

References

1. Fast, A.; Shapiro, D.; Ducommun, E.J.; Friedmann, L.W.; Bouklas, T.; Floman, Y. Low-back Pain in Pregnancy. *Spine* **1987**, *12*, 368. [CrossRef]
2. Olsson, C.; Lena, N.W. Health-related quality of life and physical ability among pregnant women with and without back pain in late pregnancy. *Acta Obs. Gynecol. Scand.* **2004**, *83*, 351–357. [CrossRef]
3. Yamada, T.; Yamato, Y.; Hasegawa, T.; Yoshida, G.; Yasuda, T.; Banno, T.; Arima, H.; Oe, S.; Mihara, Y.; Ushirozako, H.; et al. Association between Pelvic Parameters and Vaginal Delivery. *Asian Spine J.* **2021**, *16*, 248–253. [CrossRef]
4. Oura, P.; Paananen, M.; Auvinen, J.; Niinimäki, J.; Niinimäki, M.; Karppinen, J.; Junno, J.-A. Gravidity, Parity, and Vertebral Dimensions in the Northern Finland Birth Cohort 1966. *Spine* **2018**, *43*, E1102. [CrossRef]
5. Calguneri, M.B.H.A.; Bird, H.A.; Wright, V. Changes in joint laxity occurring during pregnancy. *Ann. Rheum. Dis.* **1982**, *41*, 126–128. [CrossRef]
6. Hartmann, S.; Bung, P. Physical exercise during pregnancy—Physiological considerations and recommendations. *J. Perinat. Med.* **1999**, *27*, 204–215. [CrossRef]
7. Artal, R.; O'Toole, M. Guidelines of the American College of Obstetricians and Gynecologists for exercise during pregnancy and the postpartum period. *Br. J. Sports Med.* **2003**, *37*, 6–12. [CrossRef]
8. Dehghan, F.; Haerian, B.S.; Muniandy, S.; Yusof, A.; Dragoo, J.L.; Salleh, N. The effect of relaxin on the musculoskeletal system. *Scand. J. Med. Sci. Sports* **2014**, *24*, e220–e229. [CrossRef]
9. Conder, R.; Zamani, R.; Akrami, M. The Biomechanics of Pregnancy: A Systematic Review. *J. Funct. Morphol. Kinesiol.* **2019**, *4*, 72. [CrossRef]
10. Branco, M.; Santos-Rocha, R.; Vieira, F. Biomechanics of Gait during Pregnancy. *Sci. World J.* **2014**, *2014*, 527940. [CrossRef]
11. Pauk, J.; Swinarska, D. The impact of body mass on spine alterations in pregnant women: A preliminary study. *Technol. Health Care* **2018**, *26* (Suppl. S2), 665–669. [CrossRef]
12. Romero-Vargas, S.; Zárate-Kalfópulos, B.; Otero-Cámara, E.; Rosales-Olivarez, L.; Alpízar-Aguirre, A.; Morales-Hernández, E.; Reyes-Sánchez, A. The impact of body mass index and central obesity on the spino-pelvic parameters: A correlation study. *Eur. Spine J.* **2013**, *22*, 878–882. [CrossRef]

13. Whitcome, K.K.; Shapiro, L.J.; Lieberman, D.E. Fetal load and the evolution of lumbar lordosis in bipedal hominins. *Nature* **2007**, *450*, 1075–1078. [CrossRef]
14. Yoo, H.; Shin, D.; Song, C. Changes in the spinal curvature, degree of pain, balance ability, and gait ability according to pregnancy period in pregnant and nonpregnant women. *J. Phys. Ther. Sci.* **2015**, *27*, 279–284. [CrossRef]
15. Wang, T.; Wang, H.; Liu, H.; Ma, L.; Liu, F.Y.; Ding, W.Y. Sagittal spinopelvic parameters in 2-level lumbar degenerative spondylolisthesis: A retrospective study. *Medicine* **2016**, *95*, e5417. [CrossRef]
16. Foti, T.; Davids, J.R.; Bagley, A. A Biomechanical Analysis of Gait During Pregnancy. *J. Bone Jt. Surg. Am.* **2000**, *82*, 625. [CrossRef]
17. Wu, W.; Meijer, O.G.; Lamoth, C.J.; Uegaki, K.; van Dieën, J.H.; Wuisman, P.I.; de Vries, J.I.; Beek, P.J. Gait coordination in pregnancy: Transverse pelvic and thoracic rotations and their relative phase. *Clin. Biomech.* **2004**, *19*, 480–488. [CrossRef]
18. Borg-Stein, J.; Dugan, S.A.; Gruber, J. Musculoskeletal aspects of pregnancy. *Am. J. Phys. Med. Rehabil.* **2005**, *84*, 180–192. [CrossRef]
19. Fast, A.; Hertz, G. Nocturnal low back pain in pregnancy: Polysomnographic correlates. *Am. J. Reprod. Immunol.* **1992**, *28*, 251–253. [CrossRef]
20. Mens, J.M.A.; Pool-Goudzwaard, A.; Stam, H.J. Mobility of the Pelvic Joints in Pregnancy-Related Lumbopelvic Pain: A Systematic Review. *Obs. Gynecol. Surv.* **2009**, *64*, 200. [CrossRef]
21. Ritchie, J.R. Orthopedic considerations during pregnancy. *Clin. Obs. Gynecol.* **2003**, *46*, 456–466. [CrossRef]
22. Gutke, A.; Östgaard, H.C.; Öberg, B. Predicting persistent pregnancy-related low back pain. *Spine* **2008**, *33*, E386–E393. [CrossRef]
23. Michoński, J.; Walesiak, K.; Pakuła, A.; Glinkowski, W.; Sitnik, R. Monitoring of spine curvatures and posture during pregnancy using surface topography—Case study and suggestion of method. *Scoliosis Spinal Disord.* **2016**, *11* (Suppl. S2), 31. [CrossRef]
24. McCrory, J.L.; Chambers, A.J.; Daftary, A.; Redfern, M.S. The pregnant "waddle": An evaluation of torso kinematics in pregnancy. *J. Biomech.* **2014**, *47*, 2964–2968. [CrossRef]
25. Bird, A.R.; Menz, H.B.; Hyde, C.C. The effect of pregnancy on footprint parameters: A prospective investigation. *J. Am. Podiatr. Med. Assoc.* **1999**, *89*, 405–409. [CrossRef]
26. Krkeljas, Z. Changes in gait and posture as factors of dynamic stability during walking in pregnancy. *Hum. Mov. Sci.* **2018**, *58*, 315–320. [CrossRef]
27. Matsuda, N.; Kitagaki, K.; Perrein, E.; Tsuboi, Y.; Ebina, A.; Kondo, Y.; Murata, S.; Isa, T.; Okumura, M.; Kawaharada, R.; et al. Association Between Excessive Weight Gain During Pregnancy and Persistent Low Back and Pelvic Pain After Delivery. *Spine* **2020**, *45*, 319. [CrossRef]
28. Vermani, E.; Mittal, R.; Weeks, A. Pelvic Girdle Pain and Low Back Pain in Pregnancy: A Review. *Pain Pract.* **2010**, *10*, 60–71. [CrossRef]
29. Casagrande, D.; Gugala, Z.; Clark, S.M.; Lindsey, R.W. Low Back Pain and Pelvic Girdle Pain in Pregnancy. *J. Am. Acad. Orthop. Surg.* **2015**, *23*, 539–549. [CrossRef]
30. Mens, J.M.A.; Vleeming, A.; Snijders, C.J.; Koes, B.W.; Stam, H.J. Understanding peripartum pelvic pain: Implications of a patient survey. *Spine* **1996**, *21*, 1363–1369. [CrossRef]
31. Gilleard, W.L.; Brown, J.M. Structure and function of the abdominal muscles in primigravid subjects during pregnancy and the immediate postbirth period. *Phys. Ther.* **1996**, *76*, 750–762. [CrossRef]
32. Ireland, M.L.; Ott, S.M. The effects of pregnancy on the musculoskeletal system. *Clin. Orthop. Relat. Res.* **2000**, *372*, 169–179. [CrossRef]
33. Szkwara, J.M.; Hing, W.; Pope, R.; Rathbone, E. Compression shorts reduce prenatal pelvic and low back pain: A prospective quasi-experimental controlled study. *PeerJ* **2019**, *7*, e7080. [CrossRef]
34. Nascimento, S.L.; Surita, F.G.; Cecatti, J.G. Physical exercise during pregnancy: A systematic review. *Curr. Opin. Obs. Gynecol.* **2012**, *24*, 387–394. [CrossRef]
35. Mørkved, S.; Bø, K. Effect of pelvic floor muscle training during pregnancy and after childbirth on prevention and treatment of urinary incontinence: A systematic review. *Br. J. Sports Med.* **2014**, *48*, 299–310. [CrossRef]
36. Woodley, S.J.; Lawrenson, P.; Boyle, R.; Cody, J.D.; Mørkved, S.; Kernohan, A.; Hay-Smith, E.J.C. Pelvic floor muscle training for preventing and treating urinary and faecal incontinence in antenatal and postnatal women. *Cochrane Database Syst. Rev.* **2020**, *5*, CD007471. [CrossRef]
37. Wang, X.Q.; Zheng, J.J.; Yu, Z.W.; Bi, X.; Lou, S.-J.; Liu, J.; Cai, B.; Hua, Y.-H.; Wu, M.; Wei, M.-L.; et al. A meta-analysis of core stability exercise versus general exercise for chronic low back pain. *PLoS ONE* **2012**, *7*, e52082. [CrossRef]
38. Pennick, V.; Liddle, S.D. Interventions for preventing and treating pelvic and back pain in pregnancy. *Cochrane Database Syst. Rev.* **2013**, *8*, CD001139. [CrossRef]
39. Black, R.A.; Hill, D.A. Over-the-counter medications in pregnancy. *Am. Fam. Physician* **2003**, *67*, 2517–2524.
40. Rathmell, J.P.; Viscomi, C.M.; Ashburn, M.A. Management of nonobstetric pain during pregnancy and lactation. *Anesth. Analg.* **1997**, *85*, 1074–1087. [CrossRef]

41. Betsch, M.; Wehrle, R.; Dor, L.; Rapp, W.; Jungbluth, P.; Hakimi, M.; Wild, M. Spinal posture and pelvic position during pregnancy: A prospective rasterstereographic pilot study. *Eur. Spine J.* **2015**, *24*, 1282–1288. [CrossRef]
42. Sanderson, P.L.; Fraser, R.D. The influence of pregnancy on the development of degenerative spondylolisthesis. *J. Bone Jt. Surg. Br.* **1996**, *78*, 951–954. [CrossRef]
43. Cholewicki, J.; Lee, A.S.; Popovich, J.M., Jr.; Mysliwiec, L.W.; Winkelpleck, M.D.; Flood, J.N.; Pathak, P.K.; Kaaikala, K.H.; Reeves, N.P.; Kothe, R. Degenerative Spondylolisthesis Is Related to Multiparity and Hysterectomies in Older Women. *Spine* **2017**, *42*, 1643–1647. [CrossRef]
44. Çevik, S.; Yılmaz, H.; Kaplan, A.; Yetkinel, S.; Evran, Ş.; Çalış, F.; Akkaya, E.; Katar, S.; Baygül, A.; Hanımoğlu, H. Association between parity and lumbar spine degenerative disorders in young women. *Br. J. Neurosurg.* **2020**, *34*, 172–175. [CrossRef]
45. Güngör, E.; Karakuzu Güngör, Z. Obstetric-related lower back pain: The effect of number of pregnancy on development of chronic lower back pain, worsening of lumbar disc degeneration and alteration of lumbar sagittal balance. *J. Orthop. Surg. Res.* **2024**, *19*, 174. [CrossRef]
46. Schuller, S.; Charles, Y.P.; Steib, J.P. Sagittal spinopelvic alignment and body mass index in patients with degenerative spondylolisthesis. *Eur. Spine J.* **2011**, *20*, 713–719. [CrossRef]
47. Barrey, C.; Roussouly, P.; Le Huec, J.C.; D'Acunzi, G.; Perrin, G. Compensatory mechanisms contributing to keep the sagittal balance of the spine. *Eur. Spine J.* **2013**, *22* (Suppl. S6), 834–841. [CrossRef]
48. Bailey, J.F.; Sparrey, C.J.; Williams, F.M.K.; Curran, P.F.; Lotz, J.C.; Kramer, P.A. The Effect of Parity on Age-Related Degenerative Changes in Sagittal Balance. *Spine* **2020**, *45*, E210. [CrossRef]
49. Weis, C.A.; Pohlman, K.; Draper, C.; daSilva-Oolup, S.; Stuber, K.; Hawk, C. Chiropractic Care for Adults with Pregnancy-Related Low Back, Pelvic Girdle Pain, or Combination Pain: A Systematic Review. *J. Manip. Physiol. Ther.* **2020**, *43*, 714–731. [CrossRef]
50. Shiri, R.; Coggon, D.; Falah-Hassani, K. Exercise for the prevention of low back and pelvic girdle pain in pregnancy: A meta-analysis of randomized controlled trials. *Eur. J. Pain* **2018**, *22*, 19–27. [CrossRef]
51. Sapsford, R. Rehabilitation of pelvic floor muscles utilizing trunk stabilization. *Man. Ther.* **2004**, *9*, 3–12. [CrossRef]
52. Fitzgerald, C.M.; Bennis, S.; Marcotte, M.L.; Shannon, M.B.; Iqbal, S.; Adams, W.H. The impact of a sacroiliac joint belt on function and pain using the active straight leg raise in pregnancy-related pelvic girdle pain. *PM&R* **2022**, *14*, 19–29. [CrossRef]
53. Field, T. Prenatal exercise research. *Infant. Behav. Dev.* **2012**, *35*, 397–407. [CrossRef]

bioengineering

MDPI

Review

Advances in the Measurement and Interpretation of Intervertebral Motion in the Lumbar Spine: A Scoping Review

Alan Breen [1,*], Alexander Breen [1], Jonathan Branney [2], Alister du Rose [3] and Mehdi Nematimoez [4]

1 Faculty of Science and Technology, Bournemouth University, Poole BH12 5B, UK; alexbreen@bournemouth.ac.uk
2 Faculty of Health, Environment and Medical Sciences, Bournemouth University, Poole BH8 8GP, UK; jbranney@bournemouth.ac.uk
3 AECC School of Chiropractic, Health Sciences University, 13–15 Parkwood Road, Bournemouth BH5 2DF, UK; adurose@aecc.ac.uk
4 Institute for Advanced Biomechanics and Motion Studies, Offenburg University, Max Planck Str. 1, 77656 Offenburg, Germany; nemati80@gmail.com
* Correspondence: breenalan@gmail.com

Abstract

Background: Intervertebral motion is a fundamental aspect of spinal biomechanics, crucial for understanding lumbar spine function, pain mechanisms, and surgical outcomes. Various methods exist for measuring and interpreting it, each with its own advantages, limitations, and specific applications. However, a comprehensive and standard taxonomy of study types for the measurement and interpretation of in vivo intervertebral motion in the lumbar spine is lacking. Objectives: This review aimed to systematically identify, characterise, and categorise the diverse study types deposited in the literature. Eligibility criteria: Only studies in English and of lumbar spine intervertebral motion in living subjects were considered, and only those that employed objective measurement of motion sequences were included. Sources of evidence: A comprehensive literature search was performed in PubMed, CINAHL, and SCOPUS for articles published between January 2000 and October 2025. Charting methods: After removal of duplicates, all studies were subjected to Title and abstract screening, followed by full-text screening of potentially eligible studies. Data selected were charted into tables under the headings: author, year, country, purpose, technology, participants, measurement, interpretation, radiation dosage, and significance of findings. Results: Forty-nine studies were abstracted and are described under 11 study types. These formed a taxonomy constituting the following six categories: normal biomechanical mechanisms, pathological and injury mechanisms, direct kinematic measurement, spinal stabilisation, dynamic radiography, and clinical markers. The resulting taxonomy will serve as a resource for researchers, clinicians, and policymakers by facilitating a more coherent understanding of the field and promoting standardisation in research design and reporting.

Keywords: intervertebral motion; measurement; interpretation; lumbar spine; human

1. Introduction

The scientific literature pertaining to low back pain is large and heterogeneous. It contains many studies relating to intervertebral motion, yet these sometimes seem to lack coherence. Researchers who study intervertebral motion have increasingly noticed discrepancies in the design and reporting of studies resulting from an inadequate definition of the phenomenon. This may ignore the temporal component [1], and/or ignore the path that the motion has taken [2], and/or use only the global start and stop positions to calculate the range of motion [3]. As an illustration, Figure 1 shows intervertebral rotation from a healthy asymptomatic research participant in which the L5 vertebra has returned to its starting position by the end of the participant's maximum bending range, falsely indicating "no motion" at that level. The example also demonstrates the extent to which spatiotemporal interactions within a group of motion segments may appear conflicting but are not associated with any pain or disability.

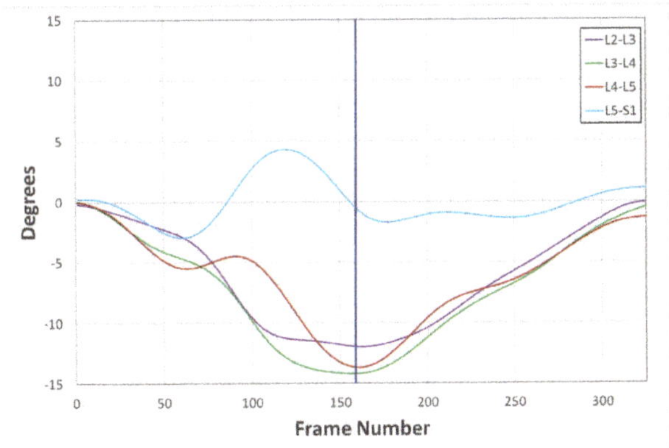

Figure 1. Intervertebral rotation graph of L2-3 to L5-S1 motion during standing flexion and return in an asymptomatic control participant measured using quantitative fluoroscopy (QF). (Blue vertical line indicates point of maximum global flexion.) L5–S1 exhibits anti-directional motion in its outward path, returning to its starting position by the end of its global range. (Source: Breen, A.; Breen, A.; Reference Database of Continuous Vertebral Flexion and Return. Open Science Framework 2022, https://doi.org/10.17605/osf.io/a27py). A video of the motion sequence depicted in this graph can be found in the Supplementary Materials.

Methods for measuring and interpreting such motion have reached higher prominence in the research literature as technological sophistication has progressed, and some expect that we are on the verge of making important advances [4–6]. However, clinical studies still tend to confine intervertebral motion to overall rotation in upright postures, ignoring, for example, incremental rotation, translation, velocity, and the sequencing of motion onsets. Passive restraint is seldom mentioned in studies of in vivo lumbar motion, but it is a necessary option in the investigation of the types of instability that could be obscured by loading and muscle activity.

Authors also sometimes imply that landmarks on the surface of the back over the lumbar spine represent the lumbar spine itself or compare (or combine) intervertebral motion findings with studies that have used different data collection protocols [7,8]. These discrepancies introduce uncontrolled variation (noise) that can obscure comparisons [9]. Thus, a comprehensive and standardised taxonomy of study types for the measurement and interpretation of in vivo intervertebral motion in the lumbar spine is needed if we are to compare findings across studies, synthesise evidence, and guide future research.

In particular, we wish to point out at the outset of this review that motion is the process of an object changing position over time, while displacement specifically refers to the change in position from an initial point to a final point, including direction. Thus, motion can be described using concepts like velocity and acceleration. By contrast, displacement is the change from an initial to a final position (spatial only) and the distance and direction between these positions. It follows that intervertebral range of motion is the maximum

https://doi.org/10.3390/bioengineering13020239

displacement between a pair of vertebral segments during a specific motion in a given plane and not, as the long-held definition states, "the difference between the two points of physiologic extent of motion" [10]. Instead, we contend that it can occur anywhere in the motion and not necessarily at the end of the path of any given section of the spine that contains the vertebral pair. Thus, "motion" is defined as "change in location from a spatial position A to a different position B, whereby the moving figure was located at position A at time T1 and then located at position B at another time T2" [11]. Talmy (1984) considers the path of motion to be the fundamental component of a motion event because, without a path, there is no motion [11]. In the case of motion events, a change in spatiotemporal location means a change in the spatial and temporal configuration of the path that is the result of the motion [12].

This scoping review aimed to address these issues by systematically identifying the study types employed in the literature to measure and interpret intervertebral motion. The intention was to map the prevailing concepts and create a taxonomy of studies that have involved the measurement and interpretation of in vivo lumbar spine intervertebral motion over the past 25 years. This information will hopefully serve as a resource for clinicians, researchers, and policymakers to help facilitate a more coherent understanding of practice in the field and promote high standards in research design and reporting [13].

2. Materials and Methods

This scoping review was conducted in accordance with the Arksey and O'Malley (2005) framework and following the PRISMA-ScR (Preferred Reporting Items for Systematic Review and Meta-Analyses (extension for Scoping Reviews) guidelines [14,15]. A protocol was preregistered on the OSF database (https://osf.io/dnwua/overview, Center for Open Science, Charlottsville, VA, USA) and a literature search was conducted by one author using the electronic databases: PubMed, SCOPUS and CINAHL Ultimate, with Boolean operators combining MeSH terms and key words relating to: "lumbar spine", "intervertebral motion", "measurement", "interpretation", "human" and "in vivo". Using the bibliographic search software: (https://rayyan.ai/, v5.4.0, Rayyan Systems Inc., Cambridge, MA, USA) reports were identified and imported into a ReadCube reference management database (https://papersapp.com, Digital Science & Research Solutions Ltd., Cambridge, MA, USA) for screening.

Eligible studies were published between January 2000 and October 2025 and involved human participants of any age group, where in vivo intervertebral motion of the lumbar spine (L1–S1) was measured and interpreted. Systematic reviews during this period were accessed and screened for additional eligible studies. Studies of interest generally encompass kinematics (e.g., range of motion (RoM), instantaneous centre of rotation (ICR), coupled motions), dynamics (e.g., motion patterns under load), and the methodologies used to acquire, analyse, and/or interpret these data. The context was any clinical or research setting in which in vivo lumbar intervertebral motion was studied. This included, but was not limited to, studies of healthy individuals, patients with low back pain, post-surgical patients, athletes, and people in occupational settings. An example search string for PubMed might include:

> "(lumbar spine[MeSH Terms] OR lumbar vertebra* OR lumbosacral OR low back) AND (intervertebral disc[MeSH Terms] OR intervertebral motion OR spinal motion OR kinematics OR dynamics OR range of motion OR coupled motion) AND (in vivo OR living OR human) AND (measurement OR imaging OR fluoroscopy OR MRI OR CT OR X-ray OR radiography OR optical tracking OR motion capture OR video analysis OR wearable sensor*)"

https://doi.org/10.3390/bioengineering13020239

All empirical study designs were considered (e.g., observational studies, experimental studies, case series, cohort studies, cross-sectional studies). Narrative reviews, theoretical papers, animal studies, in vitro studies, cadaveric studies, and biomechanical models without in vivo validation were excluded. Studies were eligible if they satisfied all the following: Quantitative measurement, Continuous intervertebral motion (>3 Hz), Lumbar spine, In vivo, Human, and Article in English. Studies considered ineligible were: Reviews, Book chapters, Commentaries, Case studies, Qualitative studies, Discussions, Editorials, Letters, Guidelines, Protocols, and Conference papers.

After removal of duplicates, all studies were subjected to title and abstract screening for eligibility by the lead author and one other, and a subset was identified for full-text screening. Cases of disagreement were arbitrated by a third author. This procedure was also followed for the charting process, in which information from studies eligible for full-text screening was initially extracted into a spreadsheet under the headings: lead author, year, short title, qualifying descriptions, include/exclude (with reasons), and access information (e.g., DOI). The results were considered by a different author and, in the event of disagreement, arbitrated by a further one before being allocated to their final charting position. Any additional articles chosen from private databases that did not appear in the search were considered in the same way at this stage.

Data items from studies that were accepted for inclusion were charted into tables according to their type and under the descriptive headings: author/year, country of lead author, purpose of study, technology, participants/gender, measurement, interpretation, radiation dose (if applicable), and significance of findings (see Supplementary Materials). Studies were not assessed for methodological quality beyond requiring correct nomenclature. However, studies were only charted if they represented peer-reviewed and published research that met the criteria for eligibility and inclusion as independently verified by three authors. The charting process strove to present information from the studies as distinctive types rather than to evaluate them. All authors approved the results of the charting process.

3. Results

Database searching identified 793 studies, of which 216 were duplicates and were removed, leaving 577 for consideration (Figure 2). After title and abstract screening, 473 of these were excluded, leaving 104 for full-text screening. Of these, 60 were excluded, mainly because they did not study intervertebral motion (11), did not study motion (14), or did not study continuous motion (13). After the addition of five studies from other sources known to the authors, and following full-text screening, 49 studies were found to be eligible for review. The 5-year distribution of publication dates of the 104 that underwent full-text screening shows an approximate doubling of articles that were published 5 years after the first decade of the study period (Figure 3).

Table 1. List of study types.

	Study Type	Number of Studies
1	Intervertebral motion level interactions	4
2	Disc degeneration kinematics	5
3	Implanted markers and internal sensors	5
4	Assessment of lumbar orthoses	2
5	Technology development	10
6	Back pain biomarkers	7
7	Post stabilisation dynamics	2
8	Normative lumbar kinematics	5
9	Intervertebral force deformation	4
10	Aerospace lumbar kinematics	2
11	Soft tissue artefact measurement	3

https://doi.org/10.3390/bioengineering13020239

The 49 studies that met the criteria for eligibility and inclusion are represented as 11 types and shown in Table 1. These had a total of 2151 participants with lead authors from 11 countries: USA (17), UK (15), Japan (3), China (3), Canada (3), Switzerland (2), Italy (2), Australia (1), Finland (1), Germany (1), and Iran (1).

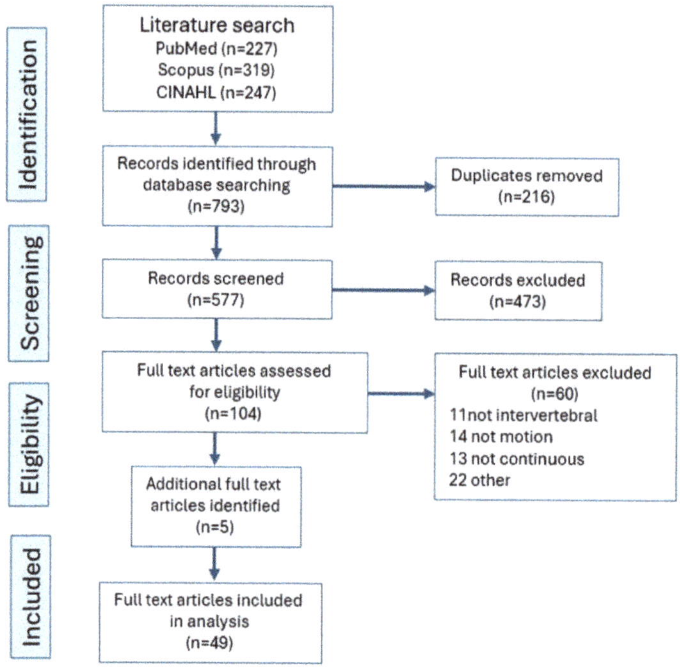

Figure 2. PRISMA flowchart.

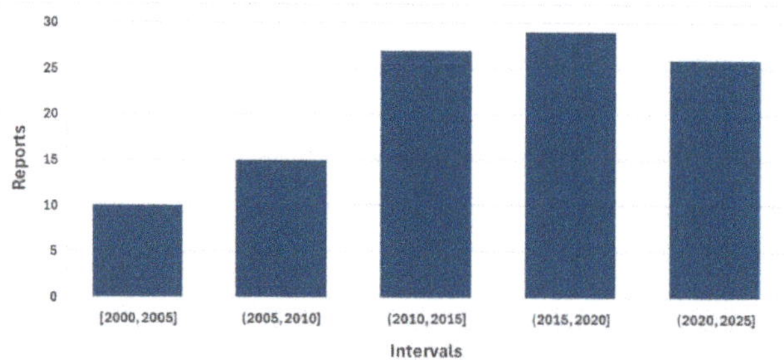

Figure 3. 5-year distribution of publication dates (2000–2025) of the 104 studies that underwent full-text screening.

3.1. Intervertebral Motion Interactions

Among the first phenomena to be studied in relation to the interactions between the intervertebral motion segments (and that occurred either prior to the review period, and not therefore part of the results [16], or during it [17]), were the relative onsets of angular intervertebral motion during bending. Like many of the studies in this review, motion X-rays were used to register the kinematics, which initially presented some technological barriers to measurement that had to be overcome to progress the work. It was found that after the initial Japanese studies, some 20 years had elapsed before automated vertebral image tracking and digital imaging improved the analysis process sufficiently to encourage

https://doi.org/10.3390/bioengineering13020239

phase lag studies to recommence. When it did, the value of measuring motion contributions for making comparisons between individuals or groups, rather than raw values that were prone to wide variation, became apparent [18]. Then, to further explore apparent kinematic biomarkers for chronic, nonspecific back pain, principal component analysis was used [19]. This was followed by the classification of the motion path types originally discovered by Harada et al. [17], using cineradiography, but now incorporating the first derivatives (velocity) of sagittal plane intervertebral rotation [20]. This demonstrated, through cluster analysis, that the top-down cascade pattern of lumbar flexion referred to by Harada in 2000 [17] appeared to be the most common pattern of "phase lag". However, clinical studies are needed to understand the relevance of this in symptomatic states.

3.2. Disc Degeneration Kinematics

The chronology of the use of intervertebral motion in the study of disc degeneration began with Takayagani et al., 2001 [21], comparing flexion and return for rotation and translation in 41 patients with L4 degenerative spondylolisthesis and 20 controls. To the best of our knowledge, this was the first study to report on spondylolisthesis kinematics. It found that rotation and translation were simultaneous in controls, but not in patients, where the more severe cases exhibited decreased translation. Following this, 7 years later, Hasegawa et al. [22] used an intraoperative spinous distraction apparatus to measure restraint in terms of absorption energy and the neutral zone (NZ), taking disc height loss into account. The authors found that "Stiffness demonstrated a significant negative and NZ a significant positive relationship with disc height". However, there were no significant differences between spines with "collapsed" discs than those without, although the NZ value was higher in those without the collapsed types and more sensitive to this measure in degenerative segments with preserved disc height. The authors went on to suggest that "degenerative segments with preserved disc height have a latent instability compared to segments with collapsed discs". These two studies illustrate an interesting comparison of the kinematic versus the loading effects of disc degeneration.

Following the appearance of quantitative fluoroscopic systems, Breen et al. [23,24] compared the kinematics of non-spondylolisthesis patients who had early-to-moderate disc degeneration to pain-free controls in both standing and lying flexion-extension. The authors' intention was to test the hypothesis of Farfan and Kirkaldy-Willis that early disc degeneration is associated with instability [25]. This study found that patients with disc degeneration exhibited more disrupted interactions between segments in terms of motion sharing, but only in passive recumbent motion. There was also only a weak-to-moderate negative correlation between disc-height loss and weight-bearing RoM, and no correlation with translation or laxity. Meanwhile, Dombrowski et al. [2], using standing 3D fluoroscopy and CT-generated bone models to study continuous dynamic sagittal rotation in degenerative spondylolisthesis, compared their findings with those from static flexion-extension radiographs. Their study found that 42% of the spondylolisthesis group demonstrated aberrant mid-range motion on intervertebral dynamic motion analysis and greater flexion translation on dynamic than static imaging. Caution is advised, however, in the interpretation of these latest studies, none of which assessed more than 10 patient-control pairs each.

3.3. Implanted Markers and Internal Sensors

Implanted marker systems in the spine are generally considered too invasive for other than serious spinal disorders, however they are still in use, although more usually for the measurement of displacement using static radiographs than for the measurement of actual intervertebral motion in disc replacement surgery [26]. However, to capture motion

https://doi.org/10.3390/bioengineering13020239

and avoid radiation, implanted wires and screws have also been used to mount sensors percutaneously for tracking by motion capture systems.

Five studies were found in this category, two of chronic back pain patients [27,28] and three of healthy controls [29–31]. All used optical sensors were attached percutaneously to pedicle screws or K-wires implanted into the spines of chronic back pain patients, and no further ones seem to have been published since 2013. The patient studies used 3D and 2D analysis, respectively, and did not involve control groups. Their authors suggest that they could be useful in testing internal fixation systems where the spine is being surgically exposed.

One system [29] used ultrasound tracking of LEDs from L12–L2 and found within-subject variation of 0-6-42% during flexion-extension, lateral bending, and axial rotation, respectively. The last two studied normal ranges of the same motions from L1–S1 (the first study to compare the whole lumbar spine in vivo using bone pins) [30] and reported that the main intervertebral motion during gait is mid-lumbar and coronal [31].

3.4. Assessment of Lumbar Orthoses

Two early US studies [32,33] were found to have used 2D quantitative fluoroscopy in small groups of controls (4 and 10, respectively). Both used free (uncontrolled) bending during flexion/extension. The first tested a custom-fitted thoracolumbosacral orthosis and found that it reduced flexion range by 2/3, while the second compared no orthotic, a soft orthotic, a semirigid orthotic, and a semirigid thoracolumbosacral orthotic and found that all three devices limited L3–4 and L4–5 flexion, but not L5–S1.

3.5. Technology Development

The 10 studies identified mainly challenged the technical viability of the technologies in this review for reliability, accuracy, and radiation dosage during controlled versus uncontrolled weight bearing and recumbent flexion, extension, and lateral bending. These mainly employed 2D quantitative fluoroscopy [7,34–41] with one study of the reliability of an implanted Kirshner wire system for measurement of 3D L1–S1 motion during gait [42]. This study reported similar findings to Mac Williams et al. [31] (see above). Studies were mainly conducted by US and UK groups, also with representation from Italy and Canada.

3.6. Back Pain Biomarkers

In their review of biomarker types in relation to chronic pain, Reckziegel et al. [43] concluded that the field has been advancing in terms of the investigation of hypotheses pertaining to underlying mechanisms. For chronic, nonspecific low back pain, the tendency has often been to frame these hypotheses around disordered intervertebral motion. The chief suspects have been excessive angular and/or linear RoM, laxity (as reflected by a steep starting motion gradient or attainment rate), or an increased attainment rate around the mid-range of the intervertebral motion path—suggestive of instability.

In our review, we found no clear evidence for excessive RoM, whether using static or dynamic radiographs [44]. However, the attainment rate during standing flexion was assessed by Teyhen et al. [45] using 2D quantitative fluoroscopy, who found it to be increased, although still with no evidence of increased RoM. A subsequent subgroup study comparing chronic nonspecific back pain patients and controls in flexion and return [46] found a significantly smaller rotational motion contribution during the return phase from standing flexion at L5–S1 in patients.

Similar 2D fluoroscopy studies using passive recumbent flexion did not detect this [23,47,48] but instead found greater inequality of motion sharing (MSI) in patients with chronic nonspecific back pain. A further passive recumbent flexion and return study probed this further by measuring the timing of peak velocity during intervertebral flexion. This was found to occur earlier in the motion path at L5–S1 in back pain patients than in controls [49]. Taken together, these studies may indicate MSI as a biomarker, linking disruption of passive restraint and chronic back pain.

Meanwhile, a recent 3D quantitative fluoroscopy study with CT models that assessed standing flexion-extension and lateral bending in patients with chronic back pain measured intervertebral rotations, translations, and coupling and found heterogeneity in these variables [50]. This study did not have a control group, and further research is required. Another recent biplanar fluoroscopy study using 3D CT models examined the movement of endplate centres with respect to the sacrum before and after lifting in patients with recurrent back pain and healthy controls [51]. This study found significant inter-test differences, with and without fatigue, in translation and z-axis rotation in both groups. This occurred slightly later in flexion-extension motion after fatigue. The authors suggest that this may indicate a protective mechanism or a role in dysfunction.

3.7. Post Stabilisation Dynamics

Only two studies were found that addressed this. The first presented preliminary data and used biplanar fluoroscopy to track implanted metal markers in five patients, 2, 3, and 6 months post-fusion, comparing intervertebral displacement at the beginning and end of trunk motion with minimum/maximum intervertebral displacement over the motion path, however, these did not correspond [3].

The other study [52] used 2D fluoroscopy in 24 patients with flexible total disc replacements (TDRs) 6 weeks and 5 years post-surgery. The upper and lower TDR endplates were tracked, and the continuous RoM with 10 repeated cycles was measured in the coronal and sagittal planes during treadmill walking, indicating partial preservation of motion.

3.8. Normative Lumbar Kinematics

Five studies were eligible for the reporting of in vivo normative intervertebral motion. All were from the authors' own labs, and all used the same 2D fluoroscopy technology protocols [53]. They cannot, therefore, be extrapolated to other systems. A normative database is presented, consisting of 127 anonymized pain-free controls imaged during controlled weight bearing with recumbent left, right, flexion, and extension lumbar motion from vertebral midplane angles throughout the motion sequences as captured at 15 fps. These data are publicly available from the Open Science Framework database (https://osf.io/a27py/) [46]. They can be used to display and manipulate the motion sequence data with the appropriate transformations for comparison with users' own studies, if desired. For weight-bearing flexion and return, transformation of intervertebral angles to proportional contributions has revealed the level-by-level normative motion contribution patterns as a phenotype for comparison with clinical and experimental studies [46,54] (Figure 4).

The other studies present the lumbar L2–S1 IV-RoM, translation, laxity, MSI, MSV, regularity, and symmetry from individual studies of healthy control populations in different configurations. All studies present their effective radiation dosage. Two studies present the minimal detectable change and intrasubject repeatability in healthy controls of some of the studies over a 6-week period [55,56]. These show evidence that measurements of translation and MSV are not suitable for longitudinal studies and that surface EMG measurements of longissimus thoracis, lumborum, and multifidus amplitudes correlate with maximum RoM at L4–5 and L5–S1 during controlled standing flexion and return motion but are essentially silent during passive recumbent flexion and return.

https://doi.org/10.3390/bioengineering13020239

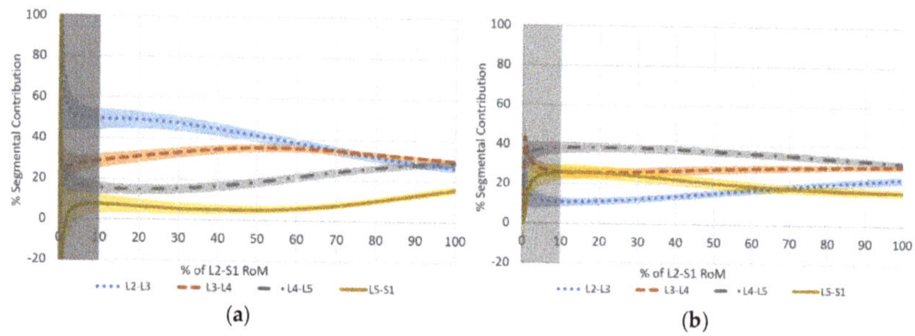

Figure 4. Average segmental contributions (L2–3 to L5–S1) during (**a**) standing flexion and (**b**) return in 127 healthy controls. Shaded areas indicate 95% confidence intervals. (from: Breen, A.; Breen, A. Reference Database of Continuous Vertebral Flexion and Return. Open Science Framework 2022, https://doi.org/10.17605/osf.io/a27py). (Grey shading over the first 10% of the motion in both graphs indicates high error areas to be ignored due to small values in the percentage calculations).

3.9. Intervertebral Force-Deformation

Four studies were found. All were of healthy controls and involved bending and lifting functional loads. All used 3D fluoroscopy linked to CT-generated bone models, one measuring sagittal intervertebral rotation and translation [57] and another measuring average ICR locations and migration ranges [58]. These two found linear relationships between load, rotation, translation, and level-specific changes in ICR location and dispersion with increased loading. In the future, this information could contribute to the establishment of biomechanical models and normative values with which to compare conditions and interventions [59].

One of the other studies investigated changes in disc height and shear strain patterns and distributions during lifting [60], finding that these restraint patterns can be mapped and could contribute to models. The last study tested facet joint kinematics during lifting and found greater translation at L3 and L4 than at other levels in response to loads, which may help to explain the pathogenesis of structural changes associated with load bearing [61].

3.10. Aerospace Lumbar Kinematics

Two studies addressed the problem of post-spaceflight disc herniation in relation to lumbar spine kinematics using 2D quantitative fluoroscopy and MRI. The first examined relationships between prolonged exposure to microgravity and weight-bearing flexion kinematics in 12 ISS crew members, six of whom developed disc hernia symptoms post-flight [62]. This study found, with MRI, that post-spaceflight disc herniation was associated with compromised multifidus quality and, using 2D fluoroscopy, that symptomatic returning astronauts had reduced flexion-extension RoM at their L3–4 and L4–5 levels.

The second study tested a microgravity countermeasure skinsuit that compressed the spine overnight to limit the intervertebral stiffness that follows a period of absence of axial loading and preserves motion in the spine. Twenty healthy controls were tested with and without skinsuit use, finding significantly more lumbar intervertebral flexion and translation range after the skinsuit was used [63]. These studies may have implications for longer space missions, including interplanetary travel.

3.11. Soft Tissue Artefact (STA) Measurement

Three recent studies attempting to measure soft tissue artefacts were eligible for this review. The first [64] compared optical motion capture in six controls for flexion and extension, with one landmark's motion measured with 3D quantitative fluoroscopy driven by CT models. With this protocol, STAs for flexion ranged from 4.0 mm for L1–3 to 13.5 mm for L4–5, and for extension, 2.7 mm for L4–5 and 6.1 mm for L1–3. Errors varied with anatomical direction, marker location, vertebral level, and bending phase.

The second study [8] aimed to evaluate static placement errors as well as STA from MoCap optical marker clusters during flexion-extension and lateral bending in 39 low back pain patients, taking patient characteristics into account. Data from surface markers positioned over L1 and L5 spinous processes were compared to continuous data from 3D fluoroscopy with CT modelling using a volumetric tracking process. Static placement errors were greatest in a superior-inferior direction (29.5 mm), and the L1–L5 RMS STA error in the sagittal plane ranged from 1.7° to 23.6°. This was larger in flexion-extension than in side-bending. Errors were participant-dependent and unrelated to age and BMI.

The third study [65] compared skin-based MoCap with accelerometers attached over L2, L4, and S1, and 2D quantitative fluoroscopy (QF) at L2–3, L3–4, L4–5 and L5–S1 for contemporaneous recording during flexion and extension in 20 controls. The 95% limits of agreement for L2–S1 between technologies for flexion were: QF vs. MoCap −10.1° to +10.1°, QF vs. accelerometer −9.8° to +9.8°, and accelerometer vs. MoCap −1.0° to 1.1°. For extension, they were: QF vs. MoCap −5.7°–+5.5°, QF vs. accelerometer −6.3° to +6.6°, and for accelerometer vs. MoCap −0.9°–+0.9°. Differences from QF tended to be greater in the lower lumbar spine, and in flexion compared to extension. However, differences between participants were highly variable, confirming the results of the other studies.

Despite the difficulty of placing external markers on the skin over adjacent vertebrae, preventing adjacent intervertebral comparisons from being made between technologies, it can probably be concluded that skin-based marker systems do not measure underlying intervertebral motion accurately. Conversion factors that would improve this do not currently seem to be within reach.

3.12. A Taxonomy of Study Types

The need for a taxonomy of study types suggested by these topics is firmly planted in the unknowns of spinal biomechanics and sustained by our erstwhile inability to measure intervertebral motion in vivo. A suggested taxonomy is presented in Table 2.

Table 2. Taxonomy of study types.

Normal biomechanical mechanisms	**Pathological and injury mechanisms**
Normative lumbar kinematics	Disc degeneration kinematics
Intervertebral motion level interactions	Aerospace lumbar kinematics
Intervertebral force deformation	
Direct kinematic measurement	**Spinal stabilisation**
Implanted markers and internal sensors	Assessment of lumbar orthoses
	Post stabilisation dynamics
Dynamic radiography	**Clinical markers**
Technology development	Nonspecific back pain biomarkers
Soft tissue artefact measurement	

4. Discussion

This review presents the results and trajectory of investigation choices by biomechanics researchers over the past quarter century by identifying the main study types that have addressed the measurement and interpretation of intervertebral motion, as defined by the motion paths and between-level interactions over time, as well as by displacement. As shown in Figure 1, intervertebral motion paths cannot always be expected to be parallel. If they were, the proportional motion that each level contributes would be constant. In real life, they vary during bending tasks, and what keeps trunk motion smooth during a bending task is continuous compensatory interactions between the levels. Access to this

https://doi.org/10.3390/bioengineering13020239

continuous motion information makes possible the measurement and interpretation of data from a full spatiotemporal suite of kinematic measures, including laxity, dynamic motion share, velocity, and acceleration. So far, the role of these factors in the clinical management of lumbar spine disorders has only begun to be explored, whereas the documented practical applications of systems have been mainly related to the aims of the 49 studies listed above. Clinicians, especially surgeons and physical therapists, now request investigations to guide treatment for suspected mechanical disorders (such as instability, fusion loosening, adjacent segment disorder, or abnormal post-injury movement) and to aid the planning of cognitive therapy, movement training, manual therapies, and occupational interventions.

In this scoping review, we have attempted to map the key areas of current study and create a taxonomy of vivo lumbar spine intervertebral motion measurement and interpretation from studies published over the past 25 years. The results indicate that research outputs in this field have increased, and the Study types suggest the headline topics. For example, the need to better understand Normal lumbar kinematics is seen as paramount and should encourage the standardisation of available technologies, which at present appear to favour imaging. However, although there are state-of-the-art techniques such as Statistical Parametric Mapping (SPM) for exploring between-level motion path relationships using vector field analysis [66], motor control strategies are inherently interconnected over time. Therefore, a point-by-point evaluation may not be sufficient to fully capture the temporal dependencies underlying coordinated movement. Nevertheless, intervertebral force deformation is also accessible via the kinematic measurement of strains, as demonstrated by Byrne et al. [60].

The attention given to Disc degeneration kinematics is unsurprising given the ubiquitousness of this condition, while the appearance of microgravity's effects on intervertebral kinematics in space has made an unexpected appearance as Aerospace lumbar kinematics. By contrast, interest in direct intervertebral kinematic measurements using implanted markers and internal sensors placed in situ during surgery may be waning, as, despite the opportunity for greater precision in motion measurement, they did not appear after 2013 in our review.

Given the progress made in disc replacement surgery, including around motion preservation systems, their rare appearance in in vivo kinematics research under post-stabilisation dynamics is surprising and is perhaps attributable to the complexity, scarcity, and cost of fluoroscopic systems for measurement [6,67]. These systems were dominant in the articles found in this review, where technology development has been heavily focused on reducing inaccuracies, imprecision, and, to some extent, radiation dosage. Research into soft tissue artefact measurement, however, has been slow to appear, but now seems largely complete, and some of the questions surrounding the assessment of lumbar orthoses appear to have been addressed.

This leaves nonspecific back pain biomarkers to complete the present taxonomy. Given the heterogeneity and complexity of this condition across biopsychosocial domains, mechanical biomarkers may best contribute to back pain assessment and treatment if considered alongside non-biomechanical mechanisms and interventions such as cognitive behavioural therapy [68,69].

Limitations and Further Work

Some studies may have been missed or erroneously excluded by our literature search, especially with respect to the recording and analysis of continuous intervertebral motion data. Non-English language studies that might have contributed were nonetheless excluded. A major weakness for interpretation was the number of studies with small samples, and we recommend replication of those that presented promising results for the discovery

https://doi.org/10.3390/bioengineering13020239

of nonspecific back pain biomarkers. Recent scoping reviews relating to the kinematics of the cervical spine and other body joints have called for greater standardisation and sophistication of data recording, image tracking, analysis, and interpolation, and we strongly support this call [6,67]. There is also a particular need to embed automated image registration as a way to reduce the laboriousness of analysis.

Radiation dosage is a cause for concern with some studies, and greater efforts to provide 3D MRI models for biplanar fluoroscopy investigations have been recommended where 3D analysis is judged necessary [70]. We also recommend kinematic studies of the performance of total disc replacements and investigations of adjacent segment disorder, utilising a more complete choice of kinematic measures, including the inclusion of sagittal alignment for structural derangements and extending image acquisition to include recumbent passive motion.

5. Conclusions

The research literature in respect of lumbar intervertebral motion is substantial, variable in subject matter, and burdened by a legacy of semantic discrepancies. However, there have been important advances around clinically useful measurement and interpretation, as well as the prospect of others to come. The result of this review challenges the traditional standard of care based on static radiographs, but with no simple, inexpensive, or readily available option to replace it. Hopefully, it will help to clarify available options and provide incentives to pursue the promising ones.

Supplementary Materials: The following supporting information can be downloaded at: https://www.mdpi.com/article/10.3390/bioengineering13020239/s1, Table S1: Supplementary file for: Advances in the measurement and interpretation of intervertebral motion in the lumbar spine: A scoping review; Video S1: Digital fluoroscopic video of a standing flexion and return motion sequence in an asymptomatic male showing the angular motion paths of intervertebral levels L2–3 to L5–S1 in Figure 1.

Author Contributions: Conceptualization, A.B. (Alan Breen), A.B. (Alexander Breen), M.N., J.B. and A.d.R.; Investigation, A.B. (Alan Breen) and A.B. (Alexander Breen); Methodology, A.B. (Alan Breen), A.B. (Alexander Breen) and J.B.; Literature search, A.B.(Alan Breen), A.B. (Alexander Breen) and J.B.; Data extraction, A.B. (Alan Breen), A.B. (Alexander Breen) and J.B.; Software, A.B. (Alexander Breen); Adjudication, A.d.R., J.B. and M.N.; Data curation, A.B. (Alan Breen) and A.B. (Alexander Breen); Writing—original draft preparation, A.B. (Alan Breen); Writing—review and editing, A.B. (Alan Breen), A.B. (Alexander Breen), M.N., A.d.R. and J.B. All authors have read and agreed to the published version of the manuscript.

Funding: This research received no external funding.

Institutional Review Board Statement: Not applicable.

Data Availability Statement: The original contributions presented in this study are included in the article/supplementary material. Further inquiries can be directed to the corresponding author.

Acknowledgments: During the preparation of this study, the authors used Gemini AI (V2.5) in formulating its context in the protocol. The authors have reviewed and edited the output and take full responsibility for the content of this publication.

Conflicts of Interest: The authors declare no conflicts of interest.

Abbreviations

The following abbreviations are used in this manuscript:

RoM	Range of motion
ICR	Instantaneous centre of rotation

MSI	Motion sharing inequality
MSV	Motion sharing variability
QF	Quantitative fluoroscopy
CT	Computed tomography
STA	Soft Tissue Artefact
NZ	Neutral Zone
TDR	Total disc replacement
SPM	Statistical parametric mapping
LED	Light-emitting diode
ISS	International Space Station

References

1. Zwambag, D.P.; Beaudette, S.M.; Gregory, D.E.; Brown, S.H.M. Distinguishing between Typical and Atypical Motion Patterns amongst Healthy Individuals during a Constrained Spine Flexion Task. *J. Biomech.* **2019**, *86*, 89–95. [CrossRef]
2. Dombrowski, M.E.; Rynearson, B.; LeVasseur, C.; Adgate, Z.; Donaldson, W.F.; Lee, J.Y.; Aiyangar, A.; Anderst, W.J. ISSLS Prize in Bioengineering Science 2018: Dynamic Imaging of Degenerative Spondylolisthesis Reveals Mid-Range Dynamic Lumbar Instability Not Evident on Static Clinical Radiographs. *Eur. Spine J.* **2018**, *27*, 752–762. [CrossRef]
3. Anderst, W.J.; Vaidya, R.; Tashman, S. A Technique to Measure Three-Dimensional in Vivo Rotation of Fused and Adjacent Lumbar Vertebrae. *Spine J.* **2008**, *8*, 991–997. [CrossRef]
4. Auerbach, J.D.; Wills, B.P.D.; McIntosh, T.C.; Balderston, R.A. Evaluation of Spinal Kinematics Following Lumbar Total Disc Replacement and Circumferential Fusion Using In Vivo Fluoroscopy. *Spine* **2007**, *32*, 527–536. [CrossRef]
5. Widmer, J.; Fornaciari, P.; Senteler, M.; Roth, T.; Snedeker, J.G.; Farshad, M. Kinematics of the Spine Under Healthy and Degenerative Conditions: A Systematic Review. *Ann. Biomed. Eng.* **2019**, *47*, 1491–1522. [CrossRef]
6. Aurangzeb, A.; Zailani, M.H.B.M.; Tan, Q.H.; Leong, D.; Kumar, D.S.; Janssen, M.E.; Chew, Z. From Static X-Rays to Dynamic 3D Tracking: A Scoping Review of Cervical Spine Imaging-Based Motion Assessment. *Eur. Spine J.* **2025**, 1–9. [CrossRef]
7. Cheng, B.; Castellvi, R.J.; Davis, D.C.; Lee, M.P.; Lorio, R.E.; Prosko, D.; Wade, C. Variability in Flexion Extension Radiographs of the Lumbar Spine: A Comparison of Uncontrolled and Controlled Bending. *Int. J. Spine Surg.* **2016**, *10*, 20. [CrossRef] [PubMed]
8. Johnson, M.E.; LeVasseur, C.; Gale, T.; Megherhi, S.; Shoemaker, J.; Pellegrini, C.; Gray, E.C.; Smith, P.; Anderst, W.J. Lumbar Spine Marker Placement Errors and Soft Tissue Artifact during Dynamic Flexion/Extension and Lateral Bending in Individuals with Chronic Low Back Pain. *J. Biomech.* **2024**, *176*, 112356. [CrossRef] [PubMed]
9. Malakoutian, M.; Volkheimer, D.; Street, J.; Dvorak, M.F.; Wilke, H.-J.; Oxland, T.R. Do in Vivo Kinematic Studies Provide Insight into Adjacent Segment Degeneration? A Qualitative Systematic Literature Review. *Eur. Spine J.* **2015**, *24*, 1865–1881. [CrossRef] [PubMed]
10. Panjabi, M.M.; White, A.A. *Clinical Biomechanics of the Spine*; JB Lippincott: Philadelphia, PA, USA, 1990.
11. Talmy, L. Lexicalization Patterns. In *Language Typology and Synchronic Description*; Schopen, T., Ed.; Cambridge University Press: Cambridge, UK, 1984; Volume 3, pp. 47–159.
12. Filipović, L.; Ibarretxe-Antuñano, I. Semantic Typology. In *Cognitive Linguistics; Key Topics*; Dąbrowska, E., Divjak, D., Eds.; De Gruyter Mouton: Berlin, Germany; Boston, MA, USA, 2007; p. 6. ISBN 978-3-11-062299-7.
13. Peters, M.D.J.; Godfrey, C.M.; Khalil, H.; McInerney, P.; Parker, D.; Soares, C.B. Guidance for Conducting Systematic Scoping Reviews. *Int. J. Evid.-Based Healthc.* **2015**, *13*, 141–146. [CrossRef]
14. Arksey, H.; O'Malley, L. Scoping Studies: Towards a Methodological Framework. *Int. J. Soc. Res. Methodol.* **2005**, *8*, 19–32. [CrossRef]
15. Tricco, A.C.; Lillie, E.; Zarin, W.; O'Brien, K.K.; Colquhoun, H.; Levac, D.; Moher, D.; Peters, M.D.J.; Horsley, T.; Weeks, L.; et al. PRISMA Extension for Scoping Reviews (PRISMA-ScR): Checklist and Explanation. *Ann. Intern. Med.* **2018**, *169*, 467–473. [CrossRef]
16. Kanayama, M.; Abumi, K.; Kaneda, K.; Tadano, S.; Ukai, T. Phase Lag of the Intersegmental Motion in Flexion-Extension of the Lumbar and Lumbosacral Spine. An in Vivo Study. *Spine* **1996**, *21*, 1416–1422. [CrossRef] [PubMed]
17. Harada, M.; Abumi, K.; Ito, M.; Kaneda, K. Cineradiographic Motion Analysis of Normal Lumbar Spine During Forward and Backward Flexion. *Spine* **2000**, *25*, 1932–1937. [CrossRef] [PubMed]
18. Breen, A.; Breen, A. Dynamic Interactions between Lumbar Intervertebral Motion Segments during Forward Bending and Return. *J. Biomech.* **2020**, *102*, 109603. [CrossRef]
19. Brownhill, K.; Mellor, F.; Breen, A.; Breen, A. Passive Intervertebral Motion Characteristics in Chronic Mid to Low Back Pain: A Multivariate Analysis. *Med. Eng. Phys.* **2020**, *84*, 115–125. [CrossRef] [PubMed]
20. Nematimoez, M.; Breen, A.; Breen, A. Spatio-Temporal Clustering of Lumbar Intervertebral Flexion Interactions in 127 Asymptomatic Individuals. *J. Biomech.* **2023**, *154*, 111634. [CrossRef]

https://doi.org/10.3390/bioengineering13020239

21. Takayanagi, K.; Takahashi, K.; Yamagata, M.; Moriya, H.; Kitahara, H.; Tamaki, T. Using Cineradiography for Continuous Dynamic-Motion Analysis of the Lumbar Spine. *Spine* **2001**, *26*, 1858–1865. [CrossRef]
22. Hasegawa, K.; Kitahara, K.; Hara, T.; Takano, K.; Shimoda, H.; Homma, T. Evaluation of Lumbar Segmental Instability in Degenerative Diseases by Using a New Intraoperative Measurement System. *J. Neurosurg. Spine* **2008**, *8*, 255–262. [CrossRef]
23. Breen, A.; Breen, A. Uneven Intervertebral Motion Sharing Is Related to Disc Degeneration and Is Greater in Patients with Chronic, Non-Specific Low Back Pain: An in Vivo, Cross-Sectional Cohort Comparison of Intervertebral Dynamics Using Quantitative Fluoroscopy. *Eur. Spine J.* **2018**, *27*, 145–153. [CrossRef]
24. Breen, A.; Mellor, F.; Morris, A.; Breen, A. An in Vivo Study Exploring Correlations between Early-to-Moderate Disc Degeneration and Flexion Mobility in the Lumbar Spine. *Eur. Spine J.* **2020**, *29*, 2619–2627. [CrossRef] [PubMed]
25. Farfan, H.F.; Kirkaldy-Willis, W.H. Instability of the Lumbar Spine. *Clin. Orthop. Relat. Res.* **1982**, *165*, 110–123. [CrossRef]
26. Ordway, N.R.; Fayyazi, A.H.; Abjornson, C.; Calabrese, J.; Park, S.-A.; Fredrickson, B.; Yonemura, K.; Yuan, H.A. Twelve-Month Follow-up of Lumbar Spine Range of Motion Following Intervertebral Disc Replacement Using Radiostereometric Analysis. *SAS J.* **2008**, *2*, 9–15. [CrossRef] [PubMed]
27. Dickey, J.P.; Pierrynowski, M.R.; Bednar, D.A.; Yang, S.X. Relationship between Pain and Vertebral Motion in Chronic Low-Back Pain Subjects. *Clin. Biomech.* **2002**, *17*, 345–352. [CrossRef] [PubMed]
28. Lund, T.; Nydegger, T.; Oxland, T.; Schlenzka, D. Three-Dimensional Motion Patterns During Active Bending in Patients with Chronic Low Back Pain. *Spine* **2002**, *27*, 1865–1874. [CrossRef]
29. Gercek, E.; Hartmann, F.; Kuhn, S.; Degreif, J.; Rommens, P.M.; Rudig, L. Dynamic Angular Three-Dimensional Measurement of Multisegmental Thoracolumbar Motion In Vivo. *Spine* **2008**, *33*, 2326–2333. [CrossRef]
30. Rozumalski, A.; Schwartz, M.H.; Wervey, R.; Swanson, A.; Dykes, D.C.; Novacheck, T. The in Vivo Three-Dimensional Motion of the Human Lumbar Spine during Gait. *Gait Posture* **2008**, *28*, 378–384. [CrossRef]
31. MacWilliams, B.A.; Rozumalski, A.; Swanson, A.N.; Wervey, R.A.; Dykes, D.C.; Novacheck, T.F.; Schwartz, M.H. Assessment of Three-Dimensional Lumbar Spine Vertebral Motion During Gait with Use of Indwelling Bone Pins. *J. Bone Jt. Surg.* **2013**, *95*, e184. [CrossRef]
32. Kooi, D.V.; Abad, G.; Basford, J.R.; Maus, T.P.; Yaszemski, M.J.; Kaufman, K.R. Lumbar Spine Stabilization With a Thoracolumbosacral Orthosis. *Spine* **2004**, *29*, 100–104. [CrossRef]
33. Basford, U.; Anderson, M.L.; Cunniff, J.G.; Kaufman, K.R.; Jelsing, E.J.; Patrick, T.A.; Magnuson, D.J.; Maus, T.P.; Yaszemski, M.J.; Basford, J. Video Fluoroscopic Analysis of the Effects of Three Commonly-Prescribed off-the-Shelf Orthoses on Vertebral Motion. *Spine* **2010**, *15*, E525–E529.
34. Nagel, T.M.; Zitnay, J.L.; Barocas, V.H.; Nuckley, D.J. Quantification of Continuous in Vivo Flexion–Extension Kinematics and Intervertebral Strains. *Eur. Spine J.* **2014**, *23*, 754–761. [CrossRef]
35. Breen, A.C.; Muggleton, J.M.; Mellor, F.E. An Objective Spinal Motion Imaging Assessment (OSMIA): Reliability, Accuracy and Exposure Data. *BMC Musculoskelet. Disord.* **2006**, *7*, 1. [CrossRef]
36. Bifulco, P.; Cesarelli, M.; Cerciello, T.; Romano, M.A. Continuous Description of Intervertebral Motion by Means of Spline Interpolation of Kinematic Data Extracted by Videofluoroscopy. *J. Biomech.* **2012**, *45*, 634–641. [CrossRef] [PubMed]
37. Davis, R.J.; Lee, D.C.; Wade, C.; Cheng, B. Measurement Performance of a Computer Assisted Vertebral Motion Analysis System. *Int. J. Spine Surg.* **2015**, *9*, 36. [CrossRef] [PubMed]
38. Breen, A.; Breen, A. Accuracy and Repeatability of Quantitative Fluoroscopy for the Measurement of Sagittal Plane Translation and Finite Centre of Rotation in the Lumbar Spine. *Med. Eng. Phys.* **2016**, *38*, 607–614. [CrossRef] [PubMed]
39. Andreozzi, E.; Pirozzi, M.A.; Fratini, A.; Cesarelli, G.; Bifulco, P. Quantitative Performance Comparison of Derivative Operators for Intervertebral Kinematics Analysis. In Proceedings of the 2020 IEEE International Symposium on Medical Measurements and Applications (MeMeA), Bari, Italy, 1 June–1 July 2020; pp. 1–6. [CrossRef]
40. To, D.; Breen, A.; Breen, A.; Mior, S.; Howarth, S. Investigator Analytic Repeatability of Two New Intervertebral Motion Biomarkers for Chronic, Nonspecific Low Back Pain in a Cohort of Healthy Controls. *Chiropr. Man. Ther.* **2020**, *28*, 62. [CrossRef]
41. Teyhen, D.S.; Flynn, T.W.; Bovik, A.C.; Abraham, L.D. A New Technique for Digital Fluoroscopic Video Assessment of Sagittal Plane Lumbar Spine Motion. *Spine* **2005**, *30*, E406–E413. [CrossRef]
42. Wang, S. In Vivo Lumbar Spine Biomechanics: Vertebral Kinematics, Intervertebral Disc Deformation, and Disc Loads. Ph.D. Thesis, Department of Mechanical Engineering, Massachusetts Institute of Technology, Cambridge, MA, USA, 2012.
43. Reckziegel, D.; Vachon-Presseau, E.; Petre, B.; Schnitzer, T.J.; Baliki, M.N.; Apkarian, A.V. Deconstructing Biomarkers for Chronic Pain- Context- and Hypothesis-Dependent Biomarker Types in Relation to Chronic Pain. *Pain* **2019**, *160*, S37–S48. [CrossRef]
44. Cheng, J.S.; Carr, C.B.; Wong, C.; Sharma, A.; Mahfouz, M.R.; Komistek, R.D. Altered Spinal Motion in Low Back Pain Associated with Lumbar Strain and Spondylosis. *Evid. -Based Spine-Care J.* **2013**, *04*, 006–012. [CrossRef]
45. Teyhen, D.S.; Flynn, T.W.; Childs, J.D.; Kuklo, T.R.; Rosner, M.K.; Polly, D.W.; Abraham, L.D. Fluoroscopic Video to Identify Aberrant Lumbar Motion. *Spine* **2007**, *32*, E220–E229. [CrossRef]

46. Breen, A.; Carvalho, D.D.; Funabashi, M.; Kawchuk, G.; Pagé, I.; Wong, A.Y.L.; Breen, A. A Reference Database of Standardised Continuous Lumbar Intervertebral Motion Analysis for Conducting Patient-Specific Comparisons. *Front. Bioeng. Biotechnol.* **2021**, *9*, 745837. [CrossRef] [PubMed]
47. Breen, A.; Mellor, F.; Breen, A. Aberrant Intervertebral Motion in Patients with Treatment-Resistant Nonspecific Low Back Pain: A Retrospective Cohort Study and Control Comparison. *Eur. Spine J.* **2018**, *27*, 2831–2839. [CrossRef] [PubMed]
48. Mellor, F.E. An Evaluation of Passive Recumbent Quantitative Fluoroscopy to Measure Mid-Lumbar Intervertebral Motion in Patients with Chronic Non-Specific Low Back Pain and Healthy Volunteers. Ph.D. Thesis, Bournemouth University, Poole, UK, 2014.
49. Breen, A.; Nematimoez, M.; Branney, J.; Breen, A. Passive Intervertebral Restraint Is Different in Patients with Treatment-Resistant Chronic Nonspecific Low Back Pain: A Retrospective Cohort Study and Control Comparison. *Eur. Spine J.* **2024**, *33*, 2405–2419. [CrossRef] [PubMed]
50. Anderst, W.; Kim, C.J.; Bell, K.M.; Gale, T.; Gray, C.; Greco, C.M.; LeVasseur, C.; McKernan, G.; Megherhi, S.; Patterson, C.G.; et al. Intervertebral Lumbar Spine Kinematics in Chronic Low Back Pain Patients Measured Using Biplane Radiography. *JOR Spine* **2025**, *8*, e70069. [CrossRef]
51. Xi, X.; Zhang, L.; Yu, H.; Qin, Y.; Jia, L.; Tsai, T.-Y.; Yu, Y.; Cheng, L. Different Spatial Characteristic Changes in Lumbopelvic Kinematics Before and After Fatigue: Comparison Between People with and Without Low Back Pain. *Bioengineering* **2025**, *12*, 214. [CrossRef]
52. Barrett, R.S.; Lichtwark, G.A.; Armstrong, C.; Barber, L.; Scott-Young, M.; Hall, R.M. Fluoroscopic Assessment of Lumbar Total Disc Replacement Kinematics During Walking. *Spine* **2015**, *40*, 436–442. [CrossRef]
53. Breen, A.C.; Teyhen, D.S.; Mellor, F.E.; Breen, A.C.; Wong, K.W.N.; Deitz, A. Measurement of Intervertebral Motion Using Quantitative Fluoroscopy: Report of an International Forum and Proposal for Use in the Assessment of Degenerative Disc Disease in the Lumbar Spine. *Adv. Orthop.* **2012**, *2012*, 802350. [CrossRef]
54. Breen, A.; Breen, A. Reference Database of Continuous Vertebral Flexion and Return. 2022. Available online: https://osf.io/a27py/ (accessed on 16 March 2022).
55. Breen, A.; Claerbout, E.; Hemming, R.; Ayer, R.; Breen, A. Comparison of Intra Subject Repeatability of Quantitative Fluoroscopy and Static Radiography in the Measurement of Lumbar Intervertebral Flexion Translation. *Sci. Rep.* **2019**, *9*, 19253. [CrossRef]
56. Breen, A.; Hemming, R.; Mellor, F.; Breen, A. Intrasubject Repeatability of in Vivo Intervertebral Motion Parameters Using Quantitative Fluoroscopy. *Eur. Spine J.* **2019**, *28*, 450–460. [CrossRef]
57. Aiyangar, A.K.; Zheng, L.; Tashman, S.; Anderst, W.J.; Zhang, X. Capturing Three-Dimensional In Vivo Lumbar Intervertebral Joint Kinematics Using Dynamic Stereo-X-Ray Imaging. *J. Biomech. Eng.* **2014**, *136*, 011004. [CrossRef]
58. Aiyangar, A.; Zheng, L.; Anderst, W.; Zhang, X. Instantaneous Centers of Rotation for Lumbar Segmental Extension in Vivo. *J. Biomech.* **2017**, *52*, 113–121. [CrossRef] [PubMed]
59. Oxland, T.R. Fundamental Biomechanics of the Spine—What We Have Learned in the Past 25 Years and Future Directions. *J. Biomech.* **2016**, *49*, 817–832. [CrossRef] [PubMed]
60. Byrne, R.M.; Aiyangar, A.K.; Zhang, X. A Dynamic Radiographic Imaging Study of Lumbar Intervertebral Disc Morphometry and Deformation In Vivo. *Sci. Rep.* **2019**, *9*, 15490. [CrossRef]
61. Song, Y.; Wen, W.; Xu, J.; Zhang, Z.; Han, Y.; Li, K.; Wang, X.; Xu, H.; Liu, J.; Miao, J. Kinematic Characteristics and Biomechanical Changes of Lower Lumbar Facet Joints Under Different Loads. *Orthop. Surg.* **2021**, *13*, 1047–1054. [CrossRef] [PubMed]
62. Bailey, J.F.; Nyayapati, P.; Johnson, G.T.A.; Dziesinski, L.; Scheffler, A.W.; Crawford, R.; Scheuring, R.; O'Neill, C.W.; Chang, D.; Hargens, A.R.; et al. Biomechanical Changes in the Lumbar Spine Following Spaceflight and Factors Associated with Post spaceflight disc herniation. *Spine J.* **2022**, *2*, 197–206. [CrossRef]
63. Breen, A.; Carvil, P.; Green, D.; Russomano, T.; Breen, A. Effects of a Microgravity SkinSuit on Lumbar Geometry and Kinematics. *Eur. Spine J.* **2023**, *32*, 839–847. [CrossRef]
64. Xi, X.; Ling, Z.; Wang, C.; Gu, C.; Zhan, X.; Yu, H.; Lu, S.; Tsai, T.-Y.; Yu, Y.; Cheng, L. Lumbar Segment-Dependent Soft Tissue Artifacts of Skin Markers during in Vivo Weight-Bearing Forward–Backward Bending. *Front. Bioeng. Biotechnol.* **2022**, *10*, 960063. [CrossRef]
65. Frey, M.; Breen, A.; Rix, J.; Carvalho, D.D. Concurrent Validity of Skin-Based Motion Capture Systems in Measuring Dynamic Lumbar Intervertebral Angles. *J. Biomech.* **2025**, *180*, 112503. [CrossRef]
66. Pataky, T.C.; Robinson, M.A.; Vanrenterghem, J. Vector Field Statistical Analysis of Kinematic and Force Trajectories. *J. Biomech.* **2013**, *46*, 2394–2401. [CrossRef]
67. Setliff, J.C.; Anderst, W.J. A Scoping Review of Human Skeletal Kinematics Research Using Biplane Radiography. *J. Orthop. Res.* **2024**, *42*, 915–922. [CrossRef]
68. Cholewicki, J.; Breen, A.; Popovich, J.J.M.; Reeves, N.P.; Sahrmann, S.A.; van Dillen, L.R.; Vleeming, A.; Hodges, P.W. Can Biomechanics Research Lead to More Effective Treatment of Low Back Pain? A Point-Counterpoint Debate. *J. Orthop. Sports Phys. Ther.* **2019**, *49*, 425–436. [CrossRef]

https://doi.org/10.3390/bioengineering13020239

69. Cholewicki, J.; Hodges, P.W.; Popovich, J.M.; Aminpour, P.; Gray, S.A.; Lee, A.S.; Breen, A.; Brumagne, S.; van Dieën, J.H.; Dillen, L.R.V.; et al. A Meta-Model of Low Back Pain to Examine Collective Expert Knowledge of the Effects of Treatments and Their Mechanisms. *Res. Sq.* **2025**. [CrossRef]
70. Morbée, L.; Chen, M.; Herregods, N.; Pullens, P.; Jans, L.B.O. MRI-Based Synthetic CT of the Lumbar Spine: Geometric Measurements for Surgery Planning in Comparison with CT. *Eur. J. Radiol.* **2021**, *144*, 109999. [CrossRef]

https://doi.org/10.3390/bioengineering13020239

bioengineering

MDPI

Article

Biomechanical Comparisons between One- and Two-Compartment Devices for Reconstructing Vertebrae by Kyphoplasty

Oliver Riesenbeck [1,*], Niklas Czarnowski [1], Michael Johannes Raschke [1], Simon Oeckenpöhler [1,†] and René Hartensuer [1,2,†]

1 Department of Trauma, Hand and Reconstructive Surgery, University Hospital Münster, Albert-Schweitzer-Campus 1, Building W1, Waldeyerstraße 1, 48149 Münster, Germany

2 Center for Orthopaedic, Traumatology, Handsurgery and Sports Medicine, Klinikum Aschaffenburg-Alzenau, 63739 Aschaffenburg, Germany

* Correspondence: oliver.riesenbeck@ukmuenster.de

† These authors contributed equally to this work.

Abstract: Background: This biomechanical in vitro study compared two kyphoplasty devices for the extent of height reconstruction, load-bearing capacity, cement volume, and adjacent fracture under cyclic loading. Methods: Multisegmental (T11–L3) specimens were mounted into a testing machine and subjected to compression, creating an incomplete burst fracture of L1. Kyphoplasty was performed using a one- or two-compartment device. Then, the testing machine was used for a cyclic loading test of load-bearing capacity to compare the two groups for the amount of applied load until failure and subsequent adjacent fracture. Results: Vertebral body height reconstruction was effective for both groups but not statistically significantly different. After cyclic loading, refracture of vertebrae that had undergone kyphoplasty was not observed in any specimen, but fractures were observed in adjacent vertebrae. The differences between the numbers of cycles and of loads were not statistically significant. An increase in cement volume was strongly correlated with increased risks of adjacent fractures. Conclusion: The two-compartment device was not substantially superior to the one-compartment device. The use of higher cement volume correlated with the occurrence of adjacent fractures.

Keywords: spinal biomechanics; spinal fracture; vertebral body; kyphoplasty; cement volume; load bearing; lumbar spine; cyclic loading

1. Introduction

Balloon kyphoplasty can result in reconstructing vertebral body height in the anterior middle region of vertebrae and in stabilizing vertebrae against compression load [1]. As well as simply reconstructing vertebral body height, restoring the endplate, which is often compressed, seems to prevent posttraumatic disc degeneration in traumatic vertebral body fractures [2]. Kyphoplasty was initially used to treat simple compression fractures, but given its clinical success, its use was extended to treat incomplete burst fractures [3].

In conventional biportal kyphoplasty, a catheter containing one compartment containing one balloon is inserted into each pedicle for a total of two balloons. After a balloon is inflated to reconstruct the vertebra, the balloon is deflated and removed, and the remaining cavity is filled with cement, typically polymethylmethacrylate. Kyphoplasty has undergone refinements ever since an inflatable balloon tamp required for its use received US Food and Drug Administration marketing clearance during 1998 [4].

For achieving the goal of preventing posttraumatic disc degeneration in traumatic vertebral body fractures, different tools for intravertebral reconstruction, all of which are refinements of kyphoplasty, have been developed [5–14].

However, anatomical endplate reconstruction with elevation remains difficult [15]. Some studies report on specific directional instruments or devices to directly manipulate the endplate but fail to detect a significant benefit considering vertebral body restoration [12–14]. A recent development in kyphoplasty potentially addresses this shortcoming by using a device that has a catheter with two compartments, each of which has a balloon [16]. A catheter is inserted into each pedicle for a total of four balloons. Each balloon could be separately filled with contrast medium, making it able to separately elevate the endplate anteriorly and posteriorly.

Considering that incomplete burst fractures vary tremendously and that fracture morphology is crucial to clinical decision-making, we developed a technique to reliably and reproducibly inflict incomplete burst fractures in cadaveric spinal specimens [17,18]. The technique consists of mounting and embedding a specimen into a custom-built frame, inserting the frame into a servo-hydraulic testing machine (Instron 8874, Instron Structural Testing Systems GmbH, Darmstadt, Germany), and applying compression. In all specimens, superior incomplete burst fractures (AO spine classification type A3) were identified [19].

This biomechanical study used this approach to compare the one-compartment catheter (two balloons) device and the two-compartment catheter (two balloons in each compartment) device for the extent of vertebral height reconstruction and load-bearing capacity. The hypothesis was that the two-compartment device was superior to the one-compartment device in both respects. Also, this study sought to evaluate the volume of cement in relation to the risk of subsequent adjacent fractures, a complication of kyphoplasty.

2. Materials and Methods

2.1. Cadaveric Specimens

In this study, 13 fresh, frozen, female lumbar spine specimens (T11–L3) were used. Specimens were thawed slowly to room temperature and kept moist throughout. All soft tissue and muscles were carefully dissected while preserving the osseous and ligamentous structures in accordance with the recommendations of Wilke et al. [20]. Vertebra L1, which was to be subject to fracture and testing, underwent measurement of the bone mineral density by quantitative computed tomography, as described [18]. Bone mineral densities were categorized according to the 2008 American College of Radiology practice parameter, which was the same as in the 2023 revision, as follows: less than 80 mg/cm^3, osteoporosis; from 80 mg/cm^3 to 120 mg/cm^3, osteopenia; and above 120 mg/cm^3, normal [21].

2.2. Mounting and Embedding Specimens

The specimen's cranial (T11) vertebra and caudal (L3) vertebrae were mounted into the top and bottom, respectively, of a custom-built mounting and embedding frame (Figure 1A–C), where the specimen remained fixed in place for the entire experiment. As described previously, an osteotomy-like procedure to weaken the cranial endplate of the target vertebra, L1, and thus, assure its facture subsequently during compression fracture creation, was performed by using a surgical chisel [18]. To prevent cement in the vertebrae from oozing out through the vertebral lesions resulting from chiseling and so to enable the correct recording of the cement volume, the lesions were sealed on the outside with plastic modeling material (Play-Doh, Hasbro, Pawtucket, RI, USA). The use of this modeling material did not have any effect on stabilizing the vertebrae.

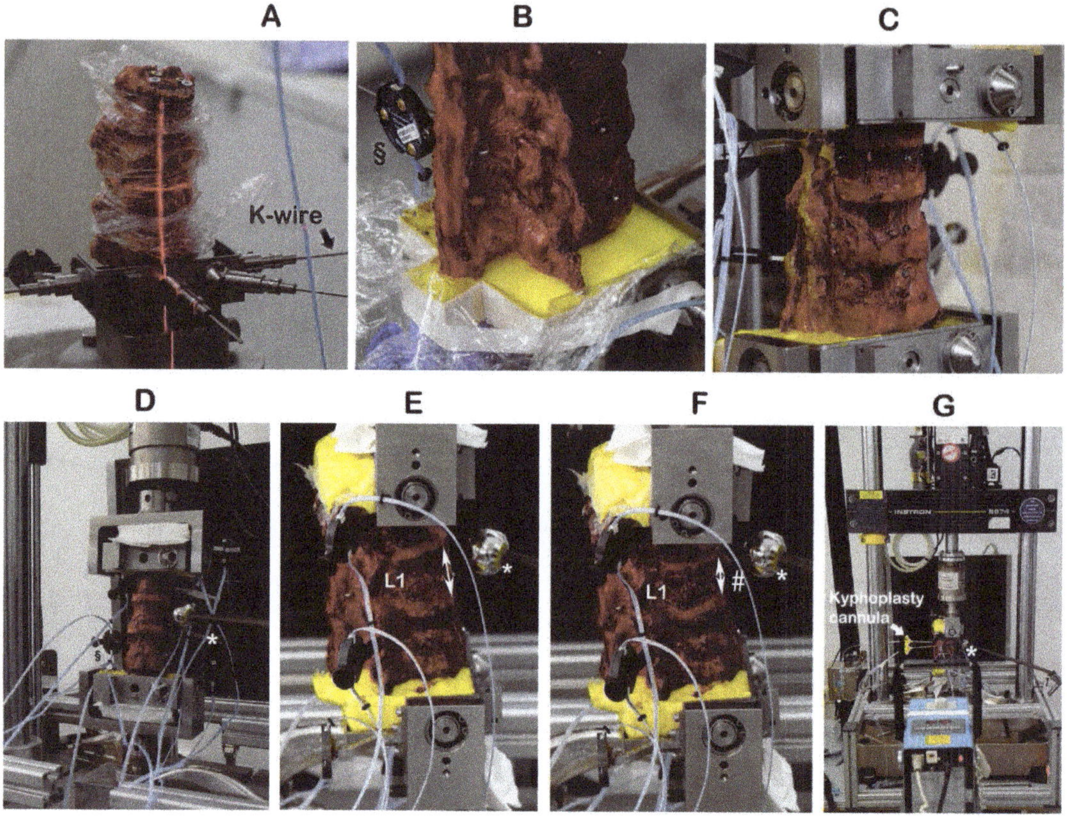

Figure 1. Embedding the specimens and creating incomplete burst fractures. § Motion trackers were not used in this work. ↕ The height of the intact vertebra L1 was measured by a ruler. * Radiographic reference ball; # incomplete burst fracture in L1.

2.3. Creating Incomplete Burst Fractures

The cranial endplate of L1 was oriented by laser into a custom-built embedding and mounting frame and held in place by K-wires (Figure 1A,B). The cranial (T11) and caudal vertebrae were embedded with polymethylmethacrylate (Technovit® 3040, Kulzer GmbH, Hanau, Germany) cement (Figure 1C). With the specimen having been mounted and embedded into the frame and the K-wires removed, the frame was inserted upright into a servo-hydraulic testing machine (Instron 8874, Instron Structural Testing Systems GmbH, Darmstadt, Germany) (Figure 1D). The height of the intact vertebra L1 was measured by a ruler (Figure 1E). The combination of osteotomy-like lesions of the target vertebra L1 and of compression at a speed of 300 mm/s to a defined distance of 67% of the intact L1 height resulted in reproducibly creating an incomplete burst fracture of L1 (Figure 1F), classified as A3 according to the AO spine classification [19].

2.4. Kyphoplasty

Kyphoplasty was always carried out according to the manufacturer's recommendations by the same experienced spine surgeon (RH) in a way that simulates actual surgery. Kyphoplasty cannulas were inserted into vertebra L1 (Figure 1G). Under radiographic guidance (Portable X-Ray Unit AJEX 9020H, JPI Healthcare Solutions, Ronkonkoma, NY, USA), balloons were inflated (Figure 2A,C).

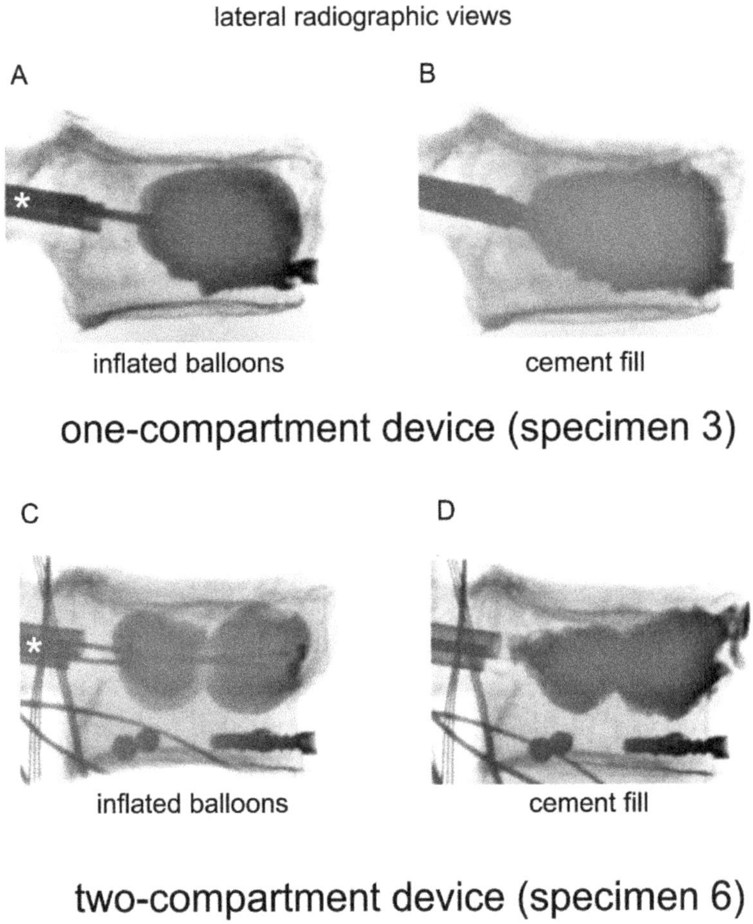

Figure 2. Lateral radiographic views of a one-compartment device with (**A**) inflated balloons and (**B**) cement fill, and a two-compartment device with (**C**) inflated balloons and (**D**) cement fill. The catheters were inserted through the cannulas, which are indicated by asterisks (*).

Both balloons in the one-compartment were inflated, and all four balloons in the two-compartment device were inflated (Figure 3). The one-compartment device (Joline S9403) contained one balloon of 16 mm diameter and 22 mm length with a volume of 6 mL and a maximum pressure of 27 bar. The two-compartment device (Joline S9420) contained two balloons of 16 mm diameter and 8 mm length each, with an individual volume of 3 mL and a maximum pressure of 27 bar.

Reconstruction of the vertebra was monitored by calibrated radiographic examinations according to clinical practice. The extent of inflation depended on surgical judgment. Once the vertebral height was determined to be sufficient, inflation was discontinued, and pressure applied to the balloons was recorded. Then, the balloons were deflated and removed, fully releasing compressive pressure in the apparatus. Having had vertebra T11 fixed into position on the servo-hydraulic machine avoided the collapse of vertebra L1 after the balloons were removed. Polymethylmethacrylate was inserted into vertebrae L1 according to the manufacturer's recommendations (Figure 2, lower row). The volume of injected polymethylmethacrylate was determined by lateral radiography to simulate clinical practice. The applicated volume was measured using a scale of 0.5 mL increments on the filling device. Vertebral height reconstruction was deemed effective if the surgeon's impression of height elevation was as observed in actual surgery.

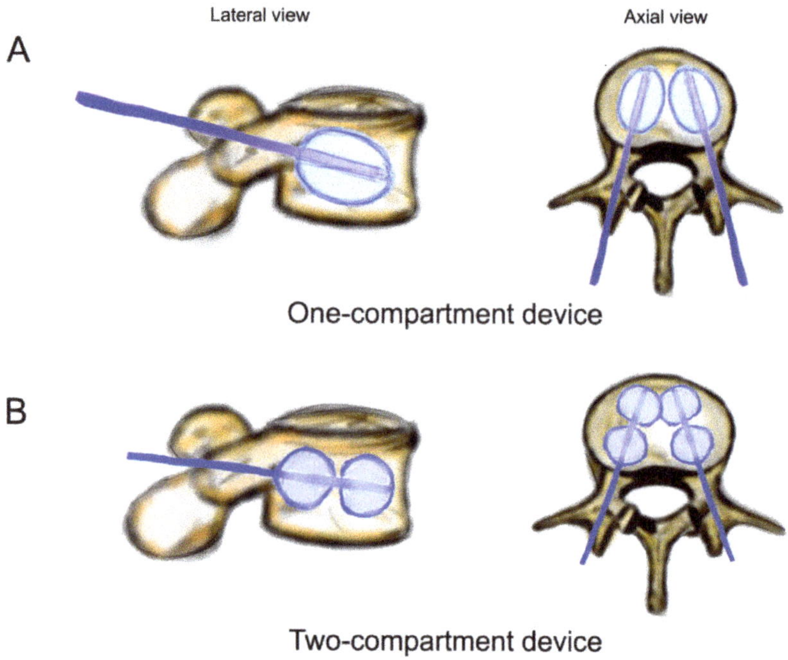

Figure 3. Schematic drawing of catheters and balloons. (**A**) One-compartment device (Joline S9403); (**B**) two-compartment device (Joline S9420).

In our model, kyphoplasty was carried out with a specimen mounted upright. Clinically, however, kyphoplasty is undertaken with the patient placed in the prone position. In this position, an axial load acts on the spine from intraabdominal and muscle tension and pressure. Therefore, this pressure needed to be accounted for in our testing. The in vivo studies by Sato et al. and by Wilke et al. published intradiscal pressure in a prone position of 91 kPa and of 0.1 MPa [22,23]. Sato et al. calculated an average spinal load of 144 N. Belkoff et al. calculated a range of 108–212 N from the original data and tested a low axial load (111 N) versus a high axial load (222 N) kyphoplasty scenario, resulting in no significant difference attributed to the axial load [24]. The low load scenario has been used several times in biomechanical testing of kyphoplasty (e.g., [25,26]). In our testing setup, the upper part of the embedding and mounting frame weighs 1.12 kg, which results in an axial compression force of 11 N by gravity. To match the published conditions of 111 N, the servo-hydraulic machine was set to maintain a minimum pressure of 100 N through axial compression on the specimen after fracture and before kyphoplasty.

During balloon inflation in kyphoplasty, the specimen's height increased, and the pressure started increasing. The pressure was regulated to maintain a constant pressure of 100 N during balloon inflation. When the balloons were deflated, as determined radiographically, and removed, the position of the testing machine was fixed to stop any additional compression, and pressure applied to the balloons was recorded, as already mentioned.

2.5. Radiographic Assessment of Vertebral Height Reconstruction

Radiographs were taken before and after a compression fracture and after kyphoplasty (Figure 4). Measurements of L1 vertebral height were taken at four places (Figure 5) using ImageJ (version 1.53s), an electronic distance measurement tool modified from McKiernan et al. [27,28].

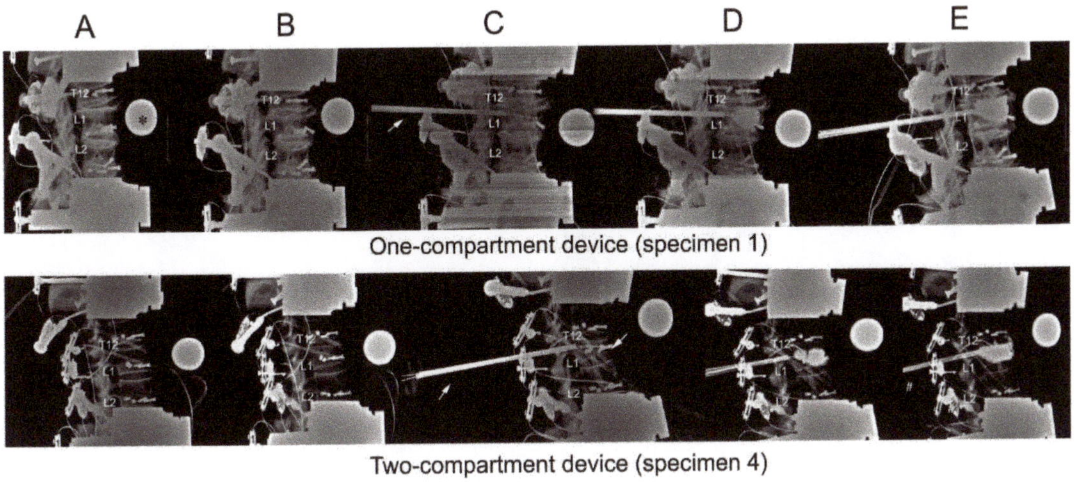

Figure 4. Lateral radiographs taken of (**A**) initial specimens before compression; (**B**) after compression; (**C**) after catheter insertion; (**D**) after balloon inflation; (**E**) and after kyphoplasty. The asterisk (*) in frame (**A**) of the one-compartment group refers to the radiographic reference ball. An arrow in both groups (**C**) points to the cannula.

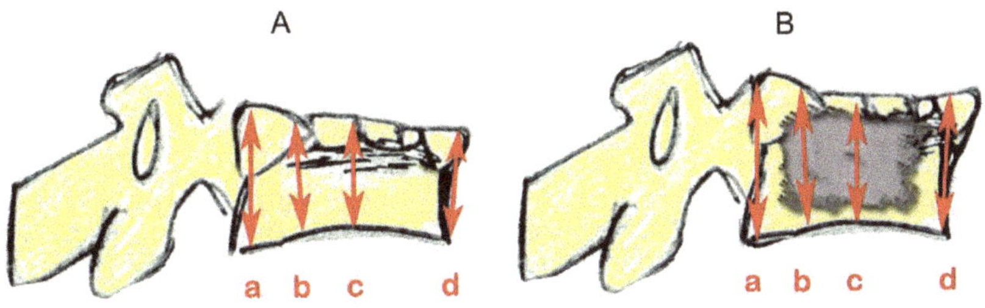

Figure 5. Schematics based on lateral vertebral radiographic projections showing vertebral height measurements (mm) by referring to the referencing ball, as shown in Figure 4. (**A**) after compression and fracture and before catheter insertion and (**B**) after cement insertion. Three of the four measurements were taken in places as per McKiernan et al. [28]: (a) posterior; (b) one-third distance from posterior to anterior; and (d) anterior. The additional location (c), centered between anterior and posterior, was added here.

2.6. Cyclic Loading Test of Load-Bearing Capacity

As a test of load-bearing capacity, cyclic loading by a stepwise cyclic compression protocol (Figure 6) was carried out using a servo-hydraulic testing machine and load until failure was recorded. Failure was defined by the radiological appearance of fractures adjacent to vertebra L1 (Figure 7) and/or by a compression depth of 7 cm underneath the starting position. All radiographic images were also examined retrospectively by R. H. to detect any earlier evidence of an adjacent fracture.

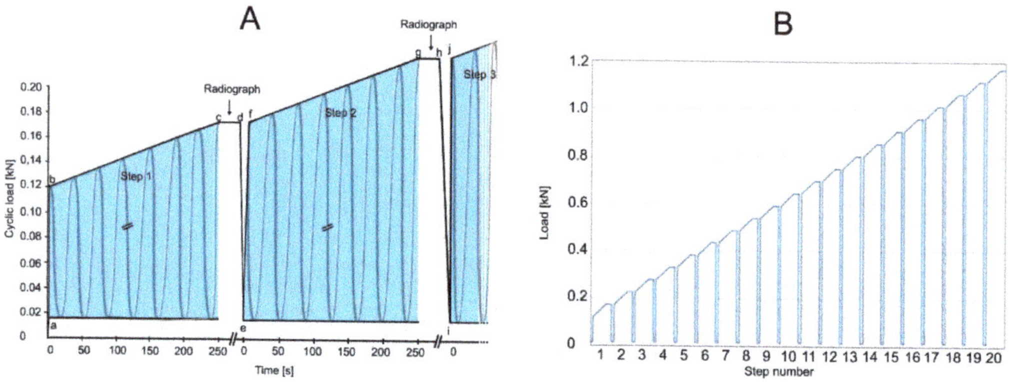

Figure 6. Test of load-bearing capacity by cyclic loading in a stepwise cyclic compression protocol. (**A**) At the beginning of a step (a; e; i), a baseline load of 0.015 kN was applied. Then, through 1.5 s, the load was applied to reach the minimum load for that step (b; f; j). Cyclic loading for that step was applied (b to c; f to g). Each complete 250 s step (the blue areas) consisted of 500 cycles (2 Hz frequency) for 52.5 N per step (slope, 0.21 N/s). With cyclic loading load of a step (c; g), the load was held constant (c to d; g to h) for the taking of a radiograph to examine for failure. When failure occurred, loading was discontinued, and cumulative load was recorded at the moment of failure. If fracture or compression was not observed, then cyclic loading was resumed by reducing to the minimum force of 0.015 kN (d to e; h to i), and a new step began. Cyclic loading was resumed (f; j) by applying the maximum force that had been reached in the previous step (c; g), and the next step began if necessary. (**B**) The test, as depicted in A, was designed for 20 steps. In the event that failure had not occurred by the end of 20 steps, testing was continued by adding additional steps by increasing cyclic loading to 70 N per step for an increased slope of 0.28 N/s.

Results from the cyclic loading tests were used for two purposes: (1) to compare the two groups for the amount of applied load until failure, and (2) to measure the amount of load applied until failure in relation to the volume of cement that had been used during kyphoplasty.

2.7. Statistics

The Mann–Whitney U test was used for group comparisons with a significance level of 0.05. To investigate the relationship between the volume of cement and the appearance of an adjacent fracture, a Spearman correlation analysis with a significance level of 0.025 was used to take into account the effects of multiple testing. Statistical analyses were carried out using SPSS (SPSS® Statistics 27; IBM, Armonk, NY, USA).

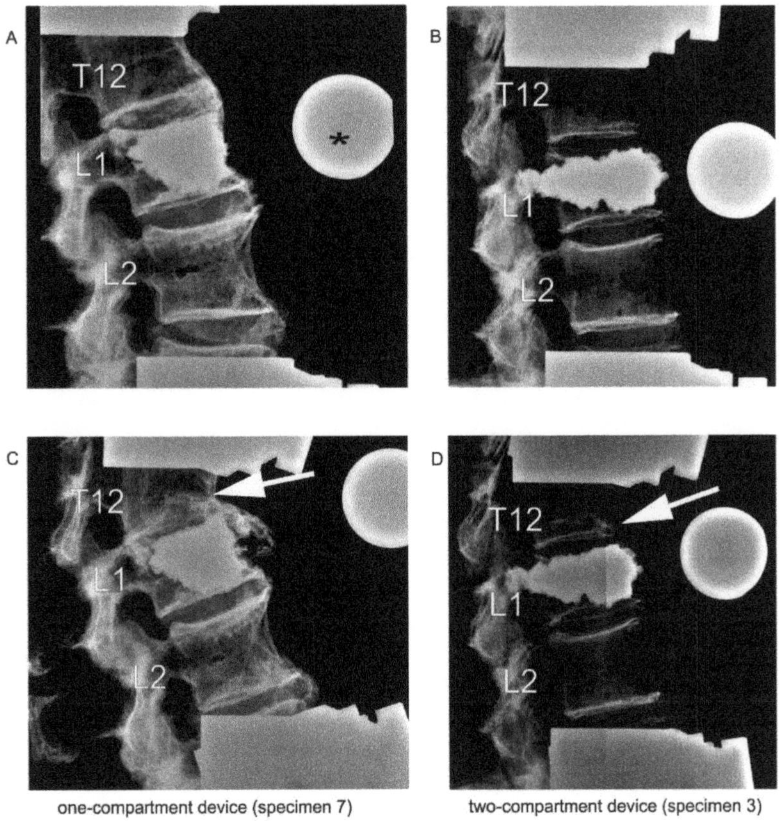

Figure 7. Lateral radiographic images upon kyphoplasty of L1 (**A**,**B**) and after adjacent fracture of T12 (**C**,**D**) resulting from the cyclic loading test of load-bearing capacity. The asterisk (*) in (**A**) refers to the radiographic reference ball. The arrows in (**C**,**D**) point to the fractured T12.

3. Results

3.1. Specimens

Of the 13 specimens, seven were allocated to the one-compartment group and six to the two-compartment group. The median (range) age in years for the one-compartment group was 79 (63 to 94), and for the two-compartment group, 82.5 (67 to 94). The median (range) bone mineral density for the one-compartment group was 71.2 (52.5 to 105.0) g/cm^3, and for the two-compartment group, 84.8 (68.6 to 120.8) g/cm^3, which was not a statistically significant difference ($p = 0.445$) All samples except for one in two-compartment group, which had a bone mineral density of 120.8 g/cm^3, were osteopenic or osteoporotic.

3.2. Decreased Vertebral Height after Compression

Compression resulted in decreasing heights of all four places in vertebra L1 (Table 1 and corresponding percentages in Figure 8), except that one increase each was noted in Place A and Place D, in the one-compartment group, as shown in the figure. These increases probably resulted from measuring error, and we inferred that compression did not occur at these two places. Decreases in height were not statistically significantly different between the two groups at any of the four places.

Table 1. Decreases in height (mm) of vertebra L1 after compression.

Place * in Vertebra L1		**a**		**b**		**c**		**d**	
		Height (mm)		**Height (mm)**		**Height (mm)**		**Height (mm)**	
		Before	**After**	**Before**	**After**	**Before**	**After**	**Before**	**After**
Group	Specimen	compression		compression		compression		compression	
one compartment	1	23.84	24.47 †	24.87	19.31	24.36	17.63	20.68	21.25 †
	2	26.91	26.49	24.85	22.28	24.26	20.57	22.07	19.99
	3	28.12	27.22	27.21	21.86	26.52	20.95	24.26	18.70
	4	28.32	26.03	25.62	22.92	25.31	22.43	27.52	27.02
	5	28.84	27.12	22.97	19.86	21.80	18.31	22.47	18.80
	6	28.97	23.88	27.41	20.01	26.28	18.47	27.02	18.26
	7	29.10	27.06	27.22	21.17	27.23	20.90	26.80	19.86
two compartment	1	27.91	23.24	24.19	18.52	24.82	18.56	26.11	19.06
	2	28.31	24.87	27.44	22.11	26.98	21.97	25.77	21.94
	3	28.37	24.83	27.09	20.60	26.26	19.13	27.77	16.94
	4	28.63	26.59	26.50	22.37	25.72	21.45	25.10	19.81
	5	28.84	26.07	26.95	19.03	26.26	17.29	26.44	16.48
	6	30.29	28.34	26.93	22.00	25.86	19.92	28.42	19.64
Comparison between before and after compression by using Mann–Whitney U-Test		$p = 0.86$		$p = 0.32$		$p = 0.39$		$p = 0.86$	

* Heights were measured at the four places shown in Figure 6. † As noted in the text, the increases in length after compression at these two places were caused by measuring error.

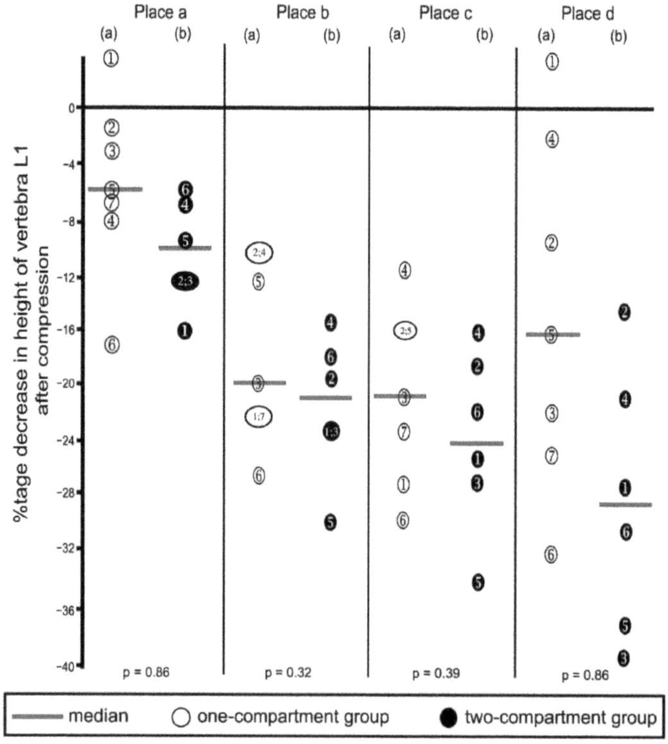

Figure 8. Percentage decreases in vertebra L1 height after compression calculated from the values shown in Table 1. Heights were measured at the four places shown in Figure 5. (a) One-compartment group; (b) two-compartment group. Numbers refer to specimens; p-values were calculated by using the Mann–Whitney U-Test.

3.3. Balloon-Filling and Cement Volumes

The combined median (range) balloon-filling volumes of the two balloons in the one-compartment group was 9.0 (7.0 to 10.0), and of the four balloons in the two-compartment group was 12.5 (11.0 to 14.5) ml. The difference between the two groups was statistically significant ($p = 0.001$). After deflation and balloon removal, the cavity remaining in the vertebra was filled with cement. The median (range) of bipedicular applied cement volume of the one-compartment group was 9.0 (9.0 to 12.0) mL. The median (range) of bipedicular applied cement volume of the two-compartment group was 11.3 (9.0 to 13.5 mL). The difference between the two groups was not statistically significant ($p = 0.138$).

Kyphoplasty resulted in increasing height (endplate elevation) in all four places in vertebra L1 (Table 2 and corresponding percentages in Figure 9), except that decreases were noted in three instances, as shown in the figure. These decreases probably resulted from measuring error. Increases in height were not/ statistically significantly different between the two groups at any of the four places. The surgeon deemed vertebral height reconstruction effective for all specimens.

Table 2. Increases in height (mm) of vertebra L1 from after compression to after kyphoplasty.

Place * in Vertebra L1		**a**		**b**		**c**		**d**	
		Height (mm)		**Height (mm)**		**Height (mm)**		**Height (mm)**	
		Before	**After**	**Before**	**After**	**Before**	**After**	**Before**	**After**
Group	Specimen	kyphoplasty		kyphoplasty		kyphoplasty		kyphoplasty	
One compartment	1	24.47	24.76	19.31	21.88	17.63	20.87	21.25	21.91
	2	26.49	27.36	22.28	23.49	20.57	23.45	19.99	21.99
	3	27.22	27.12 ‡	21.86	22.30	20.95	21.91	18.70	20.39
	4	26.03	27.22	22.92	23.25	22.43	23.19	27.02	26.81 ‡
	5	27.12	28.64	19.86	23.14	18.31	20.97	18.80	22.47
	6	23.88	24.75	20.01	20.49	18.47	19.71	18.26	23.53
	7	27.06	27.54	21.17	23.37	20.90	23.50	19.86	26.02
Two compartment	1	23.24	25.26	18.52	20.71	18.56	21.11	19.06	22.45
	2	24.87	27.58	22.11	24.32	21.97	24.44	21.94	23.24
	3	24.83	26.80	20.60	23.17	19.13	23.20	16.94	22.59
	4	26.59	27.19	22.37	23.49	21.45	22.78	19.81	20.96
	5	26.07	26.00 ‡	19.03	21.18	17.29	19.72	16.48	19.21
	6	28.34	28.86	22.00	25.13	19.92	23.83	19.64	25.70
Comparison between before and after compression by using Mann–Whitney U-Test		$p = 0.317$		$p = 0.317$		$p = 0.317$		$p = 0.568$	

* Heights were measured at the four places shown in Figure 6. ‡ As noted in the text, the decreases in length after kyphoplasty at these three places were caused by measuring error.

3.4. Cyclic Loading

After cyclic loading, refracture of vertebrae that had undergone kyphoplasty was not observed in any specimen, but fractures were observed in adjacent vertebrae. Although the median load to cause fracture of adjacent vertebrae was about 20% less in the two-compartment group than in the one-compartment group, this difference was not statistically significantly different (Figure 10A). A similar difference could be observed in the number of applied cycles that resulted in failure. Although the median number of applied cycles to result in failure was about 15% less in the two-compartment group than in the one-compartment group, this difference was also not significantly different statistically (Figure 10B).

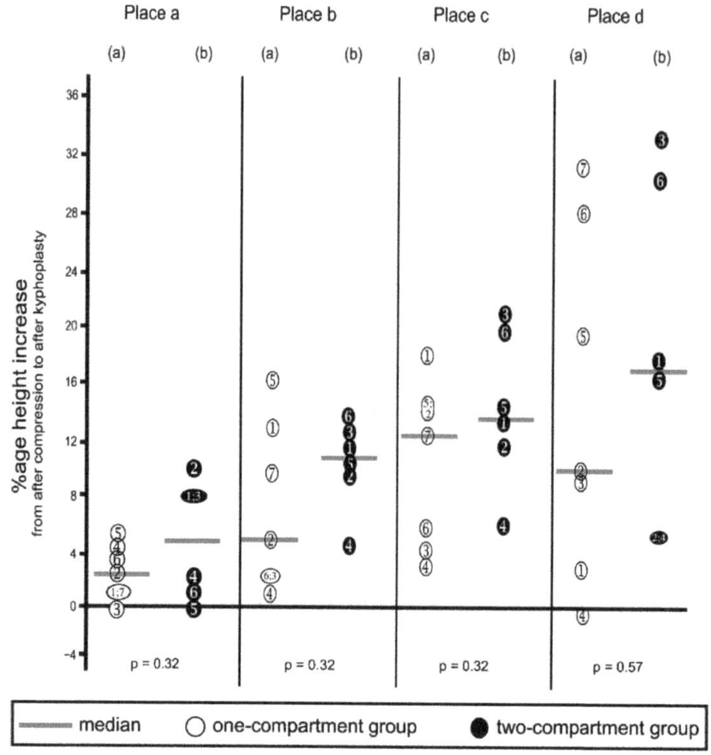

Figure 9. Percentage increases in vertebra L1 height from after compression to after kyphoplasty. Heights were measured at the four places shown in Figure 5. (a) One-compartment group; (b) two-compartment group. Numbers refer to specimens; p-values were calculated by using the Mann–Whitney U-Test.

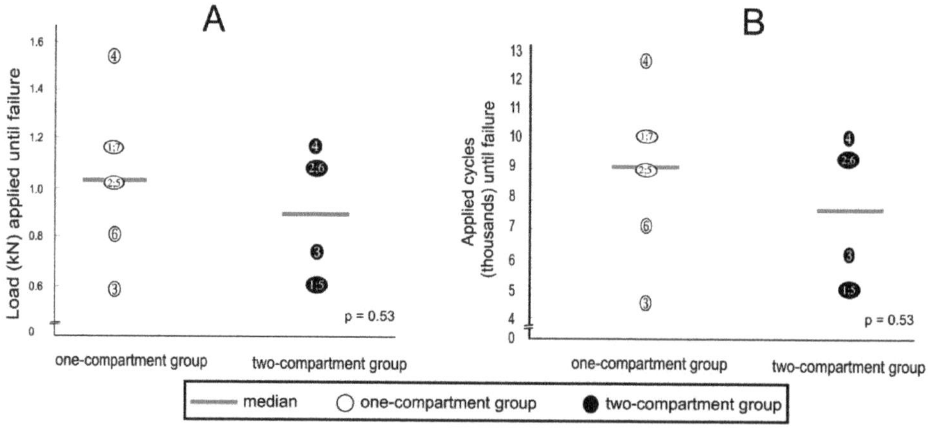

Figure 10. (**A**) applied loads and (**B**) numbers of applied cycles until failure (defined by the radiological appearance of fractures adjacent to vertebra L1 (Figure 7) and/or by a compression depth of 7 cm underneath the starting position. Numbers refer to specimens; p-values were calculated by using the Mann–Whitney U-Test.

3.5. Correlations between Applied Load and Applied Cycles at Time of Failure with Cement Volume

Spearman correlation coefficients between the applied load at the time of failure with the cement volume for the one-compartment group and for the two-compartment group were −0.805 (p = 0.029) and −0.806 (p = 0.053), respectively, and the correlation coefficients between the number of applied cycles at failure with the cement volume for the one-compartment group and for the two-compartment group were also −0.805 (p = 0.029) and −0.806 (p = 0.053), respectively. Considering that the correlation coefficients of the two groups were virtually identical, the values for the two groups were combined to increase statistical power and to shift the focus from the differences between the two groups to focus on the correlation between cement volume and failure. The Spearman correlation coefficients between the applied load at the time of failure and the cement volume (Figure 11A) and between the number of applied cycles until failure and the cement volume (Figure 11B) were both strongly negative and nearly linear.

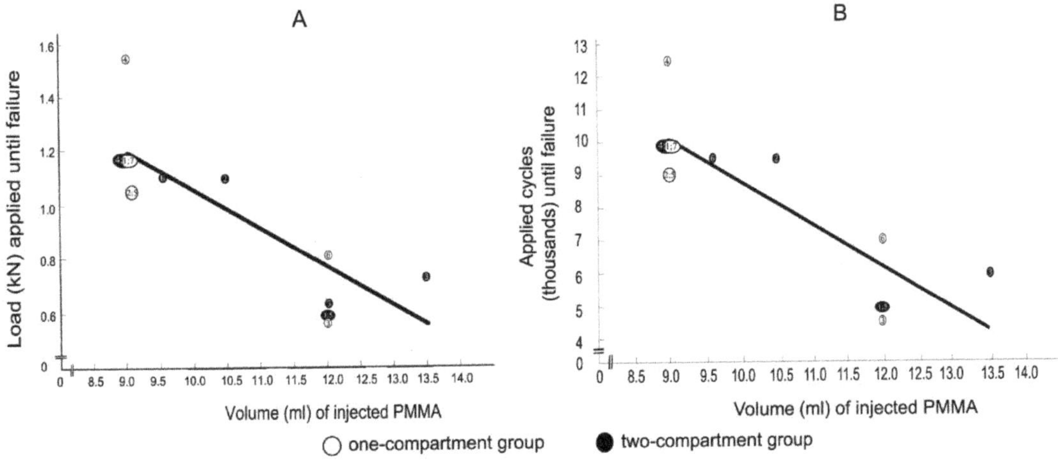

Figure 11. (**A**) correlations between applied load at failure and (**B**) between the number of applied cycles at failure with cement volume. Failure was defined by the radiological appearance of fractures adjacent to vertebra L1 (Figure 7) and/or by a compression depth of 7 cm underneath the starting position. Numbers refer to specimens. The Spearman correlation coefficients were −0.8 (p = 0.001) for both (**A**,**B**).

4. Discussion

The goal of the new protocol was to simulate the early postoperative period. We applied up to 12,500 cycles, which would correspond to a week postoperative period in a young person and perhaps 10 to 14 days in an elderly person [1]. We were unaware of standardized protocols to evaluate the biomechanical properties of cadaveric spine samples subject to experimental loads. Some protocols used constant compression force [29,30], whereas others used increasing compression force [31,32], as did this study. We believe that the model reported here has several advantages compared with others. The biomechanical model that we developed combined increasing loading with cyclic loading to reduce testing time until failure. We envisaged that this approach could be used to facilitate further studies on spine fracture treatment of incomplete burst fractures.

Another difference in published protocols is the number of segments investigated. Multisegmented models of load-bearing capacity [33,34], such as the one used in this study, better simulate the actual spine than single-segmented models [29,35]. A multisegmented specimen can flex and extend. Plus, this study combined the advantage of using multisegmented specimens with increasing loading through cyclic loading until failure. In addition, this study's compression protocol was based on radiographs taken in between each step.

However, even though multisegmented specimens were used, the effects of changes in sagittal balance could not be addressed by our model.

The most important finding of this study was that the two-compartment device was not clearly superior to the one-compartment device.

4.1. Vertebral Height Reconstruction

In particular, we had expected to find greater vertebral height reconstruction in the two places that we measured in the posterior region of the vertebra, the posterior itself (Place a), and at one-third of the distance from the posterior of the vertebra to the anterior (Place b). Such a finding would have been interesting because Verlaan et al. found that increased disc survival was associated with anatomical endplate restoration [2]. In our study, kyphoplasty resulted in increasing height (endplate elevation) in all four measured places in vertebra L1. Therefore, vertebral height reconstruction by kyphoplasty was effective in both groups, the one-compartment group and the two-compartment group. However, increases in height were not statistically significantly different between the two groups at any of the four measured places. A possible explanation for not finding a significant difference in our study was that the study was not statistically powered to detect small differences. Availability of samples is limited in most biomechanical studies.

Despite the manufacturer's instructions to use the same amount of water to fill the balloons in both the one- and two-compartment devices, the filling volume was significantly higher for the two-compartment device. The difference might have resulted from the surgeon's effort to attain optimal reconstruction in view of the manufacturer's statement that reconstruction of vertebral body height is more effectively accomplished using the two-compartment device.

Nonetheless, the extent of height reconstruction in our study was comparable with those of other studies using conventional kyphoplasty in cadaveric specimens and with clinical outcomes [15,29,36,37].

Holyoak et al. also failed to show any benefit for kyphoplasty with directional instruments considering the vertebral height restoration in comparison to conventional kyphoplasty [14].

Even a recently developed tool with the ability to directly manipulate the endplate did not show better reconstruction of the vertebral body height in comparison to conventional kyphoplasty, nor was any relevant reconstruction of the posterior wall found [12,13].

4.2. Vertebral Height Preservation

We had also hypothesized that kyphoplasty with the two-compartment device was superior to that with the one-compartment device as to the extent of load-bearing capacity until failure. Although the two-compartment device was inferior in this respect, the difference was not statistically significant.

Furthermore, although the two-compartment group had earlier adjacent fractures after cyclic loading than did the one-compartment group, the difference was not statistically significant. Therefore, we cannot conclude that the use of the one-compartment device would result in fewer fractures. However, the finding of lack of statistical significance might have resulted from a lack of significant power. The study of Holyoak et al. showed comparable results considering vertebral height preservation with no significant difference between conventional kyphoplasty and kyphoplasty with directional instruments, but a significant difference between kyphoplasty with directional instruments and augmented Kyphoplasty in favor of the first mentioned. However, there was also a correlating difference in the volume of cement used [14]. This is consistent with the findings of a recently published study by Sun et al. in which higher cement volume corresponds to better vertebral body height reconstruction [38]. Two other biomechanical studies of cadaveric vertebrae reported that height preservation depended on cement volume, in which a higher cement volume meant less height loss of the treated vertebra [39,40]. Thus, an increase in cement volume is possibly needed in an extensive height restoration to maintain the endplate.

4.3. Volume of Cement Correlates with Risk of Adjacent Fracture

Using a volume of cement appropriate for each kyphoplasty on a particular specimen rather than using a fixed volume, the surgeon was able to simulate an extent of height reconstruction similar to that observed in actual surgery. This was important because our studies on load-bearing capacity and cement volume were based on simulated clinical judgment. The volume of cement was higher for the two-compartment device, although not statistically significant. Compared with other experimental studies and with clinical data, the volume of cement was higher in this study, higher than the recommended volume of greater than 4.5 mL from Röder and colleagues [36,41–43]. However, the cement volumes reported in our study might have been slightly overestimated because of leakage resulting from lesions in the vertebra resulting from fracture.

Another key result we found is that increasing volume of cement was strongly correlated with an increasing risk of adjacent fractures after kyphoplasty.

There is controversy in the clinical literature regarding this result. Some clinical studies report that cement volume is not a risk factor for adjacent vertebral fractures after kyphoplasty. In a study by Ko et al., 123 patients underwent kyphoplasty, and 20 (16%) had early adjacent vertebral fractures (new fractures that had developed within 3 months after surgery). There was no difference in cement volume between these patients and those who did not have early adjacent vertebral fractures [44]. Teuber et al., and more recently, Essibayi et al., reported no difference in the risk of adjacent fractures after vertebroplasty or kyphoplasty compared to the natural cause of disease considering osteoporotic fractures [45,46]. Other clinical studies reported that higher cement volumes lead to a greater risk of unfavorable consequences after kyphoplasty. Yang et al. reported that larger cement volume is a risk factor for adjacent vertebral compression fracture after kyphoplasty [47]. A study by Yi et al. reported that two (2.5%) of the 79 kyphoplasty patients had subsequent adjacent fractures [48]. In these two patients, subsequent adjacent fractures occurred 20 days after kyphoplasty with 5 mL cement and at 2 months after kyphoplasty with 1 mL cement. Adida et al. were able to show a correlation between cement volume and adjacent level fracture in the thoracic spine but not in the lumbar spine [41]. Additionally, the use of less injected cement significantly boosted the potential risk of refracture [49]. A similar controversy exists considering the analgesic effect. Some studies report the use of higher cement volume to have a better analgesic effect, whereas others favor a reduced cement volume [38,43].

From a biomechanical perspective, the correlation between cement volume and adjacent fracture seems very plausible and in agreement with the previously reported literature data. Berlemann et al. reported that in vertebroplasty, the treated vertebra is fairly stiff, causing increased stress on the relatively soft adjacent osteoporotic vertebra [50]. This explanation also applies to kyphoplasty. Another study reported that strength and stiffness are weakly correlated with the percentage fill volume of cement injected [51]. A biomechanical study employing finite element analysis on an L1 vertebra reported that only a small amount of bone cement, about 15% fill volume, was necessary to restore the stiffness of the damaged vertebral body to the pre-damaged value [52]. The use of a 30% fill increased stiffness by more than 50% compared with the pre-damaged value. A greater filling can result in a substantial increase in stiffness well beyond the pre-damaged value and might not be the most biomechanically optimal. Another study using finite element analysis described that the cement distribution within the vertebral body corresponds to stress on the adjacent segment [53]. Macciachera reported in a review that mechanical percutaneous vertebral body augmentation (MPVA) reduced the rates of adjacent fractures compared to conventional kyphoplasty [54]. This is consistent with a difference in stiffness between solid cement and an MPVA. Therefore, future studies should report the biomechanical parameters of the cement used, and patient-specific adjustment of the kyphoplasty cement could reduce the risk of an adjacent fracture [55]. Currently, the intravertebral devices for reconstruction of the vertebral endplate reported in literature and this study do not significantly influence vertebral height preservation and risk of adjacent fracture.

4.4. Limitations

Biomechanical studies are based on a model and do not necessarily reflect clinical conditions. Vertebral reconstruction by kyphoplasty in this study is undertaken with upright specimens, whereas clinically, the patient is in the prone position. As mentioned, we attempted to account for differences in pressure.

The protocol in this study was intended to simulate physiologic conditions during the early postoperative period after kyphoplasty to determine the accumulated load until failure.

To develop our model simulating the accumulated load until failure, we considered previously reported observations. Wilke et al. applied 100,000 cycles of constant load in one loading test to cadaveric specimens to measure the loss of height [1]. They stated that applying such a load simulated a 5-to 6-week postoperative period in a young patient and a 3-month period in an elderly patient. Decomposition of thawed postmortem specimens limits the amount of time to apply cyclic loading until failure. This is a limitation of studies on postmortem specimens, and we were not aware of a specific time limitation. Therefore, in our study, continually increasing loading until failure was applied in one cyclic loading test. Otherwise, either failure might not have occurred, or testing until failure might have taken too long. Therefore, as with previous work, the time until failure in our work was not comparable with a patient's experience [1].

Another important factor of cyclic testing is the frequency of cyclic loading. Hansson et al. used a frequency of 0.5 Hz to simulate physiologic conditions [56]. However, there was no standard in the frequency used in previous studies, which used frequencies ranging from 0.25 to 5 Hz [29–32,57]. We used an in-between value of 2 Hz because we were unaware of validated recommendations. The values of increasing compressive load used in this study were based on published values during standing (600 N) and normal walking (900 N) [23] up to maximal loads until vertebral fracture, 1040 N based on calculation [58] and about 2500 to 3500 based on measurement of fracture load [32].

We developed a method to simulate repetitive loads to a multisegmental specimen in an effort to attempt to assure that load until failure was reached within the time range of usage from freshly frozen cadaveric specimens. Previous studies used either cyclic loading [1,29–31,42,56,57] or steady compressive force [7,36] to simulate repetitive loads onto a multisegmental specimen and to ensure that load until failure was reached within the time range of usage from freshly frozen cadaveric specimens. Our method combined these factors. Inasmuch as decomposition is an ongoing process, we sought to measure endpoint values as soon as possible to simulate physiological conditions as much as possible. However, to our knowledge, a time range of limitations on making such observations has not been reported. All specimens were tested under the same conditions, so we believe that it is reasonable to compare findings from one specimen to another. However, applying these findings to patients would not be accurate. This is a limitation of biomechanical studies of postmortem specimens.

5. Conclusions

In accordance with the intravertebral devices reported in the literature, no advantage for the two-compartment device with regard to reconstruction and preservation of the vertebral body height and the vertebral endplate could be demonstrated in the present study. Due to high stiffness, the cement volume and not the intravertebral device seem to have the decisive influence on vertebral height preservation and the risk of adjacent fractures, creating a clinical dilemma in spine surgery that has yet to be resolved.

Author Contributions: Conceptualization, O.R., M.J.R., S.O. and R.H.; methodology, O.R., N.C. and R.H.; software, O.R. and N.C.; validation, O.R., N.C. and R.H.; formal analysis, O.R., N.C. and R.H.; investigation, O.R., N.C., M.J.R. and R.H.; resources, O.R. and R.H.; data curation, O.R., N.C. and R.H.; writing—original draft preparation, O.R. and R.H.; writing—review and editing, O.R., N.C., M.J.R., S.O. and R.H.; visualization, O.R., N.C. and R.H.; supervision, M.J.R., S.O. and R.H.; project administration, O.R., M.J.R. and R.H.; funding acquisition, M.J.R. and R.H. All authors have read and agreed to the published version of the manuscript.

Funding: This study was supported in part by funds received from Joline (Joline GmbH & Co. KG, 72379 Hechingen, Germany) to cover material costs of EUR 26,031.25. There was no influence by the funder on this article. We acknowledge support from the Open Access Publication Fund of the University of Münster.

Institutional Review Board Statement: The Ethics Committee of the Westphalia-Lippe Medical Association (Ethikkommission der Ärztekammer Westfalen-Lippe) and the University of Munster approved the study (2018–2021 672-f-S).

Informed Consent Statement: Informed consent was obtained from all subjects involved in the study, consistently with the setting of collection/donation of the anatomical material.

Data Availability Statement: Dataset available on request from the authors.

Acknowledgments: We thank Martin Schulze and Markus Müller of the University of Muenster for their help in managing the robot-based testing setup, and Alexander Stump for helping in data acquisition. Michael S. Altus, ELS of Intensive Care Communications, Inc., Baltimore, MD, USA, provided professional English-language editing of our manuscript.

Conflicts of Interest: The authors declare no conflicts of interest. The funders had no role in the design of the study; in the collection, analyses, or interpretation of data; in the writing of the manuscript; or in the decision to publish the results.

References

1. Wilke, H.-J.; Mehnert, U.; Claes, L.E.; Bierschneider, M.M.; Jaksche, H.; Boszczyk, B.M. Biomechanical evaluation of vertebroplasty and kyphoplasty with polymethyl methacrylate or calcium phosphate cement under cyclic loading. *Spine* **2006**, *31*, 2934–2941. [CrossRef] [PubMed]
2. Verlaan, J.-J.; Dhert, W.J.A.; Oner, F.C. Intervertebral disc viability after burst fractures of the thoracic and lumbar spine treated with pedicle screw fixation and direct end-plate restoration. *Spine J.* **2013**, *13*, 217–221. [CrossRef] [PubMed]
3. Krüger, A.; Zettl, R.; Ziring, E.; Mann, D.; Schnabel, M.; Ruchholtz, S. Kyphoplasty for the treatment of incomplete osteoporotic burst fractures. *Eur. Spine J.* **2010**, *19*, 893–900. [CrossRef] [PubMed]
4. Centers for Medicare & Medicaid Services. Percutaneous Kyphoplasty for Vertebral Fractures Caused by Osteoporosis and Malignancy: Technology Assessment. 2005. Available online: https://www.cms.gov/medicare-coverage-database/view/technology-assessments.aspx?taid=25 (accessed on 26 February 2024).
5. Otten, L.A.; Bornemnn, R.; Jansen, T.R.; Kabir, K.; Pennekamp, P.H.; Wirtz, D.C.; Stuwe, B.; Pflugmacher, R. Comparison of balloon kyphoplasty with the new Kiva® VCF system for the treatment of vertebral compression fractures. *Pain Physician* **2013**, *16*, E505–E512. [CrossRef] [PubMed]
6. Vanni, D.; Galzio, R.; Kazakova, A.; Pantalone, A.; Grillea, G.; Bartolo, M.; Salini, V.; Magliani, V. Third-generation percutaneous vertebral augmentation systems. *J. Spine Surg.* **2016**, *2*, 13–20. [CrossRef] [PubMed]
7. Sietsma, M.S.; Hosman, A.J.F.; Verdonschot, N.J.J.; Aalsma, A.M.M.; Veldhuizen, A.G. Biomechanical evaluation of the vertebral jack tool and the inflatable bone tamp for reduction of osteoporotic spine fractures. *Spine* **2009**, *34*, E640–E644. [CrossRef] [PubMed]
8. Wang, D.; Zheng, S.; Liu, A.; Xu, J.; Du, X.; Wang, Y.; Wang, L. The Role of Minimally Invasive Vertebral Body Stent on Reduction of the Deflation Effect After Kyphoplasty: A Biomechanical Study. *Spine* **2018**, *43*, E341–E347. [CrossRef] [PubMed]
9. Kim, J.-M.; Shin, D.A.; Byun, D.-H.; Kim, H.-S.; Kim, S.; Kim, H.-I. Effect of bone cement volume and stiffness on occurrences of adjacent vertebral fractures after vertebroplasty. *J. Korean Neurosurg. Soc.* **2012**, *52*, 435–440. [CrossRef]
10. Oberkircher, L.; Struewer, J.; Bliemel, C.; Buecking, B.; Eschbach, D.-A.; Ruchholtz, S.; Krueger, A. Height restoration and preservation in osteoporotic vertebral compression fractures: A biomechanical analysis of standard balloon kyphoplasty versus radiofrequency kyphoplasty in a cadaveric model. *J. Spinal Disord. Tech.* **2014**, *27*, 283–289. [CrossRef]
11. Hoffmann, J.; Preston, G.; Whaley, J.; Khalil, J.G. Vertebral Augmentation in Spine Surgery. *J. Am. Acad. Orthop. Surg.* **2023**, *31*, 477–489. [CrossRef]
12. Krüger, A.; Bäumlein, M.; Knauf, T.; Pascal-Moussellard, H.; Ruchholtz, S.; Oberkircher, L. Height and volume restoration in osteoporotic vertebral compression fractures: A biomechanical comparison of standard balloon kyphoplasty versus Tektona® in a cadaveric fracture model. *BMC Musculoskelet. Disord.* **2021**, *22*, 76. [CrossRef] [PubMed]
13. Marie-Hardy, L.; Mohsinaly, Y.; Pietton, R.; Stencel-Allemand, M.; Khalifé, M.; Bonaccorsi, R.; Barut, N.; Pascal-Moussellard, H. Efficiency of a novel vertebral body augmentation system (Tektona™) in non-osteoporotic spinal fractures. *BMC Musculoskelet. Disord.* **2022**, *23*, 356. [CrossRef] [PubMed]
14. Holyoak, D.T.; Andreshak, T.G.; Hopkins, T.J.; Brook, A.L.; Frohbergh, M.E.; Ong, K.L. Height restoration and sustainability using bilateral vertebral augmentation systems for vertebral compression fractures: A cadaveric study. *Spine J.* **2022**, *22*, 2072–2081. [CrossRef] [PubMed]

15. Liang, L.; Chen, X.; Jiang, W.; Li, X.; Chen, J.; Wu, L.; Zhu, Y. Balloon kyphoplasty or percutaneous vertebroplasty for osteoporotic vertebral compression fracture? An updated systematic review and meta-analysis. *Ann. Saudi Med.* **2016**, *36*, 165–174. [CrossRef] [PubMed]
16. Joline GmbH & Co. KG. Spine Portfolio 2024: Quattroplasty™. 2024. Available online: https://www.joline.de/wp-content/uploads/2019/11/JOL755-Spine-maincat-GB-2024_WTK_LR.pdf (accessed on 22 July 2024).
17. Spiegl, U.J.; Josten, C.; Devitt, B.M.; Heyde, C.-E. Incomplete burst fractures of the thoracolumbar spine: A review of literature. *Eur. Spine J.* **2017**, *26*, 3187–3198. [CrossRef] [PubMed]
18. Hartensuer, R.; Gasch, A.; Gehweiler, D.; Schanz, S.; Schulze, M.; Matuszewski, L.; Langer, M.; Raschke, M.J.; Vordemvenne, T. Experimentally induced incomplete burst fractures—A novel technique for calf and human specimens. *BMC Musculoskelet. Disord.* **2012**, *13*, 45. [CrossRef]
19. Vaccaro, A.R.; Oner, C.; Kepler, C.K.; Dvorak, M.; Schnake, K.; Bellabarba, C.; Reinhold, M.; Aarabi, B.; Kandziora, F.; Chapman, J.; et al. AOSpine thoracolumbar spine injury classification system: Fracture description, neurological status, and key modifiers. *Spine* **2013**, *38*, 2028–2037. [CrossRef] [PubMed]
20. Wilke, H.J.; Wenger, K.; Claes, L. Testing criteria for spinal implants: Recommendations for the standardization of in vitro stability testing of spinal implants. *Eur. Spine J.* **1998**, *7*, 148–154. [CrossRef] [PubMed]
21. American College of Radiology. ACR–SPR–SSR Practice Parameter for the Performance of Quantitative Computed Tomography (QCT) Bone Mineral Density. 2024. Available online: https://www.acr.org (accessed on 22 July 2024).
22. Sato, K.; Kikuchi, S.; Yonezawa, T. In vivo intradiscal pressure measurement in healthy individuals and in patients with ongoing back problems. *Spine* **1999**, *24*, 2468–2474. [CrossRef]
23. Wilke, H.J.; Neef, P.; Caimi, M.; Hoogland, T.; Claes, L.E. New in vivo measurements of pressures in the intervertebral disc in daily life. *Spine* **1999**, *24*, 755–762. [CrossRef]
24. Belkoff, S.M.; Jasper, L.E.; Stevens, S.S. An ex vivo evaluation of an inflatable bone tamp used to reduce fractures within vertebral bodies under load. *Spine* **2002**, *27*, 1640–1643. [CrossRef] [PubMed]
25. Rotter, R.; Martin, H.; Fuerderer, S.; Gabl, M.; Roeder, C.; Heini, P.; Mittlmeier, T. Vertebral body stenting: A new method for vertebral augmentation versus kyphoplasty. *Eur. Spine J.* **2010**, *19*, 916–923. [CrossRef] [PubMed]
26. Gordon, M.; Peppelman, W.C.; Beutler, W.; O'Halloran, D.; Chinthakunta, S.R.; Bucklen, B. An In Vitro Evaluation of Fracture Reduction Achieved by Inflatable Bone Tamps Under Simulated Physiological Load. *Clin. Spine Surg.* **2017**, *30*, E31–E37. [CrossRef] [PubMed]
27. Rasband, W. *ImageJ*; U.S. National Institutes of Health: Bethesda, MD, USA, 2018; Available online: https://imagej.net/ij/ (accessed on 1 July 2024).
28. McKiernan, F.; Faciszewski, T.; Jensen, R. Reporting height restoration in vertebral compression fractures. *Spine* **2003**, *28*, 2517–2521. [CrossRef] [PubMed]
29. Achatz, G.; Riesner, H.-J.; Friemert, B.; Lechner, R.; Graf, N.; Wilke, H.-J. Biomechanical in vitro comparison of radiofrequency kyphoplasty and balloon kyphoplasty. *Eur. Spine J.* **2017**, *26*, 3225–3234. [CrossRef] [PubMed]
30. Huber, G.; Nagel, K.; Skrzypiec, D.M.; Klein, A.; Püschel, K.; Morlock, M.M. A description of spinal fatigue strength. *J. Biomech.* **2016**, *49*, 875–880. [CrossRef] [PubMed]
31. Kolb, J.P.; Weiser, L.; Kueny, R.A.; Huber, G.; Rueger, J.M.; Lehmann, W. Zementaugmentation an der Wirbelsäule : Was ist biomechanisch zu beachten? *Orthopade* **2015**, *44*, 672–680. [CrossRef] [PubMed]
32. Schulte, T.L.; Keiler, A.; Riechelmann, F.; Lange, T.; Schmoelz, W. Biomechanical comparison of vertebral augmentation with silicone and PMMA cement and two filling grades. *Eur. Spine J.* **2013**, *22*, 2695–2701. [CrossRef] [PubMed]
33. Disch, A.C.; Schmoelz, W. Cement augmentation in a thoracolumbar fracture model: Reduction and stability after balloon kyphoplasty versus vertebral body stenting. *Spine* **2014**, *39*, E1147–E1153. [CrossRef]
34. Germaneau, A.; Vendeuvre, T.; Saget, M.; Doumalin, P.; Dupré, J.C.; Brémand, F.; Hesser, F.; Couvertier, M.; Brèque, C.; Maxy, P.; et al. A novel approach for biomechanical spine analysis: Mechanical response of vertebral bone augmentation by kyphoplasty to stabilise thoracolumbar burst fractures. *J. Mech. Behav. Biomed. Mater.* **2016**, *59*, 291–303. [CrossRef]
35. Ananthakrishnan, D.; Berven, S.; Deviren, V.; Cheng, K.; Lotz, J.C.; Xu, Z.; Puttlitz, C.M. The effect on anterior column loading due to different vertebral augmentation techniques. *Clin. Biomech.* **2005**, *20*, 25–31. [CrossRef] [PubMed]
36. Wong, E.K.Y.; Whyne, C.M.; Singh, D.; Ford, M. A Biomechanical Assessment of Kyphoplasty as a Stand-Alone Treatment in a Human Cadaveric Burst Fracture Model. *Spine* **2015**, *40*, E808–E813. [CrossRef] [PubMed]
37. Yao, R.; Yin, P.; Zhang, Y.; Li, D.; Zhang, X.-N.; Tao, L.; Zhou, J.; Su, Q. Is there a risk of further displacement of the fractured fragment into the spinal canal in osteoporotic vertebral burst fractures after unilateral PKP? A prospective study. *Spine J.* **2023**, *24*, 1077–1086. [CrossRef] [PubMed]
38. Sun, N.; Zhang, Y.; Xie, D.; Chen, Y.; Liu, Y. Enhancing percutaneous kyphoplasty efficacy in elderly osteoporotic fractures through optimal cement filling ratio. *Front. Endocrinol.* **2024**, *15*, 1359550. [CrossRef] [PubMed]
39. Krüger, A.; Baroud, G.; Noriega, D.; Figiel, J.; Dorschel, C.; Ruchholtz, S.; Oberkircher, L. Height restoration and maintenance after treating unstable osteoporotic vertebral compression fractures by cement augmentation is dependent on the cement volume used. *Clin. Biomech.* **2013**, *28*, 725–730. [CrossRef] [PubMed]
40. Alamin, T.; Kleimeyer, J.P.; Woodall, J.R.; Agarwal, V.; Don, A.; Lindsey, D. Improved biomechanics of two alternative kyphoplasty cementation methods limit vertebral recollapse. *J. Orthop. Res.* **2018**, *36*, 3225–3230. [CrossRef]

41. Adida, S.; Taori, S.; Wong, V.R.; Tang, A.; Sefcik, R.K.; Zhang, X.; Gerszten, P.C. Analysis of injected cement volume and clinical outcomes following balloon-assisted kyphoplasty in a series of 368 patients. *Clin. Neurol. Neurosurg.* **2024**, *243*, 108367. [CrossRef] [PubMed]
42. He, X.; Li, H.; Meng, Y.; Huang, Y.; Hao, D.-J.; Wu, Q.; Liu, J. Percutaneous Kyphoplasty Evaluated by Cement Volume and Distribution: An Analysis of Clinical Data. *Pain Physician* **2016**, *19*, 495–506. [PubMed]
43. Röder, C.; Boszczyk, B.; Perler, G.; Aghayev, E.; Külling, F.; Maestretti, G. Cement volume is the most important modifiable predictor for pain relief in BKP: Results from SWISSspine, a nationwide registry. *Eur. Spine J.* **2013**, *22*, 2241–2248. [CrossRef]
44. Ko, B.-S.; Cho, K.-J.; Park, J.-W. Early Adjacent Vertebral Fractures after Balloon Kyphoplasty for Osteoporotic Vertebral Compression Fractures. *Asian Spine J.* **2019**, *13*, 210–215. [CrossRef]
45. Essibayi, M.A.; Mortezaei, A.; Azzam, A.Y.; Bangash, A.H.; Eraghi, M.M.; Fluss, R.; Brook, A.; Altschul, D.J.; Yassari, R.; Chandra, R.V.; et al. Risk of adjacent level fracture after percutaneous vertebroplasty and kyphoplasty vs natural history for the management of osteoporotic vertebral compression fractures: A network meta-analysis of randomized controlled trials. *Eur. Radiol.* **2024**. [CrossRef] [PubMed]
46. Teuber, H.; Tiziani, S.; Halvachizadeh, S.; Frey, D.; Sprengel, K.; Pape, H.-C.; Osterhoff, G. Single-level vertebral kyphoplasty is not associated with an increased risk of symptomatic secondary adjacent osteoporotic vertebral compression fractures: A matched case-control analysis. *Arch. Osteoporos.* **2018**, *13*, 82. [CrossRef]
47. Yang, S.; Liu, Y.; Yang, H.; Zou, J. Risk factors and correlation of secondary adjacent vertebral compression fracture in percutaneous kyphoplasty. *Int. J. Surg.* **2016**, *36*, 138–142. [CrossRef] [PubMed]
48. Yi, X.; Lu, H.; Tian, F.; Wang, Y.; Li, C.; Liu, H.; Liu, X.; Li, H. Recompression in new levels after percutaneous vertebroplasty and kyphoplasty compared with conservative treatment. *Arch. Orthop. Trauma Surg.* **2014**, *134*, 21–30. [CrossRef] [PubMed]
49. Li, Y.-X.; Guo, D.-Q.; Zhang, S.-C.; Liang, D.; Yuan, K.; Mo, G.-Y.; Li, D.-X.; Guo, H.-Z.; Tang, Y.; Luo, P.-J. Risk factor analysis for re-collapse of cemented vertebrae after percutaneous vertebroplasty (PVP) or percutaneous kyphoplasty (PKP). *Int. Orthop.* **2018**, *42*, 2131–2139. [CrossRef] [PubMed]
50. Berlemann, U.; Ferguson, S.J.; Nolte, L.P.; Heini, P.F. Adjacent vertebral failure after vertebroplasty. A biomechanical investigation. *J. Bone Jt. Surg. Br.* **2002**, *84*, 748–752. [CrossRef]
51. Molloy, S.; Mathis, J.M.; Belkoff, S.M. The Effect of Vertebral Body Percentage Fill on Mechanical Behavior During Percutaneous Vertebroplasty. *Spine* **2003**, *28*, 1549–1554. [CrossRef]
52. Liebschner, M.A.; Rosenberg, W.S.; Keaveny, T.M. Effects of bone cement volume and distribution on vertebral stiffness after vertebroplasty. *Spine* **2001**, *26*, 1547–1554. [CrossRef]
53. Zhang, X.; Chen, T.; Meng, F.; Li, S.; Xu, G.; Yan, J.; Zhao, W. A finite element analysis on different bone cement forms and injection volumes injected into lumbar vertebral body in percutaneous kyphoplasty. *BMC Musculoskelet. Disord.* **2022**, *23*, 621. [CrossRef]
54. Macciacchera, M.; McDonnell, J.M.; Amir, A.; Sowa, A.; Cunniffe, G.; Darwish, S.; Murphy, C.; Butler, J.S. Mechanical Vertebral Body Augmentation Versus Conventional Balloon Kyphoplasty for Osteoporotic Thoracolumbar Compression Fractures: A Systematic Review and Meta-Analysis of Outcomes. *Glob. Spine J.* **2024**, 21925682241261988. [CrossRef]
55. Fada, R.; Shahgholi, M.; Azimi, R.; Babadi, N.F. Estimation of Porosity Effect on Mechanical Properties in Calcium Phosphate Cement Reinforced by Strontium Nitrate Nanoparticles: Fabrication and FEM Analysis. *Arab. J. Sci. Eng.* **2024**, *49*, 1815–1825. [CrossRef]
56. Hansson, T.H.; Keller, T.S.; Spengler, D.M. Mechanical behavior of the human lumbar spine. II. Fatigue strength during dynamic compressive loading. *J. Orthop. Res.* **1987**, *5*, 479–487. [CrossRef] [PubMed]
57. Oakland, R.J.; Furtado, N.R.; Wilcox, R.K.; Timothy, J.; Hall, R.M. Preliminary biomechanical evaluation of prophylactic vertebral reinforcement adjacent to vertebroplasty under cyclic loading. *Spine J.* **2009**, *9*, 174–181. [CrossRef] [PubMed]
58. Kim, M.J.; Lindsey, D.P.; Hannibal, M.; Alamin, T.F. Vertebroplasty versus kyphoplasty: Biomechanical behavior under repetitive loading conditions. *Spine* **2006**, *31*, 2079–2084. [CrossRef] [PubMed]

bioengineering

MDPI

Article

Primary Stability of Kyphoplasty in Incomplete Vertebral Body Burst Fractures in Osteoporosis: A Biomechanical Investigation

Oliver Riesenbeck [1,*], Niklas Czarnowski [1], Michael Johannes Raschke [1], Simon Oeckenpöhler [1,†] and René Hartensuer [1,2,†]

1 Department of Trauma, Hand and Reconstructive Surgery, University Hospital Münster, Albert-Schweitzer-Campus 1, Building W1, Waldeyerstraße 1, 48149 Münster, Germany
2 Center for Orthopaedic, Traumatology, Handsurgery, and Sportsmedicine, Klinikum Aschaffenburg-Alzenau, 63739 Aschaffenburg, Germany
* Correspondence: oliver.riesenbeck@ukmuenster.de
† These authors contributed equally to this work.

Abstract: Background: The objective of our study was to biomechanically evaluate the use of kyphoplasty to stabilize post-traumatic segmental instability in incomplete burst fractures of the vertebrae. **Methods:** The study was performed on 14 osteoporotic spine postmortem samples (Th11–L3). First, acquisition of the native multisegmental kinematics in our robot-based spine tester with three-dimensional motion analysis was set as a baseline for each sample. Then, an incomplete burst fracture was generated in the vertebral body L1 with renewed kinematic testing. After subsequent kyphoplasty was performed on the fractured vertebral body, primary stability was examined again. **Results:** Initially, a significant increase in the range of motion after incomplete burst fracture generation in all three directions of motion (extension–flexion, lateral tilt, axial rotation) was detected as proof of post-traumatic instability. There were no significant changes to the native state in the adjacent segments. Radiologically, a significant loss of height in the fractured vertebral body was also shown. Traumatic instability was significantly reduced by kyphoplasty. However, native kinematics were not restored. **Conclusions:** Although post-traumatic segmental instability was significantly reduced by kyphoplasty in our in vitro model, native kinematics could not be reconstructed, and significant instability remained.

Keywords: kyphoplasty; incomplete burst fracture; biomechanics; primary stability; vertebral body; lumbar spine; spinal instability; spinal trauma; osteoporotic compression fracture

1. Introduction

Kyphoplasty was introduced in 1998 by Mark Reiley [1] and has become widely accepted as a treatment of osteoporotic vertebral compression fractures (OCFs) [2–4]. Kyphoplasty offers advantages over vertebroplasty for restoring vertebral body height and kyphosis angle. Although its short-term effects, especially regarding pain relief, seem to be similar to those of vertebroplasty, kyphoplasty may have advantages in safety and long-term effects in OCFs [2].

The safety and successful outcomes of kyphoplasty have led to more liberal indications for the procedure. When kyphoplasty was first introduced, involvement of the middle column of the vertebral body was considered to be a contraindication for the procedure [3]. However, today it is considered safe to treat these fractures with cement augmentation techniques [5–7]. Some researchers have proposed that kyphoplasty is safe and effective as a stand-alone treatment even for burst fractures, but no studies have provided strong evidence for this [8].

However, biomechanical investigations have confirmed that cement can stabilize osteoporotic vertebrae under cyclic loading (axial compression). Therefore, vertebral augmentation techniques such as vertebroplasty and kyphoplasty are considered to be effective and minimally invasive surgical methods for the stabilization of fractured vertebrae [9]. These biomechanical results, restoring vertebral resistance to compression forces, might partially explain the reported success of kyphoplasty in pure compression fractures. Correction of kyphosis, vertebral body height, and resistance against compression addresses the main pathology mechanism. In traumatic incomplete burst fractures, additional injuries to the disc and ligaments may influence the stability of the motion segment, the functional spinal unit (FSU).

Previous studies have shown that augmentation without correction of the compressed fracture (vertebroplasty) did not restore the stability of the FSU in a human cadaveric incomplete burst fracture model [10].

Considering the stabilization of wedge-compression fractures—A1 according to AO spine classification—by kyphoplasty in a multisegmental, biomechanical model, contradictive results have been reported [11]. Disch and Schmoelz showed that vertebral body height can be restored, but they found that kyphoplasty could not restore the stability of an intact segment. They found that the initial gain in stability after kyphoplasty was markedly reduced to the level of the fractured specimen with increased cyclic load [12]. Achatz et al. reported that kyphoplasty was neither able to restore the initial vertebral body height, nor could it restore the kinematics of the intact spinal segment, which deteriorated under further cyclic loading [13]. Contrary to this, Holyoak et al. reported that kyphoplasty was able to restore vertebral body height close to the intact status and subsequent cyclic loading did not deteriorate height relevantly [14]. To our knowledge, reports of only two additional studies of the biomechanics of kyphoplasty for complete burst fractures—A4 according to AO spine classification—are available: Wong et al. reported that kyphoplasty failed to sufficiently restore stability as a stand-alone treatment after high-energy burst fracture [15]. Germaneau et al. concluded that percutaneous kyphoplasty offers good primary stability in burst fractures, but that its success is limited by potential lesions in adjacent discs or ligaments [16].

Because, to the best of our knowledge, no study has presented information on the stabilization of incomplete burst fractures—A3 according to the AO spine classification and OF 3 according to the OF classification—by kyphoplasty, we conducted a human cadaveric study using a robot-based spine tester and performed three-dimensional motion analysis [11,17]. The spine tester has previously been evaluated for single and multilevel testing [18].

We hypothesized that kyphoplasty can restore primary stability in a traumatic incomplete burst fracture model.

2. Materials and Methods

2.1. Specimens

We used 13 human fresh-frozen cadaveric spine samples (Th11–L3). The median age of the specimen donors was 82 years (Q1 [first quartile] = 75 years; Q3 [third quartile] = 83 years), and all donors were female. In all samples, bone mineral density (BMD) was measured using quantitative computed tomography. The median BMD was 75.63 mg/cm^3 (Q1 = 70.32 mg/cm^3; Q3 = 91.18 mg/cm^3). In comparison, a BMD of >120 mg/cm^3 is considered normal, one between 80 and 120 mg/cm^3 indicates osteopenia, and one of <80 mg/cm^3 indicates osteoporosis [19]. Therefore, all samples except one were from donors who had osteopenia or osteoporosis. Samples with relevant morphologic changes beyond age-related degeneration (e.g., tumor, fracture, deformity, fusion) were excluded.

Prior to testing, all specimens were thawed slowly to room temperature and all soft tissue and muscles were dissected carefully to preserve osseous and ligamentous structures.

The caudal and cranial vertebral bodies were rigidly fixed in a standardized manner in a custom-made embedding frame filled with a two-component resin (Technovit 3040, Heraeus Kulzer GmbH, Hanau, Germany). This setup was then attached to customized tools to mount the samples into the servo-hydraulic testing machine and the testing robot.

All samples were kept moist during the dissection and testing processes, and the whole procedure was performed in accordance with the process outlined by Wilke et al. [20].

2.2. Fracture Creation

We used a previously validated and reported protocol for the fracture creation, adding a novel mounting frame in the servo-hydraulic testing machine and in the robot for kinematic testing (Figures 1 and 2) [21].

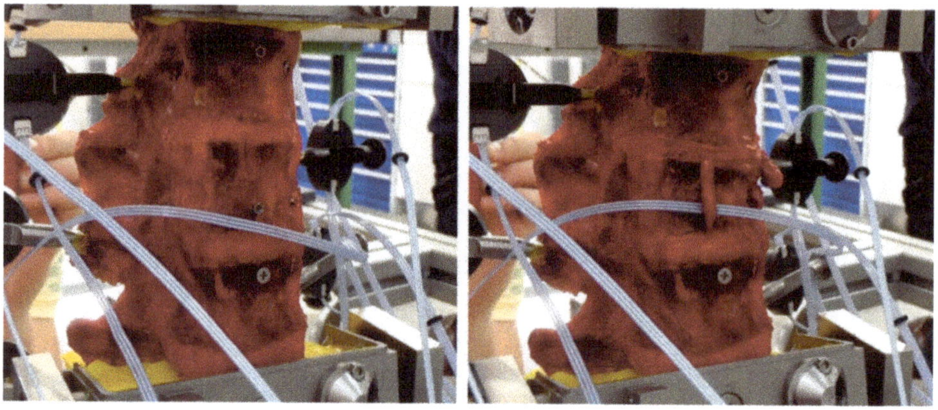

Figure 1. Modified fracture creation by distance-controlled compression after osteotomy-like weakening of the upper endplate L1. Specimen before compression (**left**) and after fracture induction by axial compression (**right**).

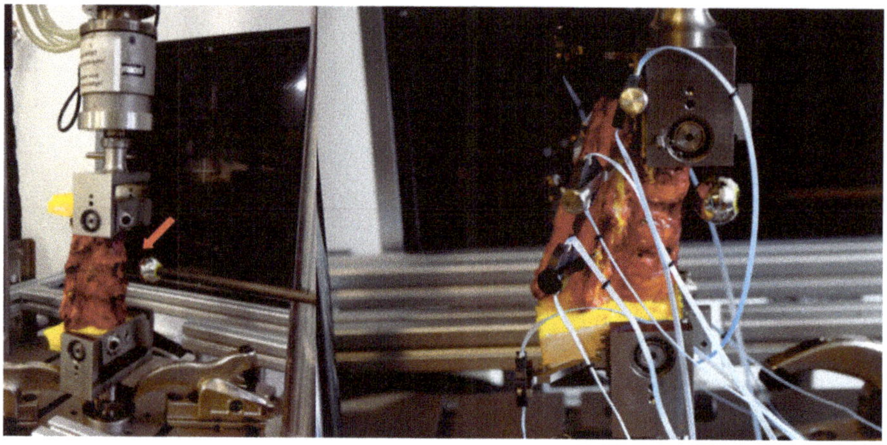

Figure 2. Hydraulic material testing machine used to create standardized incomplete burst fractures and obtain radiographs. At the **left**, the mounted motion capture marker (rigid bodies) and radiographic reference (arrow) are shown. At the **right** is a magnified lateral view of a mounted sample.

The combination of an osteotomy and a distance-controlled compression using a hydraulic testing machine (Instron 8874, Instron, Norwood, MA, USA) resulted in the reproducible creation of type A3 incomplete burst fractures according to the AO spine classification and an OF 3 according to the OF classification [11,17].

2.3. Kyphoplasty

Kyphoplasty (Figure 3) was performed by a single experienced spine surgeon (RH) who followed the manufacturer's recommendations. To simulate permanent pressure even in the supine position in vitro, a constant compressive pressure of 100 N was applied during balloon inflation, and balloon pressure was recorded. After the balloon was fully inflated, the position of the hydraulic testing machine was then preserved. After balloon deflation, there was no compressive pressure in the upright testing setup. Polymethylmethacrylate was loaded into the vertebral body according to the manufacturer's recommendations. The amount of polymethylmethacrylate was assessed via a lateral radiograph in imitation of clinical practice, and the volume of cement intrusion was recorded.

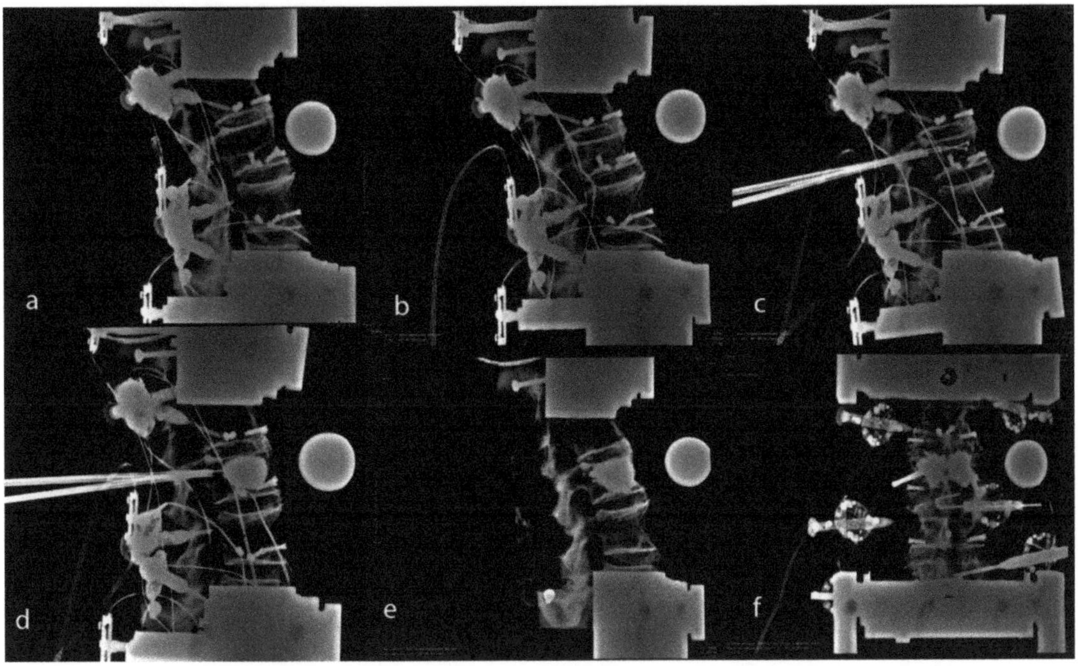

Figure 3. Radiographs of the native sample (**a**), fractured specimen (**b**), balloon in position (**c**), inflated balloon (**d**), inserted cement (**e**), and after cement insertion using the anteroposterior technique (**f**).

2.4. Kinematic Testing

A first set of kinematic tests was conducted with the intact specimens, both with and without follower load, using a robot-based system combined with a custom-made cardan drive that ensured the application of pure moments (7.5 Nm) for extension–flexion, lateral flexion, and axial rotation (Figure 4) [22,23]. All further tests were performed under follower load conditions (350 N).

Intersegmental movement was additionally recorded by the optical motion tracking system to evaluate the kinematic behavior of each FSU in the multisegmental test setup.

After fracture creation and after kyphoplasty, kinematic testing was repeated to compare the individual effects for each specimen.

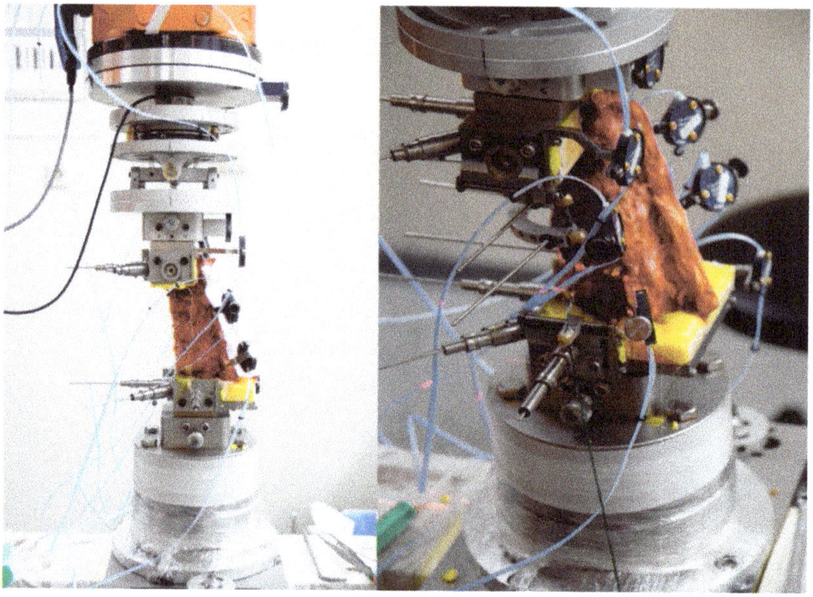

Figure 4. Mounted specimen in the robot-based spine tester combined with active optical motion tracking to record each single segmental kinematic behavior in a multisegmental setup: overview (**left**) and magnified view with rigid bodies and follower-load applications (**right**).

2.5. *Radiological Assessment*

Reconstruction of the vertebral body was monitored by calibrated radiographic examinations, in accordance with clinical practice.

Height for intact vertebral bodies, fractured bodies, and reconstructed bodies was monitored using lateral radiographs (Figure 5). Qualitative monitoring of height restoration was performed by modifying the method described by McKiernan et al. [24]. Measurements obtained included posterior vertebral body height (AB), as shown on lateral radiographs; anterior vertebral body height (CD); central height (height of the middle of the vertebral body (EF)); and the height between the posterior one-third and the anterior two-thirds of the vertebral body (GH).

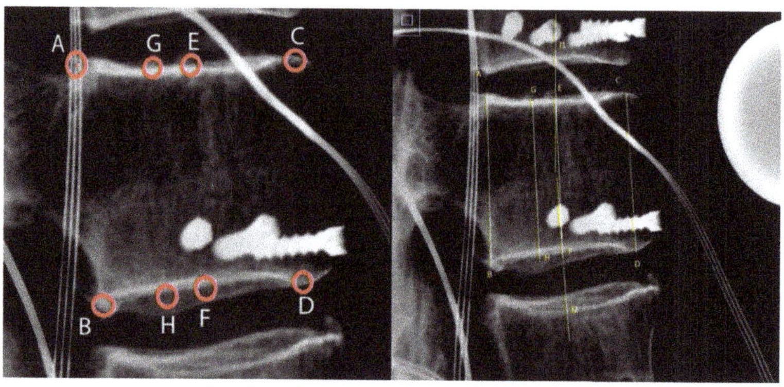

Figure 5. Schematic presentation of height measurement via lateral radiographic projection: posterior (AB), anterior (CD), middle (EF), and posterior two-thirds (GH). I1 and I2 are perpendicular midline intersections for the construction of the points E and F [24].

2.6. Groups

Each specimen was tested intact without follower load (group 1), intact with follower load (group 2), fractured (group 3), and after kyphoplasty (group 4).

2.7. Statistics

Statistical analysis was performed using the Wilcoxon signed-rank test and Bonferroni correction using SPSS (SPSS® Statistics 27; IBM, Endicott, NY, USA).

3. Results

3.1. Fracture Creation and Vertebral Body Reconstruction

The vertebral body height was decreased by the standardized fracture creation procedure to the following percentages of intact values: AB, 91.9%; CD, 75.1%; EF, 76.6%; and GH, 76.6%.

The median balloon inflation pressure was 13.5 bar (Q1 = 12; Q3 = 14.25). The median balloon volume was 10 mL (Q1 = 9; Q3 = 12.5). The median cement volume was 9.6 mL (Q1 = 9; Q3 = 12).

We were able to reconstruct the vertebral body height, using balloon kyphoplasty, to the following percentages of intact values: posterior (AB), 95.8%; anterior (CD), 86.2%; middle (EF), 88%; and posterior two-thirds (GH), 85.9%. Table 1 provides details of the losses of height after fracture creation and after kyphoplasty; Figure 6 provides an overview of the lateral vertebral body heights.

Table 1. Vertebral body height (millimeters) after fracture creation and after kyphoplasty.

Specimen	1	2	3	4	5	6	7	8	9	10	11	12	13	Median	Q1	Q3
Native																
Anterior (CD)	22.5	26.4	27.5	25.8	20.7	25.1	27	26.1	24.3	27.8	26.8	28.4	22.1	26.1	23.4	27.3
Middle (EF)	21.8	26.3	25.3	27	24.4	25.7	26.3	24.8	26.5	26.3	27.2	25.9	24.3	25.9	24.6	26.4
Posterior two-thirds (GH)	23	27	25.6	27.4	24.9	26.5	27.4	24.2	27.2	27.1	27.2	26.9	24.9	26.9	24.9	27.2
Posterior (AB)	28.8	28.8	28.3	28.3	23.8	28.6	29	27.9	28.1	28.4	29.1	30.3	26.9	28.4	28	28.9
Fractured																
Anterior (CD)	18.8	16.5	27	21.9	21.2	19.8	18.3	19.1	18.7	16.9	19.9	19.6	20	19.6	18.5	20.6
Middle (EF)	18.3	17.3	22.4	22	17.6	21.5	18.5	18.6	21	19.1	20.9	19.9	20.6	19.9	18.4	21.3
Posterior two-thirds (GH)	19.9	19	22.9	22.1	19.3	22.4	20	18.5	21.9	20.6	21.2	22	22.3	21.2	19.6	22.2
Posterior (AB)	27.1	26.1	26	24.9	24.5	26.6	23.9	23.2	27.2	24.8	27.1	28.3	26.5	26.1	24.7	27.1
Reconstructed																
Anterior (CD)	22.5	19.2	26.8	23.2	21.9	21	23.5	22.5	20.4	22.6	26	25.7	22	22.5	21.5	24.6
Middle (EF)	21	19.7	23.2	24.4	20.9	22.8	19.7	21.1	21.9	23.2	23.5	23.8	23.5	22.8	21.0	23.5
Posterior two-thirds (GH)	23.1	21.2	23.3	24.3	21.9	23.5	20.5	20.7	22.3	23.2	23.4	25.1	23.5	23.2	21.6	23.5
Posterior (AB)	28.6	26	27.2	27.6	24.8	27.2	24.7	25.3	27.1	26.8	27.5	28.9	27.4	27.2	25.7	27.6

AB, posterior vertebral body height; CD, anterior vertebral body height; EF, central vertebral body height, aka height of the middle of the vertebral body; GH, height between the posterior one-third and the anterior two-thirds of the vertebral body; Q1 and Q3, median values for first and third quartiles, respectively.

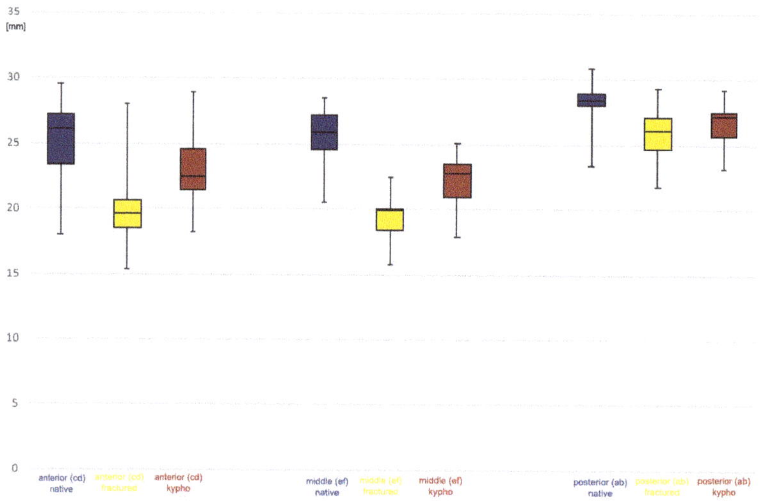

Figure 6. Boxplot of lateral vertebral body height in millimeters. Anterior (cd), middle (ef), and posterior (ab) values are presented by condition: intact, blue; fractured, yellow; reconstructed by kyphoplasty (kypho), red.

3.2. *Kinematics of the Injured Segment (Th12–L1)*

An increase in the range of motion (ROM), in the size of the neutral zone, and in the size of the elastic zone after fracture induction was obvious for axial rotation, extension–flexion, and lateral flexion.

In Figures 7 and 8, we considered the intact condition with (light blue) and without follower load (blue) and the fractured condition (yellow) as reference points for estimating the effect of kyphoplasty (red).

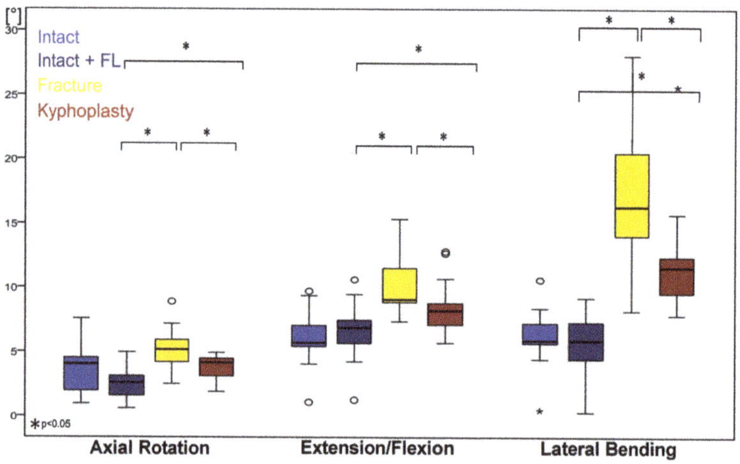

Figure 7. Boxplot of kinematic median values of functional spinal unit Th12–L1 for axial rotation, extension–flexion, and lateral flexion. Intact values without (light blue) and with follower load (blue), fractured values with follower load (yellow), and values after kyphoplasty with follower load (red). Circle represents outliers and five-pointed asterisk represents extreme outliers.

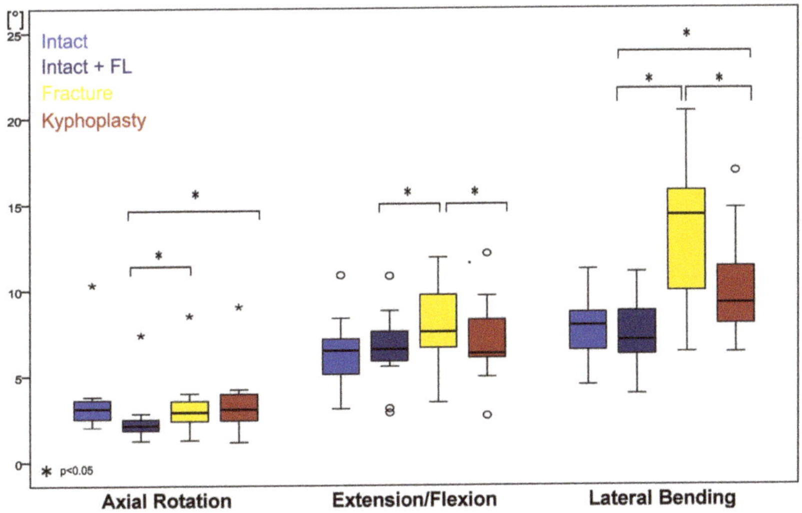

Figure 8. Boxplot of kinematic median values of functional spinal unit L1–L2 for axial rotation, extension–flexion, and lateral flexion. Intact values without (light blue) and with follower load (blue), values for fractures with follower load (yellow), and values after kyphoplasty with follower load (red). Circle represents outliers and five-pointed asterisk represents extreme outliers.

3.3. Extension–Flexion

In extension–flexion in the intact condition, ROM without follower load was 5.7° (Q1 = 5.2°; Q3 = 8.0°); under follower load, it was 6.8° (Q1 = 5.4°; Q3 = 8.4°). ROM increased after fracture by 132% ($p < 0.05$) to 9.1° (Q1 = 8.8°; Q3 = 12.4°). This change can be interpreted as traumatic segmental instability. After kyphoplasty, ROM decreased by 90% ($p < 0.05$). However, in comparison with the intact state, a significant increase of 120% ($p < 0.05$) still remained.

3.4. Axial Rotation

Axial rotation in the intact condition without follower load was 4.0° (Q1 = 1.8°; Q3 = 4.6°); under follower load, it was 2.6° (Q1 = 1.2°; Q3 = 3.2°). ROM increased after fracture to 5.1° (150%; $p < 0.05$; Q1 = 3.7; Q3 = 6.1°). These changes can be interpreted as traumatic segmental instability for rotation. After kyphoplasty, ROM significantly decreased to 4.1° (Q1 = 2.6°; Q3 = 4.5°; $p < 0.05$). However, in comparison with the intact state, a significant increase in ROM in axial rotation after kyphoplasty remained: 161% ($p < 0.05$).

3.5. Lateral Flexion

In lateral flexion, intact ROM without follower load was 5.9° (Q1 = 5.2°; Q3 = 7.8°); under follower load, it was 5.9° (Q1 = 4.0°; Q3 = 7.4°). ROM increased after fracture by 277% ($p < 0.05$) to 16.3° (Q1 = 13.9; Q3 = 20.7°). These changes can also be interpreted as traumatic segmental instability for lateral flexion. After kyphoplasty, ROM decreased to 11.6° (Q1 = 9.5°; Q3 = 12.4°; $p < 0.05$). However, in comparison with the intact state, a significant increase in ROM (197%; $p < 0.05$) remained.

3.6. Kinematics of L1–L2

The kinematics of levels L1–L2 show effects similar to those at the experimentally injured level. This effect of partial restoration of segmental stability was detectable for all movement directions: axial rotation, extension–flexion, and lateral flexion (Figure 8; Table 2).

Table 2. Kinematics (range of motion [°]), Q1 = 1st quartile; Q3 = 3rd quartile.

	Group 1: Intact without Follower Load			Group 2: Intact with Follower Load			Group 3: Fracture with Follower Load			Group 4: Kyphoplasty with Follower Load		
	Median	Q1	Q3	Median	Q1	Q3	Median	Q1	Q3	Median	Q1	Q3
Axial rotation												
Th11–Th12	5.2	2.0	5.7	3.3	1.7	4.6	3.4	1.8	4.5	3.1	1.8	5.0
Th12–L1	4.0	1.8	4.7	2.6	1.2	3.2	5.1	3.7	6.1	4.1	2.6	4.5
L1–L2	3.1	2.5	3.8	2.1	1.8	2.6	2.9	2.1	3.6	3.1	2.1	4.1
L2–L3	4.4	3.4	6.1	2.6	2.2	4.2	2.7	2.4	4.2	2.9	2.4	4.3
Extension-Flexion												
Th11–Th12	4.8	3.8	5.6	4.6	3.5	5.6	4.7	4.2	6.6	5.0	3.0	6.7
Th12–L1	5.7	5.2	8.05	6.8	5.4	8.4	9.1	8.8	12.4	8.2	6.8	9.7
L1–L2	6.5	5.1	7.2	6.5	5.7	7.8	7.5	6.0	10.0	6.3	5.5	8.9
L2–L3	8.2	6.3	8.9	8.6	6.3	9.7	9.0	6.2	9.9	9.6	6.3	9.9
Lateral flexion												
Th11–Th12	4.2	3.5	6.4	3.1	1.7	4.5	3.6	1.8	4.6	3.8	1.4	4.7
Th12–L1	5.9	5.2	7.8	5.9	4.0	7.4	16.3	13.9	20.7	11.6	9.3	14.0
L1–L2	7.9	6.3	8.9	7.1	6.1	8.9	14.3	9.2	16.0	9.2	7.4	12.4
L2– L3	9.9	6.7	11.7	9.0	5.9	11.2	9.8	5.9	14.9	9.6	6.6	13.2

3.7. Adjacent Segments

Each segment was evaluated independently using optical three-dimensional motion analysis. No significant changes were detected in the segments except FSUs involving fractured vertebra (Th12–L1 and L1–L2).

4. Discussion

In the common understanding of spinal instability, (incomplete) burst fractures are considered to be unstable. However, clinical treatment options do not necessarily reflect this assessment under the current understanding of biomechanics.

Previous studies have shown that vertebroplasty without reconstruction of vertebral body height could not restore the stability of the FSU in a human cadaveric burst fracture model [10].

These findings must be discussed within the framework of conflicting clinical findings regarding treatment success with cement augmentation. According to Germaneau et al., kyphoplasty can stabilize a traumatic fractured segment [16]. Therefore, they concluded that percutaneous kyphoplasty offers sufficient primary stability in burst fractures.

This is consistent with some other case reports noting that kyphoplasty should be a reliable and successful stand-alone option for treating traumatic burst fractures [8]. However, Wong et al. reported that kyphoplasty failed to sufficiently restore stability as a stand-alone treatment after high-energy burst fracture due to the compromised intervertebral discs [15]. Their biomechanical results are supported by multiple clinical reports, including those of Oner et al. [25], Zaryanov et al. [26], Josten et al. [27], and Spiegl et al. [28] of the need to use both posterior instrumentation and kyphoplasty to achieve vertebral body restoration and segmental stabilization.

Our findings add to the controversy by showing that kyphoplasty has some potential to increase segmental stability in a traumatic incomplete burst fracture model. We found that the increase in stability in extension/flexion, rotation, and bending was significant compared with that in the fractured state. This effect was evident in the injured index level (Th12–L1) as well as in the level below (L1–L2). It seems obvious that the reconstruction of height leads to some stabilizing effects in both involved FSUs. This effect might be in accordance with the flagpole principle described by Evans [29].

However, the post-surgery values we obtained did not reach the values of intact kinematic conditions, and significant segmental instability remained compared with the intact sample. Therefore, our findings show that kyphoplasty cannot reconstruct native kinematic values after incomplete burst fracture.

In fact, there is little knowledge about the resulting instability in vivo after incomplete burst fractures. This lack of knowledge is replenished by different treatment options, including conservative [30,31], vertebroplasty [32], kyphoplasty [33], instrumentation [30], different combinations of kyphoplasty [31], and instrumentation plus 360° fusion [30].

One explanation may be an inconsistent usage of the term incomplete burst fracture. This fracture type has a wide range of appearances that can lead to different levels of instability. For that reason, we advocate for the usage of differentiated classification systems to standardize the type of injury in clinical practice and experimental research. Even when using a specific classification for osteoporotic vertebral fractures and a corresponding scoring system, it remains difficult to recommend a treatment method. Therefore, additional posterior instrumentation should be evaluated in the presence of an OF 3 fracture [34,35].

Additionally, the importance of active stabilization of the FSU must be discussed. It is well known that biomechanical kinematic studies mainly investigate the passive factors of motion. Some may simulate muscle forces and have shown a glimmer of importance [36]. However, post mortem experiments cannot verify the role of the active motion system in stabilizing the spine, resulting in a limitation of our study.

Van Dieen et al. reported that the changed trunk muscle recruitment patterns in patients with low back pain enhance the stability of the lumbar spine [37]. These findings indicate the potential compensatory mechanisms of the active motion system in case of a resulting segmental instability and may partially explain why some patients gain enough passive stability to compensate after kyphoplasty and others do not. As long as we do not have the tools to evaluate the patient's active compensatory potential and calculate the required passive stability, we must rely on the findings of kinematic post mortem studies.

Another limitation of our study is the volume of cement used. The cement volumes in the published literature vary significantly, but smaller amounts of cement may be used in everyday clinical practice [14,38,39]. Therefore, our study might overestimate the effect of biomechanical stabilization by kyphoplasty in incomplete burst fractures.

5. Conclusions

Kyphoplasty is able to stabilize incomplete burst fractures by restoring vertebral body height, but significant instability remains in comparison with intact values. Therefore, successful treatment depends not only on correct execution of the procedure, but also on the individual capacity of active segmental stabilization to some extent.

Author Contributions: Conceptualization, O.R., M.J.R., S.O. and R.H.; methodology, O.R., N.C. and R.H.; software, O.R. and N.C.; validation, O.R., N.C. and R.H.; formal analysis, O.R., N.C. and R.H.; investigation, O.R., N.C., M.J.R. and R.H.; resources, O.R. and R.H.; data curation, O.R., N.C. and R.H.; writing—original draft preparation, O.R. and R.H.; writing—review and editing, O.R., N.C., M.J.R., S.O. and R.H.; visualization, O.R., N.C. and R.H.; supervision, M.J.R., S.O. and R.H.; project administration, O.R., M.J.R. and R.H.; funding acquisition, M.J.R. and R.H. All authors have read and agreed to the published version of the manuscript.

Funding: This study was supported in part by funds received from Joline (Joline GmbH & Co. KG, 72379 Hechingen, Germany) to cover material costs of EUR 26,031.25. There was no influence of the funder on this article. We acknowledge support from the Open Access Publication Fund of the University of Münster.

Institutional Review Board Statement: The Ethics Committee of the Westphalia-Lippe Medical Association (Ethikkommission der Ärztekammer Westfalen-Lippe) and the University of Munster approved the study (2018-21 672-f-S).

Informed Consent Statement: Informed consent was obtained from all subjects involved in the study, consistently with the setting of collection/donation of the anatomical material.

Data Availability Statement: Dataset available on request from the authors.

Acknowledgments: We thank Martin Schulze and Markus Müller of the University of Munster for help in managing the robot-based testing setup. and Alexander Stump for helping in data acquisition. Katharine O'Moore-Klopf, (East Setauket, NY, USA) provided professional English language editing of our manuscript.

Conflicts of Interest: The authors declare no conflicts of interest.

References

1. Garfin, S.R.; Yuan, H.A.; Reiley, M.A. New technologies in spine: Kyphoplasty and vertebroplasty for the treatment of painful osteoporotic compression fractures. *Spine* **2001**, *26*, 1511–1515. [CrossRef]
2. Zhao, G.; Liu, X.; Li, F. Balloon kyphoplasty versus percutaneous vertebroplasty for treatment of osteoporotic vertebral compression fractures (OVCFs). *Osteoporos. Int.* **2016**, *27*, 2823–2834. [CrossRef]
3. Ortiz, A.O.; Zoarski, G.H.; Beckerman, M. Kyphoplasty. *Tech. Vasc. Interv. Radiol.* **2002**, *5*, 239–249. [CrossRef]
4. Hoffmann, J.; Preston, G.; Whaley, J.; Khalil, J.G. Vertebral Augmentation in Spine Surgery. *J. Am. Acad. Orthop. Surg.* **2023**, *31*, 477–489. [CrossRef]
5. Li, C.-H.; Chang, M.-C.; Liu, C.-L.; Chen, T.-S. Osteoporotic burst fracture with spinal canal compromise treated with percutaneous vertebroplasty. *Clin. Neurol. Neurosurg.* **2010**, *112*, 678–681. [CrossRef]
6. Yao, R.; Yin, P.; Zhang, Y.; Li, D.; Zhang, X.-N.; Tao, L.; Zhou, J.; Su, Q. Is there a risk of further displacement of the fractured fragment into the spinal canal in osteoporotic vertebral burst fractures after unilateral PKP? A prospective study. *Spine J.* **2023**, *24*, 1077–1086. [CrossRef]
7. Chen, Y.; Yin, P.; Hai, Y.; Su, Q.; Yang, J. Is Osteoporotic Thoracolumbar Burst Fracture a Contraindication to Percutaneous Kyphoplasty? A Systematic Review. *Pain Physician* **2021**, *24*, E685–E692.
8. Oh, T.; Scheer, J.K.; Fakurnejad, S.; Dahdaleh, N.S.; Smith, Z.A. Minimally invasive spinal surgery for the treatment of traumatic thoracolumbar burst fractures. *J. Clin. Neurosci.* **2015**, *22*, 42–47. [CrossRef]
9. Wilke, H.-J.; Mehnert, U.; Claes, L.E.; Bierschneider, M.M.; Jaksche, H.; Boszczyk, B.M. Biomechanical evaluation of vertebroplasty and kyphoplasty with polymethyl methacrylate or calcium phosphate cement under cyclic loading. *Spine* **2006**, *31*, 2934–2941. [CrossRef] [PubMed]
10. Hartensuer, R.; Gehweiler, D.; Schulze, M.; Matuszewski, L.; Raschke, M.J.; Vordemvenne, T. Biomechanical evaluation of combined short segment fixation and augmentation of incomplete osteoporotic burst fractures. *BMC Musculoskelet. Disord.* **2013**, *14*, 360. [CrossRef] [PubMed]
11. Vaccaro, A.R.; Oner, C.; Kepler, C.K.; Dvorak, M.; Schnake, K.; Bellabarba, C.; Reinhold, M.; Aarabi, B.; Kandziora, F.; Chapman, J.; et al. AOSpine thoracolumbar spine injury classification system: Fracture description, neurological status, and key modifiers. *Spine* **2013**, *38*, 2028–2037. [CrossRef] [PubMed]
12. Disch, A.C.; Schmoelz, W. Cement augmentation in a thoracolumbar fracture model: Reduction and stability after balloon kyphoplasty versus vertebral body stenting. *Spine* **2014**, *39*, E1147–E1153. [CrossRef] [PubMed]
13. Achatz, G.; Riesner, H.-J.; Friemert, B.; Lechner, R.; Graf, N.; Wilke, H.-J. Biomechanical in vitro comparison of radiofrequency kyphoplasty and balloon kyphoplasty. *Eur. Spine J.* **2017**, *26*, 3225–3234. [CrossRef] [PubMed]
14. Holyoak, D.T.; Andreshak, T.G.; Hopkins, T.J.; Brook, A.L.; Frohbergh, M.E.; Ong, K.L. Height restoration and sustainability using bilateral vertebral augmentation systems for vertebral compression fractures: A cadaveric study. *Spine J.* **2022**, *22*, 2072–2081. [CrossRef] [PubMed]
15. Wong, E.K.Y.; Whyne, C.M.; Singh, D.; Ford, M. A Biomechanical Assessment of Kyphoplasty as a Stand-Alone Treatment in a Human Cadaveric Burst Fracture Model. *Spine* **2015**, *40*, E808–E813. [CrossRef] [PubMed]
16. Germaneau, A.; Vendeuvre, T.; Saget, M.; Doumalin, P.; Dupré, J.C.; Brémand, F.; Hesser, F.; Couvertier, M.; Brèque, C.; Maxy, P.; et al. A novel approach for biomechanical spine analysis: Mechanical response of vertebral bone augmentation by kyphoplasty to stabilise thoracolumbar burst fractures. *J. Mech. Behav. Biomed. Mater.* **2016**, *59*, 291–303. [CrossRef]
17. Schnake, K.J.; Blattert, T.R.; Hahn, P.; Franck, A.; Hartmann, F.; Ullrich, B.; Verheyden, A.; Mörk, S.; Zimmermann, V.; Gonschorek, O.; et al. Classification of Osteoporotic Thoracolumbar Spine Fractures: Recommendations of the Spine Section of the German Society for Orthopaedics and Trauma (DGOU). *Glob. Spine J.* **2018**, *8*, 46S–49S. [CrossRef] [PubMed]
18. Schulze, M.; Hartensuer, R.; Gehweiler, D.; Hölscher, U.; Raschke, M.J.; Vordemvenne, T. Evaluation of a robot-assisted testing system for multisegmental spine specimens. *J. Biomech.* **2012**, *45*, 1457–1462. [CrossRef] [PubMed]
19. American College of Radiology. ACR–SPR–SSR Practice Parameter for the Performance of Quantitative Computed Tomography (QCT) Bone Mineral Density. Available online: https://www.acr.org (accessed on 22 July 2024).
20. Wilke, H.J.; Wenger, K.; Claes, L. Testing criteria for spinal implants: Recommendations for the standardization of in vitro stability testing of spinal implants. *Eur. Spine J.* **1998**, *7*, 148–154. [CrossRef]
21. Hartensuer, R.; Gasch, A.; Gehweiler, D.; Schanz, S.; Schulze, M.; Matuszewski, L.; Langer, M.; Raschke, M.J.; Vordemvenne, T. Experimentally induced incomplete burst fractures—A novel technique for calf and human specimens. *BMC Musculoskelet. Disord.* **2012**, *13*, 45. [CrossRef]

22. Patwardhan, A.G.; Havey, R.M.; Carandang, G.; Simonds, J.; Voronov, L.I.; Ghanayem, A.J.; Meade, K.P.; Gavin, T.M.; Paxinos, O. Effect of compressive follower preload on the flexion-extension response of the human lumbar spine. *J. Orthop. Res.* **2003**, *21*, 540–546. [CrossRef] [PubMed]
23. Rohlmann, A.; Zander, T.; Rao, M.; Bergmann, G. Applying a follower load delivers realistic results for simulating standing. *J. Biomech.* **2009**, *42*, 1520–1526. [CrossRef] [PubMed]
24. McKiernan, F.; Faciszewski, T.; Jensen, R. Reporting height restoration in vertebral compression fractures. *Spine* **2003**, *28*, 2517–2521. [CrossRef] [PubMed]
25. Oner, F.C.; Verlaan, J.-J.; Verbout, A.J.; Dhert, W.J.A. Cement augmentation techniques in traumatic thoracolumbar spine fractures. *Spine* **2006**, *31*, S89–S95, discussion S104. [CrossRef] [PubMed]
26. Zaryanov, A.V.; Park, D.K.; Khalil, J.G.; Baker, K.C.; Fischgrund, J.S. Cement augmentation in vertebral burst fractures. *Neurosurg. Focus* **2014**, *37*, E5. [CrossRef] [PubMed]
27. Josten, C.; Schmidt, C.; Spiegl, U. Osteoporotische Wirbelkörperfrakturen der thorakolumbalen Wirbelsäule. *Diagn. Behandlungsstrategien. Chir.* **2012**, *83*, 866–874. [CrossRef] [PubMed]
28. Spiegl, U.; Jarvers, J.-S.; Heyde, C.-E.; Josten, C. Osteoporotic vertebral body fractures of the thoracolumbar spine: Indications and techniques of a 360°-stabilization. *Eur. J. Trauma Emerg. Surg.* **2017**, *43*, 27–33. [CrossRef] [PubMed]
29. Evans, J.H. Biomechanics of lumbar fusion. *Clin. Orthop. Relat. Res.* **1985**, *193*, 38–46. [CrossRef]
30. Uchida, K.; Kobayashi, S.; Matsuzaki, M.; Nakajima, H.; Shimada, S.; Yayama, T.; Sato, R.; Baba, H. Anterior versus posterior surgery for osteoporotic vertebral collapse with neurological deficit in the thoracolumbar spine. *Eur. Spine J.* **2006**, *15*, 1759–1767. [CrossRef]
31. Marco, R.A.W.; Kushwaha, V.P. Thoracolumbar burst fractures treated with posterior decompression and pedicle screw instrumentation supplemented with balloon-assisted vertebroplasty and calcium phosphate reconstruction. *J. Bone Jt. Surg. Am.* **2009**, *91*, 20–28. [CrossRef]
32. Shin, J.J.; Chin, D.K.; Yoon, Y.S. Percutaneous vertebroplasty for the treatment of osteoporotic burst fractures. *Acta Neurochir.* **2009**, *151*, 141–148. [CrossRef] [PubMed]
33. Hartmann, F.; Gercek, E.; Leiner, L.; Rommens, P.M. Kyphoplasty as an alternative treatment of traumatic thoracolumbar burst fractures Magerl type A3. *Injury* **2012**, *43*, 409–415. [CrossRef] [PubMed]
34. Ullrich, B.W.; Schenk, P.; Scheyerer, M.J.; Bäumlein, M.; Katscher, S.; Schnake, K.J.; Zimmermann, V.; Schwarz, F.; Schmeiser, G.; Scherer, M.; et al. Georg Schmorl prize of the German spine society (DWG) 2022: Current treatment for inpatients with osteoporotic thoracolumbar fractures-results of the EOFTT study. *Eur. Spine J.* **2023**, *32*, 1525–1535. [CrossRef] [PubMed]
35. Blattert, T.R.; Schnake, K.J.; Gonschorek, O.; Gercek, E.; Hartmann, F.; Katscher, S.; Mörk, S.; Morrison, R.; Müller, M.; Partenheimer, A.; et al. Nonsurgical and Surgical Management of Osteoporotic Vertebral Body Fractures: Recommendations of the Spine Section of the German Society for Orthopaedics and Trauma (DGOU). *Glob. Spine J.* **2018**, *8*, 50S–55S. [CrossRef] [PubMed]
36. Wilke, H.J.; Wolf, S.; Claes, L.E.; Arand, M.; Wiesend, A. Stability increase of the lumbar spine with different muscle groups. A biomechanical in vitro study. *Spine* **1995**, *20*, 192–198. [CrossRef] [PubMed]
37. van Dieën, J.H.; Cholewicki, J.; Radebold, A. Trunk muscle recruitment patterns in patients with low back pain enhance the stability of the lumbar spine. *Spine* **2003**, *28*, 834–841. [CrossRef] [PubMed]
38. Adida, S.; Taori, S.; Wong, V.R.; Tang, A.; Sefcik, R.K.; Zhang, X.; Gerszten, P.C. Analysis of injected cement volume and clinical outcomes following balloon-assisted kyphoplasty in a series of 368 patients. *Clin. Neurol. Neurosurg.* **2024**, *243*, 108367. [CrossRef]
39. Sun, N.; Zhang, Y.; Xie, D.; Chen, Y.; Liu, Y. Enhancing percutaneous kyphoplasty efficacy in elderly osteoporotic fractures through optimal cement filling ratio. *Front. Endocrinol.* **2024**, *15*, 1359550. [CrossRef]

bioengineering

MDPI

Article

Evaluation of Load on Cervical Disc Prosthesis by Imposing Complex Motion: Multiplanar Motion and Combined Rotational–Translational Motion

Hossein Ansaripour [1,*], Stephen J. Ferguson [1] and Markus Flohr [2]

[1] Institute for Biomechanics, D-HEST, ETH Zurich, Gloriastrasse 37/39, 8092 Zurich, Switzerland; sferguson@ethz.ch
[2] CeramTec GmbH, CeramTec-Platz 1-9, 73207 Plochingen, Germany; m.flohr@ceramtec.de
* Correspondence: hansaripour@student.ethz.ch or h.ansaripour@web.de

Abstract: (1) Background: The kinematic characteristics of disc prosthesis undergoing complex motion are not well understood. Therefore, examining complex motion may provide an improved understanding of the post-operative behavior of spinal implants. (2) Methods: The aim of this study was to develop kinematic tests that simulate multiplanar motion and combined rotational–translational motion in a disc prosthesis. In this context, five generic zirconia-toughened alumina (BIOLOX®*delta*, CeramTec, Germany) ball and socket samples were tested in a 6 DOF spine simulator under displacement control with an axial compressive force of 100 N in five motion modes: (1) flexion–extension (FE = ± 7.5°), (2) lateral bending (LB = ± 6°), (3) combined FE-LB (4) combined FE and anteroposterior translation (AP = 3 mm), and (5) combined LB and lateral motion (3 mm). For combined rotational–translational motion, two scenarios were analyzed: excessive translational movement after sample rotation (scenario 1) and excessive translational movement during rotation (scenario 2). (3) Results: For combined FE-LB, the resultant forces and moments were higher compared to the unidirectional motion modes. For combined rotational–translational motion (scenario 1), subluxation occurred at FE = 7.5° with an incremental increase in AP translation = 1.49 ± 0.18 mm, and LB = 6° with an incremental increase of lateral translation = 2.22 ± 0.16 mm. At the subluxation point, the incremental increase in AP force and lateral force were 30.4 ± 3.14 N and 40.8 ± 2.56 N in FE and LB, respectively, compared to the forces at the same angles during unidirectional motion. For scenario 2, subluxation occurred at FE = 4.93° with an incremental increase in AP translation = 1.75 mm, and LB = 4.52° with an incremental increase in lateral translation = 1.99 mm. At the subluxation point, the incremental increase in AP force and lateral force were 39.17 N and 38.94 N in FE and LB, respectively, compared to the forces in the same angles during the unidirectional motion. (4) Conclusions: The new test protocols improved the understanding of in vivo-like behavior from in vitro testing. Simultaneous translation–rotation motion was shown to provoke subluxation at lower motion extents. Following further validation of the proposed complex motion testing, these new methods can be applied future development and characterization of spinal motion-preserving implants.

Keywords: in vitro kinematics; cervical disc prosthesis; multiplanar motion; subluxation

1. Introduction

Anterior cervical discectomy and fusion (ACDF) is the gold standard for treatment of degenerative disc disease causing myelopathy or radiculopathy [1]. There are outstanding short- and long-term clinical results for ACDF [2–5]. However, Baba et al. found the development of new dynamic spinal canal stenosis in 25% of patients after a mean of 8.5 years of follow-up [6]. In another study, an adjacent segment disease (ASD) rate of 2.9% per year was reported after fusion [7]. Several in vitro biomechanical tests have also proposed that the considerable increase in range of motion (ROM) and intradiscal pressure (IDP) at the adjacent segments may contribute to the development of ASD following ACDF [8].

Concern about adjacent segment disease has prompted a surge in the development of motion-preserving disc prostheses.

Disc prostheses has been developed to preserve motion across the spine, alter IDP less at the adjacent segments, provide pain relief, and prevent the development of late ASD [8]. A 5-year meta-analysis revealed a relatively low rate of complications for cervical disc prostheses (0–4.0%) and lumbar disc prostheses (0–16.7%) [9]. Nevertheless, disc arthroplasty is a relatively new treatment, and long-term data on issues including fatigue failure, wear debris accumulation, implant luxation, and heterotopic ossification are insufficient [10–13]. Moreover, a cross-sectional analysis reported complications mainly related to implant migration, insertion problems, neck pain, heterotopic ossification, and radiculopathy following cervical disc arthroplasty [14]. These issues might be traced back to design, material selection, or surgical error [15].

Biomechanical in vitro testing is useful for the preclinical assessment of spinal implant performance and the success of operative procedures. Nonetheless, the majority of research has focused on observing simple arcs of motion in a single plane (i.e., flexion-extension (FE), lateral bending (LB), and axial rotation (AR)), making it challenging to address clinical issues such as implant luxation and migration [8]. A previous study examined cervical spine kinematics in multiplanar motion after disc replacement and ACDF [16]. They hypothesized that examining multiplanar motion could provide a better understanding of the in vivo behavior of spinal implants [16]. Penning et al. also suggested that the segmental motion of the cervical spine was not simply a rocking movement but accompanied by a displacement between vertebrae [17]. Hence, the absence of translational motion in in vitro testing can prevent the complete understanding of disc prosthesis function. For instance, abnormal or excessive vertebral translations may serve as clinically important indicators for the evaluation of disc subluxation and dislocation risk.

This study aimed to develop kinematic test methods for a cervical disc prosthesis that simulate multiplanar motion and combined rotational–translational motion. The specific research objectives were as follows:

(1) Quantifying the forces and moments generated during the multiplanar motion (combined FE-LB). In this case, three component forces (i.e., Fx, Fy, and Fz) and three component moments (i.e., Tx, Ty, and Tz) were calculated, and the resultant values for forces and moments were then compared between the unidirectional motion test and the multiplanar motion test.
(2) Determining the conditions under which subluxation or dislocation occurs during combined rotational–translational movements. In this case, the lateral force, the degree of the rotation, and the translation were measured until the samples reached the subluxation point.

By focusing on these specific metrics, this research provides detailed insights into the in vitro behavior of cervical disc prostheses under complex loading conditions.

2. Materials and Methods

2.1. Experimental Components

Five generic zirconia-toughened alumina (BIOLOX®*delta*, CeramTec, Plochingen, Germany) ball and socket samples (shape: cylindrical body; height: 20 mm; diameter: 11 mm) were tested in a 6 DOF spine simulator (material testing system MTS-370.02 Bionix, Eden Prairie, MN USA) (Figure 1). The rotations in FE and LB are achieved by a rotary gimbal drive around the y and x axis, respectively (Figure 1). A 6-component load cell (FT 15954, mini-45/SI-580–20) was attached between the upper fixture and the gimbal for data acquisition at a 100 Hz sampling frequency to record reaction moments and forces. Alignment of the sample and setup components was required during the test run to assure unconstrained rotation around the desired axis and to avoid excessive forces and moments and, as a result, damage to the sample. The X/Y table can provide appropriate translational compensation, which corrects for the offset between the testing machine's center of rotation (COR) and the sample's COR. In this study, the offset measurement was 95 mm. Therefore, a translational compensation of 12.40 mm for FE (7.5°) and 9.93 mm for LB (6°) were required ($Offset \times \sin\alpha$). The orientation of the coordinate system was the same as that suggested in the literature [18].

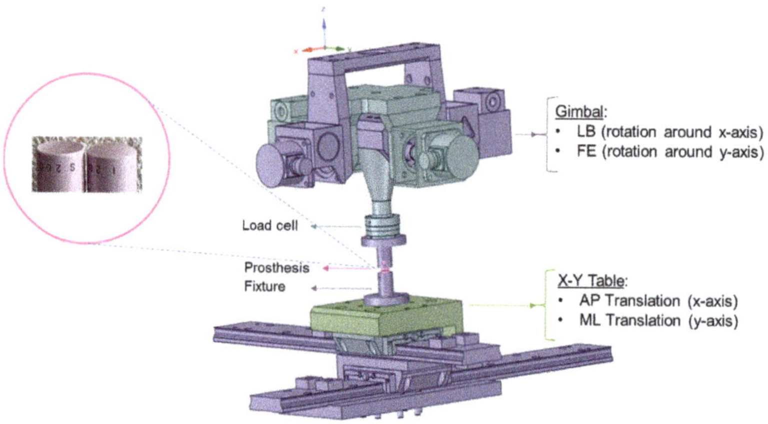

Figure 1. The spine simulator is equipped with a gimbal drive (providing FE and LB rotations) and an X/Y table (providing unconstrained translational motions). Zirconia-toughened alumina samples (pink ball and socket) were tested in a material testing machine.

2.2. Study Design

The fixtures for the specimens were attached to the gimbal (upper part) and the X/Y table (lower part) of the test rig (Figure 1). The samples were tested under displacement control with a constant axial compressive force of 100 N in five motion modes: (1) FE (±7.5°), (2) LB (±6°), (3) combined FE-LB (Figure 2a), (4) combined FE and AP (3 mm), and (5) combined LB and lateral motion (3 mm). Two scenarios were studied for the combined rotational–translational motion:

- Scenario 1: The application of an excessive translational movement (3 mm) subsequent to the complete rotation of the sample (Figure 2b). Before applying the excessive translation, the samples remained in a fully flexed condition for 10 seconds to reach the required values for rotation and axial force (i.e., the dwelling phase). During the rotation step, the applied translation was due to the offset between the testing machine's COR and the sample's COR (12.40 mm for FE and 9.93 mm for LB) in order to adjust the alignment.
- Scenario 2: The concurrent application of excessive translational motion (3 mm) and rotation (Figure 2c). A translational adjustment (command normal translation in Figure 2c) was made during the rotation to prevent excessive forces and moments due to the offset between the sample's COR and the testing machine's COR. The 3 mm of extra translation was then superimposed on the translational adjustment.

A ramp load profile and a sinusoidal load profile were applied for scenario 1 and scenario 2, respectively (Figure 2b,c) to assess its potential impact on results. It is noteworthy that both ramp and sinusoidal load profiles are prevalent in the literature and exhibit negligible difference between them [8]. Nevertheless, the sinusoidal load profile bears closer resemblance to physiological conditions.

The objective of these excessive translations was to illustrate the response of the disc prosthesis under severe loading conditions.

For FE, LB, and combined FE-LB, the tests were conducted over five cycles, with the first four cycles serving as preconditioning cycles and not being included in later analysis. For multiplanar motion and unidirectional motion tests, an X/Y table with passive linear bearings (Figure 3) was employed, allowing for the sample to move freely while rotating.

For the rotational–translational motion test, a hydraulically controlled X/Y table (Figure 4) was employed to apply the required lateral movement. In earlier laboratory procedures, tests were undertaken to optimize setup design by decreasing setup friction's effect on the outcomes. Furthermore, the interface of the samples was lubricated with Ringer's solution (i.e., an isotonic solution analogous to an animal's body fluids) to reduce the friction between articulating surfaces.

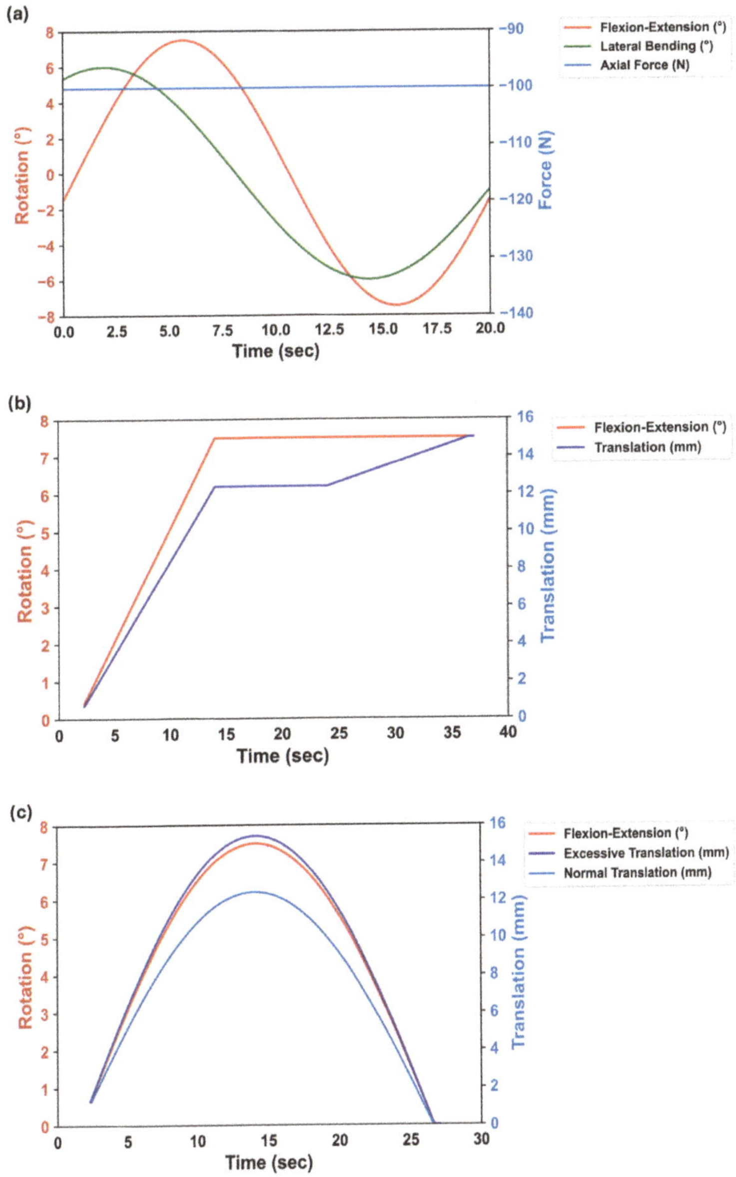

Figure 2. Illustration of load profiles: (**a**) Multiplanar motion profile (combined FE-LB) with 100 N of axial compressive force during the test. (**b**) Excessive translation after flexion. (**c**) The application of the excessive translation while the sample flexed. During rotational–translational motion experiments, a constant 100 N axial compressive force was also applied.

Figure 3. Passive linear bearing X/Y table.

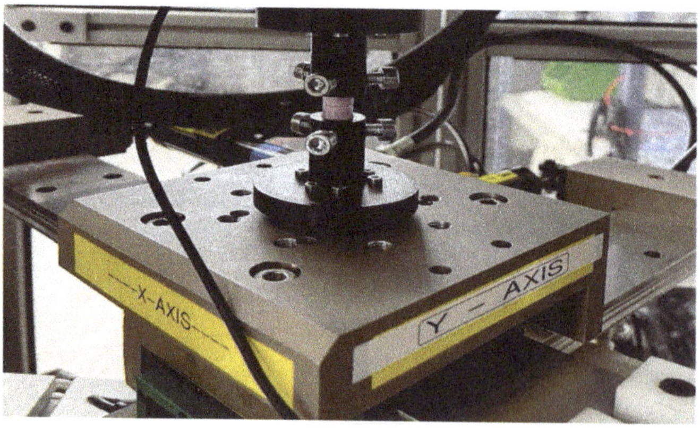

Figure 4. Hydraulic X/Y table.

2.3. *Postprocessing*

No data transformation was required since the rotation center of the sample was aligned with the vertical axis of the load cell during the test. The load cell data were filtered using a Savitzky–Golay filter to smooth the data without distorting the signal tendency. Customized Python code (Python 3.8.5 programming language) was used for data analysis to derive the moments and forces of the system.

2.3.1. Unidirectional and Multiplanar Motion

In the absence of facet joints and ligaments, the reaction shear forces and moments are primarily a function of disc friction, setup friction, and the applied axial force (both magnitude and orientation). In other words, the combination of angular motion and eccentric axial force results in anteroposterior (AP) and mediolateral (ML) forces and moments in FE and LB, respectively (Figure 5). For each test, 3 component forces (i.e., Fx, Fy, and Fz) and 3 component moments (i.e., Tx, Ty, and Tz) of the system were calculated. The resultant values for forces and moments were then compared between the unidirectional motion test and the multiplanar motion test.

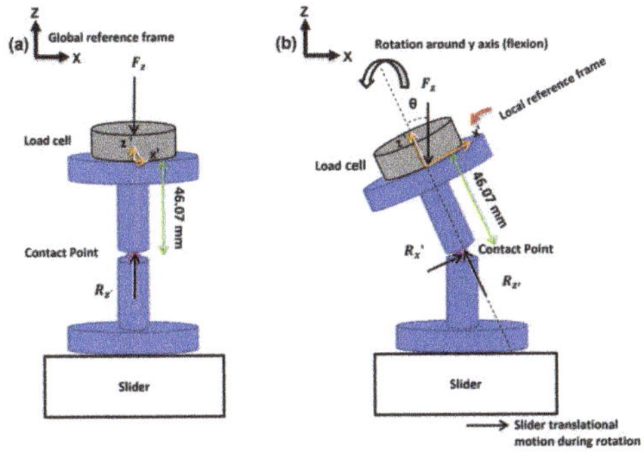

Figure 5. (**a**) A pure axial force is present in the neutral position and (**b**) the prosthesis flexes around the y-axis and the axial force becomes eccentric. Therefore, it results in lateral force and moments. The distance between the load cell's origin and the sample's contact point is 46.07 mm. $R_{X'}$ and $R_{Z'}$ show the shear force and axial force, respectively, in the local reference frame (i.e., the implanted related axis). F_Z shows the axial force in the global reference frame.

2.3.2. Rotational–Translational Motion

For the rotational–translational motion experiments, the lateral force and translation were measured until the samples reached the subluxation point, followed by micro-separation.

3. Results

3.1. Reaction Moments and Forces for Combined and Unidirectional Motion Tests

In coupled motion consisting of FE-LB, the absolute mean of the maximum (std) Tx (LB moment), Ty (FE moment), and Tz (axial torque) were 0.56 (0.01) Nm, 0.82 (0.1) Nm, and 0.01 (0.003) Nm, respectively (Figure 6a and Table 1). For the simple arcs of FE, the absolute mean of the maximum (std) Tx, Ty, and Tz were 0.15 (0.06) Nm, 0.74 (0.12) Nm, and 0.01 (0.01) Nm, respectively (Figure 6a and Table 1). For the simple arcs of LB, the absolute mean of the maximum (std) Tx, Ty, and Tz were 0.67 (0.08) Nm, 0.1 (0.07) Nm, and 0.01 (0.003) Nm, respectively (Figure 6a and Table 1).

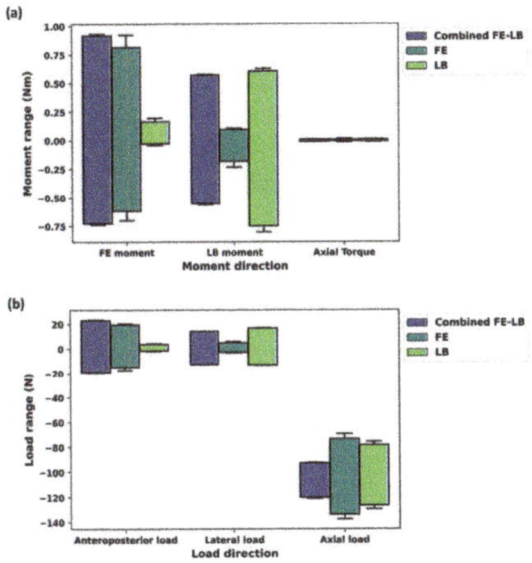

Figure 6. (**a**) Moment range (Nm) and (**b**) force range (N) for combined FE-LB, FE, and LB. In this graph, FE moment, LB moment, and axial torque represent Ty, Tx, and Tz, respectively. Anteroposterior force, lateral force, and axial force represent Fx, Fy, and Fz, respectively.

Table 1. The reaction moments and forces for combined FE-LB, FE, and LB.

Motion Type	Tx (Std) Nm	Ty (Std) Nm	Tz (Std) Nm	Fx (Std) N	Fy (Std) N	Fz (Std) N
Combined FE-LB	0.56 (0.01)	0.82 (0.1)	0.01 (0.003)	20.92 (1.84)	13.40 (0.60)	106.61 (13.61)
FE	0.15 (0.06)	0.74 (0.12)	0.01 (0.01)	17.15 (2.14)	3.86 (1.09)	103.38 (29.94)
LB	0.67 (0.08)	0.1 (0.07)	0.01 (0.003)	2.57 (1.004)	14.87 (1.65)	102.46 (24.33)

For combined FE-LB, the absolute mean of the maximum (std) Fx (anteroposterior force), Fy (lateral force), and Fz (axial force) were 20.92 (1.84) N, 13.40 (0.60) N, and 106.61 (13.61) N, respectively (Figure 6b and Table 1). For the simple arcs of FE, the absolute mean of the maximum (std) Fx, Fy, and Fz were 17.15 (2.14) N, 3.86 (1.09) N, and 103.38 (29.94) N, respectively (Figure 6b and Table 1). For the simple arcs of LB, the absolute mean of the maximum (std) Fx, Fy, and Fz were 2.57 (1.004) N, 14.87 (1.65) N, and 102.46 (24.33) N, respectively (Figure 6b and Table 1).

Table 2 indicates that the combined FE-LB has a greater resultant moment and force compared to FE and LB individually. The resultant moment and force obtained from the FE and LB are similar.

Table 2. The resultant moment and force for combined FE-LB, FE, and LB.

Measurement Type	Combined FE-LB (Std)	FE (Std)	LB (Std)
Resultant moment (Nm)	1.0 (0.08)	0.75 (0.2)	0.68 (0.08)
Resultant force (N)	109.46 (13.26)	104.86 (29.51)	103.57 (24.07)

3.2. *Shear Forces, Translations, and Degree of Rotations for Rotational–Translational Motion Tests*

At 12.4 mm and 9.9 mm, the mean (std) shear forces in disc-related axes for FE and LB, respectively, were 47.4 (6.40) N and 40.1 (4.27) N (Figure 7). Then, the excessive lateral translation was applied to reach the micro-separation or subluxation point. After complete flexion, the samples reached the subluxation point at an absolute mean (std) displacement of 13.8 (0.2) mm and absolute mean (std) shear force of 77.0 (7.1) N (Figure 7a). Thus, incremental increases in AP translation = 1.49 ± 0.18 mm and shear force = 30.4 ± 3.14 N were observed at subluxation point. After completing lateral bending rotation, the samples reached the subluxation point at an absolute mean (std) displacement of 12.1 (0.2) mm and absolute mean (std) shear force of 80.9 (6.3) N (Figure 7b). Thus, an incremental increase in lateral translation = 2.22 ± 0.16 mm and shear force = 40.8 ± 2.56 N were observed at the subluxation point.

Due to the complex physiological condition, we hypothesized that the subluxation may also occur during rotation. Thus, the excessive lateral translation was superimposed on the translational compensation while the sample was rotating. During the flexion, the sample reached the subluxation point at FE = 4.9°, lateral displacement = 9.9 mm, and shear force = 64.9 N (Figure 8). During LB, the sample reached the subluxation point at LB = 4.5°, lateral displacement = 9.5 mm, and shear force = 72.8 N (Figure 9). For FE = 4.9°, lateral displacement and shear force were 8.1 mm and 25.7 N, respectively, under normal condition (i.e., motion to a predefined physiological angle, without additional translation). For LB = 4.5°, lateral displacement and shear force were 7.5 mm and 33.8 N, respectively, under normal condition. Therefore, an incremental increase in AP translation = 1.75 mm and lateral translation = 1.99 mm was observed at the subluxation point for FE and LB, respectively. Moreover, the incremental increase in AP shear force and lateral shear force was 39.17 N and 38.94 N for FE and LB, respectively, at the subluxation point.

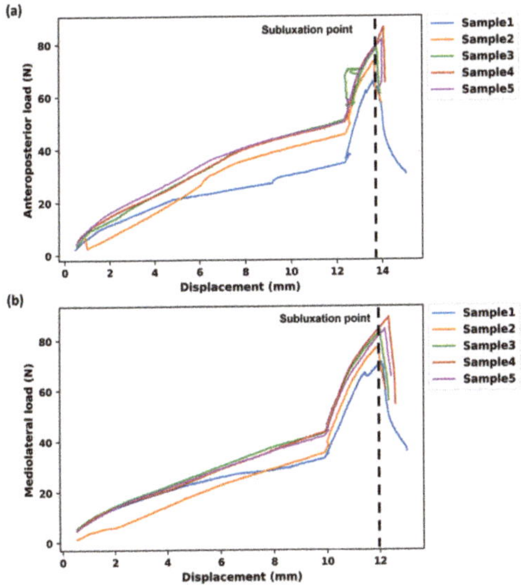

Figure 7. Subluxation test after (**a**) complete flexion and (**b**) complete lateral bending. Anteroposterior load and mediolateral load represent shear loads in the implanted related axes. The excessive translation of 3 mm was applied after the sample was fully rotated. The shear load and excessed translation were determined at the subluxation point or micro-separation site.

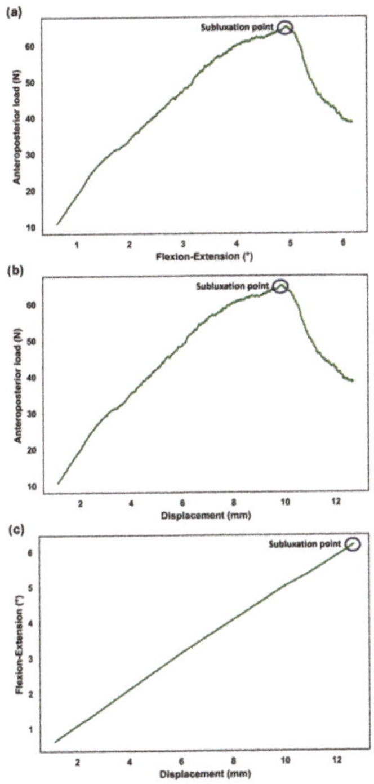

Figure 8. Subluxation test during flexion. Anteroposterior load represents shear load in the implanted related axis. The excessive translation of 3 mm was superimposed on lateral compensation, which corrects for the offset between the sample's COR and the gimbal's COR. The shear load, excessed translation, and the degree of rotation were determined at the subluxation point (**a**–**c**).

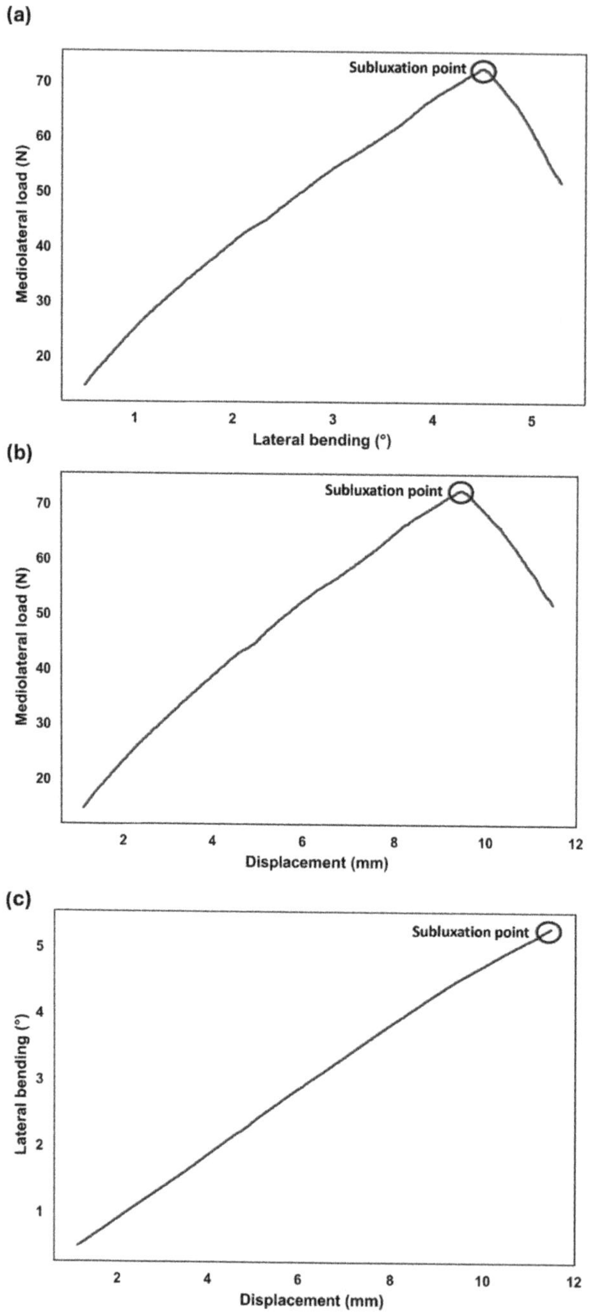

Figure 9. Subluxation test during lateral bending. Mediolateral load represents shear load in the implanted related axis. The excessive translation of 3 mm was superimposed on lateral compensation, which corrects for the offset between the sample's COR and the gimbal's COR. The shear load, excessed translation, and the degree of rotation were determined at the subluxation point (**a**–**c**).

4. Discussion

The first objective of this work was to compare the loads and moments in multiplanar motion tests (i.e., FE-LB) to those obtained in unidirectional motion tests (i.e., FE and LB). The results revealed that the resultant loads and moments for multiplanar motion were greater than those obtained for unidirectional testing. Since the samples were generic ball and socket components with no design constraints, one would expect that multiplanar motion test findings should be comparable to unidirectional motion test results. The slight discrepancy could be attributed to the fact that the friction regime is different in multiplanar motion test. It is recommended that the test be conducted on different designs (such as non-symmetrical designs in different planes) and non-articulating disc prostheses. This would comprehensively evaluate disc prosthesis response to complex motion, which is likely to occur in vivo.

In a previous study, cervical spine kinematics were examined in multiplanar motion after disc replacement utilizing human cadaver spines [16]. They found that the stiffnesses in multiplanar motion were higher than those in unidirectional motion. Due to the use of cadaver specimens in prior studies, our findings were not directly comparable, however consistent. The present study also provides insight into the influence of the testing protocol on common laboratory simulator results. Furthermore, the length of the evaluated spinal segment could affect the moment arm, resulting in greater moments for the same range of motion (ROM). In this study, rotation combined with a constant axial force was directly applied to the disc prosthesis, whereas in the literature, a pure bending moment was applied at the cranial level of C4–7 [16]. It would be beneficial to simulate facet joint loads and ligament contributions in the study. One approach is to use synthetic specimens that replicate cadaveric biomechanics and provide advantages for cross-laboratory validation studies while lowering costs and disease transmission [19,20].

This study also investigated subluxation, which is one of the most important clinical concerns following disc replacement. There have been two reports of disc dislocation in the literature [21]. The first report indicated that the prosthesis was implanted too anteriorly. The second was a technical error during implantation when the polyethylene inlay was not completely snapped into the inferior endplate. The improper snapping of the inlay into the metallic endplate, followed by the presence of shear forces, increases the likelihood of subluxation and dislocation. Furthermore, the incision of ligaments during surgery or muscle dysfunction can lead to hypermobility of the spine and an increase in shear loads on disc prosthesis, which increases the probability of migration. It is important to note that unstable fusion at the bone–implant interface may also contribute to disc dislocation. In this study, kinematic tests were prescribed to replicate subluxation by applying excessive translation during and after rotation. In both instances, the shear load increased continuously until the subluxation point. Furthermore, the tests revealed an increased risk of subluxation for complex, combined motion than for sequential movements.

In normal conditions (i.e., motion to a predefined physiological angle, without additional translation), the results indicated a linear relationship between the increase in shear loads and the rotation angle. However, the presence of excessive translation after full rotation caused a sharp increase in shear loads because the ball and socket samples had three degrees of freedom, permitting 3D rotations without translation (Figure 7). The same trend was also observed when excessive translation was applied during rotation (Figures 8 and 9). It can be inferred that mobile core prostheses can provide a more effective resistance to aberrant situations. In addition, previous research demonstrated other advantages of mobile core prostheses, in that they can disperse pressure at the bone–prosthesis interface due to the capability of anteroposterior translation as well as having less variation in the instantaneous axis of rotation [22,23]. However, the long-term clinical outcome of mobile-core prostheses is still debated.

In subluxation testing, the value of shear loads and excess translation at the micro-separation point for LB were somewhat greater than those for FE. Furthermore, in the event of simultaneous rotational and translational motion, the subluxation occurred at comparable degrees of rotation for FE and LB because the ball and socket contact point did not shift during the kinematic tests. To determine whether there are significant differences between FE and LB for other devices, further studies are required to develop subluxation tests for non-symmetrical and non-articulating designs.

It is currently challenging to pinpoint what shear force magnitude leads to subluxation of discs in an in vivo condition. Based on the present results, even a modest increase in shear loads is enough to cause subluxation in this design. The development of subluxation tests using spinal loading simulators (with or without cadaver specimens) and finite element simulations in the future is crucial for a better understanding multi-directional loading conditions and their consequences.

The wear analysis of certain materials, such as ceramic, is difficult due to their remarkable resistance to low friction under lubricated conditions. Thus, the development of severe loading conditions such as subluxation or edge loading could be beneficial for a tribology investigation of the materials for TDR applications. At present, there is an absence of prior research that investigates the effects of edge loading on disc prosthesis deterioration. Notwithstanding this, a number of research studies examined the effects of edge loading on hip replacements, which may shed light on analogous phenomena affecting spinal disc prostheses. In metal-on-metal articulations for total hip replacements, edge loading resulted in pseudotumor formation, metallosis, aseptic lymphocytic vasculitis-associated lesions, and accelerated wear of the entire articulation [24–28]. In ceramic-on-ceramic articulations for total hip replacements, edge loading led to squeaking, stripe wear, and accelerated wear of the whole joint [29–32].

5. Conclusions

This study developed new test protocols for determining kinetic parameters of disc prosthesis during multiplanar motion (i.e., FE-LB) and subluxation tests. For the multiplanar motion test, the resultant forces and moments were slightly higher but comparable to the unidirectional motion test (i.e., FE and LB). For the subluxation test (scenario 1), the application of excessive translation led to a 62.44% and 101.74% increase in shear forces for FE and LB, respectively. For the subluxation test (scenario 2), the application of excessive translation resulted in a 152.53% and 115.38% increase in shear forces for FE and LB, respectively. The highest shear force was found at the subluxation point, followed by a sharp reduction due to complete dislocation. Multiplanar motion and subluxation testing will help develop and evaluate disc prosthesis function, longevity, and safety, as well as address clinical issues.

Author Contributions: Conceptualization, H.A.; methodology, H.A.; investigation, H.A.; writing—original draft preparation, H.A.; writing—review and editing, S.J.F., M.F. and H.A.; supervision, M.F. and S.J.F.; All authors have read and agreed to the published version of the manuscript.

Funding: This project has received funding from the European Union's Horizon 2020 research and innovation program under the Marie Sklodowska-Curie grant agreement No 812765.

Institutional Review Board Statement: Not applicable.

Informed Consent Statement: Not applicable.

Data Availability Statement: The data are available within the article.

Conflicts of Interest: The authors declare no conflicts of interest.

References

1. Patel, V.V.; Wuthrich, Z.R.; McGilvray, K.C.; Lafleur, M.C.; Lindley, E.M.; Sun, D.; Puttlitz, C.M. Cervical facet force analysis after disc replacement versus fusion. *Clin. Biomech.* **2017**, *44*, 52–58. [CrossRef]
2. Gore, D.R.; Sepic, S.B. Anterior discectomy and fusion for painful cervical disc disease. A report of 50 patients with an average follow-up of 21 years. *Spine* **1998**, *23*, 2047–2051. [CrossRef]

3. Matsumoto, M.; Okada, E.; Ichihara, D.; Watanabe, K.; Chiba, K.; Toyama, Y.; Fujiwara, H.; Momoshima, S.; Nishiwaki, Y.; Iwanami, A.; et al. Anterior cervical decompression and fusion accelerates adjacent segment degeneration: Comparison with asymptomatic volunteers in a ten-year magnetic resonance imaging follow-up study. *Spine* **2010**, *35*, 36–43. [CrossRef]
4. Nabhan, A.; Ahlhelm, F.; Pitzen, T.; Steudel, W.I.; Jung, J.; Shariat, K.; Steimer, O.; Bachelier, F.; Pape, D. Disc replacement using Pro-Disc C versus fusion: A prospective randomised and controlled radiographic and clinical study. *Eur. Spine J.* **2007**, *16*, 423–430. [CrossRef] [PubMed]
5. Grob, D.; Porchet, F.; Kleinstück, F.S.; Lattig, F.; Jeszenszky, D.; Luca, A.; Mutter, U.; Mannion, A.F. A comparison of outcomes of cervical disc arthroplasty and fusion in everyday clinical practice: Surgical and methodological aspects. *Eur. Spine J.* **2010**, *19*, 297–306. [CrossRef] [PubMed]
6. Baba, H.; Furusawa, N.; Imura, S.; Kawahara, N.; Tsuchiya, H.; Tomita, K. Late radiographic findings after anterior cervical fusion for spondylotic myeloradiculopathy. *Spine* **1993**, *18*, 2167–2173. [CrossRef] [PubMed]
7. Hilibrand, A.S.; Carlson, G.D.; Palumbo, M.A.; Jones, P.K.; Bohlman, H.H. Radiculopathy and myelopathy at segments adjacent to the site of a previous anterior cervical arthrodesis. *J. Bone Jt. Surg. Am.* **1999**, *81*, 519–528. [CrossRef] [PubMed]
8. Ansaripour, H.; Ferguson, S.; Flohr, M. In-vitro Biomechanics of the Cervical Spine: A Systematic Review. *J. Biomech. Eng.* **2022**, *144*, 100801. [CrossRef] [PubMed]
9. Formica, M.; Divano, S.; Cavagnaro, L.; Basso, M.; Zanirato, A.; Formica, C.; Felli, L. Lumbar total disc arthroplasty: Outdated surgery or here to stay procedure? A systematic review of current literature. *J. Orthop. Traumatol.* **2017**, *18*, 197–215. [CrossRef]
10. Tu, T.-H.; Wu, J.-C.; Huang, W.-C.; Guo, W.-Y.; Wu, C.-L.; Shih, Y.-H.; Cheng, H. Heterotopic ossification after cervical total disc replacement: Determination by CT and effects on clinical outcomes. *J. Neurosurg. Spine* **2011**, *14*, 457–465. [CrossRef]
11. Cavanaugh, D.A.; Nunley, P.D.; Kerr, E.J., 3rd; Werner, D.J.; Jawahar, A. Delayed hyper-reactivity to metal ions after cervical disc arthroplasty: A case report and literature review. *Spine* **2009**, *34*, E262–E265. [CrossRef] [PubMed]
12. Guyer, R.D.; Shellock, J.; MacLennan, B.; Hanscom, D.; Knight, R.Q.; McCombe, P.; Jacobs, J.J.; Urban, R.M.; Bradford, D.; Ohnmeiss, D.D. Early failure of metal-on-metal artificial disc prostheses associated with lymphocytic reaction: Diagnosis and treatment experience in four cases. *Spine* **2011**, *36*, E492–E497. [CrossRef]
13. Amoretti, N.; Iannessi, A.; Lesbats, V.; Marcy, P.Y.; Hovorka, E.; Bronsard, N.; Fonquerne, M.E.; Hauger, O. Imaging of intervertebral disc prostheses. *Diagn. Interv. Imaging* **2012**, *93*, 10–21. [CrossRef]
14. Virk, S.; Phillips, F.; Khan, S.; Qureshi, S. A cross-sectional analysis of 1347 complications for cervical disc replacements from medical device reports maintained by the United States Food and Drug Administration. *Spine J.* **2020**, *21*, 265–272. [CrossRef] [PubMed]
15. Ansaripour, H.; Haeussler, K.L.; Ferguson, S.J.; Flohr, M. Prioritizing biomaterials for spinal disc implants by a fuzzy AHP and TOPSIS decision making method. *Sci. Rep.* **2023**, *13*, 21531. [CrossRef]
16. Daniels, A.H.; Paller, D.J.; Feller, R.J.; Thakur, N.A.; Biercevicz, A.M.; Palumbo, M.A.; Crisco, J.J.; Madom, I.A. Examination of cervical spine kinematics in complex, multiplanar motions after anterior cervical discectomy and fusion and total disc replacement. *Int. J. Spine Surg.* **2012**, *6*, 190–194. [CrossRef]
17. Penning, L. Normal movements of the cervical spine. *AJR Am. J. Roentgenol.* **1978**, *130*, 317–326. [CrossRef] [PubMed]
18. Wilke, H.J.; Wenger, K.; Claes, L. Testing criteria for spinal implants: Recommendations for the standardization of in vitro stability testing of spinal implants. *Eur. Spine J.* **1998**, *7*, 148–154. [CrossRef] [PubMed]
19. Sherrill, J.T.; Siddicky, S.F.; Davis, W.D.; Chen, C.; Bumpass, D.B.; Mannen, E.M. Validation of a custom spine biomechanics simulator: A case for standardization. *J. Biomech.* **2020**, *98*, 109470. [CrossRef]
20. Wang, T.; Ball, J.R.; Pelletier, M.H.; Walsh, W.R. Biomechanical evaluation of a biomimetic spinal construct. *J. Exp. Orthop.* **2014**, *1*, 3. [CrossRef]
21. Aunoble, S.; Donkersloot, P.; le Huec, J.C. Dislocations with intervertebral disc prosthesis: Two case reports. *Eur. Spine J.* **2004**, *13*, 464–467. [CrossRef] [PubMed]
22. Lee, S.-H.; Im, Y.-J.; Kim, K.-T.; Kim, Y.-H.; Park, W.-M.; Kim, K. Comparison of Cervical Spine Biomechanics after Fixed- and Mobile-Core Artificial Disc Replacement: A Finite Element Analysis. *Spine* **2011**, *36*, 700–708. Available online: https://journals.lww.com/spinejournal/Fulltext/2011/04200/Comparison_of_Cervical_Spine_Biomechanics_After.4.aspx (accessed on 20 April 2011). [CrossRef] [PubMed]
23. Rousseau, M.A.; Bradford, D.S.; Bertagnoli, R.; Hu, S.S.; Lotz, J.C. Disc arthroplasty design influences intervertebral kinematics and facet forces. *Spine J.* **2006**, *6*, 258–266. [CrossRef] [PubMed]
24. Hussain, A.; Counsell, L.; Kamali, A. Clinical effects of edge loading on metal-on-metal hip resurfacings. *Orthop. Proc.* **2010**, *92-B*, 399.
25. Langton, D.J.; Joyce, T.J.; Jameson, S.S.; Lord, J.; Van Orsouw, M.; Holland, J.P.; Nargol, A.V.F.; De Smet, K.A. Adverse reaction to metal debris following hip resurfacing. *J. Bone Jt. Surg. Br.* **2011**, *93-B*, 164–171. [CrossRef]
26. Willert, H.-G.; Buchhorn, G.H.; Fayyazi, A.; Flury, R.; Windler, M.; Köster, G.; Lohmann, C.H. Metal-on-Metal Bearings and Hypersensitivity in Patients with Artificial Hip Joints: A Clinical and Histomorphological Study. *JBJS* **2005**, *87*, 28–36. Available online: https://journals.lww.com/jbjsjournal/fulltext/2005/01000/metal_on_metal_bearings_and_hypersensitivity_in.6.aspx (accessed on 2 January 2005). [CrossRef]
27. Langton, D.J.; Jameson, S.S.; Joyce, T.J.; Hallab, N.J.; Natu, S.; Nargol, A.V.F. Early failure of metal-on-metal bearings in hip resurfacing and large-diameter total hip replacement. *J. Bone Jt. Surg. Br.* **2010**, *92-B*, 38–46. [CrossRef]

28. Harris, W.H. Edge Loading Has a Paradoxical Effect on Wear in Metal-on-Polyethylene Total Hip Arthroplasties. *Clin. Orthop. Relat. Res.* **2012**, *470*, 3077–3082. Available online: https://journals.lww.com/clinorthop/fulltext/2012/11000/edge_loading_has_a_paradoxical_effect_on_wear_in.17.aspx (accessed on 30 November 2012). [CrossRef]
29. Jarrett, C.A.; Ranawat, A.S.; Bruzzone, M.; Blum, Y.C.; Rodriguez, J.A.; Ranawat, C.S. The Squeaking Hip: A Phenomenon of Ceramic-on-Ceramic Total Hip Arthroplasty. *JBJS* **2009**, *91*, 1344–1349. Available online: https://journals.lww.com/jbjsjournal/fulltext/2009/06000/the_squeaking_hip__a_phenomenon_of.7.aspx (accessed on 1 June 2009). [CrossRef]
30. Keurentjes, J.C.; Kuipers, R.M.; Wever, D.J.; Schreurs, B.W. High Incidence of Squeaking in THAs with Alumina Ceramic-on-ceramic Bearings. *Clin. Orthop. Relat. Res.* **2008**, *466*, 1438–1443. Available online: https://journals.lww.com/clinorthop/fulltext/2008/06000/high_incidence_of_squeaking_in_thas_with_alumina.28.aspx (accessed on 26 February 2008). [CrossRef]
31. Stewart, T.; Tipper, J.; Streicher, R.; Ingham, E.; Fisher, J. Long-term wear of HIPed alumina on alumina bearings for THR under microseparation conditions. *J. Mater. Sci. Mater. Med.* **2001**, *12*, 1053–1056. [CrossRef] [PubMed]
32. Lusty, P.J.; Watson, A.; Tuke, M.A.; Walter, W.L.; Walter, W.K.; Zicat, B. Orientation and wear of the acetabular component in third generation alumina-on-alumina ceramic bearings. *J. Bone Jt. Surg. Br.* **2007**, *89-B*, 1158–1164. [CrossRef] [PubMed]

bioengineering

MDPI

Article

Bilateral Iliosacral and Transsacral Screws Are Biomechanically Favorable and Reduce the Risk for Fracture Progression in Fragility Fractures of the Pelvis—A Finite Element Analysis

Moritz F. Lodde *,†, Matthias Klimek †, Elmar Herbst, Christian Peez, Oliver Riesenbeck, Michael J. Raschke and Steffen Roßlenbroich

Department of Trauma, Hand and Reconstructive Surgery, University Hospital Münster, 48149 Münster, Germany
* Correspondence: moritz.lodde@ukmuenster.de
† These authors contributed equally to this work.

Abstract: (1) Background: The incidence of fragility fractures of the pelvis (FFP) has increased significantly over the past decades. Unilateral non-displaced fractures, defined as FFP II, are the most common type of fracture. When conservative treatment fails, surgical treatment is indicated. We hypothesize that the use of bilateral SI screws (BSIs) or a transsacral screw (TSI) is superior compared to a unilateral screw (USI) because of a significant reduction in the risk of adjacent fractures and a reduction in fracture progression. (2) Methods: A finite element model of a female pelvic ring was constructed. The ligaments were simulated as tension springs. The load was applied through the sacrum with the pelvis fixed to both acetabula. An FFP IIc was simulated and fixed with either a USI or BSI or TSI. The models were analyzed for a quantitative statement of stress and fracture dislocation. (3) Results: The BSI and TSI resulted in less dislocation compared to the USI. The stress distribution on both sides of the sacrum was favorable in the BSI and TSI groups. The BSI resulted in a higher rotational stability compared to the TSI. (4) Conclusions: The use of either a BSI or TSI for fixation of unilateral FFP is biomechanically favorable compared to the use of a USI. In addition, the use of a BSI or TSI reduces the stress on the contralateral uninjured side of the sacrum. This may reduce the risk of an adjacent fracture or fracture progression.

Keywords: FE analysis; biomechanics; SI screw; transsacral SI screw; fragility fracture of the pelvis (FFP); fracture progression of FFP (FP)

1. Introduction

Clinically, fragility fractures of the pelvis (FFP) are associated with a high one-year mortality rate, ranging from 11–27% and a substantial loss of quality of life or function [1–4]. FFP result from low-energy falls or occur spontaneously. These fragility fractures are associated with a reduced bone quality or osteoporosis [5]. FFP II—unilateral non-displaced fractures of the anterior and posterior pelvic ring—are the most commonly observed fracture type [6–10]. The results of conservative and the surgical treatment techniques for FFP II are controversially discussed and reported in the literature [8,9,11]. When conservative treatment fails, surgical treatment is indicated [8,9,11,12]. While several other techniques have been described in the literature, the SI screw is widely accepted as the standard of care treatment [13–17]. However, several biomechanical studies have been conducted to further improve the fixation of the posterior pelvic ring using percutaneous SI screws [18–20]. A recently published systematic review and meta-analysis showed that

 https://doi.org/10.3390/bioengineering12010027

fracture progression is relatively high in patients with FFP [17,21]. The authors suggested that a TSI could reduce the risk of fracture progression [17].

Finite element analysis (FE analysis) is a well-established method for investigating the biomechanics of different percutaneous screw fixation techniques for unstable pelvic ring fractures, even allowing for the simulation of bone cement [22–32].

The aim of the present FE analysis was to investigate the biomechanical behavior of a USI, BSI, and TSI for the fixation of FFP IIc in terms of fracture dislocation and stress distribution.

We hypothesize that the use of bilateral SI screws (BSIs) or a transsacral screw (TSI) in unilateral fractures of the pelvic ring is biomechanically favorable compared to a unilateral SI screw (USI). We further hypothesize that the risk of fracture progression is reduced with a BSI and a TSI.

2. Materials and Methods

A finite element model of a female pelvic ring was created using the CT data from the Visible Human Project. The model was segmented in Slicer 3D [33].

Segmentation was performed with a threshold between 220 and 2000 Hounsfield units (HU) based on the CT scan [34].

Voxels outside the pelvic ring were deleted, and the segmentation was smoothed. The surface of the segmentation was equalized to a faceted model with 0.7 mm triangles to reduce the extrusions from the segmentation while still smoothing the surface and maintaining the computational complexity. This model was transformed into a full-body model. On the full-body model, the surface for each ligament insertion point was manually created. The symphysis and articular disc were manually created as an automatically attached body to the bone surface. The pelvic ring was meshed using linear tetrahedron elements.

For the convergence analysis, the pelvic ring was fixed at the upper level of the sacrum, and a bipedal stance was simulated.

For the mesh analysis, the change in the maximum tension and compression was used as the reference. A change of less than 5% was considered accurate. At this stage, the bone was modeled as a linear elastic material with a Young's modulus of 6000 megapascals (MPa). After convergence analysis, a maximum size of 5 mm and a minimum size of 1 mm were chosen for the tetrahedrons to ensure sufficient accuracy and speed of execution.

The resulting mesh was imported into bonemat (BIC Lab, Bologna, Italy). In bonemat, each element was assigned an individual Young's modulus based on the CT scan [35]. The maximum HU value was sorted by the apparent density $\rho_{app} = 1.8\ \mathrm{g/cm^3}$ [36]. The Dalstra formula was used to correlate the apparent density. Values of Young's modulus, E, were determined as follows [37]:

$$\rho_{app} = 0.00032362 * HU + 1 \quad (1)$$

$$E = 1958.6 * {\rho_{app}}^{2.33} \quad (2)$$

The Young's modulus values ranged from 200 MPa to 6530 MPa, simulating the bone conditions of an elderly woman [32]. The range was divided into increments of 50 MPa. A total of 1266 different materials were defined and processed in the mesh. After importing into Ansys Mechanical (v24.1, Ansys, Inc., Canonsburg, PA, USA), contacts between the cartilage and bone were determined as boned.

The load was applied to the first sacral body of the sacrum. The pelvis was fixed to both acetabular bones. The load value corresponded to the validation of Miller et al. [38].

A load of 294 N was applied in the anterior, posterior, superior, and inferior directions. Sacral motion and the validation parameters were measured. In the present study, the ligament configurations from the previously published study by Eichenseer et al. were used for all ligaments except the superior pubic ligament (SP) and inferior pubic ligament (IP) (Table 1) [31]. The values used for SP and IP were taken from the study by Shi et al. [24].

Table 1. Spring stiffnesses from previously published studies are shown. ASL: anterior sacroiliac ligament, SPSL: short posterior sacroiliac ligament, LPSL: long posterior sacroiliac ligament, ISL: interosseous sacroiliac ligament, SS: sacrospinous ligament, ST: sacrotuberous ligament, SP: superior pubic ligament, IP: inferior pubic ligament.

	Shi et al. [24]	Yao et al. [23]	Eichenseer et al. [31]
	N/mm	N/mm	N/mm Defined by the Percentage Elongation
ASL	700	18.9	39–103
SPSL	400	21.0	200–525
LPSL	1.000	21.0	29–75
ISL	2.800	22.4	13–34
SS	1.400	12.6	26–68
ST	1.500	22.5	17–45
SP	500	12.0	/
IP	500	12.0	/

The next step was to simulate an FFP IIc [12]. The SI screw insertions were placed by a board-certified trauma surgeon and independently validated by another board-certified trauma surgeon. The design of the USI, BSI, and TSI was adapted from commonly used implants (Marquardt Aaxomed, ISG System, Freiburg, Germany). The screws were simulated without threads and bonded to the bone (Figure 1).

Figure 1. This figure shows the SI screw with a washer used to simulate the USI, BSI, and TSI.

A maximum walking force of 2048 N was applied separately in each acetabulum during the single-legged stance, and consequently, half of the maximum force was applied separately during the bipedal stance [39].

Normal stress and fracture dislocation were investigated in the bone aligned with the anatomical directions, taking into account that the von Mises stress is only applicable to ductile material and bone behavior is brittle. All the different fixation techniques were analyzed in order to make a quantitative statement about the stresses and dislocations in the sacrum. In this way, it was possible to assess the load on the uninjured contralateral part of the sacrum. Stress and dislocation were measured at the posterior pelvic ring at the superior and inferior fracture gap in zone II according to the classification of Denis et al. [40]. Dislocation at the anterior pelvic ring was measured at the fracture gap at the superior pubic ramus.

3. Results

The BSI and TSI provided more stability and less dislocation compared to the USI (Figures 1–6, Table 2). The fracture gap at the posterior pelvic ring was lowest with the TSI (0.69 mm) compared to the BSI (1.21 mm) and USI (1.84 mm). The stress distribution on both sides of the sacrum was favorable with the BSI and TSI. It should be noted that the TSI was even more stable than the BSI. The posterior dislocation was less with the TSI (Table 2). However, the BSI resulted in a smaller fracture gap distance in the inferior–superior direction compared to the TSI (Figures 2–7). Figure 8 shows the increased fracture gap distance in the inferior–superior direction with the TSI from a more inferior view for better visualization. When the load was applied via a bipedal stance, the distance of the fracture gap at the anterior pelvic ring was 1.53 mm with the BSI and −0.20 mm with the TSI (Table 2). With the TSI, the medial and lateral parts of the superior pubic ramus slid over each other (Figures 2–8). The same was true when the load was applied via the right-sided one-legged stance (fracture gap distance: 3.31 mm for BSI vs. −1.19 mm for TSI, Table 2).

The present FE analysis shows that the use of a USI to fix a unilateral FFP IIc led to a higher stress concentration on the contralateral uninjured side of the sacrum (Figures 2–7, Table 3). This was true for both the single-legged and bipedal stance (Table 3). The BSI and TSI had similar stress distributions (Table 3). The highest stress concentration was observed at the uninjured side of the sacrum in the simulated single-legged stance of the uninjured side (stress: USI at S1: 15.71 MPa, BSI at S1: 9.98 MPa, TSI at S1: 8.91 MPa) (Figures 2–7, Table 3). Correspondingly, the BSI and TSI resulted in a lower stress concentration on the contralateral side of the sacrum.

Table 2. The distance of the fracture gap in the lateral–medial direction (mm) is shown depending on the load application and fixation technique.

Load Application	USI	BSI	TSI
Bipedal stance			
Posterior pelvic ring	1.84	1.21	0.69
Anterior pelvic ring	4.13	1.53	−0.20
Right one-legged stance			
Posterior pelvic ring	3.22	2.05	0.86
Anterior pelvic ring	5.63	3.31	−1.19
Left one-legged stance			
Posterior pelvic ring	0.01	0	0.05
Anterior pelvic ring	3.01	0.13	0.13

Table 3. Observed stress in the lateral–medial direction (MPa) is shown depending on the load application and fixation technique.

Load Application	USI	BSI	TSI
Bipedal stance			
S1 right side	9.95	7.22	5.23
S1 left side	7.55	4.26	4.88
S2 right side	5.11	2.51	3.27
S2 left side	4.76	2.49	3.26
Right one-legged stance			
S1 right side	19.99	16.23	9.27
S1 left side	−1.37	−0.56	0.28
S2 right side	8.77	6.77	4.92
S2 left side	1.33	0.41	2.08
Left one-legged stance			
S1 right side	−0.52	−0.42	0.20
S1 left side	15.71	9.98	8.91
S2 right side	1.09	0.04	0.67
S2 left side	10.58	9.28	7.94

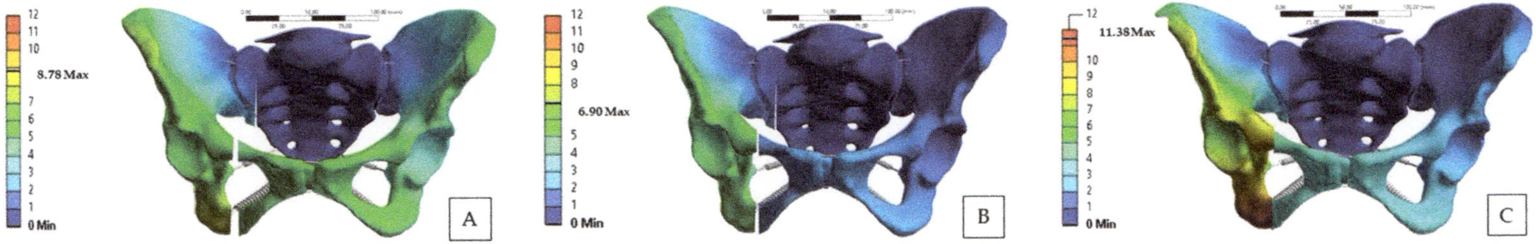

Figure 2. This figure shows the dislocation [mm] in the simulated FFP IIc. The load was applied via a bipedal stance. (**A**) Fixation with the USI. (**B**) Fixation with the BSI. (**C**) Fixation with the TSI. Dislocation of the fracture at the posterior pelvic ring was less with the BSI or TSI compared to the USI. Dislocation at the anterior pelvic ring was less with the BSI compared to the TSI or USI.

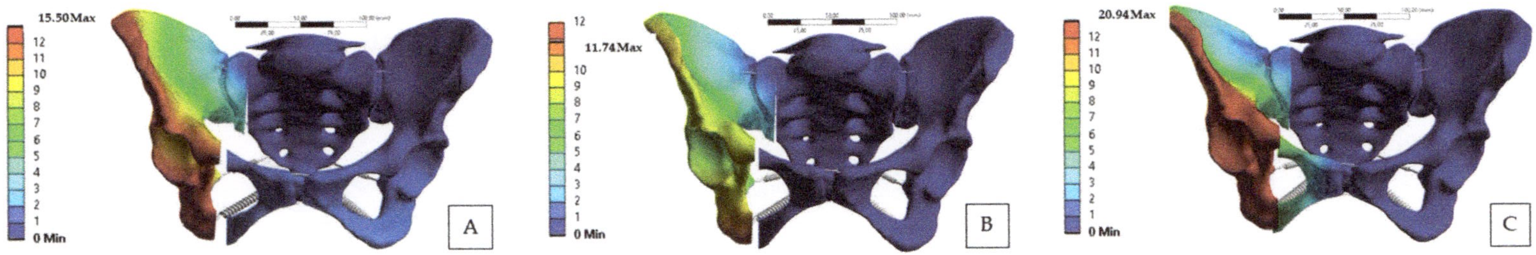

Figure 3. This figure shows the dislocation [mm] in the simulated FFP IIc. The load was applied via a right one-legged stance. (**A**) Fixation with the USI. (**B**) Fixation with the BSI. (**C**) Fixation with the TSI. Dislocation at the posterior pelvic ring was less with the TSI compared to the BSI and USI. Dislocation at the anterior pelvic ring was less with the BSI.

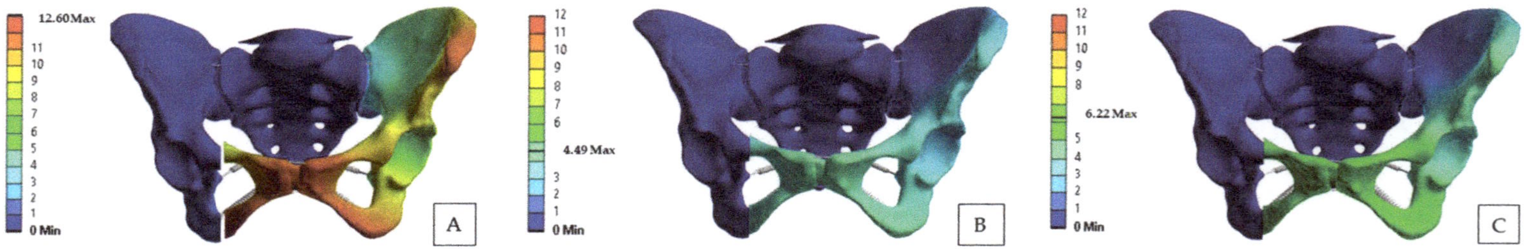

Figure 4. This figure shows the dislocation [mm] in the simulated FFP IIc. The load was applied via a left one-legged stance. (**A**) Fixation with the USI. (**B**) Fixation with the BSI. (**C**) Fixation with the TSI. Dislocation was less with the BSI. Dislocation was less with the TSI compared to the USI. Dislocation at the anterior pelvic ring was less with the BSI.

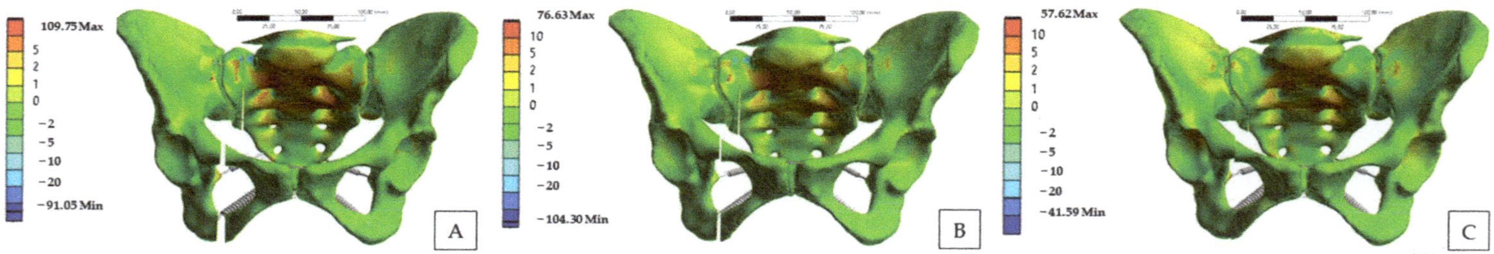

Figure 5. This figure shows the normal stress [MPa] in the mediolateral direction in the simulated FFP IIc. The load was applied via a bipedal stance. (**A**) Fixation with the USI. (**B**) Fixation with the BSI. (**C**) Fixation with the TSI. The observed stress at the posterior pelvic ring was less with the BSI or TSI. The stress at the anterior pelvic ring was less with the BSI compared to the TSI or USI.

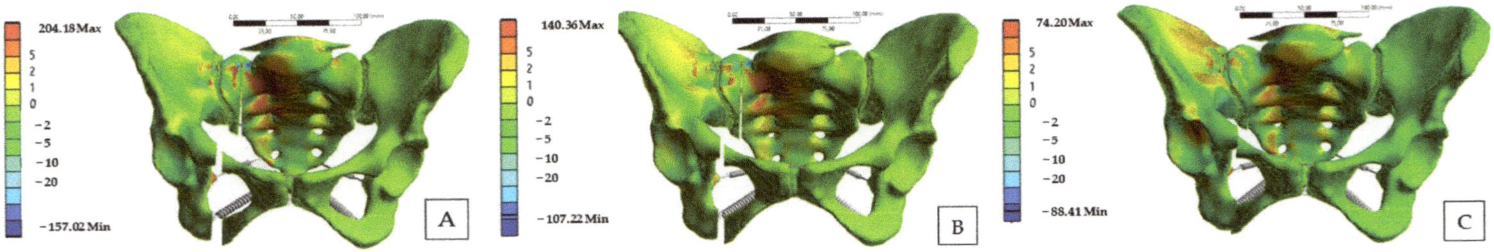

Figure 6. This figure shows the normal stress [MPa] in the mediolateral direction in the simulated FFP IIc. The load was applied via a right one-legged stance. (**A**) Fixation with the USI. (**B**) Fixation with the BSI. (**C**) Fixation with the TSI. The observed stress at the posterior pelvic ring was less with the TSI compared to the BSI or USI. The stress at the anterior pelvic ring was less with the BSI.

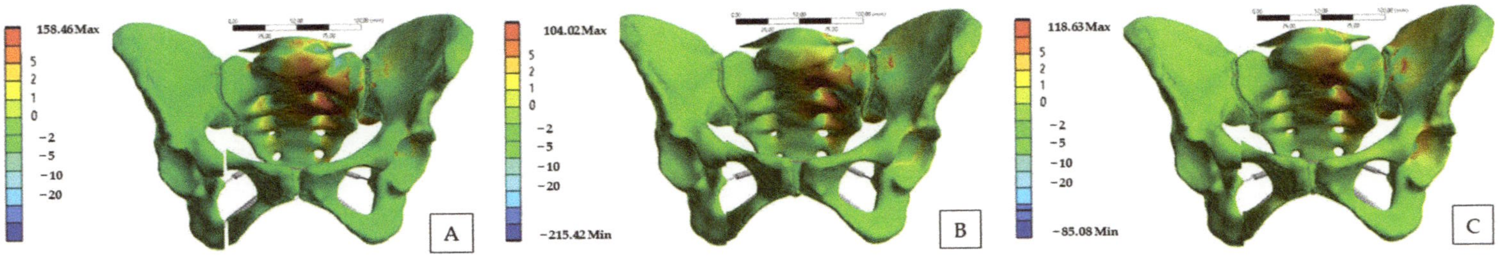

Figure 7. This figure shows the normal stress [MPa] in the mediolateral direction in the simulated FFP IIc. The load was applied via a left one-legged stance. (**A**) Fixation with the USI. (**B**) Fixation with the BSI. (**C**) Fixation with the TSI. The stress was less with the BSI or TSI compared to the USI. The stress on the anterior pelvic ring was less with the BSI.

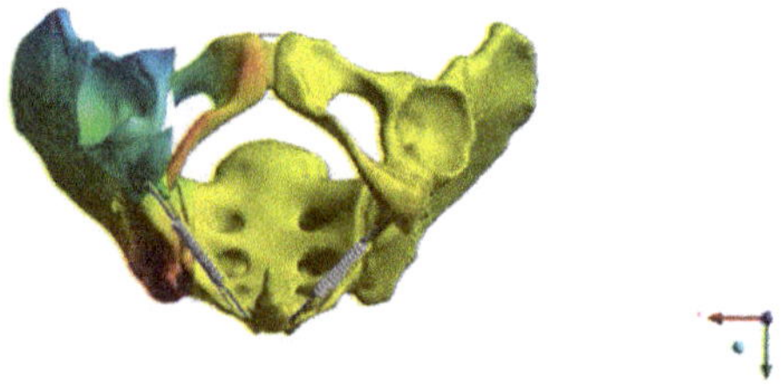

Figure 8. This figure shows, from a more inferior view, that the fracture gap in the inferior–superior direction was larger with the TSI.

4. Discussion

The present study focused on different types of fixation of FFP IIc and the risk of an adjacent fracture of the uninjured contralateral part of the sacrum. The main findings of the present study are that the BSI and TSI were biomechanically favorable compared to the USI in terms of unilateral fragility fractures of the pelvis (FFP IIc). Furthermore, the BSI and TSI reduced the risk of an adjacent fracture of the contralateral uninjured side of the sacrum. The TSI was even more stable than the BSI. However, the BSI resulted in higher rotational stability compared to the TSI. Our hypotheses were confirmed.

In the present study, an FFP IIc was simulated [12]. According to the comprehensive classification of fragility fractures of the pelvic ring, FFP II are defined as non-displaced fractures [12]. FFP IIa are isolated non-displaced fractures of the posterior pelvic ring without involvement of the anterior pelvic ring. FFP IIb are defined as non-displaced sacral fractures combined with anterior disruption [12]. More specifically, FFP IIb are pubic and ischial rami fractures combined with a crush zone of the sacral ala without displacement [12]. In contrast to this, FFP IIc lesions are pubic and ischial rami fractures combined with a non-displaced sacral ala fracture [12]. These subtle differences are important for the results of the present study because FFP IIc are more unstable than FFP IIa and FFP IIb. In particular, the rotational instability is higher in FFP IIc compared to FFP IIa and FFP IIb.

As percutaneous SI screw fixation has become widely accepted as the standard of care, the type of fracture and, thus, the degree of instability and the rotational instability influence the risk of screw loosening [19]. The present study shows that TSIs and BSIs are superior compared to USIs. TSIs result in a smaller fracture gap distance and less stress on the posterior pelvic ring compared to BSIs. TSIs may prevent screw loosening and reduce pain better than BSIs. Biomechanical studies investigating this issue are lacking. The distance of the fracture gap in the inferior–superior direction at the anterior pelvic ring is smaller with BSIs. The use of BSIs with slightly different axes provides greater rotational stability. This is consistent with another biomechanical study analyzing a screw-in-screw fixation for fragility fractures of the pelvis [19].

Adjacent fractures have long been recognized in the field of osteoporotic vertebral compression fractures. Data on adjacent fractures of the sacrum are lacking in the literature. However, a recently published systematic review and meta-analysis showed that fracture progression is relatively high in patients with an FFP [17,21]. Sensitivity analysis showed that the pooled prevalence of fracture progression in patients with an FFP was 10% (95% CI, 4–18%) [17]. A pooled prevalence >10% is relevant [41]. Yamamoto et al. suggest that fixation of a unilateral sacral fracture with a TSI may prevent subsequent fracture progression of the contralateral side of the sacrum [17]. The results of the present study support this previously published hypothesis. Clinically, the use of BSIs and TSIs is safe and feasible [16,42,43]. In FFP IIc, BSIs and TSIs are biomechanically superior due to a higher overall construct stability, a better load distribution, and a reduced risk of an adjacent fracture and fracture progression, respectively [17]. The stress and strain on the contralateral side of the sacrum are particularly high when the uninjured side is loaded. This emphasizes the need for BSIs or TSIs, as elderly patients may not be able to comply with post-operative weight-bearing restrictions [44,45]. In their retrospective study, Heydemann et al. showed that TSIs or USIs had no positive or negative effect on pain and functional outcome at a minimum follow-up of one year [46]. However, young patients were also included, and the mean ISS (Injury Severity Score) was 23.4 and 23.6, respectively. In addition, TSIs are generally likely to be used in patients with unstable injuries or in osteopenic patients [46]. Therefore, these inclusion criteria and methods may explain the findings. A clinical trial comparing USIs and TSIs or BSIs for the fixation of FFP is lacking.

Several biomechanical studies and FE analyses have examined fixation techniques of the posterior pelvic ring after high-energy trauma. The more distant and parallel the implant used is to the sacroiliac axis of rotation, the more favorable the fixation [25]. Zhang et al. found in simulated Tile B and C pelvic ring injuries that a single SI screw in S1 (first sacral body) was sufficient to stabilize type B injuries and that an additional SI screw in S2 (second sacral body) increased the biomechanical stability [26]. In another FE analysis, Wu et al. investigated six different combinations of SI screws and TSIs for the fixation of Tile B or C pelvic ring fracture models (Dennis-II-type fracture) [27]. In their study, the highest stability was achieved with an oblique SI screw in S1 and a TSI in S2 [27]. In a simulation of a Tile C pelvic ring fracture, Hu et al. showed that stress shielding was favorable with the fixation of two SI screws and a minimally invasive adjustable plate [29]. In a simulated bilateral Tile C pelvic ring fracture, the risk of fracture was low with bilateral symmetric double-segmental screws [28]. The authors strongly recommend the use of bilateral symmetric screw fixation [28]. Ma et al. observed in a unilateral Tile C pelvic ring fracture that the risk of screw breakage was lower with double-segmental screw fixation than with single-segmental screw fixation [30]. The studies published to date have evaluated fixation techniques for Tile B and Tile C pelvic ring fractures following high-energy trauma [26–30]. The main findings are that bilateral fixation techniques or the use of an additional screw result in increased biomechanical stability and a favorable stress distribution. These findings are consistent with the results of the present study.

A biomechanical study investigating the competence of a TSI and USI for the fixation of an FFP IIc showed that the TSI resulted in a significantly higher stability for the gap angle, flexion, vertical motion, and overall stability compared to the USI [18]. The authors suggest that TSIs should be used clinically for the fixation of FFP IIc [18]. These findings and conclusions are again consistent with the present study.

Strengths and limitations: The present FE analysis contains known approximations and limitations that are comparable to the limitations of previously published studies [26–30]. The FE analysis presented is an in vitro model. In the present model, the effects of the ligaments were taken into account. The effects of the pelvic muscles, fascia, and organs on pelvic stability were not considered.

All FE analyses in the present study were performed under the same in vitro experimental conditions. Biomechanical differences between the different fixation techniques could be reliably investigated without the confounding factors that occur in cadaveric biomechanical testing. Another advantage of the present study is the global anistropic bone behavior and the use of a hyperelastic symphysis. The pelvic ring was meshed using linear tetrahedral elements. Due to the complex design of the bone, tetrahedral elements are less rigid and more accurate in modeling the bone surface than hexahedral elements.

In conclusion, BSIs and TSIs are biomechanically advantageous for unilateral FFP IIc fixation. BSIs and TSIs may reduce the risk of FFP fracture progression, leading to a better clinical outcome. Further clinical studies analyzing the prevention of fracture progression are needed.

5. Conclusions

The use of BSIs or TSIs for the fixation of unilateral FFPc is biomechanically favorable compared to the use of USIs. In addition, BSIs and TSIs reduce the stress concentration and total dislocation on the uninjured contralateral side of the sacrum. This reduces the risk of an adjacent fracture on the uninjured side of the sacrum and fracture progression and may be associated with a better clinical outcome and faster recovery in the elderly population.

Author Contributions: Conceptualization, M.F.L., M.K. and S.R.; methodology, M.F.L., M.K., O.R. and S.R.; software, M.F.L. and M.K.; validation, M.F.L., E.H., C.P., O.R. and M.J.R.; formal analysis, M.F.L., E.H., O.R., M.J.R. and S.R.; investigation, M.F.L., M.K., O.R. and S.R.; resources, M.F.L., M.K., M.J.R. and S.R.; data curation, M.F.L., M.K., E.H., C.P., O.R. and S.R.; writing—original draft preparation, M.F.L.; writing—review and editing, M.F.L., M.K., E.H., C.P., O.R., M.J.R. and S.R.; visualization, M.F.L. and M.K.; supervision, M.J.R. and S.R.; project administration, M.F.L., M.K. and S.R.; funding acquisition, M.F.L., M.K., M.J.R. and S.R. All authors have read and agreed to the published version of the manuscript.

Funding: This research received no external funding.

Institutional Review Board Statement: No ethical approval was needed.

Informed Consent Statement: Not applicable.

Data Availability Statement: All data are available on request.

Conflicts of Interest: The authors declare no conflicts of interest.

References

1. Andrich, S.; Haastert, B.; Neuhaus, E.; Neidert, K.; Arend, W.; Ohmann, C.; Grebe, J.; Vogt, A.; Jungbluth, P.; Thelen, S.; et al. Excess Mortality After Pelvic Fractures Among Older People. *J. Bone Miner. Res.* **2017**, *32*, 1789–1801. [CrossRef] [PubMed]
2. Benzinger, P.; Riem, S.; Bauer, J.; Jaensch, A.; Becker, C.; Büchele, G.; Rapp, K. Risk of institutionalization following fragility fractures in older people. *Osteoporos. Int.* **2019**, *30*, 1363–1370. [CrossRef] [PubMed]
3. Krappinger, D.; Struve, P.; Schmid, R.; Kroesslhuber, J.; Blauth, M. Fractures of the pubic rami: A retrospective review of 534 cases. *Arch. Orthop. Trauma Surg.* **2009**, *129*, 1685–1690. [CrossRef] [PubMed]
4. Marrinan, S.; Pearce, M.S.; Jiang, X.Y.; Waters, S.; Shanshal, Y. Admission for osteoporotic pelvic fractures and predictors of length of hospital stay, mortality and loss of independence. *Age Ageing* **2015**, *44*, 258–261. [CrossRef] [PubMed]
5. Schönenberg, D.; Guggenberger, R.; Frey, D.; Pape, H.-C.; Simmen, H.-P.; Osterhoff, G. CT-based evaluation of volumetric bone density in fragility fractures of the pelvis—A matched case-control analysis. *Osteoporos. Int.* **2018**, *29*, 459–465. [CrossRef]
6. Lodde, M.F.; Fischer, T.M.; Riesenbeck, O.; Roßlenbroich, S.; Katthagen, J.C.; Rometsch, D.; Günes, D.; Schneider, M.; Raschke, M.J.; Liljenqvist, U. Analysis of the Spinopelvic Parameters in Patients with Fragility Fractures of the Pelvis. *J. Clin. Med.* **2023**, *12*, 4445. [CrossRef]
7. Morris, R.O.; Sonibare, A.; Green, D.J.; Masud, T. Closed pelvic fractures: Characteristics and outcomes in older patients admitted to medical and geriatric wards. *Postgrad. Med. J.* **2000**, *76*, 646–650. [CrossRef]
8. Nuber, S.; Ritter, B.; Fenwick, A.; Förch, S.; Wanzl, M.; Nuber, M.; Mayr, E. Midterm follow-up of elderly patients with fragility fractures of the pelvis: A prospective cohort-study comparing operative and non-operative treatment according to a therapeutic algorithm. *Injury* **2022**, *53*, 496–505. [CrossRef]
9. Rommens, P.M.; Arand, C.; Hofmann, A.; Wagner, D. When and How to Operate Fragility Fractures of the Pelvis? *Indian J. Orthop.* **2019**, *53*, 128–137. [CrossRef]
10. Wilson, D.G.G.; Kelly, J.; Rickman, M. Operative management of fragility fractures of the pelvis—A systematic review. *BMC Musculoskelet. Disord.* **2021**, *22*, 717. [CrossRef]
11. Schuetze, K.; Eickhoff, A.; Dehner, C.; Blidon, A.; Gebhard, F.; Richter, P.H. Short-term outcome of fragility fractures of the pelvis in the elderly treated with screw osteosynthesis and external fixator. *Eur. J. Trauma Emerg. Surg.* **2022**, *48*, 2413–2420. [CrossRef] [PubMed]
12. Rommens, P.M.; Hofmann, A. Comprehensive classification of fragility fractures of the pelvic ring: Recommendations for surgical treatment. *Injury* **2013**, *44*, 1733–1744. [CrossRef] [PubMed]
13. Matta, J.M.; Tornetta, P. Internal fixation of unstable pelvic ring injuries. *Clin. Orthop. Relat. Res.* **1996**, *329*, 129–140. [CrossRef] [PubMed]

14. Chip, M.L.C., Jr.; Simonian, P.T.; Agnew, S.G.; Mann, F.A. Radiographic Recognition of the Sacral Alar Slope for Optimal Placement of Iliosacral Screws: A Cadaveric and Clinical Study. *J. Orthop. Trauma* **1996**, *10*, 171. [CrossRef] [PubMed]
15. Choy, W.-S.; Kim, K.J.; Lee, S.K.; Park, H.J. Anterior pelvic plating and sacroiliac joint fixation in unstable pelvic ring injuries. *Yonsei Med. J.* **2012**, *53*, 422–426. [CrossRef]
16. Hartensuer, R.; Lodde, M.F.; Keller, J.; Eveslage, M.; Stolberg-Stolberg, J.; Riesenbeck, O.; Raschke, M.J. Safety, Effect and Feasibility of Percutaneous SI-Screw with and without Augmentation—A 15-Year Retrospective Analysis on over 640 Screws. *J. Clin. Med.* **2020**, *9*, 2660. [CrossRef]
17. Yamamoto, N.; Someko, H.; Tsuge, T.; Nakashima, Y.; Nakao, S. Prevalence of fracture progression in fragility fractures of the pelvis: Systematic review and meta-analysis. *Injury* **2024**, *55*, 111727. [CrossRef]
18. Cintean, R.; Fritzsche, C.; Zderic, I.; Gueorguiev-Rüegg, B.; Gebhard, F.; Schütze, K. Sacroiliac versus transiliac–transsacral screw osteosynthesis in osteoporotic pelvic fractures: A biomechanical comparison. *Eur. J. Trauma Emerg. Surg.* **2023**, *49*, 2553–2560. [CrossRef]
19. Zderic, I.; Wagner, D.; Schopper, C.; Lodde, M.; Richards, G.; Gueorguiev, B.; Rommens, P.; Acklin, Y.P. Screw-in-screw fixation of fragility sacrum fractures provides high stability without loosening-biomechanical evaluation of a new concept. *J. Orthop. Res.* **2021**, *39*, 761–770. [CrossRef]
20. Lodde, M.F.; Katthagen, J.C.; Schopper, C.O.; Zderic, I.; Richards, R.G.; Gueorguiev, B.; Raschke, M.J.; Hartensuer, R. Does Cement Augmentation of the Sacroiliac Screw Lead to Superior Biomechanical Results for Fixation of the Posterior Pelvic Ring? A Biomechanical Study. *Medicina* **2021**, *57*, 1368. [CrossRef]
21. Rommens, P.M.; Arand, C.; Hopf, J.C.; Mehling, I.; Dietz, S.O.; Wagner, D. Progress of instability in fragility fractures of the pelvis: An observational study. *Injury* **2019**, *50*, 1966–1973. [CrossRef] [PubMed]
22. Fada, R.; Shahgholi, M.; Azimi, R.; Babadi, N.F. Estimation of Porosity Effect on Mechanical Properties in Calcium Phosphate Cement Reinforced by Strontium Nitrate Nanoparticles: Fabrication and FEM Analysis. *Arab. J. Sci. Eng.* **2024**, *49*, 1815–1825. [CrossRef]
23. Yao, F.; He, Y.; Qian, H.; Zhou, D.; Li, Q. Comparison of Biomechanical Characteristics and Pelvic Ring Stability Using Different Fixation Methods to Treat Pubic Symphysis Diastasis: A Finite Element Study. *Medicine* **2015**, *94*, e2207. [CrossRef] [PubMed]
24. Shi, D.; Wang, F.; Wang, D.; Li, X.; Wang, Q. 3-D finite element analysis of the influence of synovial condition in sacroiliac joint on the load transmission in human pelvic system. *Med. Eng. Phys.* **2014**, *36*, 745–753. [CrossRef]
25. Bruna-Rosso, C.; Arnoux, P.-J.; Bianco, R.-J.; Godio-Raboutet, Y.; Fradet, L.; Aubin, C.-É. Finite Element Analysis of Sacroiliac Joint Fixation under Compression Loads. *Int. J. Spine Surg.* **2016**, *10*, 16. [CrossRef]
26. Zhang, L.; Peng, Y.; Du, C.; Tang, P. Biomechanical study of four kinds of percutaneous screw fixation in two types of unilateral sacroiliac joint dislocation: A finite element analysis. *Injury* **2014**, *45*, 2055–2059. [CrossRef]
27. Wu, C.; Zeng, B.; Deng, J.; Qin, B.; Shen, D.; Wang, X.; Hu, H.; Huang, L.; Liu, X.; Qiu, G. Finite Element Analysis and Transiliac-Transsacral Screw Fixation for Posterior Pelvic Ring with Sacrum Dysplasia. *Orthop. Surg.* **2023**, *15*, 337–346. [CrossRef]
28. Fu, S.; Zhao, Y.; Lian, W.; Zou, D.; Sun, T.; Zhao, Y.; Tan, J.; Zhang, S.; Wang, D. Comparison of the risk of breakage of two kinds of sacroiliac screws in the treatment of bilateral sacral fractures. *Eur. Spine J.* **2014**, *23*, 1558–1567. [CrossRef]
29. Hu, P.; Wu, T.; Wang, H.; Qi, X.; Yao, J.; Cheng, X.; Chen, W.; Zhang, Y. Biomechanical Comparison of Three Internal Fixation Techniques for Stabilizing Posterior Pelvic Ring Disruption: A 3D Finite Element Analysis. *Orthop. Surg.* **2019**, *11*, 195–203. [CrossRef]
30. Ma, Y.; Zhao, Y.; Zou, D.; Dong, S.; Sun, X.; Cheng, G.; Lian, W.; Zhao, Y.; Sun, T.; Wang, D.; et al. Breakage risk of different sacroiliac screws in unilateral sacral fractures a finite element analysis. *BMC Musculoskelet. Disord.* **2022**, *23*, 953. [CrossRef]
31. Eichenseer, P.H.; Sybert, D.R.; Cotton, J.R. A finite element analysis of sacroiliac joint ligaments in response to different loading conditions. *Spine* **2011**, *36*, E1446–E1452. [CrossRef] [PubMed]
32. Lipphaus, A.; Klimek, M.; Witzel, U. Comparative Finite Element Analysis of Fixation Techniques for APC II Open-Book Injuries of the Pelvis. *Biomechanics* **2021**, *1*, 321–333. [CrossRef]
33. Fedorov, A.; Beichel, R.; Kalpathy-Cramer, J.; Finet, J.; Fillion-Robin, J.-C.; Pujol, S.; Bauer, C.; Jennings, D.; Fennessy, F.; Sonka, M.; et al. 3D Slicer as an image computing platform for the Quantitative Imaging Network. *Magn. Reson. Imaging* **2012**, *30*, 1323–1341. [CrossRef] [PubMed]
34. Rubio-Pérez, I.; Díaz Lantada, A. Surgical Planning of Sacral Nerve Stimulation Procedure in Presence of Sacral Anomalies by Using Personalized Polymeric Prototypes Obtained with Additive Manufacturing Techniques. *Polymers* **2020**, *12*, 581. [CrossRef]
35. Taddei, F.; Schileo, E.; Helgason, B.; Cristofolini, L.; Viceconti, M. The material mapping strategy influences the accuracy of CT-based finite element models of bones: An evaluation against experimental measurements. *Med. Eng. Phys.* **2007**, *29*, 973–979. [CrossRef]
36. Taddei, F.; Pancanti, A.; Viceconti, M. An improved method for the automatic mapping of computed tomography numbers onto finite element models. *Med. Eng. Phys.* **2004**, *26*, 61–69. [CrossRef]

37. Dalstra, M.; Huiskes, R.; Odgaard, A.; van Erning, L. Mechanical and textural properties of pelvic trabecular bone. *J. Biomech.* **1993**, *26*, 523–535. [CrossRef]
38. Miller, J.A.; Schultz, A.B.; Andersson, G.B. Load-displacement behavior of sacroiliac joints. *J. Orthop. Res.* **1987**, *5*, 92–101. [CrossRef]
39. Bergmann, G.; Deuretzbacher, G.; Heller, M.; Graichen, F.; Rohlmann, A.; Strauss, J.; Duda, G.N. Hip contact forces and gait patterns from routine activities. *J. Biomech.* **2001**, *34*, 859–871. [CrossRef]
40. Denis, F.; Davis, S.; Comfort, T. Sacral Fractures. *Clin. Orthop. Relat. Res.* **1988**, *227*, 67–81. [CrossRef]
41. Deddens, J.A.; Petersen, M.R. Approaches for estimating prevalence ratios. *Occup. Environ. Med.* **2008**, *65*, 501–506. [CrossRef] [PubMed]
42. Wagner, D.; Kisilak, M.; Porcheron, G.; Krämer, S.; Mehling, I.; Hofmann, A.; Rommens, P.M. Trans-sacral bar osteosynthesis provides low mortality and high mobility in patients with fragility fractures of the pelvis. *Sci. Rep.* **2021**, *11*, 14201. [CrossRef] [PubMed]
43. Regenbogen, S.; Barbari, J.E.; Vetter, S.Y.; Franke, J.; Grützner, P.A.; Swartman, B. Feasibility and Safety of 3D-Navigated Trans-Sacral Bar Osteosynthesis for Fragility Fractures of the Sacrum: FIRST Clinical Experiences. *J. Clin. Med.* **2024**, *13*, 5244. [CrossRef] [PubMed]
44. Kammerlander, C.; Pfeufer, D.; Lisitano, L.A.; Mehaffey, S.; Böcker, W.; Neuerburg, C. Inability of Older Adult Patients with Hip Fracture to Maintain Postoperative Weight-Bearing Restrictions. *J. Bone Joint Surg. Am.* **2018**, *100*, 936–941. [CrossRef]
45. Merkle, T.P.; Hofmann, N.; Knop, C.; Da Silva, T. Can elderly individuals perform partial weight bearing on their lower limbs? A prospective cohort study using ambulatory real-time biofeedback. *J. Orthop. Surg. Res.* **2023**, *18*, 324. [CrossRef]
46. Heydemann, J.; Hartline, B.; Gibson, M.E.; Ambrose, C.G.; Munz, J.W.; Galpin, M.; Achor, T.S.; Gary, J.L. Do Transsacral-transiliac Screws Across Uninjured Sacroiliac Joints Affect Pain and Functional Outcomes in Trauma Patients? *Clin. Orthop. Relat. Res.* **2016**, *474*, 1417–1421. [CrossRef]

bioengineering

MDPI

Article

A Muscle-Driven Spine Model for Predictive Simulations in the Design of Spinal Implants and Lumbar Orthoses

Robin Remus [1,*], Andreas Lipphaus [2,3], Marisa Ritter [1], Marc Neumann [1] and Beate Bender [1]

1 Chair of Product Development, Department of Mechanical Engineering, Ruhr-University Bochum, 44801 Bochum, Germany
2 Biomechanics Research Group, Department of Mechanical Engineering, Ruhr-University Bochum, 44801 Bochum, Germany
3 Clinic of Pediatric Surgery, Marien Hospital Witten, Ruhr-University Bochum, 58452 Witten, Germany
* Correspondence: robin.remus@ruhr-uni-bochum.de

Abstract: Knowledge of realistic loads is crucial in the engineering design process of medical devices and for assessing their interaction with the spinal system. Depending on the type of modeling, current numerical spine models generally either neglect the active musculature or oversimplify the passive structural function of the spine. However, the internal loading conditions of the spine are complex and greatly influenced by muscle forces. It is often unclear whether the assumptions made provide realistic results. To improve the prediction of realistic loading conditions in both conservative and surgical treatments, we modified a previously validated forward dynamic musculoskeletal model of the intact lumbosacral spine with a muscle-driven approach in three scenarios. These exploratory treatment scenarios included an extensible lumbar orthosis and spinal instrumentations. The latter comprised bisegmental internal spinal fixation, as well as monosegmental lumbar fusion using an expandable interbody cage with supplementary posterior fixation. The biomechanical model responses, including internal loads on spinal instrumentation, influences on adjacent segments, and effects on abdominal soft tissue, correlated closely with available in vivo data. The muscle forces contributing to spinal movement and stabilization were also reliably predicted. This new type of modeling enables the biomechanical study of the interactions between active and passive spinal structures and technical systems. It is, therefore, preferable in the design of medical devices and for more realistically assessing treatment outcomes.

Keywords: musculoskeletal multibody model; FEM; forward dynamics simulation; lumbar fusion; interbody cage; extensible lumbar belt; soft tissue biomechanics; ArtiSynth; biomedical development process; engineering design

1. Introduction

Low back pain is one of the leading causes of disability worldwide and is predicted to increase due to population growth and aging [1]. While its prevalence rises with age, most cases occur between the ages of 50 and 55 [2], resulting in an enormous socio-economic burden, as back pain is a major contributor to absenteeism [1,3,4]. The causes of pain are manifold, and treatment can be complicated when no specific disease or structural cause is known [5,6]. Treatment is generally based on the following two pillars [7–9]: conservative treatment, such as psychotherapy, medication, injections, and physical therapy and surgical treatment. In both cases, medical devices are frequently used. However, even when all treatment options have been exhausted, it is often not possible to achieve complete relief of symptoms [9]. A significant proportion of patients often experience recurrent or new problems [10]. To improve treatment for back conditions, reduce surgery and recovery times, minimize post-operative complications, and improve overall quality of life [8,11], new medical devices are constantly being developed [12,13].

Two examples of medical devices used to stabilize the spine are lumbar orthoses for conservative treatment [14–17] and spinal implants for surgical treatment [18,19]. The latter include screws and rods, interbody cages, and plates that are implanted for the direct fixation and fusion of the spine [20–23]. A lumbar orthosis, on the other hand, is an external device that encompasses all or part of the lumbar and sacroiliac regions of the trunk and is used to modify the structural and functional characteristics of the neuromuscular and skeletal systems [16,24]. To support the engineering design process [25] and predict the treatment outcomes of new devices, biomechanical studies are conducted [26]. These provide essential quantitative measures to ensure that the design process is not based solely on expert opinion rather than by biomechanical data and clinical evidence [27]. However, even the latest experimental methods used to study spinal mechanics and the loads and effects of medical devices have their limitations [28]. As a result, virtual models are increasingly being used, with varying levels of detail and solution approaches [29–31]. These models are intended to describe the existing knowledge of the complex physiology and pathology of the human body using mathematical models in order to derive useful predictions for treatment as a quantitative hypothesis [32].

A variety of different virtual spine models have been developed, which can be categorized into the following three types of numerical modeling [29,33]: finite element (FE), musculoskeletal multibody (MB), and combined FE–MB modeling. Each type of modeling has specific advantages and disadvantages and is used to answer different scientific, clinical, and engineering questions. To support the design process of medical devices, FE modeling is most frequently used [30,34]. It enables a detailed structural mechanics analysis of products or engineering designs in interaction with parts of the healthy or pathological osteoligamentous spine [28,35]. To obtain relevant results, realistic boundary conditions must be chosen that reproduce physiologically appropriate spinal loads [36–38]. However, the most common limitation of FE models is the absence of the muscles and tissues surrounding the spinal column [34,39]. The effects of muscle forces, body weight, and external loads due to physical activity must be represented by external forces and moments as simplified boundary conditions [38,40–44]. As it is recognized that the loading of the spine is highly complex due to the multiple surrounding redundant intrinsic and extrinsic muscles [45–47], the use of simplified boundary conditions can lead to over-simplification and inaccurate predictions of in vivo loading conditions [39,48–52]. Experimental [53] and simulation [54] studies have shown differences in implant loads between simplified and realistic loading conditions, which can be critical if underestimated in the design process. Muscles are also essential for spinal stability [55–59], which is affected by treatment with medical devices such as implants or orthoses [60,61].

To overcome these limitations and retain the advantages of FE modeling, the advantages of musculoskeletal MB modeling can be exploited in the context of a forward dynamic hybrid FE–MB model of the spine [29,33,62,63] with a muscle-driven approach [64–66]. These are still rarely used in spinal biomechanics and represent an alternative to coupled FE–MB models [49,52,67–69]. A hybrid FE–MB model allows for the calculation of the stresses and strains of specific structures while using rigid bodies and muscles in a single simulation step. This avoids, for example, the technically and mechanically complex coupling of two or more separate FE and MB models, which is necessary for the two-way exchange of simulation results and model synchronization [33,67]. A muscle-driven approach also allows for the simulation of movement and posture by predicting the muscle activation patterns for a dynamic musculoskeletal spine model and its interaction with the environment [66,70]. Such a hybrid FE–MB model can be used, for example, to conduct virtual technical feasibility studies and engineering designs of medical devices to clarify their interactions with biomechanical factors and generate new knowledge to improve patient care [31,71–73].

The aim of this study is, therefore, to illustrate a novel model for predicting in vivo-like loading conditions of the lumbar spine treated conservatively or surgically with medical devices. For this purpose, a previously validated intact hybrid FE–MB spine model [74] was modified in three exploratory scenarios, fitted with implants or an orthosis. The model validity was evaluated for different physical activities, including a comparison with data from the literature for the biomechanical responses of the FE and MB components, as well as the internal loads on the spinal implants. As the entire model was muscle-driven, the stabilization mechanisms of the muscle forces were also evaluated and passive minimal models without muscle influence were extracted and validated in advance.

2. Materials and Methods

In the following, we describe our approach to modeling three scenarios for the use of medical devices in the lower back and their integration into a musculoskeletal lumbosacral spine (MLS) model for predictive simulations. The inclusion of scenarios was based on available in vivo and in vitro data for validation [27] and consisted of modifications to the intact MLS model (Figure 1). The development and validation of the active intact MLS model, which extended a passive hybrid FE–MB osteoligamentous lumbosacral spine (OLS) model [75], have been described elsewhere [74]. Modeling and simulation were performed within the open-source modeling framework ArtiSynth [76], and the implemented Tracking Controller [77,78] was used to realize the muscle-driven approach for the MLS model. A summary of the details of the intact MLS model relevant to this study is provided in Appendix A.

The predictive simulation of the active hybrid FE–MB MLS model was divided into the following six phases, of which [i], [iii], [iv], and [v] were optional depending on the scenario and physical activity:

[i] First MLS model manipulation during runtime,
[ii] activation of the Tacking Controller and all boundary conditions including gravity, with subsequent 2° flexion and 2° extension of the thorax (spinal settling phase),
[iii] second MLS model manipulation during runtime,
[iv] changing the body posture,
[v] application of external manipulators and loads,
[vi] evaluation of model responses once static equilibrium was reached.

We defined the resulting orientation of the dynamic vertebrae after the spinal settling phase [ii] as the stable, upright reference posture (in the following also abbreviated as standing) with a lumbar lordosis of 52° (COBB angle [79] in between L1 and S1). After reaching a static target posture in [vi], the biomechanical model responses extracted from the MLS model included intradiscal pressure (IDP), intra-abdominal pressure (IAP), range of motion (ROM), intervertebral displacement, and muscle force.

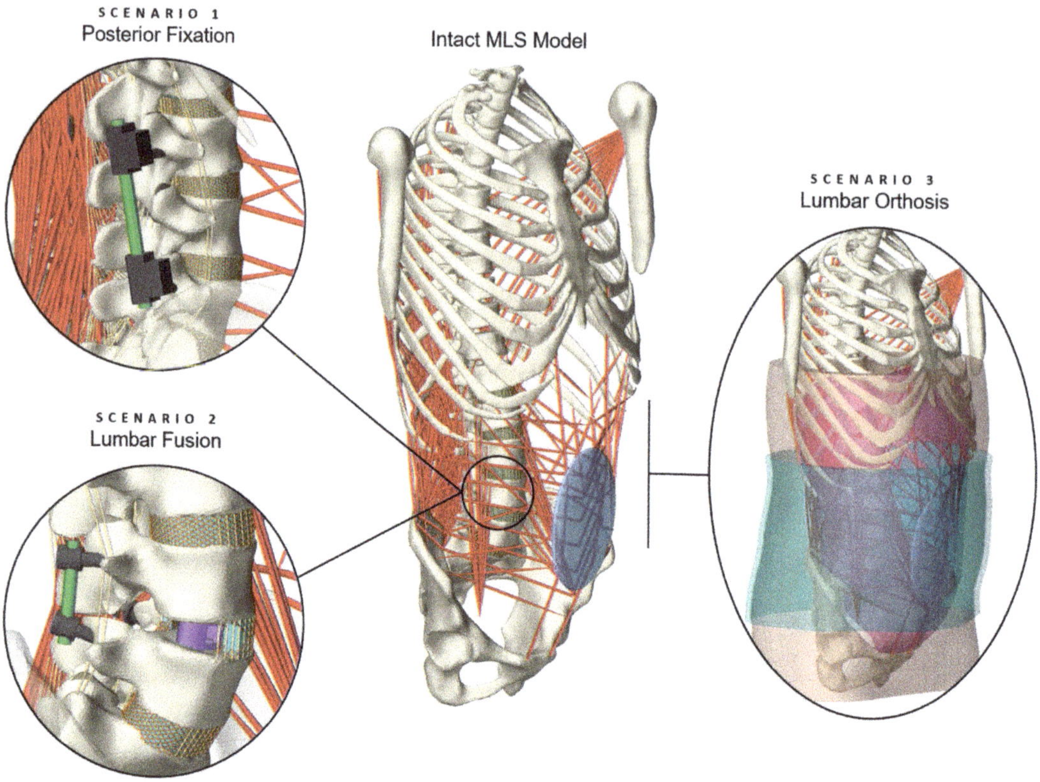

Figure 1. Visualization of the active hybrid FE–MB MLS (musculoskeletal lumbosacral spine) model with scenario overview. The intact MLS model [74] was modified to replicate three treatment scenarios involving medical devices. The procedures can be categorized into surgical treatment in the form of lumbar spinal instrumentation (scenarios 1 and 2) and conservative treatment with an orthosis (scenario 3). For better visibility, the muscles on the right side of the body are hidden in scenarios 1 and 2, and the L4/5 annulus fibrosus is shown in sectional view in scenario 2.

2.1. Spinal Instrumentation

Two types of bi- and monosegmental instrumentation were simulated for the first two scenarios. These comprised posteriorly fixed internal spinal fixators spanning from L3 to L5 and L4 to L5, and a L4/5 interbody fusion cage. All implants were implemented by 3D finite elements, geometrically simplified, and symmetrical to the sagittal plane. Their meshes were generated using ANSYS Workbench (2020 R1, ANSYS Inc., Canonsburg, PA, USA) and fed into ArtiSynth. Hexahedral elements were used for the rods. All other elements were meshed with linear tetrahedral elements. Surface nodes of the pedicle screws located within a rigid vertebra were attached to it and the nodes of the screw holes were connected to the adjacent elements of the respective rod in place. All implanted component materials were represented by the titanium alloy Ti6Al4V (Table 1), which was modeled as linear elastic.

Table 1. Material properties of the anatomical and technical FE components. *c* and *m* are material parameters for the Ogden material model. All other material data of the MLS model can be found elsewhere [74,75].

Component	Material Properties		References
	Young's Modulus	Poisson's Ratio	
Bone (cancellous and cortical)	0.1 ... 12 GPa	0.30	[34,80,81]
Diaphragm muscle tissue	5.32 MPa	0.33	[82]
Diaphragm tendon tissue	33 MPa	0.33	[82,83]
Implants (Ti6Al4V alloy)	110 GPa	0.30	[80,84]
Lumbar belt fabric	3 MPa	0.49	[50]
	c	*m*	
Abdominal and pelvic cavity region	7 kPa	17	[85]
Abdominal wall region	10 ... 19 kPa	19 ... 22	[85]
Posterior muscle region	25 ... 100 kPa	19 ... 23	[85]

For preliminary validation of the internal instrumentations, sections of the passive OLS model were used (Figure 2). Moments were applied to the cranial vertebra (L3 or L4) and axial compressive forces with a path optimized follower load [86,87]. The most inferior vertebra L5 was fixed in place. All load cases used are summarized in Table 2.

Subsequently, two instrumented OLS spine sections (*cf.* Figure 2) were transferred unchanged into the MLS model (Figure 1). Muscular structures and adjacent motion segments were not altered. To predict loads on the implants and adjacent level effects [88], sagittal postures were simulated. These were the same as those used to validate the intact MLS model and have been described elsewhere [74].

Table 2. Load cases used for OLS model validation with spinal instrumentations. Moments were applied in all spatial directions.

Used for Scenario	Moment in Nm	Follower Load in N	References
1	3.75		[89]
1	6.60		[36,90]
1		250	[36,90]
2	10.0	200	[91]

2.1.1. Scenario 1: Posterior Fixation

The *fixateur interne* described by Dick [92] was utilized to model a bilateral two-level posterior fixation (PF). Rohlmann et al. [36,93] used modified versions [94] of this posterior spinal fixation device to measure the internal loads in the longitudinal rods in vivo and in vitro. We modeled the screws, nuts, and adjustable clamps of the device as a single FE body each. The diameter of pedicle screws was 5 mm and that of longitudinal rods was 7 mm. As a result of the virtual implantation, the maximum unsupported screw length was 12 mm, and the unclamped length of a rod was 40.5 mm.

Von Mises stresses and internal loads on the rods were evaluated in the center between two screw heads. Transformation into an axial force F_Z and a bending moment in the sagittal plane $M_{b,sag}$ was carried out according to Rohlmann et al. [36,95]. Due to anatomical conditions, the alignment of the implant coordinate system deviated from that of the upper body. The *y*-axis, which pointed ventrally and ran almost parallel to the pedicle screws, was rotated 15° around the *z*-axis, which represented the longitudinal axis of a rod. Negative values indicated an axial compressive force and a bending moment in flexion.

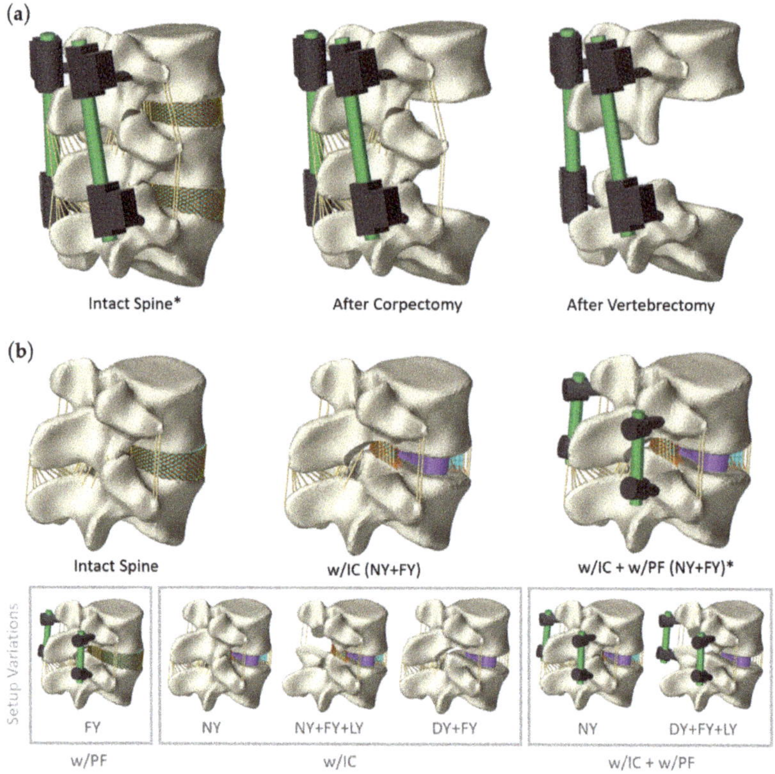

Figure 2. All validation setups for both spinal instrumentation scenarios using sections of the OLS (osteoligamentous lumbosacral spine) model [75]. Only setups marked with an asterisk (*) were used for the active MLS model. (**a**) Instrumented two-level posterior fixators [92] spanning from L3 to L5. Vertebra L4 is bridged in three different clinical scenarios: intact spine, after corpectomy, and after vertebrectomy. (**b**) Implantation of an interbody cage (IC) and bilateral PF (posterior fixation) for single-level spinal fusion. The intact L4/5 functional spinal unit was dissected as shown to replicate a complete bilateral facetectomy (FY) or a FY with laminectomy (LY). The nucleus pulposus was always removed (nucleotomy (NY)) for the IC and the annulus fibrosus remained intact or was entirely removed in a discectomy (DY). For visualization purposes, the annulus fibrosus is always shown in a sagittal section when the IC is implanted.

In a preliminary sensitivity study, we analyzed the influences of different meshes of the PF. When comparing hexahedral elements with linear and quadratic shape functions (Hex8 and Hex20), the changes in internal loads were <0.7% and those in stresses were <4.1%. For quadratic elements in both rods with the final edge length in the longitudinal direction of $\leq$2.5 mm, the increase in calculation time was, on average, 359% (validation setup after corpectomy, Figure 2a). We ended up using 722 elements for a rod and up to 2890 elements for a single screw clamp combination. The validation of the model was carried out by comparing the internal loads (axial load and bending moment) of the rods with in vitro [36,89,90] and in silico [26] studies under the pure bending moments and pure axial compression forces listed in Table 2. To increase comparability, we modeled three spinal conditions, as shown in Figure 2a.

For predictive simulations with the MLS model, the rotational components of the vertebral target frames were modified. The L4 target frame was removed, and its expected rotational contribution was equally distributed among the remaining vertebral target frames. The expected rotation from L3 to L5 was reduced by 68% (intact OLS with PF compared to no PF during flexion loading), and this reduction was also distributed.

2.1.2. Scenario 2: Lumbar Fusion

Single-level lumbar fusion was modeled, consisting of an interbody fusion cage with supplementary bilateral PF and varying degrees of resection. The diameter of the fixator pedicle screws was 6 mm and that of the straight rods was 5.5 mm. Resections were based on common surgical procedures (e.g., TLIF or PLIF) and are described and visualized in Figure 2b. The L4/5 nucleus pulposus was removed prior to cage insertion. The annulus fibrosus remained intact or was completely removed, depending on the validation setup. In accordance with standard TLIF procedures [19], the cage was banana-shaped, 34 mm wide, with an external cylinder diameter of 82 mm, and had a lordosis angle of 0°. The cage height of 15 mm was determined according to the disc space, thus providing a slight interference fit (press fit).

For a structural mechanics analysis of the cage and its contacts, the two adjacent vertebral bodies L4 and L5 were integrated into the OLS model as additional 3D FE bodies. The cranial endplate of the FE vertebral body L4 was attached to RB vertebra L4 and the caudal endplate of the FE vertebral body L5 was attached to RB vertebra L5 via their respective nodes. All components of the MLS model and vice versa interacted with these FE models via the RBs L4 and L5. Figure 3 visualizes all the components that defined the mechanical relationship between the RB vertebrae L4 and L5.

Linear tetrahedral FE meshes of the vertebral bodies L4 and L5 were automatically generated in ArtiSynth using the TetGen algorithm. The triangular surface meshes used for this were manually segmented from the CT data of the Visible Human Male [96] using 3D Slicer (Version 5.6.2). Assuming a negligible influence for this scenario, the posterior bony elements were removed in the middle of the pedicles. The surface meshes were re-triangulated with SpaceClaim (v201, ANSYS Inc., USA) using the Regularize tool to ensure a sufficient mesh quality. Consistent facet aspect ratios were chosen with maximum edge lengths ranging from 0.5 mm for the end plates to 2.0 mm for the remaining topology (see Figure 3). Automatic mapping of linearly elastic and isotropic material data onto the tetrahedral elements of the vertebral bodies was carried out using BoneMat (Build_152) [97]. Both FE meshes were not symmetrized for this purpose. Each element was assigned an average material property based on the Hounsfield unit (HU) of voxel-wise bone tissue in the CT grid (voxels resampled to 1 mm). Using HU integration with four integration steps and a gap value of 50 for the Young's modulus calculation and algorithm parameters from Schileo et al. [98] resulted in up to 204 different parameter groups. Since no mechanical tests could be performed to calibrate the material properties [99], the Young's moduli were scaled to the range from 100 to 12,000 MPa [34,80,81], maintaining their relative distances. Poisson's ratio was assumed to be 0.3 [80] for all elements. The FE data were imported into ArtiSynth using VTK files along with the material data. The heterogeneous material parameter distributions of both vertebrae are visualized in Figure 4a.

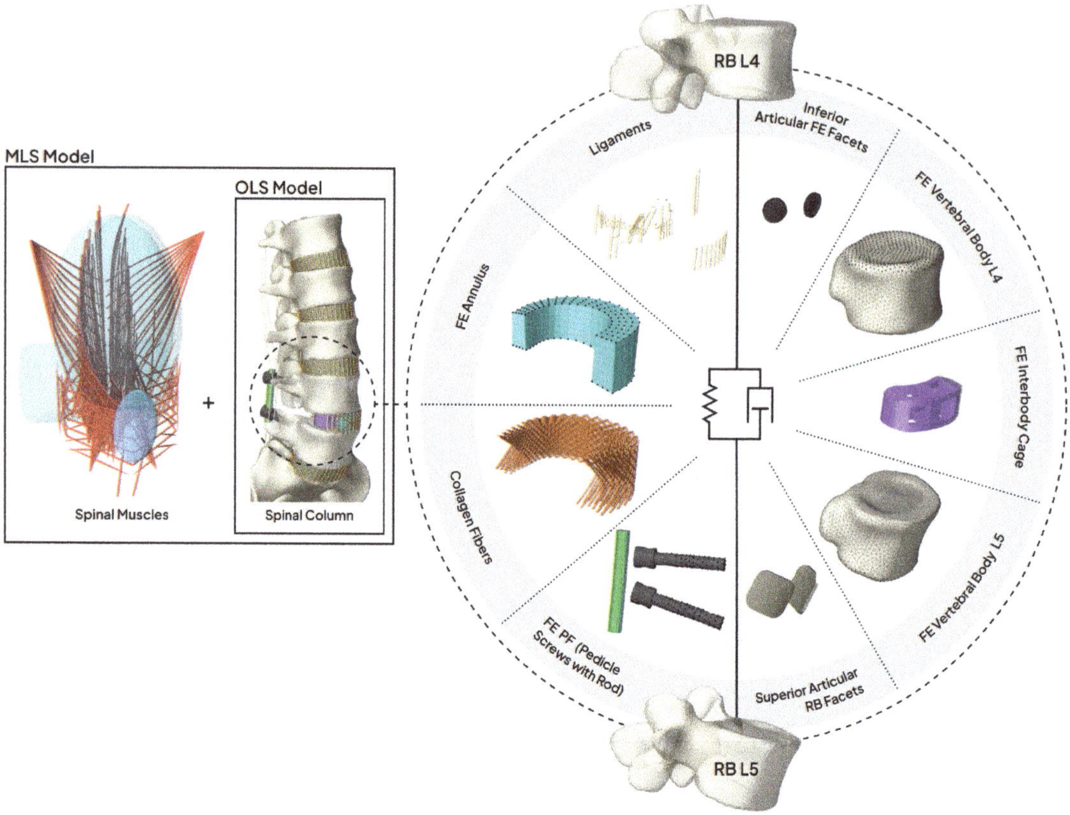

Figure 3. Detailed view of scenario 2 of the MLS (musculoskeletal lumbar spine) model with an exploded view of the L4/5 lumbar fusion. The two subsystems [74,100] spinal muscles and osteoligamentous spine (setup w/IC + w/PF (NY + FY), *cf.* Figure 2b) are shown separately on the left side. All muscles were attached to the RB bones or abdominal plate and were redirected by the cyan colored wrapping bodies in the area of the rib cage and lumbar spine. In contrast to classic musculoskeletal MB models, the mechanical relationship between RB bones were also defined by 3D FE bodies and contact conditions. All the components modeled for this purpose are shown separately on the right (setup w/IC + w/PF (NY)). Using this hybrid FE–MB modeling approach, the dynamic relationships of both RB vertebrae were highly non-linear and modularly adaptable (depending on the setup, components such as facet joints, ligaments, or PF (posterior fixation) were removed or added; *cf.* Figure 2b). All nodes of the FE bodies that were attached to the RB vertebrae are visualized as black dots. These were nodes of pedicle screws, annulus, inferior articular facets, vertebral body L4, and vertebral body L5. The FE cage was placed as a contact body between the two vertebral FE bodies and was only in contact with them. For its caudal contact, the finely meshed endplate of FE vertebral body L5 can be seen. Rods were attached to pedicle screws and start and end points of the ligaments and collagen fibers to RB L4 and L5. Superior articular RB facets were in frictionless contact with the inferior articular FE facets. Note: L4/5 annulus fibrosus is shown in a sagittal section.

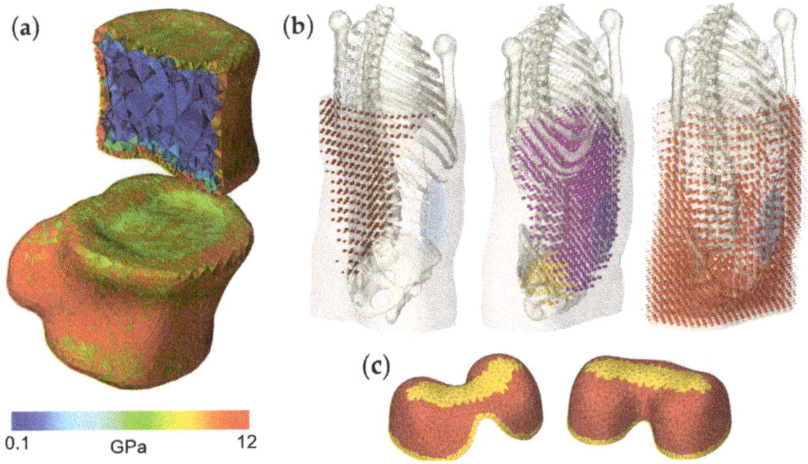

Figure 4. Visualizations of the implemented material heterogeneities. (**a**) Subject-specific Young's modulus distribution for the FE vertebral bodies L4 and L5 in anterolateral view. Vertebra L4 is cut parallel to the sagittal plane, displaying the internal elements with reduced stiffness. (**b**) Color coding of the FE elements of the embedding mesh to which different material parameters have been assigned: posterior muscle region (left), abdominal and pelvic cavity region (center), and abdominal wall region (right). In all views, the right side of the body is at the front, the left view is posterior-lateral and the central and right view is anterior-lateral. (**c**) Diaphragm shell elements with distinction between muscle (red) and tendon tissue (yellow) in anterior (left) and posterior (right) view.

The cage was implanted symmetrically to the sagittal plane in the anterior part of the removed nucleus pulposus. Initially, the cage height was reduced (anisotropic scaling), which allowed for a virtual insertion without contacts and intersections. For subsequent disc space distraction during runtime (step [i]), the cage was first moved caudally to the FE vertebral body L5, attached to it with two central nodes for stability, and then enlarged to its design height. This included physics simulation and contact friction for the cage to be temporarily deactivated. Lastly, the PF was locked and a friction coefficient of 0.5 [101,102] was set for the cage–bone surface contact to simulate the immediate post-operative time. To perform this during runtime, multiple controllers and input probes, which are data streams that can set control inputs and modulate model parameter values over time [103], were implemented.

In a preliminary FE mesh analysis of the rods, the maximum von Mises stresses were convergent with deviations of <5% for maximum element edge lengths of 1.8 mm. The element edge lengths of the screws were 2.5 mm. The element sums for a rod were 456 and were 865 for a screw. When comparing elements with linear and quadratic shape functions for the rods, differences between the results were <0.3% for ROM and < 8.5% for stresses with an increase in the calculation time of 130% (w/IC + w/PF (NY + FY)).

For validation, multidirectional ROMs of the fused L4/5 motion segment were simulated using the same load cases (Table 2) as in the in vitro study by Lund et al. [91]. The following three different conditions of the motion segment were compared: intact, with cage, and with cage and PF (the entire lumbar fusion). The follower load was applied before the pedicle screws were rigidly attached to the RB vertebrae. Different resections (see Figure 2) were simulated to evaluate influences of the OLS model's biological structures.

For load prediction with the MLS model, the rotational components of the vertebral tracking target frames were modified. The expected rotational contribution from L4 to L5 was assumed to be reduced by 90%, and this proportion was equally distributed among all other vertebral target frames. The physics simulation of vertebra L3 and the Tracking Controller were only activated from step [ii] onwards. Implantation with distraction by cage expansion was performed in the passive state of the MLS model (surgical procedure in prone position, step [i]). In addition to the internal loads of the rods (see Section 2.1.1), contact forces and pressures between the cage and vertebral bodies were also used to evaluate the instrumentation.

2.2. Scenario 3: Lumbar Orthosis

In the third scenario, the intact MLS model was extended by a 3D FE model of the surrounding soft tissue of the torso, which was in direct bidirectional interaction with the bony spinal structures. Soft tissue geometries were segmented from the image data of the Visible Human Male [96] from vertebra T9 to the pelvis using 3D Slicer. The flattening of the posterior tissue due to the ventrodorsal weight force [104] was corrected. Within the torso section, the following anatomical structures were separated: diaphragm, thoracic cavity, abdominal cavity, pelvic cavity, back muscles, and abdominal wall (anterolateral and posterior region combined). The diaphragm was derived from the adjacent structures as a surface mesh. All separate geometries were revised in SpaceClaim concerning faceting, sagittal symmetry, and conformity to the MLS model geometry. Additionally, a surface mesh was generated that fully enclosed the lumbar spine except for the spinous processes and that served as a continuous surface for a geometric skinning approach [105,106]. This approach was chosen to obtain a simplified and uniform contact surface that provided a two-way contact force coupling between the soft tissue and the underlying dynamic RB vertebrae. For this, the passive skinning mesh (Figure 5a) was attached to the vertebrae as the master bodies (linear blending), was driven by them, and deformed according to their movements.

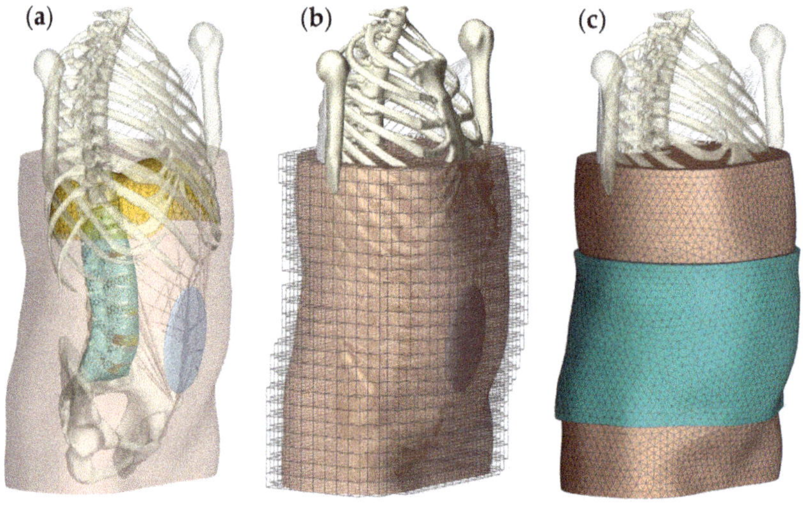

Figure 5. Third scenario, in which the MLS model was extended to include the surrounding FE soft tissue of the trunk and an extensible lumbar orthosis. Three rendering properties are used to visualize different details: (**a**) Inside the transparent torso, one can see the diaphragm (yellow), which was attached to the thorax, and the skinning mesh of the spine (cyan), which was used for the internal contact calculation (spine-soft tissue). (**b**) Surface mesh of the skin that was added to the regular embedding FE mesh as a contact surface. (**c**) Faceting of the polygonal surface mesh of the skin and the FE orthosis. The applied FE orthosis is in its initial state (not tensioned).

In contrast to the common method of meshing free-form bodies with tetrahedral elements, where achieving a high degree of conformity is usually time-consuming and often results in poorly conditioned elements, we used a mesh embedding approach [106,107]. This involved embedding the triangulated surface meshes of the segmented abdominal soft tissue structures in a comparatively coarse but regular grid of 4525 hexahedral elements (Figure 5b). The nodal and element masses and stiffnesses were calculated [107] based on a surface mesh that represented the outer contour of the trunk, excluding the skinned mesh volume of the spine and the thoracic cavity. Two additional surface meshes were embedded for the contact modeling of the skin–orthosis and soft tissue–spine. For the latter, the skinning mesh of the spine was duplicated with an offset of 0.25 mm during runtime at the end of step [ii] to minimize influences on the vertebral dynamics before settling. This contact was assumed to be frictionless [50,108,109] to prevent shear forces between the surfaces.

For an isotropic, hyperelastic, and nearly incompressible material behavior of the abdominal FE elements, the one-term Ogden material model [110,111] was used. The strain energy density function with logarithmic bulk potential is of the form

$$W(\lambda_i) = \frac{c}{m^2}\sum_{i=1}^{3}(\tilde{\lambda}_i^m - 1) + \frac{\kappa}{2}(\ln J)^2 \quad (1)$$

where $\tilde{\lambda}_i$ is the ith deviatoric principal stretch, c and m are the Ogden material parameters, and J represents the volume ratio. Based on the in vivo study by Remus et al. [85], tissue structures were combined and three regions with different material parameter sets were distinguished (Table 1) to obtain a heterogeneous distribution. The element allocations were conducted using the segmented soft tissue geometries and are visualized in Figure 4b. For the posterior muscle and the abdominal wall regions, additional linear dependencies on the excitations of the corresponding muscle groups of the MLS model were incorporated. To do this, the muscle excitations calculated by the Tracking Controller were averaged separately for both regions and the respective material parameter sets (c and m) were scaled between the minimum and maximum values given in Table 1.

The abdominal cavity was bounded cranially by the dome-shaped diaphragm. This was modeled with 1488 triangle shell elements (Tri3) with a constant thickness of 2.5 mm [112]. All inferior diaphragm boundary nodes were attached to the thorax and three regions [113] for two isotropic linear elastic materials were distinguished (Figure 4c, Table 1), as follows: (1) tendinous region cranial central (phrenic center) and at the border, and (2) the muscular region in between. All other shell nodes were linked to an embedded surface mesh representing the abdominal cavity. Any weight force influences of the FE soft tissues on the MLS model were neglected. To avoid the erroneous application of forces due to boundary conditions, the nodes of the transverse torso sections were fixed caudally to the static pelvis and cranially to an auxiliary body kinematically driven by the dynamic thorax. Further nodes within the bony structures of the thorax and between the ribs were attached to the dynamic thorax. All abdominal nodes within or close to the spinous processes were attached to them. Assuming a rigid pelvic floor, its FE nodes were attached to the pelvis.

The orthosis was an extruded copy of the skin surface facets with a constant height of 24 cm from L1 to the sacrum, assuming a continuous, skintight, extensible lumbar belt (Figure 5c). The resulting 8004 FE wedge elements in two layers of 4 mm each were assigned a linear material behavior (Table 1). The displacement of the caudal orthosis surface nodes was restricted in the vertical body direction. Because measured shear forces between an orthosis and body were low [114], we considered orthosis–skin contact to be frictionless [115].

For a nearly incompressible behavior of the abdomen, the bulk modulus κ was set to 40 kPa for all hexahedral elements. This reduced the L3/4 and L4/5 IDPs by about 25% [116] in an upright posture with a maximum pressure of 12 kPa and a mean pressure of 9 kPa under the orthosis. This is consistent with the available literature data, assuming that the orthosis was tightened more, as participants with 8 to 9.4 kPa did voluntarily [60,117,118], in order to simulate a corset inflated to the tolerance limit of the subjects in the study by Nachemson and Morris [116]. To do this, the FE orthosis was tightened during runtime at the beginning of step [v] by continuously reducing its length by up to 8% and temporarily deactivating its physics simulation. The physical simulation of the orthosis was then reactivated, causing it to stretch again, to some extent, due to its elasticity. Contact calculation was permanently enabled.

To evaluate this scenario, we predicted the influences of the soft tissue on the biomechanical responses of the MLS model in different postures. In addition, we simulated the influence of the applied orthosis in an upright posture with and without holding a 20 kg crate in both hands with arms bent in front of the upper body. In the center of the abdominal cavity [109], we examined the IAP via the hydrostatic pressure of the FE elements. We explicitly identify this component as FE–IAP, to distinguish the origin of the IAP for evaluation. Thus, the added FE soft tissue represented a second pressure force source acting cranially on the diaphragm and, hence, the thorax. The IAP resulted from muscle forces acting on the abdominal plate and was not specified in more detail for reasons of consistency. Trunk stiffness was estimated based on the sum of all muscle forces required to maintain a body posture.

3. Results

In the following, we distinguish between results using sections of the passive OLS model with simplified loading conditions and results predicted with the muscle-driven MLS model.

3.1. Scenario 1: Posterior Fixation

To validate the instrumented internal spinal fixators, the cutting conditions of a rod for the most relevant load cases were compared to data from the literature [26,36,89,90] in Figure 6. The calculated axial forces and bending moments were within the ranges of the published data depending on the spinal condition. Removing biological structures generally increased the load on the rods, which were at their maximum under axial compression. Since our model was not affected by experimental inaccuracies [26], the axial forces in the rods were approximately zero under pure moments after corpectomy. For the intact spine without external loads, F_Z was −2.4 N and $M_{b,sag}$ was −0.25 Nm. In the case of corpectomy, both values were reduced to −1.2 N and −0.02 Nm, and in the case of vertebrectomy, to 0 N and 0 Nm.

The body postures and resulting muscle activation patterns of the MLS model strongly influenced the internal loads on the PF (Figure 7a). The predicted results compared to in vivo data [36] are visualized in Figure 8a. The forces and moments measured by Rohlmann et al. varied considerably between the three patients. Our results fell almost completely within these ranges and varied depending on different thorax extension and flexion angles. The maximum axial force in a rod was −245 N in 10° extension and the minimum was −12.4 N in 30° flexion. In the latter case, we also measured the maximum bending moment of −7.4 Nm. Thus, the axial forces in extension exceeded the forces measured in vivo. All relative changes due to postural changes were plausible. However, it should be noted that neither the exact postures nor the equivalent spinal conditions of the participants could be replicated in our study.

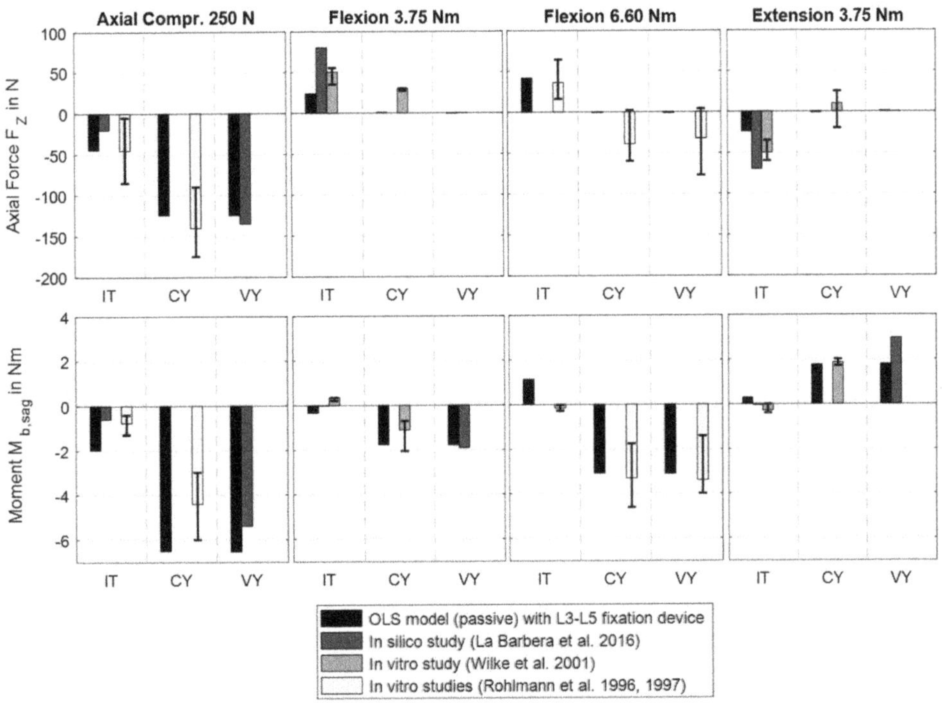

Figure 6. Excerpt from the simulated internal loads in the right rod of the posterior lumbar spinal fixation device (scenario 1) under pure axial compression force or pure bending moment. For validation, the simulation results were compared with in vitro measurements by Rohlmann et al. [36,90] and Wilke et al. [89], and in silico data by La Barbera et al. [26]. Only available comparison data were visualized. The fixators bridged vertebra L4 in three different clinical scenarios (*cf.* Figure 2a): Intact spine (IT), after corpectomy (CY), and after vertebrectomy (VY).

Further MLS model responses are visualized in Figures 9, 10 and 11a. The IDP was quantified in the nucleus pulposus of an intervertebral disc as the hydrostatic stress from the node normal stresses. Due to the PF and expected load sharing, the IDPs of levels L3/4 and L4/5 were similarly reduced. The greatest IPD reductions were found in extension (−78% and −71%, respectively). With an increasing flexion angle, the relative pressure differences between unchanged and fixed spinal levels decreased to 13% and 11% (Figure 9). In a previous in vivo study [116], PF resulted in a comparable reduction in disc loading of about 30%. Cranial to the fixed levels, IDPs were increased or unchanged. The pressure increase was maximal in extension and minimal in 20° flexion. For 30° flexion, the IDP change increased again. Except for 10° flexion, the pressure on level L5/S1 was reduced due to PF. The segmental rotation contributions of the lumbar motion segments were altered by the PF (Figure 10). The contributions of L3–L4 and L4–L5 were reduced and compensated by the adjacent levels. The rotation contribution of L2–L3 increased most in flexion. The associated changes in muscle forces (Figure 11a) were heterogeneous for the postures examined. As a result of the increased stiffness of the spine, the total muscle force in the upright posture was reduced by 22%. To generate the maximum flexion angle, more force was applied by the abdominal muscles and less by the posterior muscles. In extension, an inverse behavior was observed. The sum of all predicted muscle forces was, thus, reduced by 18% and 4%, respectively.

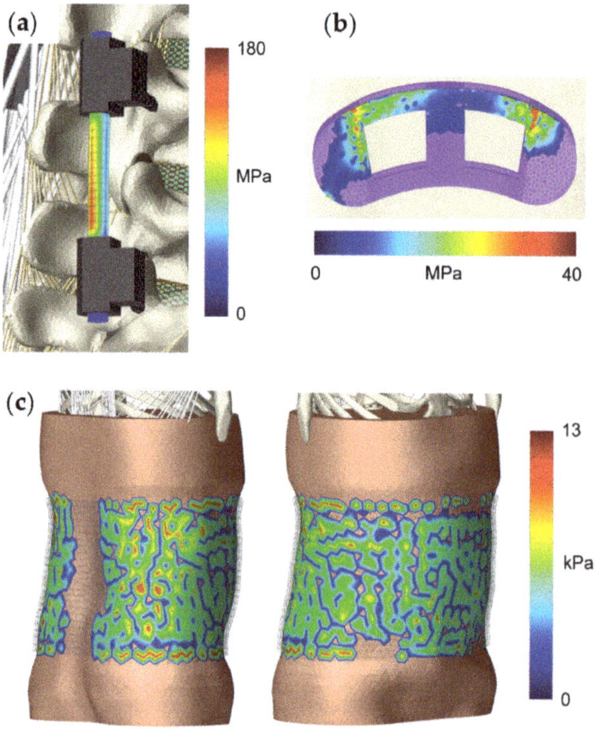

Figure 7. Exemplary visualizations of predicted stress and pressure distributions for the three scenarios: (**a**) Von Mises stress in the right rod of the fixation device in posterolateral view with the thorax in 30° flexion. Muscles on the right side are hidden. The almost stress-free sections of the rod adjacent to the clamps resulted from the attachment conditions of the rod elements. (**b**) Pressure distribution on the cranial side of the interbody cage with the thorax in 30° flexion. All components except the cage were hidden cranial to vertebra L5. (**c**) Pressure distribution under the orthosis applied with maximum tension in relaxed standing position, shown in posterolateral (left) and anterolateral (right) view. The pressure is not interpolated, and the orthosis is displayed transparently.

3.2. Scenario 2: Lumbar Fusion

The calculated ROMs of the validation load cases for the lumbar fusion are visualized in Figure 12 for a comparison with in vitro study data [91]. For flexion and axial rotation, the deviations from their median values were minimal in the intact state. In extension, our motion segment overestimated and in lateral bending it underestimated the in vitro movements. These deviations were absent compared to the in vitro study [119] used to validate the OLS model [75]. As can be seen in Figure 12, these systematic deviations also remained after the implantation of the interbody cage. On the other hand, relative ROM changes with and without the cage were consistent with the in vitro data in all spatial directions. Overall, cage implantation altered the ROMs to between −31% and +21% of the intact movement. The height expansion of the cage increased the disc space by approximately 1 mm, ensured a press fit, and pre-tensioned the structures of the motion segment. This was 6% of the mean disc height, which corresponds closely to previous studies [120–122] and was found to mimic the validation results best. The tight annulus fibrosus had the greatest influence on mechanical behavior and stability. For the setups w/IC (NY + FY) and w/IC (DY + FY), distraction resulted in contact forces of 250 N and

120 N on the anterior side of the cage, respectively. Regardless of the validation setup examined, the additional implantation of the PF reduced the ROMs to almost 0° and the stiffness of the lumbar fusion (w/IC + w/PF) was at the upper end of the study data.

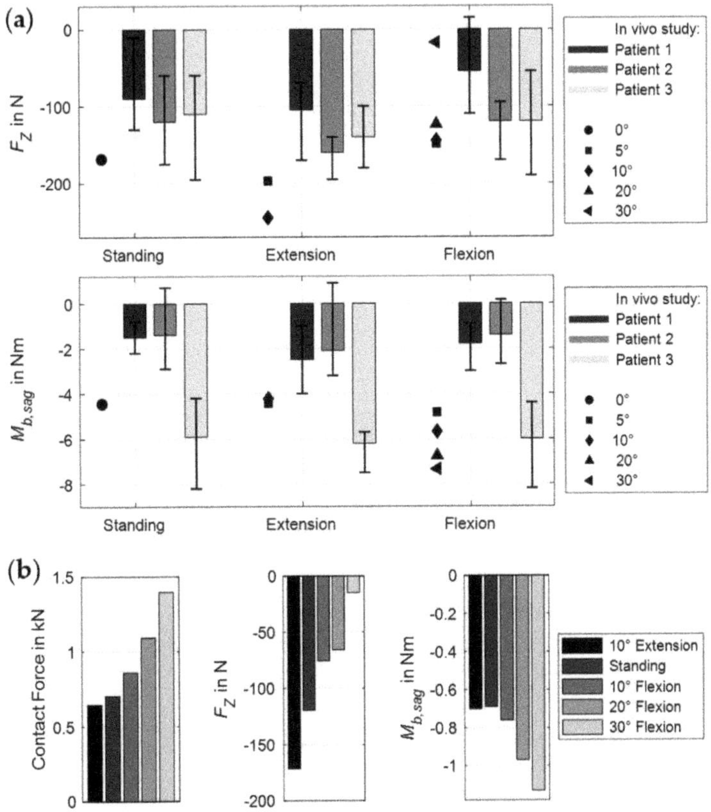

Figure 8. Predicted implant loads in the MLS model for different postures from 10° extension to 30° flexion. (**a**) Two-level posterior fixators implanted in the intact spine (scenario 1, Figure 2a). The predicted axial force components (top) and bending moments (bottom) in the left rod are shown as black symbols for the respective absolute thoracic angle. For comparison, the in vivo data from three patients with anterior fusion from the study by Rohlmann et al. [36] are visualized to the right of each of these. (**b**) For lumbar fusion (scenario 2), the contact forces between vertebral body L4 and the expanded cage (left), as well as the axial force components (center) and the bending moments (right) in the left rod are visualized.

In the modified MLS model, the ROM of the fused motion segment L4–L5 decreased to almost 0° in all load cases (Figure 10). While maintaining the total ROM of the thorax, the motion segments adjacent to the lumbar fusion showed the anticipated rotational compensation mechanisms [123]. This is consistent with previous in vivo measurements of spinal kinematics after lumbar fusion [124], in which the loss of intersegmental rotation was compensated by adjacent levels. Proportionally, sagittal intervertebral ROMs were initially compensated caudally and to a greater extent by the cranial segments at larger flexion angles. Overall, the total force in the erector spinae increased with the flexion angle, while extension slightly reduced this force. Except for the posterior muscle groups, the estimated muscle forces were slightly increased due to fusion (Figure 11a). At 30° flexion, these were reduced to 326 N for the erector spinae and to 80 N for the multifidus muscle. The estimated muscle forces were, therefore, lower than in previous in vitro [37] and FE studies [51,125]. Changes in IDPs were calculated to be at least as high or higher cranially from L4/5 (Figure 9). The pressure only decreased in the L5/S1 disc, the most in extension (−4%). This pattern of IDP changes showed a high correlation with the in silico study results of Kumaran et al. [69]. The contact force acting on the cage depended on the postures examined (Figures 7b and 8b). In the upright position, this was 703 N, which agreed with the results of an earlier FE study [126]. The lowest contact force occurred in extension and the highest in 30° flexion. In contrast, the absolute axial forces in the rods were inverted. Maximum axial compression was seen in 10° extension and minimal axial compression in 30° flexion. For the latter, the rod was almost unloaded at −14.8 N. In an upright position, F_Z was −119.5 N with a bending moment of −0.7 Nm. In extension and flexion, the bending moment increased to up to −1.13 Nm.

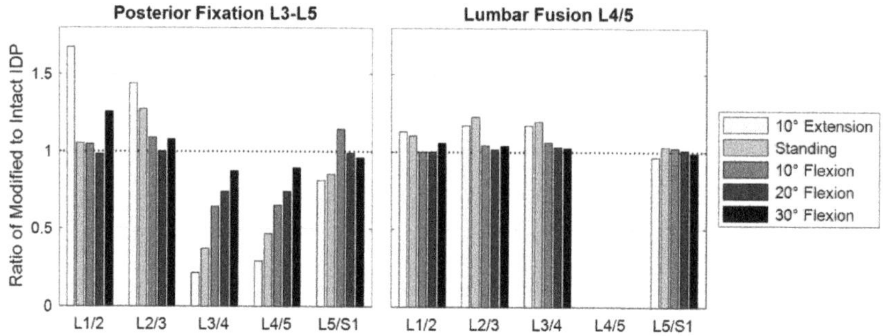

Figure 9. Summary of the predicted IDP changes for the examined postures, in each case as the ratio of instrumented spine to the results of the intact MLS model. The implanted cage replaced the nucleus of the L4/5 disc, which is why no pressure value is available. Refer to Figure 8b for contact forces acting on the cage.

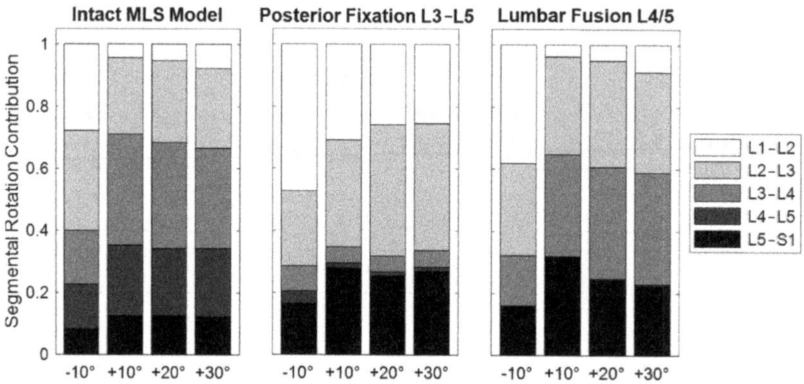

Figure 10. Summary of the segmental rotation contributions of the motion segments L1–L2 to L5–S1 in sagittal plane. The rotation contributions are given for the five postures 10° extension (−10°) to 30° flexion (+30°) for the intact MLS model and the two scenarios with spinal instrumentations. All values are given in relation to the respective upright posture (end of phase [iii]).

3.3. Scenario 3: Lumbar Orthosis

The abdominal FE soft tissue surrounding the spinal column is visualized without (Figure 5a,b) and with (Figure 5c) the orthosis applied. Using the same conditions and physical activities, we found that the soft tissue had an influence on the biomechanical responses of the MLS model. In the upright position, lumbar lordosis was reduced by 1°.

The absolute IDP changes were <7.8% (minimal for L4/5 with −2.1% and maximal for L3/4 with +7.8%). In 30° flexion, IDP was +15% at L1/2 and −16.3% at L4/5. The results were within the tolerance range of in vivo studies [127–129], but demonstrated the sensitivity of the muscle-driven model to the contact force coupling. We also measured changes in the predicted muscle activation patterns and the resulting muscle forces. The sum of all muscle forces was reduced, as follows: standing relaxed by −1.8%, standing with a 20 kg box by −13%, and in 30° flexion by −38%. While less force was primarily applied by the posterior muscles, the opposite was usually true for the abdominal muscles (Figure 11b). However, since the posterior individual muscle group forces were much greater in comparison, the sums were reduced, and the spinal stiffness was assumed to be increased.

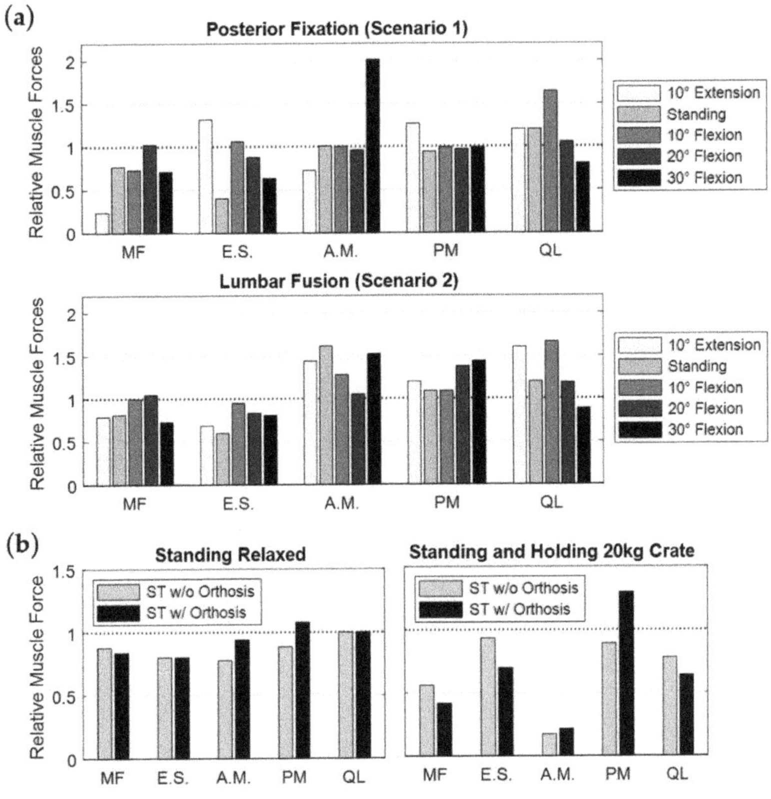

Figure 11. Predicted muscle forces given as the ratio of the modified to the intact MLS model in the same posture. Forces of iliocostalis thoracis, iliocostalis lumborum, and longissimus lumborum are combined into erector spinae (E.S.) and internus abdominis, obliquus externus abdominis, and rectus abdominis are combined into abdominal muscles (A.M.). Further individual muscle fibers were summed for multifidus (MF), psoas major (PM), and quadratus lumborum (QL). Scenarios are the spinal instrumentations (**a**) and the extension with soft tissues (ST) with (w/ orthosis) and without (w/o orthosis) lumbar orthosis (**b**).

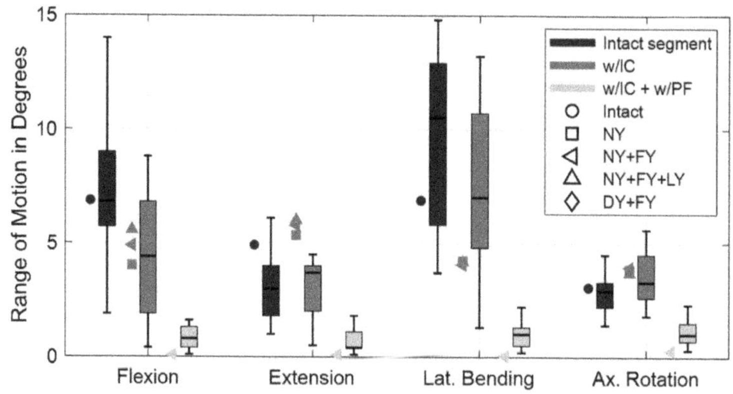

Figure 12. Validation of simulated ROMs of the L4/5 motion segment in intact condition, with interbody cage (w/IC), and with cage and posterior fixation (w/IC + w/PF). The absolute values for the four principal directions are compared with the in vitro data of Lund et al. [91] (shown as boxplots) for the same loading conditions (Table 2). Our simulation data are shown to the left of the corresponding boxplot, and the motion segment condition is color-coded. Further model variations, which are shown in Figure 2b, are illustrated by different marker symbols. No results are visualized for DY + FY without PF, because no stable state was reached, and for variations of the condition w/IC + w/PF (lightest grey), because ROM did not differ.

The calculated contact pressure on the skin and the associated soft tissue deformation after the orthosis was fully tightened while standing are shown in Figure 7c. Contact pressure varied with body region and load case. In all cases, the orthosis lifted off the skin in the lumbar region, resulting in zero pressure. This has already been observed in other studies [117,130]. Posterolateral contact pressure was maximal at up to 12 kPa. This resulted in a 23–33% reduction in IDPs and a 0.5° reduction in lumbar lordosis. The predicted FE–IAP increased from 1.1 mmHg to 30.5 mmHg when the orthosis was fully tightened, which slightly exceeded data from the literature [131,132]. The former was the pressure present without orthosis and the IAP remained almost constant at 10 mmHg. The estimated muscle force changes are visualized in Figure 11b. In both cases examined, the posterior muscle groups exerted less force to maintain an upright posture with and without a 20 kg box in both hands. The sum of muscle forces was, thus, reduced by 7.5% and 35% compared to the unmodified MLS model. For these two load cases, the muscle forces with the orthosis applied were reduced by 5.9% and 19.1% for the MLS model with soft tissue (compared to no orthosis). In vivo, trunk muscle activity was similarly little or less affected by elastic lumbar belts under similar conditions [133,134]. The maximum change in muscle force was 81 N for the erector spinae when holding the box. Thoracic rotation differences were negligible. Based on the reduced muscle forces, we assumed that this amount passively increased the stiffness of the trunk and, thus, the spinal stability, without being able to separate the influences of FE–IAP, the orthosis, and other FE soft tissue interactions. Under these simplifying assumptions, the results were in the range of an in vivo study by Cholewicki et al. [15]. In conclusion, all muscles combined contributed less to the active stabilization of the spine against external forces with the orthosis applied. However, the predicted changes in muscle excitation patterns were inconclusive except for the posterior muscle groups.

4. Discussion

Determining the internal dynamics of the human spine under realistic loading conditions is an essential step towards a better understanding of the treatment of low back disorders. Therefore, the purpose of our present study was to develop a forward dynamic hybrid FE–MB model capable of predicting the effects of the interplay between passive and active spinal structures, as well as medical devices. The modeling process involved adapting our previously validated active MLS model [74,75] to include different treatment scenarios. The back and abdominal muscles, facet joints, and all spinal ligaments were included. Intervertebral discs, implanted devices, abdominal soft tissue, and an orthosis were modeled using 3D finite elements with linear and non-linear material properties. The comparisons with available in vivo, in vitro, and in silico studies showed a high validity of the three scenarios in the context of the simplified passive OLS and the entire active MLS model.

Even if this model offers a new way to validly study clinical scenarios under enhanced loading conditions, some general limitations must be noted. Firstly, only one model was created and modified in three scenarios. The effects of subject variability could not be addressed. In order to reduce the time currently required for the largely manual creation of such a model, automated pipelines [135] using a deep learning approach [136] may be a way to generate larger cohorts of patient-specific models in the future. Individual segment masses, which are essential for MLS models, could also be more easily predicted from image data using artificial neural networks [137]. Due to the complexity of the model and the scarcity of data from the literature, we validated the model incrementally. This was under the assumption of component validation [138], assuming that models consisting of well-validated sub-models are likely to be valid. The OLS model [75] and the intact MLS model [74] based on it were extensively validated in advance. The investigations in this study were limited to sagittal tasks, the results were evaluated when the entire model reached a static state (equilibrium), and this research was not intended to address specific clinical objectives [16]. However, as a new simulation approach was realized to predict the interdependent effects of implants and orthoses on a muscle-driven musculoskeletal lumbosacral spine model, the three variants were sufficient for a proof of concept. Future studies can include different physical activities to investigate new or patient-specific implant and orthosis designs. Therefore, the proposed model can be used as a basis for a prediction tool for the internal dynamics of the spine in the virtual product design of medical devices. In addition, manufacturers could be supported in proving their compliance with regulatory requirements, necessary under the Medical Devices Regulation (MDR, EU-V 745/2017) before placing such devices on the market in the European Union. These involve demonstrating safety and efficacy, with the potential benefit to the patient increasing with the level of risk.

A key differentiator from comparable spine models is the inherent representation of muscles in the hybrid MLS model. To determine the muscle forces that produced coordinated spinal movements, an optimization routine was utilized. Using a simplifying optimization-based routine, however, may not realistically represent physiological muscle activation patterns lacking a physiological basis, leading to errors in predicted muscle forces [29,139], and internal loads. Validations of the directly correlated spinal loads are, therefore, even more relevant. As appropriate in vivo data are difficult or almost impossible to obtain and are often difficult to justify ethically, the available in vivo studies are limited. For different postures and situations, we compared the results of our model with in vivo data on lumbar IDPs [127–129] and internal loads in implanted posterior fixators [36]. There was good to very good agreement, which led us to conclude that the calculated muscle force patterns were valid predictions. It should be noted, however, that in previous studies [140], similar lumbosacral model responses were found for different muscle force predictions for the major spine extensors. As it is very difficult to measure the electrical activity of autochthonous muscles in vivo, there is a lack of data to test mathematical models in this respect [47]. Furthermore, the current optimization routine could not explicitly account for co-contractions of the abdominal and oblique antagonistic muscles [29], as used, for example, when lifting heavy loads [141], which resulted in a tendency to underestimate their activities.

The muscle-driven approach used enabled the forward dynamic MLS model to be moved and held in different body postures without prescribing complete spinal kinematics. However, as partial kinematic specifications were also part of the multi-criteria optimization, an influence is inevitable [142]. For this reason, we examined and minimized the influences of the rotational contributions of the vertebral target frames on the MLS model responses in a previous study [74]. In the first two scenarios of this study, we adopted the hypothesis proposed by Panjabi [123] that a fusion causes an intervertebral motion redistribution that enables the patient to achieve the same ROM as before the surgery. Therefore, we kept the total thoracic ROM and modified the expected rotational contributions of the vertebral tracking target frames based on preliminary examinations with the OLS model. The limitations of this approach were that it is uncertain how individual intervertebral rotation contributions break down and whether patients actually have the same lumbar ROM after fusion surgery. In the in vivo study by Rohlmann et al. [36], for example, patients were asked to move as much as possible without pain. Because the ROM varied, we calculated the internal loads on the fixators for extension and flexion angles from -10° up to $+30^\circ$ and compared the deviations (see Figure 8a). In addition, the pelvis was fixed so that the simulated flexion of the upper body could not represent a combination of hip and lumbar spine movements [125].

Further common limitations of muscle-driven and forward dynamics simulation approaches in the context of clinical biomechanics are a high computational complexity, the required model accuracy, and a *Sim2Real* gap. A predicted movement or posture is the result of a dynamic interplay of all modeled structural components. This requires a high degree of model accuracy concerning all load-bearing structures (*cf.* Figure 3) and the motor control unit, which predicts individual muscle activation patterns [65]. As a consequence, forward dynamics simulations do not generate residual forces and moments, but may continuously diverge from experimental data (e.g., kinematic tracking), which is referred to as *Sim2Real* gap [31]. The computational times were significantly increased with driving muscle excitations compared to the passive OLS model. Using a desktop PC with Intel Core i7-13700K @ 3.40 GHz, 32 GB Ram, and 1 TB SSD running Windows 10 Pro 64-bit, scenarios 1, 2, and 3 took a maximum of 48, 95, and 89 minutes, respectively. Overall, the unavailability of detailed internal loads in other methods emphasizes the importance of a forward dynamics approach as the most powerful approach for investigating changes in mechanical biological spinal structures [143].

As is usual with numerical models [34], this study used an idealized situation for spinal fixation and lumbar fusion. Factors that were disregarded included, for example, pathological bone elasticity, muscle injury, incomplete resection of the cartilage endplate, spinous process fracture, and poor blood supply. All intervertebral discs were modeled as non-degenerated, which is rarely the case for treated in vivo segments [144] and is often the reason for the surgical procedure [37,122]. As a result, both ROMs [145] and IDPs [146] may have been affected. Our internal fixators were simulated without preloading. In reality, this is hardly possible due to the fixation of the screws, results in deviations in the internal loads [89], and can have a strong effect on the mechanical behavior of the bridged region [125,126]. The bone–screw connections were assumed to be ideally bonded, with no relative movement or loosening. Even if this may represent a realistic scenario after complete osseointegration [147], a stiffer construct can be assumed for the primary stability condition tested here. For the posterior fixations and the cage, the available reference data could not be exactly replicated, which may be a factor contributing to differences in the biomechanical model responses. For the screws, this does not allow for a valid assessment of their failure or stress intensifications [147,148]. Our focus was on the rods and their internal loads, which were comparable to study data. The implanted cage was of a different type, narrower, and positioned more anteriorly compared to the cages in the reference study. However, in vitro studies showed that there were no significant differences between cage types in ROM with [149] and without [22,91] PF. Segment stiffness generally increases as the cage size increases, as shown in a numerical study [150]. A further simplification of our current modeling relates to cage insertion and the evaluation of cage subsidence. No bone grafts or fragments were placed in and around the cage, as is usually done [151], which contributes to better osseointegration and fusion [152]. Consequently, it can be assumed that the primary stability was reduced, since the cage contact area was smaller and the stress distribution was less homogeneous (*cf.* Figure 7b).

The interaction between the trunk and an orthosis is highly complex, with a variety of biological and technical factors influencing each other [24]. The modeling in the third scenario was a feasibility study for extending a detailed muscle-driven spine model to include abdominal soft tissue. Although the full validation of such a model is impossible [61], the validity of the underlying components has been confirmed in previous studies [74,75,85]. In contrast to the first two scenarios, no passive OLS model components could be compared with in vitro studies. To the best of our knowledge, there is no comprehensive study of the exact mechanism of action of lumbar orthoses [153] that would allow for the estimated biomechanical model responses to be compared under similar conditions. Only separate biomechanical responses of model components could be quantitatively compared with independent data from the literature, as follows: the IAP resulting from external radial compression [131], the pressure on the skin [60,117,118], the stiffness and deformation of the active abdominal tissue under local compression [85], and the IDP [116]. Although moderate to good correlations were found, this is the main limitation in assessing the validity of the third scenario.

Further limitations and potentials, the influences of which we consider to be particularly relevant for scenario 3, concern the diaphragm, abdominal wall, and IAP. For the diaphragm, its material properties, geometry, and function were simplified. These include orthotropic and muscle-activation-dependent non-linear material properties [82]. Furthermore, it was not taken into account that breathing and body posture influence the morphology of the diaphragm [154] and how this, together with pelvic floor contraction, contributes to an increase in IDP [155]. The abdominal wall also has a decisive influence on the IAP, the biomechanics of which depend on the variable IAP, the passive mechanical properties of fascial and muscle tissue, and the activation of the abdominal muscles [156]. The active abdominal wall is, thus, deformed both by breathing and muscle activity [157,158], as well as by the external radial compression of an orthosis [159]. One way of taking the biomechanics of the abdominal wall into account is to directly embed detailed, active fiber and aponeurosis structures into the abdominal hexahedral mesh [160]. Another option is to use a realistic active 3D FE model where the volumetric abdominal wall is modeled with separate muscle layers and muscle contractility [161]. This active compression of the abdominal content is likely to further increase the stabilizing effect of the trunk already observed in this study. In addition to the IAP, the spine compression force is expected to be increased by the co-contraction of the trunk muscles [158]. The multipoint transversal and oblique abdominal muscles of the MLS model could be replaced and the volumetric muscles could be activated by the Tracking Controller [106]. This has the advantage that the IAP acting on the thorax can only consist of the second component, as follows: (1) a surrogate force calculated from the sum of the abdominal muscle forces acting posteriorly on the abdominal plate (*cf.* Appendix A) and (2) the forces of the abdominal FE soft tissue acting directly on the diaphragm.

Any time-related effects on the abdominal FE soft tissue were neglected. We only examined the immediate effects after the orthosis was applied. The interaction between the soft tissue and the spine was realized using simplified contact conditions (geometric skinning of the spine) and nodal attachments. The results showed that trunk stiffness, quantified by a decrease in predicted muscle force, was increased for the same postures compared to the intact MLS model without soft tissue. This was inevitable with the current approach and must be taken into account when interpreting the results. An additional and expected increase in trunk stiffness occurred due to the external radial compression of the abdomen. The orthosis modeled corresponded to a tall and flexible belt that was tightened more than usual. In this form, it represented a simplified version of orthoses for conservative treatment and is not comparable to weightlifting belts. The latter are narrower, usually made of leather, and, therefore, almost inextensible [162]. Orthoses for treatment purposes are usually higher posteriorly than anteriorly, closed asymmetrically, and made of different fabrics with integrated reinforcements [60,117,153]. This allows for the targeted application of forces to the pelvis, chest, and abdomen, for example to decrease the lordosis angle [163,164] and to posteriorly shift the abdominal center of mass [24]. Studies have shown that the design of the orthosis has an influence on trunk stiffness [60,165], ROM [166], and muscle activity [167]. Consequently, our results were specific to the modeled orthosis in the applied position.

5. Conclusions

In this study, we introduced a muscle-driven model of the lumbosacral spine to examine the effects on the musculoskeletal system under realistic loading conditions during conservative and surgical treatment with medical devices. Three scenarios were realized in which the spine was stabilized with implants or an orthosis and different physical activities were simulated. The predicted loading conditions accounted for local and global musculature, body weight, intra-abdominal pressure, and forces resulting from the support of the abdominal soft tissue. The validation of the responses of biological structures and internal loads of implants showed a high degree of agreement with in vivo studies where available. In addition, the model enabled the analysis of structures adjacent to the treated segments for altered spinal loading and load sharing, which may allow hypotheses to be made about adjacent segment disease. Thus, this model represents a new approach to computational spine biomechanics by combining optimization-driven muscles, intervertebral discs, ligaments, facet joints, abdominal soft tissue, implants, and orthoses in one forward dynamic model. In the future, it can be a feasible tool for preliminary clinical evaluation and rapid assessment in the biomedical engineering design of medical devices for conservative and surgical treatment of the lumbar spine.

Author Contributions: Conceptualization, R.R.; methodology, R.R.; software, R.R.; validation, R.R., A.L. and M.R.; formal analysis, R.R.; investigation, R.R.; data curation, R.R.; writing—original draft preparation, R.R.; writing—review and editing, R.R., A.L., M.R., M.N. and B.B.; visualization, R.R.; supervision, B.B.; project administration, R.R. and M.N.; funding acquisition, M.N. All authors have read and agreed to the published version of the manuscript.

Funding: The authors R.R. and M.R. gratefully acknowledges the support of the German Federal Ministry for Economic Affairs and Climate Action through the research project KK5055218SK2 (Central Innovation Programme for small and medium-sized enterprises). This research received no external funding.

Data Availability Statement: The source code of the validated passive hybrid OLS model is openly available at https://github.com/RemusR9/artisynth_lumbosacralSpineModel, accessed on 30 January 2025. Additional data and source code supporting the conclusions of this article will be made available by the authors on request.

Acknowledgments: We acknowledge support by the DFG Open Access Publication Funds of the Ruhr-University Bochum.

Conflicts of Interest: The authors declare no conflicts of interest. The funders had no role in the design of the study; in the collection, analyses, or interpretation of data; in the writing of the manuscript; or in the decision to publish the results.

Abbreviations

The following abbreviations are used in this manuscript:

CY	Corpectomy
DY	Discectomy
IC	Interbody cage
IT	Intact spine
FE	Finite element
FY	Facetectomy
HU	Hounsfield unit
IDP	Intradiscal pressure
LY	Laminectomy
OLS	Osteoligamentous lumbosacral spine
MB	Multibody
MLS	Musculoskeletal lumbosacral spine
PF	Posterior fixation
PLIF	Posterior lumbar interbody fusion
ROM	Range of motion
TLIF	Transforaminal lumbar interbody fusion
VY	Vertebrectomy

Appendix A

The muscle-driven MLS model was created using ArtiSynth (www.artisynth.org, accessed on 30 January 2025), which allowed FE method and musculoskeletal MB dynamics to be combined in one model and solved directly. The modeling details can be described based on the three subsystems that work together in the concept of the spinal stabilization system [57,100]: (1) Passive spinal column, (2) muscles surrounding the spine, and (3) muscle coordination control unit.

(1) The passive OLS model [75] consisted of five rigid lumbar vertebrae, the rigid sacrum, five fiber-reinforced FE intervertebral discs, facet joints, and seven pre-tensioned ligaments. The discs were composed of a quasi-incompressible nucleus pulposus surrounded by the annulus fibrosus with crisscrossed collagen fiber rings. The collagen fibers were embedded in the hyperelastic ground substance and represented by tension-only springs with non-linear material properties. The facet joints were curved with an initial gap width of 0.5 mm and the inferior articular facets had a thin hyperelastic cartilage layer.

(2) The passive hybrid FE-MB OLS model was extended to include 12 muscle groups of the lower trunk, as well as four auxiliary bodies for muscle attachments and wrapping [168]. The muscle groups were modeled with 129 muscle fascicles for each lateral side of the body using a Hill-type muscle model [169] that comprehensively considered the respective lines of action. The IAP was considered as a force acting cranially on the thorax and calculated from the abdominal muscle contractions compressing the abdominal cavity. For

this purpose, the abdominal muscles were attached to a kinematically driven abdominal plate that replaced the ligamentous structures of the abdominal wall.

(3) To determine a pattern of muscle excitations that produced a coordinated movement and, thus, solved muscle and spinal redundancy, we used the Tracking Controller [77,78]. This approach made use of *forward dynamics assisted data tracking* [64,170] and enabled the optimization-based prediction of spinal movements and stable postures without prescribing accurate kinematic data. The mathematical model of the MLS fully described how the positions and velocities of model components changed due to applied forces and moments. In contrast to inverse dynamics [65,143,171], the 258 muscle forces that resulted from the musculotendon lengths, velocities, and activations (iteratively determined by the Tracking Controller) actuated or drove the dynamic MLS model and can be referred to as *muscle-driven forward dynamics simulation* [65,66,70,74,139,169,172,173]. As primary objective of the Tracking Controller with chase control as motion tracking setting, target frames were specified for all dynamic and kinematically independent RB skeletal components (thorax and vertebrae L2 to L5). When minimizing the tracking error between tracking target frames and RBs in each time step, 9 of the total 30 degrees of freedom were included. These were all rotations for the thorax and its translations, but not along the longitudinal axis, and for the vertebrae, solely their rotations in the sagittal plane. The nine degrees of freedom of the five target frames were driven by kinematic constraints, and segmental rotational contributions (spinal rhythms) were taken into account [174]. In this study, the excitation group of the lumbar multifidus, which included all fascicles, was divided into five independent excitation groups on each side of the body. This was based on the cranial muscle insertion points (L1 to L5) and allowed for segmental excitation. Details about MLS model calibration and tracking errors can be found in [74].

References

1. GBD 2021 Low Back Pain Collaborators. Global, regional, and national burden of low back pain, 1990–2020, its attributable risk factors, and projections to 2050: A systematic analysis of the Global Burden of Disease Study 2021. *Lancet Rheumatol.* **2023**, *5*, e316–e329. [CrossRef] [PubMed]
2. GBD 2019 Diseases and Injuries Collaborators. Global burden of 369 diseases and injuries in 204 countries and territories, 1990–2019: A systematic analysis for the Global Burden of Disease Study 2019. *Lancet* **2020**, *396*, 1204–1222. [CrossRef] [PubMed]
3. Andersson, G.B.J. Epidemiology of low back pain. *Acta Orthop. Scand.* **1998**, *69*, 28–31. [CrossRef]
4. Kliner, K.; Rennert, D.; Richter, M. (Eds.) *BKK Gesundheitsatlas 2019*; MWV Medizinisch Wisschenschaftliche Verlagsgesellschaft: Berlin, Germany, 2019; ISBN 978-3-95466-443-6.
5. Setchell, J.; Costa, N.; Ferreira, M.L.; Hodges, P.W. What decreases low back pain? A qualitative study of patient perspectives. *Scand. J. Pain* **2019**, *19*, 597–603. [CrossRef] [PubMed]
6. Matheve, T.; Hodges, P.W.; Danneels, L. The Role of Back Muscle Dysfunctions in Chronic Low Back Pain: State-of-the-Art and Clinical Implications. *J. Clin. Med.* **2023**, *12*, 5510. [CrossRef]
7. Duthey, B. *Background Paper 6.24 Low Back Pain*; WHO: Geneva, Switzerland, 2013.
8. Lee, S.-B.; Yoon, J.; Park, S.-J.; Chae, D.-S. Expandable Cages for Lumbar Interbody Fusion: A Narrative Review. *J. Clin. Med.* **2024**, *13*, 2889. [CrossRef] [PubMed]
9. Chou, R.; Deyo, R.; Friedly, J.; Skelly, A.; Hashimoto, R.; Weimer, M.; Fu, R.; Dana, T.; Kraegel, P.; Griffin, J.; et al. Nonpharmacologic Therapies for Low Back Pain: A Systematic Review for an American College of Physicians Clinical Practice Guideline. *Ann. Intern. Med.* **2017**, *166*, 493–505. [CrossRef]
10. Mayer, F.; Arampatzis, A.; Banzer, W.; Platen, P.; Schneider, C. (Eds.) *Rückenschmerz und Sport: Evidenzbasierte Prävention und Therapie aus dem Forschungsprojekt "RanRücken"*, 1st ed.; Springer: Berlin/Heidelberg, Germany, 2024; ISBN 978-3-662-63253-6.
11. La Barbera, L. Fixation and Fusion. In *Biomechanics of the Spine: Basic Concepts, Spinal Disorders and Treatments*, 1st ed.; Galbusera, F., Wilke, H.-J., Eds.; Academic Press: London, UK, 2018; Chapter 17, pp. 301–327, ISBN 0128128526.
12. Wilke, H.-J.; Wenger, K.; Claes, L.E. Testing criteria for spinal implants: Recommendations for the standardization of in vitro stability testing of spinal implants. *Eur. Spine J.* **1998**, *7*, 148–154. [CrossRef]
13. Phan, K.; Mobbs, R.J. Evolution of Design of Interbody Cages for Anterior Lumbar Interbody Fusion. *Orthop. Surg.* **2016**, *8*, 270–277. [CrossRef]

14. Jellema, P.; van Tulder, M.W.; van Poppel, M.N.M.; Nachemson, A.L.; Bouter, L.M. Lumbar supports for prevention and treatment of low back pain: A systematic review within the framework of the Cochrane Back Review Group. *Spine* **2001**, *26*, 377–386. [CrossRef]
15. Cholewicki, J.; Juluru, K.; Radebold, A.; Panjabi, M.M.; McGill, S.M. Lumbar spine stability can be augmented with an abdominal belt and/or increased intra-abdominal pressure. *Eur. Spine J. Off. Publ. Eur. Spine Soc. Eur. Spinal Deform. Soc. Eur. Sect. Cerv. Spine Res. Soc.* **1999**, *8*, 388–395. [CrossRef] [PubMed]
16. Hsu, J.D.; Michael, J.W.; Fisk, J.R. *AAOS Atlas of Orthoses and Assistive Devices*, 4th ed.; Mosby/Elsevier: Philadelphia, PA, USA, 2008; ISBN 0323076319.
17. Anderson, C.W.; Redford, J.B. Orthotic treatment for injuries and diseases of the spinal column. *Phys. Med. Rehabil. State Art Rev.* **2000**, *14*, 471–484.
18. Lei, W.; Yan, Y. (Eds.) *Internal Fixation of the Spine: Principles and Practice*; Springer: Berlin/Heidelberg, Germany, 2021; ISBN 978-981-16-1562-7.
19. Mobbs, R.J.; Phan, K.; Malham, G.; Seex, K.; Rao, P.J. Lumbar interbody fusion: Techniques, indications and comparison of interbody fusion options including PLIF, TLIF, MI-TLIF, OLIF/ATP, LLIF and ALIF. *J. Spine Surg.* **2015**, *1*, 2–18. [CrossRef]
20. Evans, J.H. Biomechanics of lumbar fusion. *Clin. Orthop. Relat. Res.* **1985**, 38–46. [CrossRef]
21. Campbell, R.C.; Mobbs, R.J.; Lu, V.M.; Xu, J.; Rao, P.J.; Phan, K. Posterolateral Fusion Versus Interbody Fusion for Degenerative Spondylolisthesis: Systematic Review and Meta-Analysis. *Glob. Spine J.* **2017**, *7*, 482–490. [CrossRef] [PubMed]
22. Oxland, T.R.; Lund, T. Biomechanics of stand-alone cages and cages in combination with posterior fixation: A literature review. *Eur. Spine J.* **2000**, *9* (Suppl. S1), S95–S101. [CrossRef]
23. Schwab, F.; Patel, A.; Ungar, B.; Farcy, J.-P.; Lafage, V. Adult spinal deformity-postoperative standing imbalance: How much can you tolerate? An overview of key parameters in assessing alignment and planning corrective surgery. *Spine* **2010**, *35*, 2224–2231. [CrossRef] [PubMed]
24. White, A.A.; Panjabi, M.M. *Clinical Biomechanics of the Spine*, 2nd ed.; Lippincott: Philadelphia, PA, USA, 1990; ISBN 0397507208.
25. Bender, B.; Gericke, K. *Pahl/Beitz Konstruktionslehre*; Springer: Berlin/Heidelberg, Germany, 2021; ISBN 978-3-662-57302-0.
26. La Barbera, L.; Galbusera, F.; Wilke, H.-J.; Villa, T.M.T. Preclinical evaluation of posterior spine stabilization devices: Can the current standards represent basic everyday life activities? *Eur. Spine J.* **2016**, *25*, 2909–2918. [CrossRef]
27. Augat, P.; Hast, M.W.; Schemitsch, G.; Heyland, M.; Trepczynski, A.; Borgiani, E.; Russow, G.; Märdian, S.; Duda, G.N.; Hollensteiner, M.; et al. Biomechanical models: Key considerations in study design. *OTA Int.* **2021**, *4*, e099. [CrossRef]
28. Galbusera, F.; Wilke, H.-J. (Eds.) *Biomechanics of the Spine: Basic Concepts, Spinal Disorders and Treatments*, 1st ed.; Academic Press: London, UK, 2018; ISBN 0128128526.
29. Knapik, G.G.; Mendel, E.; Bourekas, E.; Marras, W.S. Computational lumbar spine models: A literature review. *Clin. Biomech.* **2022**, *100*, 105816. [CrossRef]
30. Wilke, H.-J.; Heuer, F.; Schmidt, H. Prospective design delineation and subsequent in vitro evaluation of a new posterior dynamic stabilization system. *Spine* **2009**, *34*, 255–261. [CrossRef]
31. Wechsler, I.; Wolf, A.; Shanbhag, J.; Leyendecker, S.; Eskofier, B.M.; Koelewijn, A.D.; Wartzack, S.; Miehling, J. Bridging the sim2real gap. Investigating deviations between experimental motion measurements and musculoskeletal simulation results—A systematic review. *Front. Bioeng. Biotechnol.* **2024**, *12*, 1386874. [CrossRef] [PubMed]
32. Viceconti, M. Biomechanics-based in silico medicine: The manifesto of a new science. *J. Biomech.* **2015**, *48*, 193–194. [CrossRef]
33. Nispel, K.; Lerchl, T.; Senner, V.; Kirschke, J.S. Recent Advances in Coupled MBS and FEM Models of the Spine—A Review. *Bioengineering* **2023**, *10*, 315. [CrossRef] [PubMed]
34. Wang, R.; Wu, Z. Recent advancement in finite element analysis of spinal interbody cages: A review. *Front. Bioeng. Biotechnol.* **2023**, *11*, 1041973. [CrossRef]
35. Loenen, A.C.Y.; Noailly, J.; Ito, K.; Willems, P.C.; Arts, J.J.; van Rietbergen, B. Patient-Specific Variations in Local Strain Patterns on the Surface of a Trussed Titanium Interbody Cage. *Front. Bioeng. Biotechnol.* **2021**, *9*, 750246. [CrossRef] [PubMed]
36. Rohlmann, A.; Bergmann, G.; Graichen, F.; Weber, U. Comparison of loads on internal spinal fixation devices measured in vitro and in vivo. *Med. Eng. Phys.* **1997**, *19*, 539–546. [CrossRef]
37. Wilke, H.-J.; Rohlmann, A.; Neller, S.; Graichen, F.; Claes, L.E.; Bergmann, G. ISSLS prize winner: A novel approach to determine trunk muscle forces during flexion and extension: A comparison of data from an in vitro experiment and in vivo measurements. *Spine* **2003**, *28*, 2585–2593. [CrossRef]
38. Rohlmann, A.; Zander, T.; Rao, M.; Bergmann, G. Realistic loading conditions for upper body bending. *J. Biomech.* **2009**, *42*, 884–890. [CrossRef]
39. Abbasi-Ghiri, A.; Ebrahimkhani, M.; Arjmand, N. Novel force–displacement control passive finite element models of the spine to simulate intact and pathological conditions; comparisons with traditional passive and detailed musculoskeletal models. *J. Biomech.* **2022**, *141*, 111173. [CrossRef]

40. Fang, G.; Lin, Y.; Wu, J.; Cui, W.; Zhang, S.; Guo, L.; Sang, H.; Huang, W. Biomechanical Comparison of Stand-Alone and Bilateral Pedicle Screw Fixation for Oblique Lumbar Interbody Fusion Surgery-A Finite Element Analysis. *World Neurosurg.* **2020**, *141*, e204–e212. [CrossRef]
41. Cegoñino, J.; Calvo-Echenique, A.; Pérez-del Palomar, A. Influence of different fusion techniques in lumbar spine over the adjacent segments: A 3D finite element study. *J. Orthop. Res.* **2015**, *33*, 993–1000. [CrossRef] [PubMed]
42. Zhang, H.; Zhu, W. The path to deliver the most realistic follower load for a lumbar spine in standing posture: A finite element study. *J. Biomech. Eng.* **2019**, *141*, 031010. [CrossRef] [PubMed]
43. Rajaee, M.A.; Arjmand, N.; Shirazi-Adl, A. A novel coupled musculoskeletal finite element model of the spine—Critical evaluation of trunk models in some tasks. *J. Biomech.* **2021**, *119*, 110331. [CrossRef]
44. Ebrahimkhani, M.; Arjmand, N.; Shirazi-Adl, A. Adjacent segments biomechanics following lumbar fusion surgery: A musculoskeletal finite element model study. *Eur. Spine J.* **2022**, *31*, 1630–1639. [CrossRef]
45. Ledet, E.H.; Liddle, B.; Kradinova, K.; Harper, S. Smart implants in orthopedic surgery, improving patient outcomes: A review. *Innov. Entrep. Health* **2018**, *5*, 41–51. [CrossRef]
46. Rohlmann, A.; Bergmann, G.; Graichen, F.; Mayer, H.M. Influence of muscle forces on loads in internal spinal fixation devices. *Spine* **1998**, *23*, 537–542. [CrossRef] [PubMed]
47. Schmidt, H.; Reitmaier, S. Modelle zur Vorhersage von Kräften und der Belastung der Wirbelsäule. In *Rückenschmerz und Sport: Evidenzbasierte Prävention und Therapie aus dem Forschungsprojekt "RanRücken"*, 1st ed.; Mayer, F., Arampatzis, A., Banzer, W., Platen, P., Schneider, C., Eds.; Springer: Berlin/Heidelberg, Germany, 2024; pp. 121–139, ISBN 978-3-662-63253-6.
48. Azari, F.; Arjmand, N.; Shirazi-Adl, A.; Rahimi-Moghaddam, T. A combined passive and active musculoskeletal model study to estimate L4-L5 load sharing. *J. Biomech.* **2018**, *70*, 157–165. [CrossRef]
49. Favier, C.D.; McGregor, A.H.; Phillips, A.T.M. Maintaining Bone Health in the Lumbar Spine: Routine Activities Alone Are Not Enough. *Front. Bioeng. Biotechnol.* **2021**, *9*, 661837. [CrossRef]
50. Bernier, E.; Driscoll, M. Numerical investigation of intra-abdominal pressure and spinal load-sharing upon the application of an abdominal belt. *J. Biomech.* **2023**, *161*, 111863. [CrossRef]
51. Zander, T.; Rohlmann, A.; Calisse, J.; Bergmann, G. Estimation of muscle forces in the lumbar spine during upper-body inclination. *Clin. Biomech.* **2001**, *16* (Suppl. S1), S73–S80. [CrossRef]
52. Khoddam-Khorasani, P.; Arjmand, N.; Shirazi-Adl, A. Trunk Hybrid Passive-Active Musculoskeletal Modeling to Determine the Detailed T12-S1 Response Under In Vivo Loads. *Ann. Biomed. Eng.* **2018**, *46*, 1830–1843. [CrossRef]
53. Rohlmann, A.; Bergmann, G.; Graichen, F. Loads on an internal spinal fixation device during walking. *J. Biomech.* **1997**, *30*, 41–47. [CrossRef] [PubMed]
54. Panico, M.; Bassani, T.; Villa, T.M.T.; Galbusera, F. The Simulation of Muscles Forces Increases the Stresses in Lumbar Fixation Implants with Respect to Pure Moment Loading. *Front. Bioeng. Biotechnol.* **2021**, *9*, 745703. [CrossRef] [PubMed]
55. Wilke, H.-J.; Wolf, S.; Claes, L.E.; Arand, M.; Wiesend, A. Stability increase of the lumbar spine with different muscle groups. A biomechanical in vitro study. *Spine* **1995**, *20*, 192–198. [CrossRef]
56. Ghezelbash, F.; Shahvarpour, A.; Larivière, C.; Shirazi-Adl, A. Evaluating stability of human spine in static tasks: A combined in vivo-computational study. *Comput. Methods Biomech. Biomed. Eng.* **2022**, *25*, 1156–1168. [CrossRef]
57. Panjabi, M.M. The stabilizing system of the spine. Part I. Function, dysfunction, adaptation, and enhancement. *J. Spinal Disord.* **1992**, *5*, 383–389; discussion 397. [CrossRef]
58. Oxland, T.R. Fundamental biomechanics of the spine--What we have learned in the past 25 years and future directions. *J. Biomech.* **2016**, *49*, 817–832. [CrossRef] [PubMed]
59. Gardner-Morse, M.G.; Stokes, I.A.; Laible, J.P. Role of muscles in lumbar spine stability in maximum extension efforts. *J. Orthop. Res.* **1995**, *13*, 802–808. [CrossRef]
60. Cholewicki, J.; Lee, A.S.; Peter Reeves, N.; Morrisette, D.C. Comparison of trunk stiffness provided by different design characteristics of lumbosacral orthoses. *Clin. Biomech.* **2010**, *25*, 110–114. [CrossRef]
61. Cholewicki, J. The effects of lumbosacral orthoses on spine stability: What changes in EMG can be expected? *J. Orthop. Res.* **2004**, *22*, 1150–1155. [CrossRef] [PubMed]
62. Holzinger, T.; Cazzola, D.; Sagl, B. Development, calibration and validation of impact-specific cervical spine models: A novel approach using hybrid multibody and finite-element methods. *Comput. Methods Programs Biomed.* **2024**, 108430. [CrossRef]
63. Stavness, I.; Lloyd, J.E.; Payan, Y.; Fels, S. Coupled hard-soft tissue simulation with contact and constraints applied to jaw-tongue-hyoid dynamics. *Int. J. Numer. Meth. Biomed. Eng.* **2011**, *27*, 367–390. [CrossRef]
64. Erdemir, A.; McLean, S.; Herzog, W.; van den Bogert, A.J. Model-based estimation of muscle forces exerted during movements. *Clin. Biomech.* **2007**, *22*, 131–154. [CrossRef] [PubMed]
65. Meszaros-Beller, L.; Hammer, M.; Schmitt, S.; Pivonka, P. Effect of neglecting passive spinal structures: A quantitative investigation using the forward-dynamics and inverse-dynamics musculoskeletal approach. *Front. Physiol.* **2023**, *14*, 1135531. [CrossRef]

66. van den Bogert, A.J.; Blana, D.; Heinrich, D. Implicit methods for efficient musculoskeletal simulation and optimal control. *Procedia IUTAM* **2011**, *2*, 297–316. [CrossRef]
67. Liu, T.; Khalaf, K.; Naserkhaki, S.; El-Rich, M. Load-sharing in the lumbosacral spine in neutral standing & flexed postures—A combined finite element and inverse static study. *J. Biomech.* **2018**, *70*, 43–50. [CrossRef]
68. Honegger, J.D.; Actis, J.A.; Gates, D.H.; Silverman, A.K.; Munson, A.H.; Petrella, A.J. Development of a multiscale model of the human lumbar spine for investigation of tissue loads in people with and without a transtibial amputation during sit-to-stand. *Biomech. Model. Mechanobiol.* **2021**, *20*, 339–358. [CrossRef]
69. Kumaran, Y.; Shah, A.; Katragadda, A.; Padgaonkar, A.; Zavatsky, J.; McGuire, R.; Serhan, H.; Elgafy, H.; Goel, V.K. Iatrogenic muscle damage in transforaminal lumbar interbody fusion and adjacent segment degeneration: A comparative finite element analysis of open and minimally invasive surgeries. *Eur. Spine J.* **2021**, *30*, 2622–2630. [CrossRef]
70. Delp, S.L.; Anderson, F.C.; Arnold, A.S.; Loan, P.; Habib, A.; John, C.T.; Guendelman, E.; Thelen, D.G. OpenSim: Open-source software to create and analyze dynamic simulations of movement. *IEEE Trans. Biomed. Eng.* **2007**, *54*, 1940–1950. [CrossRef]
71. Ignasiak, D.; Turner, A.W. Corrigendum to 'A novel method for prediction of postoperative global sagittal alignment based on full-body musculoskeletal modeling and posture optimization' [J. Biomech. (2019) https://doi.org/10.1016/j.jbiomech.2019.109324]. *J. Biomech.* **2020**, *102*, 109672. [CrossRef] [PubMed]
72. Wolf, A.; Miehling, J.; Wartzack, S. Challenges in interaction modelling with digital human models—A systematic literature review of interaction modelling approaches. *Ergonomics* **2020**, *63*, 1442–1458. [CrossRef]
73. Neumann, M.; Bender, B. Challenges in the Development of Biomechatronic Systems. In *Design Methodology for Future Products*; Springer: Cham, Switzerland, 2022; pp. 105–122.
74. Remus, R.; Selkmann, S.; Lipphaus, A.; Neumann, M.; Bender, B. Muscle-driven forward dynamic active hybrid model of the lumbosacral spine: Combined FEM and multibody simulation. *Front. Bioeng. Biotechnol.* **2023**, *11*, 1223007. [CrossRef]
75. Remus, R.; Lipphaus, A.; Neumann, M.; Bender, B. Calibration and validation of a novel hybrid model of the lumbosacral spine in ArtiSynth-The passive structures. *PLoS ONE* **2021**, *16*, e0250456. [CrossRef] [PubMed]
76. Lloyd, J.E.; Stavness, I.; Fels, S. ArtiSynth: A fast interactive biomechanical modeling toolkit combining multibody and finite element simulation. In *Soft Tissue Biomechanical Modeling for Computer Assisted Surgery*; Payan, Y., Ed.; Springer: Berlin/Heidelberg, Germany, 2012; pp. 355–394, ISBN 978-3-642-29014-5.
77. Stavness, I.; Hannam, A.G.; Lloyd, J.E.; Fels, S. Predicting muscle patterns for hemimandibulectomy models. *Comput. Methods Biomech. Biomed. Eng.* **2010**, *13*, 483–491. [CrossRef] [PubMed]
78. Stavness, I.; Lloyd, J.E.; Fels, S. Automatic prediction of tongue muscle activations using a finite element model. *J. Biomech.* **2012**, *45*. [CrossRef]
79. Been, E.; Kalichman, L. Lumbar lordosis. *Spine J.* **2014**, *14*, 87–97. [CrossRef]
80. Galbusera, F.; Bellini, C.M.; Anasetti, F.; Ciavarro, C.; Lovi, A.; Brayda-Bruno, M. Rigid and flexible spinal stabilization devices: A biomechanical comparison. *Med. Eng. Phys.* **2011**, *33*, 490–496. [CrossRef]
81. Turbucz, M.; Pokorni, A.J.; Szőke, G.; Hoffer, Z.; Kiss, R.M.; Lazary, A.; Eltes, P.E. Development and Validation of Two Intact Lumbar Spine Finite Element Models for In Silico Investigations: Comparison of the Bone Modelling Approaches. *Appl. Sci.* **2022**, *12*, 10256. [CrossRef]
82. Behr, M.; Thollon, L.; Arnoux, P.-J.; Serre, T.; Berdah, S.V.; Baque, P.; Brunet, C. 3D reconstruction of the diaphragm for virtual traumatology. *Surg. Radiol. Anat.* **2006**, *28*, 235–240. [CrossRef]
83. Pato, M.P.M.; Santos, N.J.G.; Areias, P.; Pires, E.B.; de Carvalho, M.; Pinto, S.; Lopes, D.S. Finite element studies of the mechanical behaviour of the diaphragm in normal and pathological cases. *Comput. Methods Biomech. Biomed. Eng.* **2011**, *14*, 505–513. [CrossRef] [PubMed]
84. Villa, T.M.T.; La Barbera, L.; Galbusera, F. Comparative analysis of international standards for the fatigue testing of posterior spinal fixation systems. *Spine J.* **2014**, *14*, 695–704. [CrossRef] [PubMed]
85. Remus, R.; Sure, C.; Selkmann, S.; Uttich, E.; Bender, B. Soft tissue material properties based on human abdominal in vivo macro-indenter measurements. *Front. Bioeng. Biotechnol.* **2024**, *12*, 1384062. [CrossRef]
86. Renner, S.M.; Natarajan, R.N.; Patwardhan, A.G.; Havey, R.M.; Voronov, L.I.; Guo, B.Y.; Andersson, G.B.J.; An, H.S. Novel model to analyze the effect of a large compressive follower pre-load on range of motions in a lumbar spine. *J. Biomech.* **2007**, *40*, 1326–1332. [CrossRef] [PubMed]
87. Dreischarf, M.; Zander, T.; Bergmann, G.; Rohlmann, A. A non-optimized follower load path may cause considerable intervertebral rotations. *J. Biomech.* **2010**, *43*, 2625–2628. [CrossRef]
88. Volkheimer, D.; Malakoutian, M.; Oxland, T.R.; Wilke, H.-J. Limitations of current in vitro test protocols for investigation of instrumented adjacent segment biomechanics: Critical analysis of the literature. *Eur. Spine J.* **2015**, *24*, 1882–1892. [CrossRef]
89. Wilke, H.-J.; Rohlmann, A.; Neller, S.; Schultheiss, M.; Bergmann, G.; Graichen, F.; Claes, L.E. Is it possible to simulate physiologic loading conditions by applying pure moments? A comparison of in vivo and in vitro load components in an internal fixator. *Spine* **2001**, *26*, 636–642. [CrossRef]

90. Rohlmann, A.; Riley, L.H.; Bergmann, G.; Graichen, F. In vitro load measurement using an instrumented spinal fixation device. *Med. Eng. Phys.* **1996**, *18*, 485–488. [CrossRef]
91. Lund, T.; Oxland, T.R.; Jost, B.; Cripton, P.; Grassmann, S.; Etter, C.; Nolte, L.P. Interbody cage stabilisation in the lumbar spine: Biomechanical evaluation of cage design, posterior instrumentation and bone density. *J. Bone Jt. Surgery. Br. Vol.* **1998**, *80*, 351–359. [CrossRef]
92. Dick, W. *Internal Fixation of Thoracic and Lumbar Spine Fractures*; H. Huber: Toronto, ON, Canada; Lewiston, NY, USA, 1989; ISBN 978-0920887301.
93. Rohlmann, A.; Bergmann, G.; Graichen, F. Loads on internal spinal fixators measured in different body positions. *Eur. Spine J.* **1999**, *8*, 354–359. [CrossRef]
94. Rohlmann, A.; Bergmann, G.; Graichen, F. A spinal fixation device for in vivo load measurement. *J. Biomech.* **1994**, *27*, 961–967. [CrossRef]
95. Rohlmann, A.; Eick, O.; Bergmann, G.; Graichen, F. In vitro-Messung der Belastung mit instrumentierten Wirbel-Fixateur interne. *Z. Orthop. Ihre Grenzgeb.* **1995**, *133*, 136–140. [CrossRef]
96. Spitzer, V.; Ackerman, M.J.; Scherzinger, A.L.; Whitlock, D. The visible human male: A technical report. *J. Am. Med. Inform. Assoc.* **1996**, *3*, 118–130. [CrossRef] [PubMed]
97. Taddei, F.; Schileo, E.; Helgason, B.; Cristofolini, L.; Viceconti, M. The material mapping strategy influences the accuracy of CT-based finite element models of bones: An evaluation against experimental measurements. *Med. Eng. Phys.* **2007**, *29*, 973–979. [CrossRef] [PubMed]
98. Schileo, E.; Balistreri, L.; Grassi, L.; Cristofolini, L.; Taddei, F. To what extent can linear finite element models of human femora predict failure under stance and fall loading configurations? *J. Biomech.* **2014**, *47*, 3531–3538. [CrossRef] [PubMed]
99. Ashjaee, N.; Kalajahi, S.M.H.; Johnston, J.D. QCT-FE modeling of the proximal tibia: Effect of mapping strategy on convergence time and model accuracy. *Med. Eng. Phys.* **2021**, *88*, 41–46. [CrossRef]
100. Panjabi, M.M. Clinical spinal instability and low back pain. *J. Electromyogr. Kinesiol.* **2003**, *13*, 371–379. [CrossRef]
101. Shirazi-Adl, A.; Dammak, M.; Paiement, G. Experimental determination of friction characteristics at the trabecular bone/porous-coated metal interface in cementless implants. *J. Biomed. Mater. Res.* **1993**, *27*, 167–175. [CrossRef]
102. Galbusera, F.; Schmidt, H.; Wilke, H.-J. Lumbar interbody fusion: A parametric investigation of a novel cage design with and without posterior instrumentation. *Eur. Spine J.* **2012**, *21*, 455–462. [CrossRef]
103. Fels, S.; Vogt, F.; van den Doel, K.; Lloyd, J.E.; Stavness, I.; Vatikiotis-Bateson, E. ArtiSynth: A Biomechanical Simulation Platform for the Vocal Tract and Upper Airway. Technical Report. Available online: https://www.cs.ubc.ca/sites/default/files/tr/2006/TR-2006-10_0.pdf (accessed on 3 March 2025).
104. Kimura, S.; Steinbach, G.C.; Watenpaugh, D.E.; Hargens, A.R. Lumbar spine disc height and curvature responses to an axial load generated by a compression device compatible with magnetic resonance imaging. *Spine* **2001**, *26*, 2596–2600. [CrossRef]
105. Stavness, I.; Sánchez, C.A.; Lloyd, J.E.; Ho, A.; Wang, J.; Fels, S.; Huang, D. Unified skinning of rigid and deformable models for anatomical simulations. In Proceedings of the SIGGRAPH Asia 2014 Technical Briefs, Shenzhen, China, 3–6 December 2014; ACM: New York, NY, USA, 2014; pp. 1–4, ISBN 9781450328951.
106. Lloyd, J.E.; Sánchez, C.A.; Widing, E.; Stavness, I.; Fels, S.; Niroomandi, S.; Perrier, A.; Payan, Y.; Perrier, P. New Techniques for Combined FEM-Multibody Anatomical Simulation. In *New Developments on Computational Methods and Imaging in Biomechanics and Biomedical Engineering*; Springer: Cham, Switzerland, 2019; pp. 75–92.
107. Nesme, M.; Faure, F.; Payan, Y. Accurate interactive animation of deformable models at arbitrary resolution. *Int. J. Image Graph.* **2010**, *10*, 175–202. [CrossRef]
108. Snedeker, J.G.; Barnstuble, B.B.; Iaizzo, P.A.; Farshad, M.; Niederer, P.; Schmidlin, F.R. A comprehensive renal injury concept based on a validated finite element model of the human abdomen. *J. Trauma* **2007**, *62*, 1240–1249. [CrossRef]
109. Todros, S.; de Cesare, N.; Concheri, G.; Natali, A.N.; Pavan, P.G. Numerical modelling of abdominal wall mechanics: The role of muscular contraction and intra-abdominal pressure. *J. Mech. Behav. Biomed. Mater.* **2020**, *103*, 103578. [CrossRef]
110. Simo, J.C.; Taylor, R.L. Quasi-incompressible finite elasticity in principal stretches. Continuum basis and numerical algorithms. *Appl. Mech. Eng.* **1991**, *85*, 273–310. [CrossRef]
111. Lohr, M.J.; Sugerman, G.P.; Kakaletsis, S.; Lejeune, E.; Rausch, M.K. An introduction to the Ogden model in biomechanics: Benefits, implementation tools and limitations. *Philos. Trans. A Math. Phys. Eng. Sci.* **2022**, *380*, 20210365. [CrossRef]
112. Lalwani, L.K.; Govindagoudar, M.B.; Singh, P.K.; Sharma, M.; Chaudhry, D. The role of diaphragmatic thickness measurement in weaning prediction and its comparison with rapid shallow breathing index: A single-center experience. *Acute Crit. Care* **2022**, *37*, 347–354. [CrossRef]
113. Netter, F.H. *Atlas of Human Anatomy*, 7th ed.; Elsevier: Philadelphia, PA, USA, 2017; ISBN 9780323547048.
114. Petrofsky, J.S.; McLellan, K.; Prowse, M.; Bains, G.; Berk, L.; Lee, S. The effect of body fat, aging, and diabetes on vertical and shear pressure in and under a waist belt and its effect on skin blood flow. *Diabetes Technol. Ther.* **2010**, *12*, 153–160. [CrossRef]

115. Périé, D.; Aubin, C.E.; Lacroix, M.; Lafon, Y.; Labelle, H. Biomechanical modelling of orthotic treatment of the scoliotic spine including a detailed representation of the brace-torso interface. *Med. Biol. Eng. Comput.* **2004**, *42*, 339–344. [CrossRef]
116. Nachemson, A.L.; Morris, J.M. In Vivo Measurements of Intradiscal Pressure: Discometry, a method for the determination of pressure in the lower lumbar disc. *J. Bone Jt. Surg. Am.* **1964**, *46*, 1077–1092. [CrossRef]
117. Molimard, J.; Bonnaire, R.; Han, W.S.; Convert, R.; Calmels, P. In-silico pre-clinical trials are made possible by a new simple and comprehensive lumbar belt mechanical model based on the Law of Laplace including support deformation and adhesion effects. *PLoS ONE* **2019**, *14*, e0212681. [CrossRef] [PubMed]
118. Larivière, C.; Shahvarpour, A.; Gravel, C.; Gauvin, M.; Jean, A.-M.; Viau, A.; Mecheri, H. Revisiting the effect of manipulating lumbar stability with load magnitudes and positions: The effect of sex on trunk muscle activation. *J. Electromyogr. Kinesiol.* **2019**, *46*, 55–62. [CrossRef] [PubMed]
119. Heuer, F.; Schmidt, H.; Klezl, Z.; Claes, L.E.; Wilke, H.-J. Stepwise reduction of functional spinal structures increase range of motion and change lordosis angle. *J. Biomech.* **2007**, *40*, 271–280. [CrossRef] [PubMed]
120. Fantigrossi, A.; Galbusera, F.; Raimondi, M.T.; Sassi, M.; Fornari, M. Biomechanical analysis of cages for posterior lumbar interbody fusion. *Med. Eng. Phys.* **2007**, *29*, 101–109. [CrossRef]
121. Calvo-Echenique, A.; Cegoñino, J.; Del Perez Palomar, A. Is there any advantage of using stand-alone cages? A numerical approach. *Biomed. Eng. Online* **2019**, *18*, 63. [CrossRef]
122. Nibu, K.; Panjabi, M.M.; Oxland, T.R.; Cholewicki, J. Intervertebral disc distraction with a laparoscopic anterior spinal fusion system. *Eur. Spine J.* **1998**, *7*, 142–147. [CrossRef]
123. Panjabi, M.M. Hybrid multidirectional test method to evaluate spinal adjacent-level effects. *Clin. Biomech.* **2007**, *22*, 257–265. [CrossRef]
124. Lin, S.-C.; Tsai, W.-C.; Wu, S.-S.; Chen, P.-Q. Radiological and Mathematical Studies Regarding the Effects of Spinal Fixation on Kinematics and Mechanics at the Parafixed Segments. *J. Mech.* **2010**, *26*, 413–422. [CrossRef]
125. Rohlmann, A.; Bauer, L.; Zander, T.; Bergmann, G.; Wilke, H.-J. Determination of trunk muscle forces for flexion and extension by using a validated finite element model of the lumbar spine and measured in vivo data. *J. Biomech.* **2006**, *39*, 981–989. [CrossRef]
126. Zander, T.; Rohlmann, A.; Klöckner, C.; Bergmann, G. Comparison of the mechanical behavior of the lumbar spine following mono- and bisegmental stabilization. *Clin. Biomech.* **2002**, *17*, 439–445. [CrossRef]
127. Sato, K.; Kikuchi, S.; Yonezawa, T. In vivo intradiscal pressure measurement in healthy individuals and in patients with ongoing back problems. *Spine* **1999**, *24*, 2468–2474. [CrossRef]
128. Wilke, H.-J.; Neef, P.; Hinz, B.; Seidel, H.; Claes, L.E. Intradiscal pressure together with anthropometric data—A data set for the validation of models. *Clin. Biomech.* **2001**, *16*, 111–126. [CrossRef]
129. Takahashi, I.; Kikuchi, S.; Sato, K.; Sato, N. Mechanical load of the lumbar spine during forward bending motion of the trunk-a biomechanical study. *Spine* **2006**, *31*, 18–23. [CrossRef] [PubMed]
130. Bonnaire, R. Caractérisation Mécanique des Orthèses: Application aux Ceintures de Soutien Lombaire Dans le Cadre de la Lombalgie. Doctoral Dissertation, Ecole des Mines de Saint-Étienne, Saint-Étienne, France, 2015.
131. Morris, J.M.; Lucas, D.; Bresler, B. Role of the Trunk in Stability of the Spine. *J. Bone Jt. Surg.* **1961**, *43*, 327–351. [CrossRef]
132. El Bojairami, I.; Jacobson, N.; Driscoll, M. Development and evaluation of a numerical spine model comprising intra-abdominal pressure for use in assessing physiological changes on abdominal compliance and spinal stability. *Clin. Biomech.* **2022**, *97*, 105689. [CrossRef] [PubMed]
133. Jorgensen, M.J.; Marras, W.S. The effect of lumbar back support tension on trunk muscle activity. *Clin. Biomech.* **2000**, *15*, 292–294. [CrossRef]
134. van Poppel, M.N.M.; de Looze, M.P.; Koes, B.W.; Smid, T.; Bouter, L.M. Mechanisms of action of lumbar supports: A systematic review. *Spine* **2000**, *25*, 2103–2113. [CrossRef]
135. Nispel, K.; Lerchl, T.; Gruber, G.; Möller, H.; Graf, R.; Senner, V.; Kirschke, J.S. From MRI to FEM: An automated pipeline for biomechanical simulations of vertebrae and intervertebral discs. *Front. Bioeng. Biotechnol.* **2025**, *12*, 1485115. [CrossRef]
136. Möller, H.; Graf, R.; Schmitt, J.; Keinert, B.; Schön, H.; Atad, M.; Sekuboyina, A.; Streckenbach, F.; Kofler, F.; Kroencke, T.; et al. SPINEPS-automatic whole spine segmentation of T2-weighted MR images using a two-phase approach to multi-class semantic and instance segmentation. *Eur. Radiol.* **2024**, *35*, 1178–1189. [CrossRef]
137. Liu, T.; El-Rich, M. Subject-specific trunk segmental masses prediction for musculoskeletal models using artificial neural networks. *Med. Biol. Eng. Comput.* **2024**, *62*, 2757–2768. [CrossRef]
138. Lewandowski, A. *Issues in Model Validation*; Angewandte Systemanalyse; IIASA: Laxenburg, Austria, 1982; pp. 1–29.
139. Kia, M.; Stylianou, A.P.; Guess, T.M. Evaluation of a musculoskeletal model with prosthetic knee through six experimental gait trials. *Med. Eng. Phys.* **2014**, *36*, 335–344. [CrossRef]
140. Ignasiak, D.; Ferguson, S.J.; Arjmand, N. A rigid thorax assumption affects model loading predictions at the upper but not lower lumbar levels. *J. Biomech.* **2016**, *49*, 3074–3078. [CrossRef] [PubMed]

141. Hodges, P.W.; Ferreira, P.H.; Ferreira, M.L. Chapter 14—Lumbar Spine: Treatment of Motor Control Disorders. In *Pathology and Intervention in Musculoskeletal Rehabilitation*, 2nd ed.; Magee, D.J., Zachazewski, J.E., Quillen, W.S., Manske, R.C., Eds.; Elsevier Health Sciences: Amsterdam, The Netherlands, 2015; pp. 520–560, ISBN 978-0-323-31072-7.
142. Malakoutian, M.; Sánchez, C.A.; Brown, S.H.M.; Street, J.; Fels, S.; Oxland, T.R. Biomechanical Properties of Paraspinal Muscles Influence Spinal Loading-A Musculoskeletal Simulation Study. *Front. Bioeng. Biotechnol.* **2022**, *10*, 852201. [CrossRef] [PubMed]
143. Rupp, T.K.; Ehlers, W.; Karajan, N.; Günther, M.; Schmitt, S. A forward dynamics simulation of human lumbar spine flexion predicting the load sharing of intervertebral discs, ligaments, and muscles. *Biomech. Model. Mechanobiol.* **2015**, *14*, 1081–1105. [CrossRef] [PubMed]
144. Rohlmann, A.; Zander, T.; Bergmann, G.; Boustani, H.N. Optimal stiffness of a pedicle-screw-based motion preservation implant for the lumbar spine. *Eur. Spine J.* **2012**, *21*, 666–673. [CrossRef] [PubMed]
145. Kettler, A.; Rohlmann, F.; Ring, C.; Mack, C.; Wilke, H.-J. Do early stages of lumbar intervertebral disc degeneration really cause instability? Evaluation of an in vitro database. *Eur. Spine J.* **2011**, *20*, 578–584. [CrossRef]
146. McNally, D.S.; Adams, M.A. Internal intervertebral disc mechanics as revealed by stress profilometry. *Spine* **1992**, *17*, 66–73. [CrossRef]
147. La Barbera, L.; Galbusera, F.; Villa, T.M.T.; Costa, F.; Wilke, H.-J. ASTM F1717 standard for the preclinical evaluation of posterior spinal fixators: Can we improve it? *Proc. Inst. Mech. Eng. H* **2014**, *228*, 1014–1026. [CrossRef]
148. Chen, C.-S.; Chen, W.-J.; Cheng, C.-K.; Jao, S.-H.E.; Chueh, S.-C.; Wang, C.-C. Failure analysis of broken pedicle screws on spinal instrumentation. *Med. Eng. Phys.* **2005**, *27*, 487–496. [CrossRef]
149. Cho, W.; Wu, C.; Mehbod, A.A.; Transfeldt, E.E. Comparison of cage designs for transforaminal lumbar interbody fusion: A biomechanical study. *Clin. Biomech.* **2008**, *23*, 979–985. [CrossRef]
150. Calvo-Echenique, A.; Cegoñino, J.; Chueca, R.; Pérez-del Palomar, A. Stand-alone lumbar cage subsidence: A biomechanical sensitivity study of cage design and placement. *Comput. Methods Programs Biomed.* **2018**, *162*, 211–219. [CrossRef]
151. Reinshagen, C.; Ruess, D.; Walcott, B.P.; Molcanyi, M.; Goldbrunner, R.; Rieger, B. A novel minimally invasive technique for lumbar decompression, realignment, and navigated interbody fusion. *J. Clin. Neurosci.* **2015**, *22*, 1484–1490. [CrossRef] [PubMed]
152. Pao, J.-L. Biportal Endoscopic Transforaminal Lumbar Interbody Fusion Using Double Cages: Surgical Techniques and Treatment Outcomes. *Neurospine* **2023**, *20*, 80–91. [CrossRef] [PubMed]
153. Bonnaire, R.; Woo-Suck, H.; Calmels, P.; Convert, R.; Molimard, J. Parametric Study of Lumbar Belts in the Case of Low Back Pain: Effect of Patients' Specific Characteristics. In *Computational Biomechanics for Medicine: Personalisation, Validation and Therapy*; Nash, M.P., Nielsen, P.M.F., Wittek, A., Miller, K., Joldes, G.R., Eds.; Springer: Cham, Switzerland, 2020; pp. 43–59, ISBN 303015923X.
154. Guo, J.; Guo, W.; Ren, G. Embodiment of intra-abdominal pressure in a flexible multibody model of the trunk and the spinal unloading effects during static lifting tasks. *Biomech. Model. Mechanobiol.* **2021**, *20*, 1599–1626. [CrossRef]
155. Hodges, P.W.; Cresswell, A.G.; Daggfeldt, K.; Thorstensson, A. In vivo measurement of the effect of intra-abdominal pressure on the human spine. *J. Biomech.* **2001**, *34*, 347–353. [CrossRef]
156. Spadoni, S.; Todros, S.; Pavan, P.G. Numerical modeling of the abdominal wall biomechanics and experimental analysis for model validation. *Front. Bioeng. Biotechnol.* **2024**, *12*, 1472509. [CrossRef]
157. Jourdan, A.; Rapacchi, S.; Guye, M.; Bendahan, D.; Masson, C.; Bège, T. Dynamic-MRI quantification of abdominal wall motion and deformation during breathing and muscular contraction. *Comput. Methods Programs Biomed.* **2022**, *217*, 106667. [CrossRef] [PubMed]
158. Cholewicki, J.; Ivancic, P.C.; Radebold, A. Can increased intra-abdominal pressure in humans be decoupled from trunk muscle co-contraction during steady state isometric exertions? *Eur. J. Appl. Physiol.* **2002**, *87*, 127–133. [CrossRef]
159. Miyamoto, K.; Iinuma, N.; Ueki, S.; Shimizu, K. Effects of abdominal belts on the cross-sectional shape of the trunk during intense contraction of the trunk muscles observed by computer tomography. *Clin. Biomech.* **2008**, *23*, 1220–1226. [CrossRef]
160. Sánchez, C.A.; Li, Z.; Hannam, A.G.; Abolmaesumi, P.; Agur, A.; Fels, S. Constructing Detailed Subject-Specific Model of the Human Masseter. In *Imaging for Patient-Customized Simulations and Systems for Point-of-Care Ultrasound: International Workshops, BIVPCS 2017 and POCUS 2017, Held in Conjunction with MICCAI 2017, Québec City, QC, Canada, September 14, 2017, Proceedings*; Cardoso, M.J., Arbel, T., Tavares, J., Aylward, S., Li, S., Boctor, E., Fichtinger, G., Cleary, K., Freeman, B., Kohli, L., Eds.; Springer International Publishing: Berlin/Heidelberg, Germany, 2017; pp. 52–60, ISBN 9783319675527.
161. Jourdan, A.; Dhume, R.; Guérin, E.; Siegel, A.; Le Ruyet, A.; Palmer, M. Numerical investigation of a finite element abdominal wall model during breathing and muscular contraction. *Comput. Methods Programs Biomed.* **2024**, *244*, 107985. [CrossRef]
162. Miyamoto, K.; Iinuma, N.; Maeda, M.; Wada, E.; Shimizu, K. Effects of abdominal belts on intra-abdominal pressure, intramuscular pressure in the erector spinae muscles and myoelectrical activities of trunk muscles. *Clin. Biomech.* **1999**, *14*, 79–87. [CrossRef]
163. Tuong, N.H.; Dansereau, J.; Maurais, G.; Herrera, R. Three-dimensional evaluation of lumbar orthosis effects on spinal behavior. *J. Rehabil. Res. Dev.* **1998**, *35*, 34–42. [PubMed]
164. Thoumie, P.; Drape, J.-L.; Aymard, C.; Bedoiseau, M. Effects of a lumbar support on spine posture and motion assessed by electrogoniometer and continuous recording. *Clin. Biomech.* **1998**, *13*, 18–26. [CrossRef] [PubMed]

165. Ivancic, P.C.; Cholewicki, J.; Radebold, A. Effects of the abdominal belt on muscle-generated spinal stability and L4/L5 joint compression force. *Ergonomics* **2002**, *45*, 501–513. [CrossRef]
166. Jegede, K.A.; Miller, C.P.; Bible, J.E.; Whang, P.G.; Grauer, J.N. The effects of three different types of orthoses on the range of motion of the lumbar spine during 15 activities of daily living. *Spine* **2011**, *36*, 2346–2353. [CrossRef]
167. Granata, K.P.; Marras, W.S.; Davis, K.G. Biomechanical assessment of lifting dynamics, muscle activity and spinal loads while using three different styles of lifting belt. *Clin. Biomech.* **1997**, *12*, 107–115. [CrossRef]
168. Lloyd, J.E.; Roewer-Després, F.; Stavness, I. Muscle Path Wrapping on Arbitrary Surfaces. *IEEE Trans. Biomed. Eng.* **2021**, *68*, 628–638. [CrossRef]
169. Millard, M.; Uchida, T.; Seth, A.; Delp, S.L. Flexing computational muscle: Modeling and simulation of musculotendon dynamics. *J. Biomech. Eng.* **2013**, *135*, 21005. [CrossRef] [PubMed]
170. Sagl, B.; Dickerson, C.R.; Stavness, I. Fast Forward-Dynamics Tracking Simulation: Application to Upper Limb and Shoulder Modeling. *IEEE Trans. Biomed. Eng.* **2019**, *66*, 335–342. [CrossRef]
171. Byrne, R.M.; Aiyangar, A.K.; Zhang, X. Sensitivity of musculoskeletal model-based lumbar spinal loading estimates to type of kinematic input and passive stiffness properties. *J. Biomech.* **2020**, *102*, 109659. [CrossRef]
172. Meszaros-Beller, L.; Hammer, M.; Riede, J.M.; Pivonka, P.; Little, J.P.; Schmitt, S. Effects of geometric individualisation of a human spine model on load sharing: Neuro-musculoskeletal simulation reveals significant differences in ligament and muscle contribution. *Biomech. Model. Mechanobiol.* **2023**, *22*, 669–694. [CrossRef]
173. Lerchl, T.; Nispel, K.; Baum, T.; Bodden, J.; Senner, V.; Kirschke, J.S. Multibody Models of the Thoracolumbar Spine: A Review on Applications, Limitations, and Challenges. *Bioengineering* **2023**, *10*, 202. [CrossRef] [PubMed]
174. Breen, A.; de Carvalho, D.; Funabashi, M.; Kawchuk, G.; Pagé, I.; Wong, A.Y.L.; Breen, A. A Reference Database of Standardised Continuous Lumbar Intervertebral Motion Analysis for Conducting Patient-Specific Comparisons. *Front. Bioeng. Biotechnol.* **2021**, *9*, 745837. [CrossRef] [PubMed]

bioengineering

MDPI

Article

Dynamic Surface Topography for Thoracic and Lumbar Pain Patients—Applicability and First Results

Johanna Kniepert [1,†], Henriette Rönsch [2,†], Ulrich Betz [2,*], Jürgen Konradi [2], Janine Huthwelker [2], Claudia Wolf [2], Ruben Westphal [3] and Philipp Drees [1]

1 Department of Orthopedics and Trauma Surgery, University Medical Center of the Johannes Gutenberg University Mainz, Langenbeckstrasse 1, 55131 Mainz, Germany; johanna.kniepert@unimedizin-mainz.de (J.K.); philipp.drees@unimedizin-mainz.de (P.D.)

2 Institute of Physical Therapy, Prevention and Rehabilitation, University Medical Center of the Johannes Gutenberg University Mainz, Langenbeckstrasse 1, 55131 Mainz, Germany; henriette.roensch@unimedizin-mainz.de (H.R.); juergen.konradi@unimedizin-mainz.de (J.K.); janine.huthwelker@unimedizin-mainz.de (J.H.); claudia.wolf@unimedizin-mainz.de (C.W.)

3 Institute of Medical Biostatistics, Epidemiology and Informatics, University Medical Center of the Johannes Gutenberg University Mainz, Obere Zahlbacher Straße 69, 55131 Mainz, Germany

* Correspondence: ulrich.betz@unimedizin-mainz.de

† These authors contributed equally to this work.

Abstract: Current routine diagnostic procedures for back pain mainly focus on static spinal analyses. Dynamic Surface Topography (DST) is an easy-to-use, radiation-free addition, allowing spine analyses under dynamic conditions. Until now, it is unclear if this method is applicable to back pain patients, and data reports are missing. Within a prospective observational study, 32 patients suffering from thoracic and lumbar back pain were examined while walking, randomized at four speeds (2, 3, 4, 5 km/h), using a DST measuring device (DIERS 4Dmotion® Lab). The measurement results were compared with those of a healthy reference group. We calculated the intrasegmental rotation for every subject and summed up the spinal motion in a standardized gait cycle. The Mann–Whitney U Test was used to compare the painful and healthy reference groups at the four different speeds. In a subgroup analysis, the painful group was divided into two groups: one with less pain (≤3 points on the Visual Analogue Scale) and one with more pain (>3 points on the Visual Analogue Scale). The Kruskal–Wallis Test was used to compare these subgroups with the healthy reference group. Of the 32 included patients, not all could walk at the intended speeds (5 km/h: 28/32). At speeds of 2–4 km/h, our results point to greater total segmental rotation of back pain patients compared to the healthy reference group. At a speed of 3 km/h, we observed more movement in the patients with more pain. Overall, we monitored small differences on average between the groups but large standard deviations. We conclude that the DST measuring approach is eligible for back pain patients when they feel confident enough to walk on a treadmill. Initial results suggest that DST can be used to obtain interesting therapeutic information for an individual patient.

Keywords: back pain; spinal motion; rotational movement; dynamic surface topography; videorasterstereography

1. Introduction

Back pain is still a widespread health problem. Besides affecting the patients' quality of life and movement behavior, its influence constitutes a significant burden on social security finance [1]. Back pain is very complex, and various factors contribute to the situation, so it cannot be treated according to a standardized protocol. One problem concerning the determination of the correct therapy for back pain is the fact that its source is various and often unknown, leaving us with a high percentage of patients suffering from unspecific pain without any structural correlation [2,3].

Even if a specific diagnosis can be made, it is well known that a wide variety of pain conditions can be linked with different functional findings [4,5].

Current routine diagnostic methods, like X-ray and MRI, are usually static analyses, able to show structural damages and degenerations but not dysfunctions of joints or tissues (e.g., a joint blockage). That is why the focus on static structural alterations has limited explanatory power for unspecific back pain [6].

Hence, an appropriate measuring system should be able to analyze the spinal motion under dynamic conditions in a standardized way.

For example, Lamoth et al. [7] investigated the movement pattern of unspecific back pain patients with a marker-based optic measurement system. They could show that back pain patients walked at lower preferred speeds and had a more in-phase movement pattern between the thorax and pelvis. Their measuring procedure just allowed global angular measurements of the upper body and pelvis but not on a vertebral plane [7]. A recent meta-analysis revealed that subjects with persistent low back pain walked slower and with shorter stride lengths. Furthermore, they also showed a more in-phase movement pattern of the thorax and the lumbar spine [8].

However, as most motion capture systems work on a marker-capturing basis [9], until now, during gait analyses, spinal motion was examined mostly only by observing the motion of the thoracic or lumbar back as a block. Almost no segmental motion was mapped, and no relationship to the corresponding gait phases was drawn [7,10].

Dynamic surface topography (DST) might offer a possibility to derive segmental spinal motion from the surface of the back during gait. DST is an easy-to-use, radiation- and contact-free measurement opportunity, allowing further spine analyses not only under static but also under dynamic conditions. A pattern of horizontal, parallel lights is projected onto the subject's bare back and distorted by the back's curvature. That distortion is analyzed by triangulation and mathematical shape analysis due to a fixed angle between the camera and the light projector.

Using a mathematical algorithm, conclusions can be drawn from the surface topographic curvature picture to the underlying spine, providing a virtually constructed 3D position of each thoracic and lumbar vertebral body (from C7 to L4) and the pelvis during gait. The spinous process of L5 cannot be reflected adequately onto the skin; therefore, the section L4 to the pelvis is considered as one moving unit. A spine model, which was developed by Turner Smith in the 1980s, is used to create this 3D rendering [11–17]. Through comparison with X-ray or computed tomography (CT) images, the reliability [18] and validity [16,19] of static ST could be shown. A gold standard for detailed segmental motion analysis (e.g., dynamic X-ray-based measurements) is missing, so there exists no proven validity of DST. Nevertheless, based on the reliability, reproducibility [11,20], and accuracy [11,21,22] of DST, inter- and intraindividual comparisons can be made. For more detailed background information on the operation principle of DST, the reader is referred to [23,24]. Currently, there is no validation for DST in walking [25].

There are first results available, describing DST-captured spinal motion during gait in healthy reference cohorts [25]. For patients with back pain, such data are still lacking. In addition, it is not yet clear whether this method is at all suitable for patients with back pain, as it requires participants to walk on a treadmill without using the handrails. For this reason, the aim of this paper is to present our experience from a first study performing

DST measurements with back pain patients and comparing the results for segmental spinal movement of the painful area to the healthy reference group.

2. Methods

2.1. Inclusion and Exclusion Criteria of the Study Population

Subjects suffering from acute or chronic back pain (specific or non-specific) with any pain intensity (at least 1 on Visual Analogue Scale (VAS, 0–10) in rest or motion) in the thoracic or lumbar spine region were enrolled. We did not differentiate between specific and non-specific back pain, as the focus was on the applicability of DST rather than the explicit study of back pain patients. The in- and outpatient recruitment was conducted at the University Medical Center of Mainz from September 2017 till January 2018. Furthermore, patients were informed about the study via flyers and word-of-mouth recommendations. Subjects were excluded if any of the following aspects applied to them: acute fractures; balance disorders or major gait abnormalities, which prevent freehand walking on the treadmill; and illnesses, which influence gait or balance (e.g., Mb. Parkinson, Multiple sclerosis, hemiparesis or -plegia, polyneuropathy).

To ensure patient safety walking on a treadmill with back pain, subjects had to perform the Two-Minute Walk Test (2MWT) [26] beforehand. The 2MWT was used to discover any major gait abnormalities (exclusion criterion), to ensure a walking speed of at least 2 km/h, and to determine the subject's possible maximum speed on the treadmill.

Subjects were later excluded when they were unable to walk at least 2 km/h or freehand on the treadmill or due to surface alterations of the back (e.g., tattoos, big scars). The latter would lead to measuring errors of the DST.

Because of the contradictory results of other studies on the influence of BMI on the measurement results of DST [19,27], no participant was excluded concerning BMI. As the focus of the paper was on clinical applicability of DST and only in the second step on the collection of data from back pain patients in general, no additional imaging to distinguish the cause of back pain (specific or non-specific) was required to participate.

2.2. Measuring Device

For the examination of spinal motion during gait, the modified DIERS 4Dmotion® Lab was utilized, consisting of the DIERS Formetric III 4D and the DIERS pedogait, using its own software DICAM v3.7 (Diers International GmbH, Schlangenbad, Germany).

The device measures more than 100 different parameters of the spine (C7-L4) and the pelvis motion with a frequency of 60 Hz. Rotation in the transverse plane was selected from this wide spectrum for the analysis in this study, as the reference values of the healthy comparison group are also available for this parameter [25,28]. In cooperation with the manufacturer, a device interface was built, with which all measured data are exported chronologically and further evaluated. The measured values of the trunk's surface rotation at the height of each vertebral body (except for the fifth lumbar (L5)) could be mapped continuously onto the thoracic and lumbar spine. Timely synchronized data of an embedded foot pressure measuring plate (100 Hz) allowed the relation of the spinal model data to the gait cycle phases.

2.3. Measuring Procedure

Before the measurements were taken, the subjects completed a pain questionnaire and performed the 2MWT. The pain questionnaire included the location of the pain (thoracic and/or lumbar), the pain intensity (VAS (10 cm, 0–10)), and the BMI.

For the dynamic measurement, it was necessary to apply red light reflecting markers onto the anatomical landmarks of the vertebra prominens (VP) and the two dimples of

Venus. Therefore, we used the static DIERS Formetric 4D average measurement to confirm or correct the marker's accurate position. This procedure was also recommended by Betsch et al. [11] prior to obtaining dynamic measurements. To capture habitual gait, at each measurement, the participants had to walk on the treadmill for two minutes before the recording was started without an additional announcement. More information about the measuring setup and procedure can be found in the dissertation by JK [29].

With the Formetric III 4D, the subject's bare back was recorded for eight seconds (at walking speeds of 2 and 3 km/h) and six seconds (at speeds of 4 and 5 km/h) to capture at least three gait cycles per measurement.

2.4. Healthy Reference Group

A group of 134 healthy subjects (average age: 39.81 years, standard deviation (SD): 12.45 years, body mass index (BMI) $\leq$ 30.0 kg/m^2) was used for comparison [25,26]. The study on the reference data of the healthy subjects was carried out using the same measurement methodology as this study on back pain patients. Due to the interindividual variations in motion sequences described in other studies, healthy subjects were not matched in terms of age and sex in this study [30,31]. To be considered healthy, the subjects had to meet several prerequisites (no history of surgery or fracture between C7 and pelvis, no medical or therapeutic treatments between C7 and pelvis within the last 12 months and no due to musculoskeletal problems, in general, within the last 6 months) and pass several tests to demonstrate adequate gait stability (timed up-and-go test, 2MWT, back performance scale) ((WHO (INT: DRKS00010834)) [25,26].

2.5. Statistical Analysis

The number of subjects was determined in advance using a power analysis: With 29 patients, a difference of 1 standard deviation with a power of 90% to the 0.83% level can be demonstrated. In order to test the clinical applicability, the primary endpoint of the study was defined as how many subjects the measurements could be carried out on and, secondarily, at which walking speeds.

The raw data of the measurements were exported in relation to the phases of the gait cycles as separate files with DICAM v3.7 software and merged using SAS software (Version 9.4). The data of each measurement were set in relation to the duration of its gait cycles (0–100%). Afterwards, the three gait cycles per measurement were combined as one gait cycle as a smooth curve using a spline function. A total of 101 values were obtained using integer percentages (0–100%) found in the graph. This procedure of generating a standardized gait cycle per measurement allowed all the measurements of different lengths to be compared [32–36].

Starting at the VP, we calculated the segmental rotational motion for all the segments down to the pelvis with IBM SPSS Statistics (Version 23). The rotational position of a spinal segment was computed by subtracting the position value in degrees of one vertebral body in the transversal plane from the position of the vertebral body below at the same point in time. By subtracting the segment's position between two consecutive time periods, the segment's rotational motion was obtained. For each measurement and segment, the absolute values of the rotational motion per standardized gait cycle were summed up (sum of motion per gait cycle (SoMpGC)). The calculation concept for the SoMpGC is displayed in Figure 1.

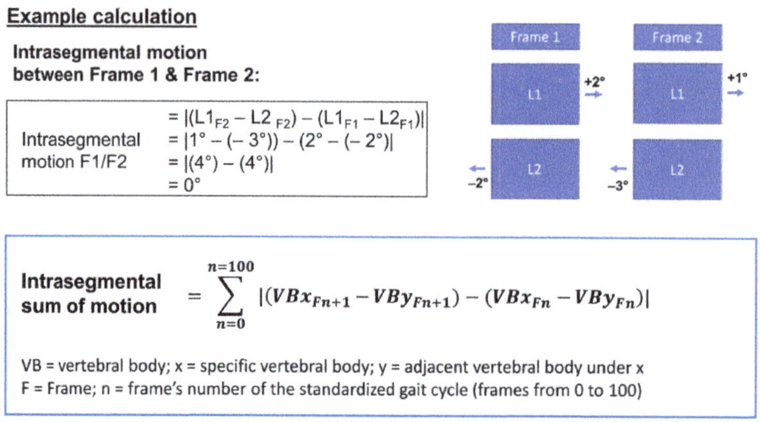

$$\text{Intrasegmental sum of motion} = \sum_{n=0}^{n=100} |(VBx_{Fn+1} - VBy_{Fn+1}) - (VBx_{Fn} - VBy_{Fn})|$$

Figure 1. Example calculation for calculating intrasegmental rotation and sum of motion per gait cycle (SoMpGC).

Subjects could mostly only name a pain area and not a single painful spinal segment, which is why a distinction was only made between thoracic and/or lumbar pain. Depending on the location of the subject's pain, the SoMpGC of the affected segments (thoracic and/or lumbar) was used to calculate the average total intrasegmental rotation for each spinal segment separately.

We collected data on pain characteristics (e.g., duration of pain, drug medication, and paresthesia) and anthropometric data such as height, weight, and age, which will be used in future analyses.

Despite the metric character of intrasegmental rotation and the sum of motion per gait cycle, we used non-parametric tests for group comparison because of the small sample size and a conservative approach to interpretation. We chose the Mann–Whitney U Test for comparison of the groups at different speeds. For subgroup analysis, the back pain group was divided by VAS values (median VAS 3.25) into a low-grade (VAS $\leq$ 3) and a high-grade (VAS > 3) pain group. The Kruskal–Wallis Test was used for group comparison. We used calculation of Cohen's d with the respective pooled standard deviation for effect sizes [37,38]. Positive values point toward the painful areas, and negative values points toward the healthy reference group. All calculations were conducted with IBM SPSS Statistics (Version 29).

3. Results

3.1. Characteristics of the Study Population

Thirty-four subjects suffering from back pain (specific and non-specific) between C7 and L5 were recruited, of which the measurements of 32 subjects (18 women, 14 men) between the ages of 19 and 68 were evaluated. The process of enrollment and dropout is displayed in Figure 2, and the participants' characteristics are displayed in Table 1. During the 2MWT, the subjects showed an average walking distance of 190.72 m (SD: 35.60). Compared to the corresponding age cohort (18–54 years) of Bohannon, Wang, and Gershon [27] for obtaining standard data (women, 183.0 m; men, 200.9 m), the participants with back pain did not show any difference in maximum possible speed. All of them had pain in rest and over 70 % reported back pain also while walking. On average, the subjects showed a pain intensity of 4/10 (see Table 2).

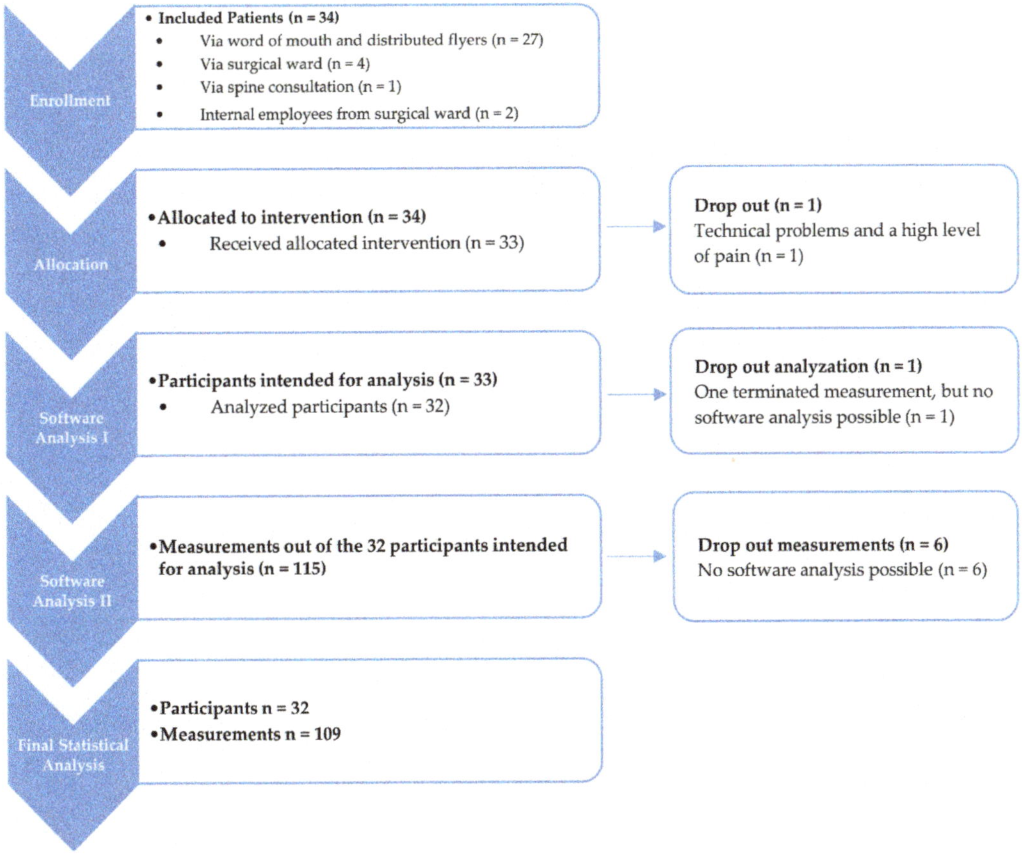

Figure 2. The process of enrollment, allocation, analysis, and reasons for dropout are displayed for the collective of subjects suffering from thoracic and/or lumbar back pain.

Table 1. Characteristics of the analyzed participants: 1a. Display of the demographic characteristics; 1b. Display of the number and percentage of subjects and location of pain (thoracic and/or lumbar).

	Characteristics	**Analyzed Participants (n = 32)**
1a	Age (years): average (SD), range	44.53 (14.84), 19–68
	Male sex n (%)	14 (44%)
	BMI average (SD), range	26.01 (4.79), 16.76–37.56
	2MWT (distance in meters): average (SD), range	190.72 (35.60), 81.28–243.84
	2MWT (speed in km/h): average (SD), range	5.72 (1.07), 2.4–7.3
1b	**Location of back pain**	**Number of subjects (percentage)**
	Thoracic spine	2 (6%)
	Lumbar spine	23 (72%)
	Thoracic and lumbar spine	7 (22%)

Abbreviations: SD, standard deviation; BMI, body mass index; SD, standard deviation; 2MWT, two-minute walk test.

Table 2. Evaluation of the pain questionnaire: percentage of the subjects with rest pain and pain when walking, and average pain in terms of the average number of the subjects' pain.

Pain Characteristics	Number of Subjects (Percentage)
Rest pain	32/32 (100%)
Pain when walking	23/32 (72%)
Pain level (VAS: 0–10)	average: 3.67/10 (minimum: 1, maximum: 8; SD: 1.83) median: 3.25; range: 1–8 SD: 1.83
Time period of pain	
<1 month	5 (15%)
1–6 month	4 (12%)
6–12 months	3 (9%)
12–24 months	6 (18%)
24–60 months	2 (6%)
>60 months	13 (39%)
Pain medication	
Yes	6 (18%)
None	27 (82%)
Radiation to the legs	
Right	5 (15%)
Left	9 (27%)
None	19 (58%)
Paresthesia/Reduction in strength	
Right	2 (6%)
Left	6 (18%)
Both sides	2 (6%)
Cervical spine	1 (3%)
None	22 (67%)

Abbreviations: VAS, Visual Analogue Scale; SD, standard deviation.

3.2. *Applicability of Dynamic Surface Topography*

The recruitment was made more difficult because not all patients with acute pain felt confident enough for the pre-examinations and later for walking on a treadmill without holding. When subjects were enrolled, only 1 of 34 enrolled subjects could not be examined with the device due to difficulty when walking on a treadmill while suffering from a high level of pain and a delay caused by technical problems of the device.

Seven single measurements at different speeds (5%) could not be analyzed by the software due to technical/software problems. For one subject, only one measurement at 2 km/h was possible. Due to technical problems, this measurement and the subject both dropped out. In total, 115 measurements of 32 subjects at speeds from 2 km/h to 5 km/h were evaluated.

The faster the treadmill speed, the more difficult it was for subjects with severe back pain to walk on it. So, 28/32 (85%) of the included subjects could be examined at a treadmill speed of 5 km/h. The evaluation of the subjects' number in relation to the treadmill's walking speeds is displayed in Table 3.

Table 3. Evaluation of the subjects' number in relation to the treadmill's walking speed.

Walking Speed on the Treadmill	Number of Subjects Examined (Percentage)
2 km/h	33 (100%)
3 km/h	31 (94%)
4 km/h	30 (91%)
5 km/h	28 (85%)

3.3. Data of Dynamic Spinal Rotation

3.3.1. Analysis of the Entire Group

The subjects' average segmental rotation in the transversal plane of the segments in the painful area is displayed as the average SoMpGC and is set in comparison to the average SoMpGC of the healthy reference group. For the patient group, only the data of the segments in the respective painful regions were included in the analysis (thoracic spine or lumbar spine/thoracic and lumbar spine).

Figure 3a–d show the SoMpGC of the back pain patients and the healthy reference group at all four speeds measured. At 2, 3, and 4 km/h, an increased total rotation of segments in the painful area is seen in relation to the same segments of the healthy reference group. In the upper thorax, this difference can be up to 2°, and in the remaining spine, the difference is less than 1°. At 5 km/h, the difference reverses in the lower spine (between T9 to Pelvis), with the total segmental rotation of subjects with back pain minimally less than the total segmental rotation of the healthy reference group. In the Mann–Whitney U Test, we found $p \leq 0.05$ only for thoracic segments.

Effect sizes for 2 km/h were large in the painful upper thoracic segments (d = 0.9 for T3/T4 and T4/T5), whereas the smallest value was detected for the painful segment L4/pelvis (d = 0.3). These effect sizes decrease at speeds of 3 and 4 km/h in the upper thorax. In the segments T6/T7, the value turns negative to indicate a greater movement in the healthy segments. At a speed of 5 km/h, there are small effect sizes in all segments, changing from positive to negative values and reverting. For further information, see Supplementary Table S1.

Because of the very small groups for the painful areas, the location of pain was not taken into account in the calculation. Hence, due to this explorative approach, we regard the results with $p \leq 0.05$ as a trend.

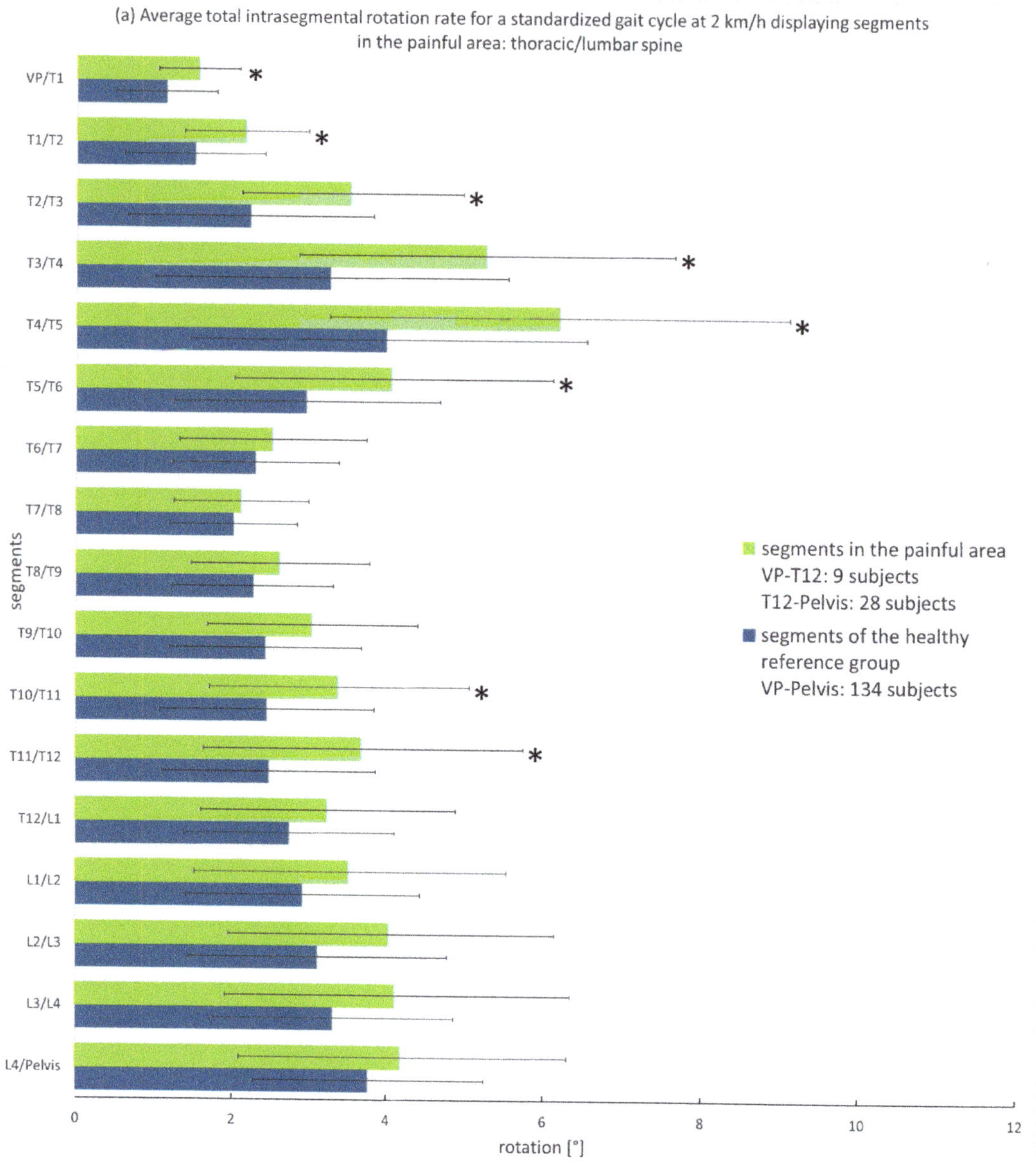

Figure 3. *Cont.*

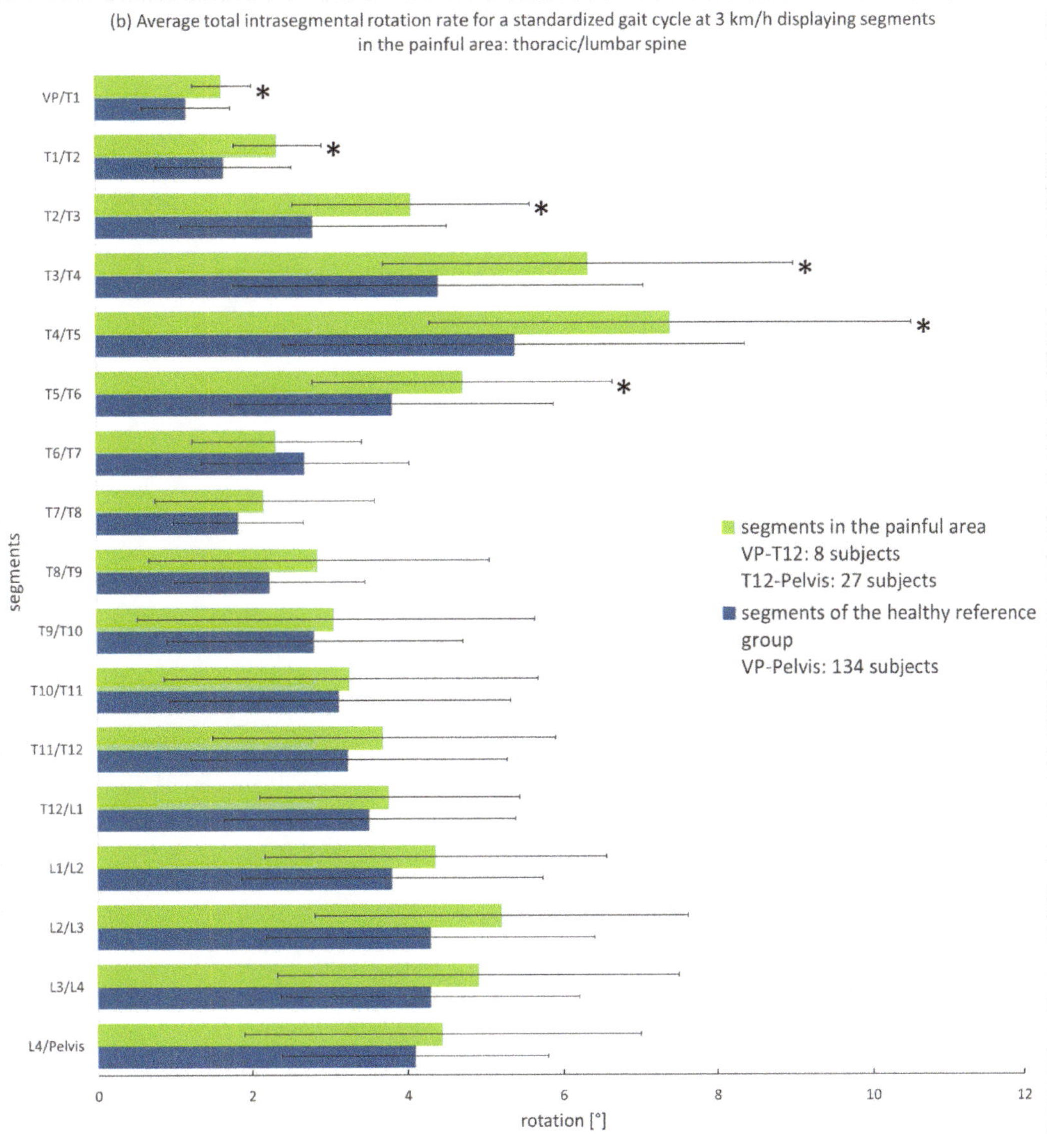

Figure 3. *Cont.*

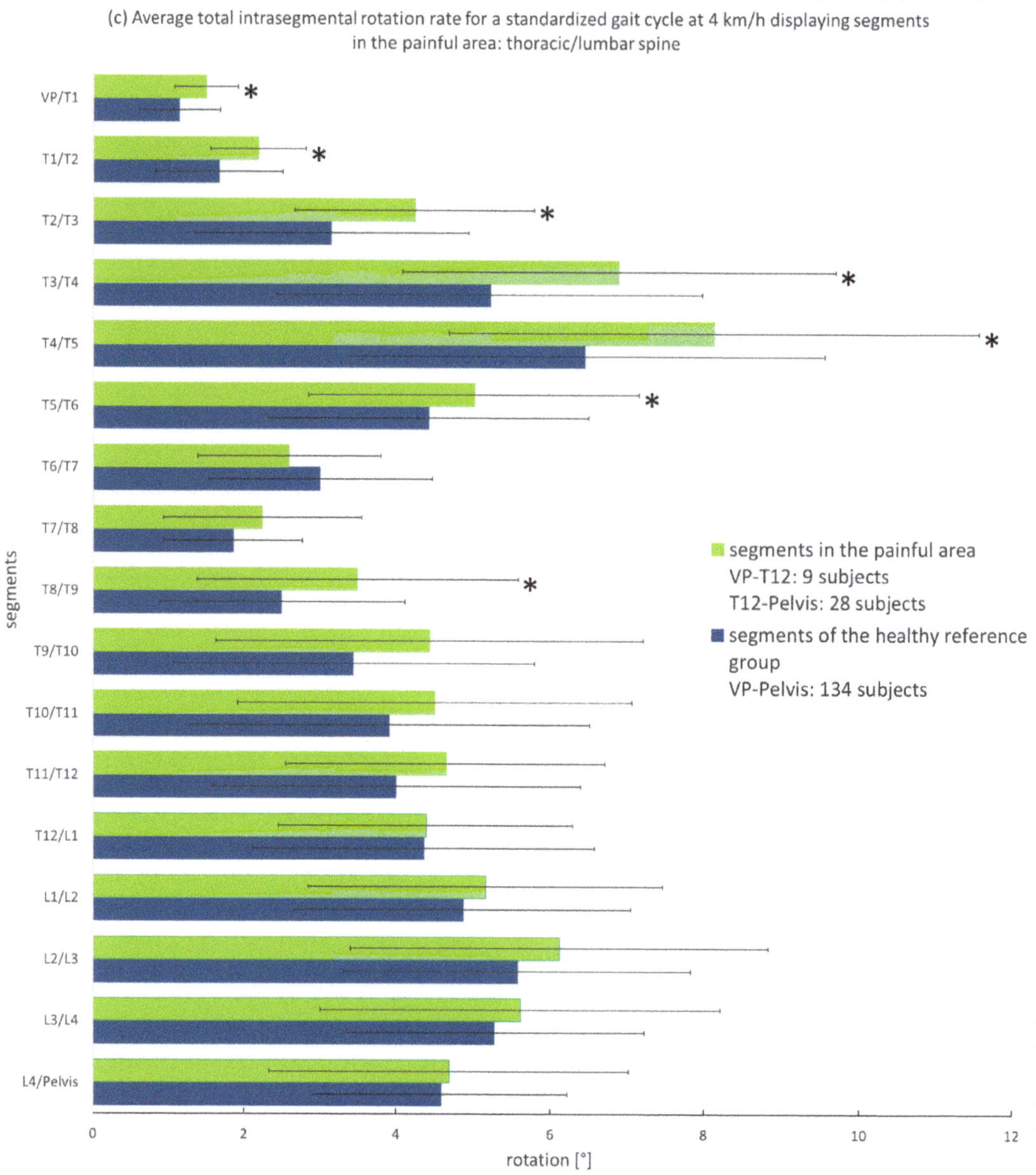

Figure 3. *Cont.*

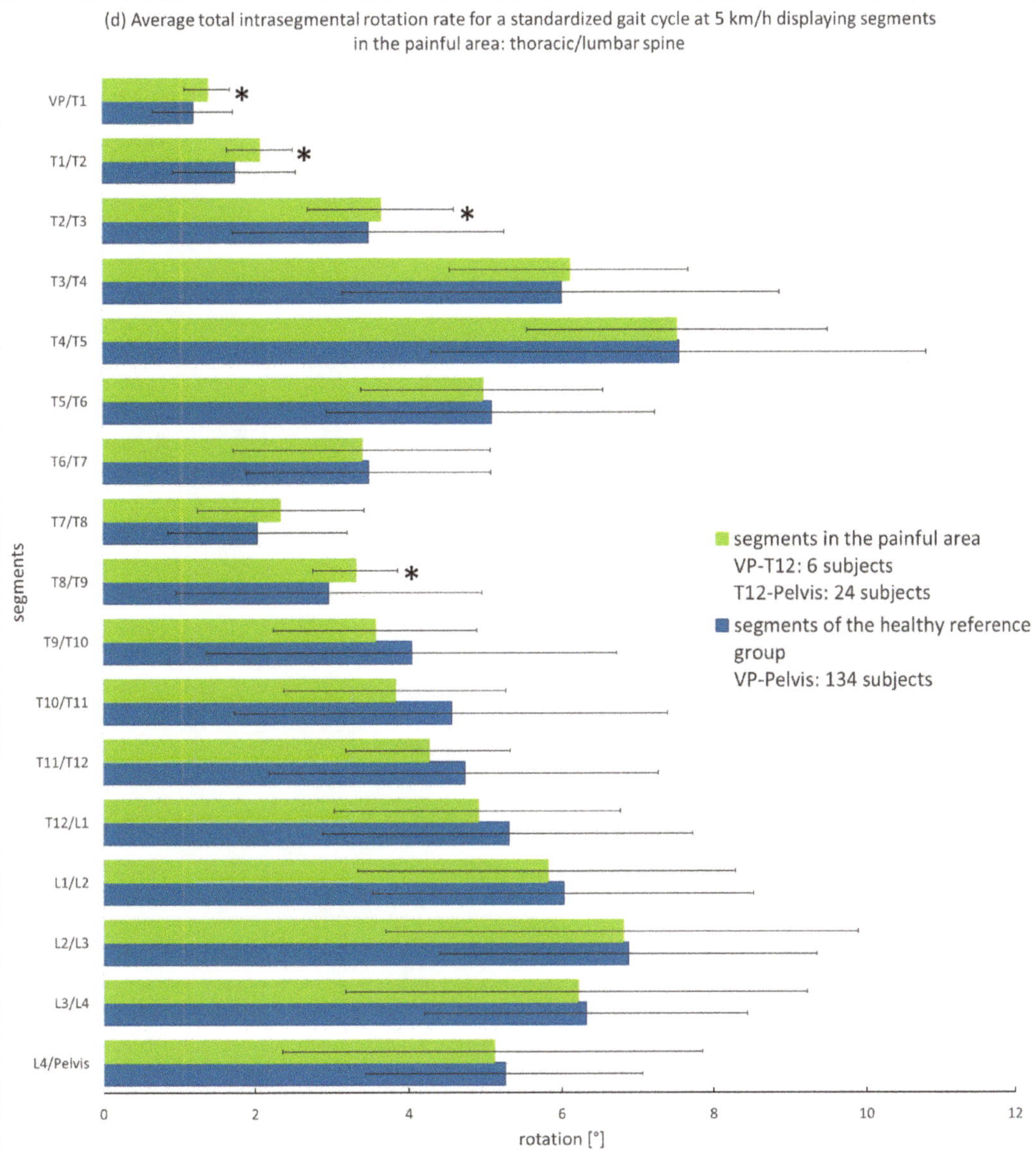

Figure 3. (**a**–**d**): Displayed are the mean and standard deviation of the SoMpGC in degrees (x-axis), displaying spinal segments (y-axis) located in the painful area thoracic or lumbar region (green) in comparison to the segments of a healthy reference group (blue): (**a**) at 2 km/h, (**b**) at 3 km/h, (**c**) at 4 km/h, (**d**) at 5 km/h; number of subjects in the painful area exceeds the maximum cohort number because subjects could have thoracic and lumbar pain. * $p \leq 0.05$ (because of the small subgroups, we consider the results as a trend).

3.3.2. Analysis According to Pain Intensity

The subjects with lumbar back pain were also examined based on their pain intensity. At the speed of 3 km/h, 27 subjects (of 30 subjects with lumbar pain, 3 measurements could not be analyzed) with a median pain of 3.25 (derived from VAS) were evaluated (see Figure 4).

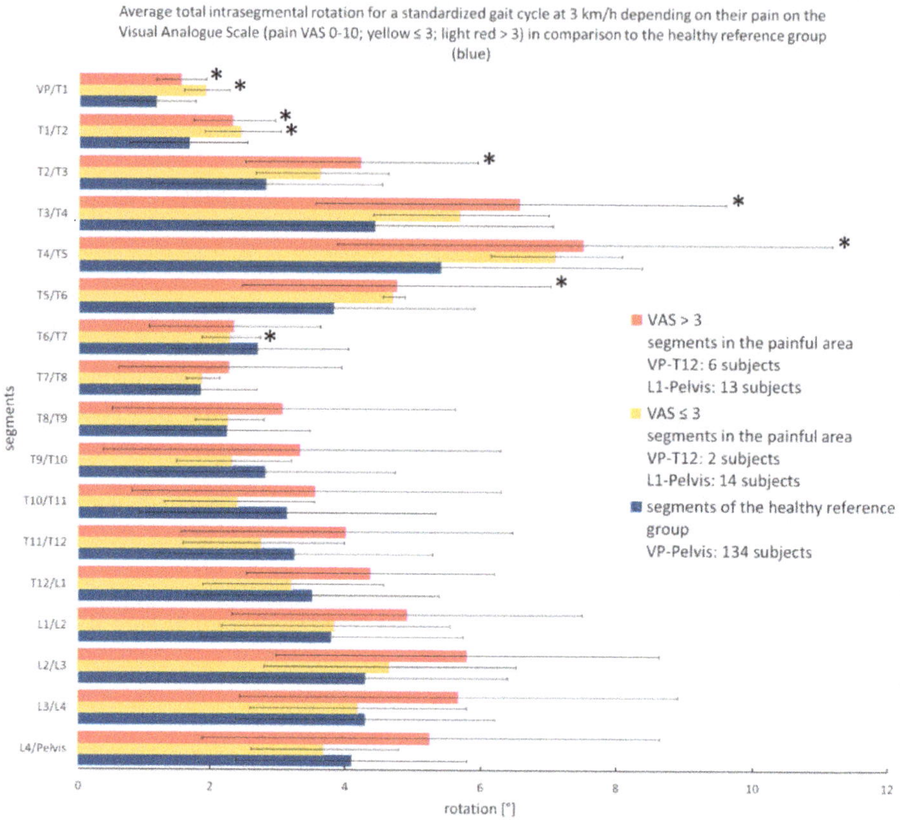

Figure 4. Displayed are the mean and standard deviation of the SoMpGC at 3 km/h in degrees (x-axis), displaying lumbar segments (y-axis) located in the painful area of low back pain depending on their pain on the VAS (Visual Analogue Scale) (pain VAS 0-10; yellow ≤ 3; red > 3) in comparison to the healthy reference group (blue); number of subjects in the painful area exceeds the maximum cohort number because subjects could have thoracic and lumbar pain. * $p \leq 0.05$, comparison between the respective pain group and healthy reference group (because of the small subgroups, we consider the results as a trend).

The low-grade pain group revealed an inhomogeneous pattern of the SoMpGC. While in the upper thoracic spine, their SoMpGC values were constantly higher than those of the healthy controls, starting at the T6/T7 segment, a strict movement pattern can no longer be recognized. Rather, the extent of movement fluctuates between more and less movement compared to the healthy comparison group.

The high-grade pain group has higher SoMpGC values compared to the healthy reference group, except the segment T6/7. The Kruskal–Wallis Test found differences between the high-grade pain and the healthy cohort in the segments VP/T1 to T5/T6 ($p < 0.05$). Because of the small subgroups, we consider the results as a trend.

In the comparison of both pain groups, the SoMpGC values of the segments T5/6 and T6/7 are nearly the same. In all other segments, the movement values of the high-grade pain group are higher than those of the low-grade pain group.

4. Discussion

The study could show that Dynamic Surface Topography can be used to examine back pain patients when they feel confident enough to walk freehand on a treadmill. However, this only affects a part of the overall group. Patients with severe pain, in particular, were often reluctant to undergo the examination. Unfortunately, it is not known how large this group is and what its exact composition is. It could be interesting to see why these patients refused to participate and if they can be convinced under better conditions. In this case, further investigations have to ensure the patients' safety, e.g., very slow walking on the treadmill while holding on, followed by a very cautious increase in demand. It is conceivable that, thereby, unconfident and anxious patients could be examined as well.

A possibility to further increase the number of suitable patients could be the use of additional safety devices such as safety straps. However, this option is limited due to the necessary view of the back during the DST measurement. If these approaches fail, the DST measurement does not appear to be a suitable method for objective and standardized functional testing. For this group, a different measurement method, for example in standing, must be found. But, if patients feel confident enough to walk on a treadmill despite having back pain, a DST measurement can be carried out.

4.1. Spinal Motion of Thoracic and Lumbar Pain Patients

Most of the literature describes an average total rotation of only 1–2° [39,40], with some studies reporting 4° to a maximum of 6° [41] in the lumbar spine. Unlike passive examinations, active examinations, such as walking, do not exhaust the maximum rotational range of spinal motion [42,43]. Feipel et al. reported more rotational movement with increasing walking speed, but this did not exceed 40% of the maximum range of motion [42]. Segment height and degeneration level also seem to be relevant for segmental motion [39,44]. Since DST measuring offers continuous observation of the rotational behavior, we decided to present the movement as SoMpGC. The limitation to the maximum range of motion would significantly reduce the value of the measurement for the functional evaluation. However, this means that our results cannot be directly compared with the results above [39,40,42]. In a study using the VICON system for investigating back pain patients during level walking at individual speeds, significant differences in the amplitude of rotation were found. Patients with low back pain used 25 to 50% less rotational ROM in the lumbar spine, depending on the corresponding reference values [45]. On the other hand, Crosbie et al. [46] reported little or no effect of back pain on the segmental motion of the lumbar spine during overground walking. In contrast, our results indicate, at least at speeds up to 4 km/h, a trend for more movement in the back pain patients compared to the healthy reference group. At 5 km/h, this effect is no longer detectable [46]. An important factor could be the walking speed. Subjects with low back pain prefer a slower walking speed, lower step length, and cadence [8], but the surprising effect was particularly noticeable at slow walking speeds. It is possible that our study shows hypermobile syndromes triggered by degeneration, for example. However, we observed intensified rotation, particularly in the more painful patients. In these patients, higher muscle tone can be expected as a result of the pain to guard the spine during walking [8,47,48]. This effect is more evident, especially at speeds higher than 4.6 km/h [7]. The study by Lamoth et al. reports even synchronous, in-phase movements between the pelvis and thorax in subjects with back pain, which may lead to less movement in the spine [7].

We observed an increased difference in rotational movement compared to the healthy reference group at lower walking speeds. Walking at lower speeds is uneconomic and more unstable than faster walking or preferred walking speed, which needs more neuromuscular activation [49,50]. In back patients, the activation of local and global muscles is altered [51], which leads to modified movement patterns [51,52]. This could be an explanation for our detected greater difference in rotational movement between the back pain and healthy participants.

4.2. Limitations of the Measuring System and the Study Design

Since the DST measurement does not directly measure segmental mobility but derives the movement from the back surface, it is conceivable that the movement does not actually change, but only the back surface, e.g., an increased muscle tone on the surface is incorrectly interpreted as the movement of the spine. Examinations of back pain patients with electromyography during gait detected higher muscle activity in the swing phase and additionally positively correlated with higher pain intensity [53]. In principle, this effect of the influence of the muscle tone is also conceivable with the other measurement methods for movement analysis. However, the DST measurement could react particularly sensitively to changes in muscle tone, as the entire width of the back contour is used for the analysis, not just the line of dorsal processes. This argument is even more validate, that as mentioned above, back pain patients activate more global than local stabilizing muscles [51]. Possibly, the movement of subcutaneous fat could also be misinterpreted by the software as segmental movement, especially in patients with a BMI >29 [54]. The combination of systematic and random mistakes makes it difficult to integrate the exact relationship between soft tissue and bony motion in data calculations [55]. The accuracy of DST measurements with respect to rotation has been determined with a mean deviation of up to 3° [16,18] compared to X-ray images and a deviation of 1–1.5% compared to the Vicon system [11]. Regarding these results and the small amount of rotational motion of the individual vertebral bodies in relation to the error indicated in the literature, the difficulty of determining the exact individual vertebral body rotation by means of DST must be considered. To our knowledge, there is no validation of the DST in walking. Measuring a cohort of pain patients with different measurement systems with comparable data processing would be very interesting in this context. In addition, the repeated measurement of patients in different stages of pain would be very informative in order to better understand the impact of back pain on the individual spine motion, in the case of this study, on the segmental rotation or SoMpGC.

4.3. Implications of the Findings for Clinical Practice

Overall, in our analysis, the mean difference in SoMpGC between the back pain patients and the healthy subjects is small, especially in relation to the standard deviation in both groups. This is to be expected for the group as a whole, as the variability is already high in the healthy group, and the group of back pain patients in our study was small and not uniform. Hence, we regard our results as an indication of maybe altered movement patterns on the vertebral plane. Nonetheless, based on the proven reliability, reproducibility [11,20], and accuracy [11,21,22] of DST measurements, useful information can be obtained from the surface of the back for individual cases as to whether the patient is currently using their spine more or less intensively. This can be a valuable hint for the therapeutic process of Clinical Reasoning. It objectively represents the movement information that is currently usually collected subjectively in the visual examination, also of the back surface. The information can become much more valuable if numerous parameters are included in the analysis at the same time in the future, resulting in a movement pattern. However, this is a challenge for the classically established analytical methods. Artificial intelligence offers an opportunity here, which has already been used in initial work in connection with DST measurement [20,21,34].

5. Conclusions

In our study, we were able to successfully apply DST measurement to back pain patients. The study population was limited by acute pain, fear, and the necessity of walking on a treadmill for the examination. Because of this, the application is limited to a subgroup of the overall back pain collective.

Until now, the influence of muscle activity and soft tissue displacement on the measurement results is unclear. On average, the differences in the measurement results between back pain patients and healthy individuals are small, but the standard deviation is large. Nonetheless, information can be obtained on the movement behavior of individual patients, therefore, could be useful in follow-up visits.

This study showed there is still a significant need for research in segmental motion analysis of the spine; yet, given the tremendous importance of back pain in society, it should be pursued further. For future investigations, precautions are conceivable to increase the subgroup (e.g., slowly increasing walking speeds, the decision of the patient when the speed increases, and to which speed limits). The integration of more parameters into the analysis, could greatly increase the value of the analysis.

Supplementary Materials: The following supporting information can be downloaded at: https://www.mdpi.com/article/10.3390/bioengineering12030289/s1, Caption Supplementary Table S1: Displayed are the effect sizes for all segments of the sum of motion per gait cycle (SoMpGC) for all speeds. Positive values points towards the painful area, negative values points towards the healthy reference group.

Author Contributions: Conceptualization, J.K. (Johanna Kniepert), C.W., U.B., J.K. (Jürgen Konradi) and J.H.; methodology, J.K. (Johanna Kniepert), U.B., J.K. (Jürgen Konradi) and J.H.; software, R.W.; validation, R.W.; formal analysis, J.K. (Johanna Kniepert); investigation, J.K. (Johanna Kniepert); resources, J.K. (Johanna Kniepert), P.D. and U.B.; data curation, J.K. (Johanna Kniepert), J.K. (Jürgen Konradi) and R.W.; writing—original draft preparation, J.K. (Johanna Kniepert) and H.R.; writing—review and editing, U.B., J.K. (Jürgen Konradi), H.R., C.W., J.H. and P.D.; visualization, J.K. (Johanna Kniepert), J.H. and H.R.; supervision, P.D. and U.B.; project administration, J.K. (Jürgen Konradi). All authors have read and agreed to the published version of the manuscript.

Funding: This research received no external funding.

Institutional Review Board Statement: The study was approved by the Ethics Committee of the medical chamber Rhineland-Palatinate Germany (837.321.17 (11155)). The collected data were pseudonymized. The trial is registered with the World Health Organization (INT: DRKS00013145). The study was conducted in accordance with the Declaration of Helsinki.

Informed Consent Statement: Informed consent was obtained from all subjects involved in the study.

Data Availability Statement: The data can be made available upon request.

Conflicts of Interest: The authors declared no conflicts of interest.

Abbreviations

The following abbreviations are used in this manuscript:

2MWT	Two-minute Walk Test
BMI	Body Mass Index
DST	Dynamic Surface Topography
SoMpGC	Sum of motion per gait cycle
VAS	Visual Analogue Scale
VP	Vertebra prominens

References

1. GBD 2021 Low Back Pain Collaborators. Global, regional, and national burden of low back pain, 1990–2020, its attributable risk factors, and projections to 2050: A systematic analysis of the Global Burden of Disease Study 2021. *Lancet Rheumatol.* **2023**, *5*, e316–e329. [CrossRef] [PubMed]
2. Maher, C.; Underwood, M.; Buchbinder, R. Non-specific low back pain. *Lancet* **2017**, *389*, 736–747. [CrossRef] [PubMed]
3. Ract, I.; Meadeb, J.M.; Mercy, G.; Cueff, F.; Husson, J.L.; Guillin, R. A review of the value of MRI signs in low back pain. *Diagn. Interv. Imaging* **2015**, *96*, 239–249. [CrossRef]
4. Bayer, T.L.; Baer, P.E.; Early, C. Situational and psychophysiological factors in psychologically induced pain. *Pain* **1991**, *44*, 45–50. [CrossRef]
5. Jensen, M.C.; Brant-Zawadzki, M.N.; Obuchowski, N.; Modic, M.T.; Malkasian, D.; Ross, J.S. Magnetic resonance imaging of the lumbar spine in people without back pain. *N. Engl. J. Med.* **1994**, *331*, 69–73. [CrossRef]
6. O'Sullivan, P. Diagnosis and classification of chronic low back pain disorders: Maladaptive movement and motor control impairments as underlying mechanism. *Man. Ther.* **2005**, *10*, 242–255. [CrossRef]
7. Lamoth, C.J.C.; Meijer, O.G.; Daffertshofer, A.; Wuisman, P.I.J.M.; Beek, P.J. Effects of chronic low back pain on trunk coordination and back muscle activity during walking: Changes in motor control. *Eur. Spine J.* **2006**, *15*, 23–40. [CrossRef] [PubMed]
8. Smith, J.A.; Stabbert, H.; Bagwell, J.J.; Teng, H.-L.; Wade, V.; Lee, S.-P. Do people with low back pain walk differently? A systematic review and meta-analysis. *J. Sport Health Sci.* **2022**, *11*, 450–465. [CrossRef]
9. Goodvin, C.; Park, E.J.; Huang, K.; Sakaki, K. Development of a real-time three-dimensional spinal motion measurement system for clinical practice. *Med. Biol. Eng. Comput.* **2006**, *44*, 1061–1075. [CrossRef]
10. Fernandes, R.S.; Armada-da-Silva, P.; Pool-Goudzwaard, A.L.; Moniz-Pereira, V.; Veloso, A.N.P. Corrigendum to "Test-retest reliability and minimal detectable change of three-dimensional gait analysis in chronic low back pain patients" [Gait Posture 42 (2015) 491-497]. *Gait Posture* **2016**, *48*, 268, Erratum in *Gait Posture* **2015**, *42*, 491–497. [CrossRef]
11. Betsch, M.; Wild, M.; Johnstone, B.; Jungbluth, P.; Hakimi, M.; Kuhlmann, B.; Rapp, W. Evaluation of a novel spine and surface topography system for dynamic spinal curvature analysis during gait. *PLoS ONE* **2013**, *8*, e70581. [CrossRef]
12. Betsch, M.; Wild, M.; Rath, B.; Tingart, M.; Schulze, A.; Quack, V. Radiation-free diagnosis of scoliosis: An overview of the surface and spine topography. *Orthopade* **2015**, *44*, 845–851. [CrossRef] [PubMed]
13. Drerup, B.; Ellger, B.; Meyer zu Bentrup, F.M.; Hierholzer, E. Functional rasterstereographic images. A new method for biomechanical analysis of skeletal geometry. *Orthopade* **2001**, *30*, 242–250. [CrossRef] [PubMed]
14. Drerup, B.; Hierholzer, E. Evaluation of frontal radiographs of scoliotic spines—Part II. Relations between lateral deviation, lateral tilt and axial rotation of vertebrae. *J. Biomech.* **1992**, *25*, 1443–1450. [CrossRef]
15. Drerup, B.; Hierholzer, E. Evaluation of frontal radiographs of scoliotic spines—Part I. Measurement of position and orientation of vertebrae and assessment of clinical shape parameters. *J. Biomech.* **1992**, *25*, 1357–1362. [CrossRef] [PubMed]
16. Drerup, B.; Hierholzer, E. Back shape measurement using video rasterstereography and three-dimensional reconstruction of spinal shape. *Clin. Biomech.* **1994**, *9*, 28–36. [CrossRef]
17. Turner-Smith, A.R. A television/computer three-dimensional surface shape measurement system. *J. Biomech.* **1988**, *21*, 515–529. [CrossRef]
18. Drerup, B.; Hierholzer, E. Assessment of scoliotic deformity from back shape asymmetry using an improved mathematical model. *Clin. Biomech.* **1996**, *11*, 376–383. [CrossRef]
19. Mohokum, M.; Mendoza, S.; Udo, W.; Sitter, H.; Paletta, J.R.; Skwara, A. Reproducibility of rasterstereography for kyphotic and lordotic angles, trunk length, and trunk inclination: A reliability study. *Spine* **2010**, *35*, 1353–1358. [CrossRef]
20. Gipsman, A.; Rauschert, L.; Daneshvar, M.; Knott, P. Evaluating the Reproducibility of Motion Analysis Scanning of the Spine during Walking. *Adv. Med.* **2014**, *2014*, 721829. [CrossRef]
21. Dindorf, C.; Konradi, J.; Wolf, C.; Taetz, B.; Bleser, G.; Huthwelker, J.; Drees, P.; Fröhlich, M.; Betz, U. General method for automated feature extraction and selection and its application for gender classification and biomechanical knowledge discovery of sex differences in spinal posture during stance and gait. *Comput. Methods Biomech. Biomed. Eng.* **2020**, *24*, 299–307. [CrossRef]
22. Dindorf, C.; Konradi, J.; Wolf, C.; Taetz, B.; Bleser, G.; Huthwelker, J.; Werthmann, F.; Bartaguiz, E.; Kniepert, J.; Drees, P.; et al. Classification and Automated Interpretation of Spinal Posture Data Using a Pathology-Independent Classifier and Explainable Artificial Intelligence (XAI). *Sensors* **2021**, *21*, 6323. [CrossRef]
23. Frobin, W.; Hierholzer, E. Automatic measurement of body surfaces using rasterstereography. Part I: Image scan and control point measurement. *Photogramm. Eng. Rem. S* **1983**, *49*, 377–384.
24. Frobin, W.; Hierholzer, E. Automatic measurement of body surfaces using rasterstereography. Part II: Analysis of the rasterstereographic line pattern and three-dimensional surface reconstruction. *Photogramm. Eng. Rem. S* **1983**, *49*, 1443–1452.

25. Huthwelker, J.; Konradi, J.; Wolf, C.; Westphal, R.; Schmidtmann, I.; Schubert, P.; Drees, P.; Betz, U. Reference values and functional descriptions of transverse plane spinal dynamics during gait based on surface topography. *Hum. Mov. Sci.* **2023**, *88*, 103054. [CrossRef]
26. Bohannon, R.W.; Wang, Y.C.; Gershon, R.C. Two-minute walk test performance by adults 18 to 85 years: Normative values, reliability, and responsiveness. *Arch. Phys. Med. Rehabil.* **2015**, *96*, 472–477. [CrossRef] [PubMed]
27. Asamoah, V.; Mellerowicz, H.; Venus, J.; Klöckner, C. Oberflächenvermessung des Rückens. *Der Orthopäde* **2000**, *29*, 480–489. [CrossRef]
28. Huthwelker, J.; Konradi, J.; Wolf, C.; Westphal, R.; Schmidtmann, I.; Drees, P.; Betz, U. Reference Values for 3D Spinal Posture Based on Videorasterstereographic Analyses of Healthy Adults. *Bioengineering* **2022**, *9*, 809. [CrossRef]
29. Kniepert, J. 4D-Bewegungsanalyse der Wirbelsäule im Gang nach mono-und Multisegmentaler Versteifung der Wirbelsäule Sowie bei Patienten mit Rückenschmerzen. Ph.D. Thesis, Johannes Gutenberg-Universität Mainz, Mainz, Germany, 2019.
30. Haimerl, M.; Nebel, I.; Linkerhagner, A.; Konradi, J.; Wolf, C.; Drees, P.; Betz, U. Comprehensive visualization of spinal motion in gait sequences based on surface topography. *Hum. Mov. Sci.* **2022**, *81*, 102919. [CrossRef]
31. Dindorf, C.; Konradi, J.; Wolf, C.; Taetz, B.; Bleser, G.; Huthwelker, J.; Werthmann, F.; Drees, P.; Frohlich, M.; Betz, U. Machine learning techniques demonstrating individual movement patterns of the vertebral column: The fingerprint of spinal motion. *Comput. Methods Biomech. Biomed. Eng.* **2021**, *25*, 821–831. [CrossRef]
32. Konradi, J. SPSS syntax script to create graphs of spinal motion relative to phases of gait. *Mendeley Data* **2022**, *V1*. [CrossRef]
33. Konradi, J. SPSS syntax script to create graphs of spinal motion for a Standardized Gait Cycle. *Mendeley Data* **2022**, *V1*. [CrossRef]
34. Konradi, J.; Betz, U. Validation of automatic detection of gait phases. *Mendeley Data* **2022**, *V1*. [CrossRef]
35. Schmidtmann, I.; Konradi, J. SAS syntax script for merging export files. *Mendeley Data* **2022**, *V1*. [CrossRef]
36. Westphal, R.; Konradi, J. SAS syntax script for creation of a Standardized Gait Cycle. *Mendeley Data* **2022**, *V1*. [CrossRef]
37. Cohen, J. *Statistical Power Analysis for the Behavioural Sciences*, 2nd ed.; Lawrence Erlbaum: Hillsdale, MI, USA, 1988.
38. Hedges, L.V.; Olkin, I. *Statistical Methods for Meta-Analysis*, 1st ed.; Academic Press: New York, NY, USA, 1985; p. 369.
39. Dickey, J.P.; Pierrynowski, M.R.; Bednar, D.A.; Yang, S.X. Relationship between pain and vertebral motion in chronic low-back pain subjects. *Clin. Biomech.* **2002**, *17*, 345–352. [CrossRef]
40. Blankenbaker, D.G.; Haughton, V.M.; Rogers, B.P.; Meyerand, M.E.; Fine, J.P. Axial rotation of the lumbar spinal motion segments correlated with concordant pain on discography: A preliminary study. *AJR Am. J. Roentgenol.* **2006**, *186*, 795–799. [CrossRef]
41. Passias, P.G.; Wang, S.; Kozanek, M.; Xia, Q.; Li, W.; Grottkau, B.; Wood, K.B.; Li, G. Segmental lumbar rotation in patients with discogenic low back pain during functional weight-bearing activities. *J. Bone Jt. Surg.* **2011**, *93*, 29–37. [CrossRef]
42. Feipel, V.; De Mesmaeker, T.; Klein, P.; Rooze, M. Three-dimensional kinematics of the lumbar spine during treadmill walking at different speeds. *Eur. Spine J.* **2001**, *10*, 16–22. [CrossRef]
43. Dvorak, J.; Vajda, E.G.; Grob, D.; Panjabi, M.M. Normal motion of the lumbar spine as related to age and gender. *Eur. Spine J.* **1995**, *4*, 18–23. [CrossRef]
44. Ochia, R.S.; Inoue, N.; Takatori, R.; Andersson, G.B.; An, H.S. In vivo measurements of lumbar segmental motion during axial rotation in asymptomatic and chronic low back pain male subjects. *Spine* **2007**, *32*, 1394–1399. [CrossRef]
45. Gombatto, S.P.; Brock, T.; DeLork, A.; Jones, G.; Madden, E.; Rinere, C. Lumbar spine kinematics during walking in people with and people without low back pain. *Gait Posture* **2015**, *42*, 539–544. [CrossRef]
46. Crosbie, J.; de Faria Negrao Filho, R.; Nascimento, D.P.; Ferreira, P. Coordination of spinal motion in the transverse and frontal planes during walking in people with and without recurrent low back pain. *Spine* **2013**, *38*, E286–E292. [CrossRef] [PubMed]
47. Schmölz, W.; Bostelmann, R. 2.1 Biomechanik der Wirbelsäule. In *Wirbelsäule Interdisziplinär: Operative und Konservative Therapie*; Börm, W., Meyer, F., Bullmann, V., Knop, C., Eds.; Schatthauer: Stuttgart, Germany, 2017; pp. 21–28.
48. van der Hulst, M.; Vollenbroek-Hutten, M.M.; Rietman, J.S.; Schaake, L.; Groothuis-Oudshoorn, K.G.; Hermens, H.J. Back muscle activation patterns in chronic low back pain during walking: A "guarding" hypothesis. *Clin. J. Pain* **2010**, *26*, 30–37. [CrossRef]
49. Beauchet, O.; Annweiler, C.; Lecordroch, Y.; Allali, G.; Dubost, V.; Herrmann, F.R.; Kressig, R.W. Walking speed-related changes in stride time variability: Effects of decreased speed. *J. Neuroeng. Rehabil.* **2009**, *6*, 32. [CrossRef] [PubMed]
50. Akl, A.R.; Gonçalves, P.; Fonseca, P.; Hassan, A.; Vilas-Boas, J.P.; Conceição, F. Muscle Co-Activation around the Knee during Different Walking Speeds in Healthy Females. *Sensors* **2021**, *21*, 677. [CrossRef]
51. Hodges, P.W.; Moseley, G.L.; Gabrielsson, A.; Gandevia, S.C. Experimental muscle pain changes feedforward postural responses of the trunk muscles. *Exp. Brain Res.* **2003**, *151*, 262–271. [CrossRef]
52. Vlaeyen, J.W.S.; Linton, S.J. Fear-avoidance model of chronic musculoskeletal pain: 12 years on. *Pain* **2012**, *153*, 1144–1147. [CrossRef] [PubMed]
53. Arendt-Nielsen, L.; Graven-Nielsen, T.; Svarrer, H.; Svensson, P. The influence of low back pain on muscle activity and coordination during gait: A clinical and experimental study. *Pain* **1996**, *64*, 231–240. [CrossRef]

54. Mörl, F.; Blickhan, R. Three-dimensional relation of skin markers to lumbar vertebrae of healthy subjects in different postures measured by open MRI. *Eur. Spine J.* **2006**, *15*, 742–751. [CrossRef]
55. Colyer, S.L.; Evans, M.; Cosker, D.P.; Salo, A.I.T. A Review of the Evolution of Vision-Based Motion Analysis and the Integration of Advanced Computer Vision Methods Towards Developing a Markerless System. *Sports Med. Open* **2018**, *4*, 24. [CrossRef] [PubMed]

bioengineering

MDPI

Article

Augmenting Screw Technique to Prevent TLIF Cage Subsidence: A Biomechanical In Vitro Study

Alina Jacob [1,*], Alicia Feist [2], Ivan Zderic [2], Boyko Gueorguiev [2], Jan Caspar [2], Christian R. Wirtz [1], Geoff Richards [2], Markus Loibl [3], Daniel Haschtmann [3] and Tamas F. Fekete [3]

[1] Department of Neurosurgery, University of Ulm, 89081 Ulm, Germany
[2] AO Research Institute Davos, 7270 Davos, Switzerland
[3] Department of Spine Surgery, Schulthess Clinic, 8008 Zurich, Switzerland
* Correspondence: alina.jacob@uni-ulm.de

Abstract: (1) Cage subsidence in spine surgery is a frequent clinical challenge. This study aimed to assess a novel screw augmentation technique for Transforaminal Lumbar Interbody Fusion in cadavers of reduced bone mineral density (BMD). (2) Forty human lumbar vertebrae (BMD 84.2 ± 24.4 mgHA/cm^3, range 51–119 mgHA/cm^3) were assigned to two groups: augmenting screw group and control group. The augmentation technique comprised placement of two additional subcortical screws. Ten constructs per group were loaded with a quasi-static load-to-failure protocol and other ten were cyclically loaded. Failure modes were documented. (3) During the quasi-static load-to-failure testing, the augmenting screw technique showed a significantly higher failure load (1426.0 ± 863.6 N) versus the conventional technique in the control group (682.2 ± 174.5 N, p = 0.032). Cyclic loading revealed higher number of cycles and corresponding load until reaching 5 mm subsidence and significantly higher number of cycles and corresponding load until reaching 10 mm subsidence for the augmenting screw technique (9645 ± 3050; 1164.5 ± 305.0 N) versus the conventional technique in the control group (5395 ± 2340; 739.5 ± 234.0 N, $p < 0.05$). Failure modes were different and showed bending of the augmenting screws, followed by cut-out. (4) The investigated augmenting screw technique demonstrated higher failure loads and cycles to failure against cage subsidence compared to conventional cage placement. Failure modes were different between the two techniques and may lead to a different kind of complications.

Keywords: subsidence; lumbar spine surgery; osteoporosis; biomechanics; TLIF

1. Introduction

Annually, 39 million individuals worldwide are diagnosed with lumbar spondylolisthesis, 103 million with lumbar spinal stenosis, and 403 million with symptomatic lumbar disc degeneration [1–3]; given the demographic trends, a further rise in incident rates is expected [4]. The increase in degenerative diseases of the lumbar spine has led to a worldwide rise in the number of lumbar fusion surgeries, in which intervertebral cages are frequently used [2,5]. The benefits of a correctly placed cage include the increase in disc and foraminal height and the correction of lordosis. Biomechanically, an intervertebral cage absorbs axial compression forces and thereby decreases the load on the screw–rod construct, which potentially reduces the risk of construct and implant failure [6,7]. Transforaminal lumbar interbody fusion (TLIF) surgery is an established treatment option for degenerative diseases of the lumbar spine including foraminal stenosis, osteochondrosis, spinal canal stenosis, and spondylolisthesis [8]. Although TLIF shows various advantages over other types of procedures and achieves decent clinical outcomes with high fusion rates [9], cage subsidence is one of the most common complications in TLIF surgery. The reported rate of cage subsidence in TLIF is up to 51.2% [10]. Subsidence is not necessarily associated with poor outcomes for the patient but can result in loss of correction and foraminal stenosis. If severe subsidence occurs, the biomechanical advantages of cages are lost. Potential long-term complications include construct failure up to kyphotic deformity, implant loosening, and pseudarthrosis [10–12]. Risk factors for cage subsidence are low bone mineral density (BMD) [13,14], specifically in the endplates [15], endplate weakening during disc space preparation [16], and cage geometry [11,17–22]. Singular cages with a smaller contact area are more likely to subside than multiple and customed cages with a larger contact area due to higher peak loads acting on the endplate [18,22,23]. Endplate resistance to subsidence is strongest in the posterolateral region, and weakest in the central and anterior regions [24]. The typical cage geometry and preferred placement for optimal correction of the sagittal profile make TLIF cages susceptible to subsidence. TLIF cages are typically singular cages and have a significantly smaller footprint coverage than anterior lumbar interbody fusion and lateral lumbar interbody fusion cages due to limited access to the disc space [25,26]. Their placement is often preferred in the anterior third of the vertebra for lordosis restoration [10,25].

In low bone quality, sinking of a cage which cannot be halted may occur. Cases have been observed where the cage eventually settles on a pedicle screw. At this stage, alignment correction is lost, and the construct has failed. This observation raises the question of whether targeted subcortical screw placement could help prevent cage subsidence at an earlier stage. Therefore, the present study aimed to biomechanically assess whether bicortical screw augmentation can mitigate the subsidence of TLIF cages in specimens of reduced BMD. To the best of our knowledge, this technique has not been evaluated biomechanically for TLIF cages so far.

2. Materials and Methods

The Null hypotheses tested were that (1) the maximum failure load of TLIF cage subsidence cannot be altered by subcortically placed augmenting screws, (2) the number of cycles to failure cannot be increased by augmenting screws, and (3) failure modes are indifferent to TLIF cages without augmentation.

2.1. Specimens and Preparation

Forty fresh-frozen human cadaveric lumbar vertebrae (two L1, twelve L2, twelve L3, six L4, eight L5) from four male and four female donors aged 82 years on average (range 70–94 years) were used. Quantitative computed tomography (CT) scanning was performed (Revolution Evo, GE Healthcare Chicago, IL, USA) to measure the trabecular BMD using visualization software (AMIRA, Visage Imaging, Berlin, Germany). Trabecular BMD was 77.9 ± 34.3 mgHA/cm^3 (mean $\pm$ standard deviation, SD). Based on BMD, the vertebrae were assigned to two clusters of twenty vertebrae each for treatment with a TLIF cage with or without additional subchondral screw augmentation. Within each cluster, ten constructs (n = 10) were further assigned for either destructive quasi-static or cyclic biomechanical loading.

2.2. Surgical Technique

Prior to preparation and biomechanical testing, all specimens were thawed at room temperature overnight, the soft tissue was carefully removed, and the vertebrae were isolated. The intervertebral discs were removed on both sides, with special care taken to preserve the integrity of the endplates.

Figure 1 schematically illustrates the surgical technique. The augmentation technique commenced with the removal of the dorsal structure using an oscillating saw to avoid compression force transmission through the posterior structures.

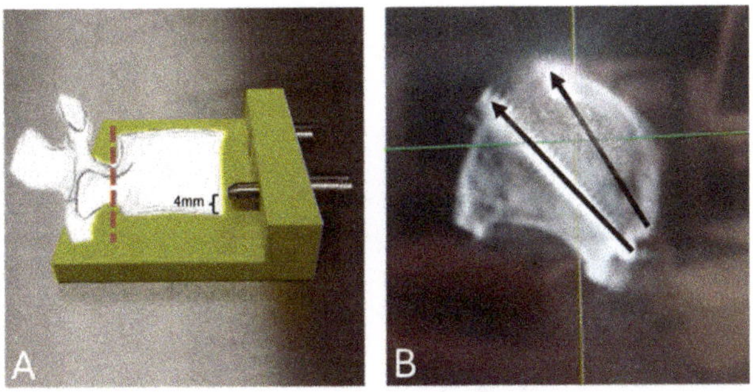

Figure 1. Visualization of the augmentation technique. (**A**) The picture shows a custom-made 3D-printed drill guide which ensured uniform screw placement 4 mm below the endplate. The red dashed line indicates the removal of the dorsal structures. A custom-made drill guide ensured uniform screw placement in all vertebrae 4 mm below the endplate. (**B**) The picture represents a 3D scan of a prepared specimen. The dorsal structures were removed. Two screws were inserted bicortically from the posterolateral third of the vertebra, the area which is accessible during surgery during a TLIF approach.

The isolated vertebral body was then placed upside down on a custom-made 3D-printed drill guide which ensured uniform screw placement 4 mm below the endplate. The entry points were set at the left posterolateral third of the vertebra in the area that is accessible during a common approach during in vivo TLIF surgery. The bony bridge between both entry points was maintained at a minimum of 5 mm to prevent the creation of a potential weak point for cortical fracture between the screws. The screw trajectories were not parallel and aimed to ensure optimal support of the TLIF cage. Following pre-drilling of the holes using a 2.7 mm drill, fully threaded, cancellous 4.0 mm screws (Johnson & Johnson MedTech, Zuchwil, Switzerland) were placed manually. The length of each screw was determined to provide bicortical anchorage. In this context, bicortical anchorage refers to the fixation of each augmenting screw to penetrate both lateral cortices of the vertebral body. The length of the screws ranged between 42 mm and 50 mm, and they were angled relative to each other within a range of 11° to 17°. Correct placement was radiographically verified.

A polyether ether ketone (PEEK) T-PAL cage (12 × 13 × 30 mm, Johnson & Johnson MedTech, MA, USA) was placed in the anterior third of the vertebra in the median sagittal plane.

2.3. Biomechanical Testing

Biomechanical testing was performed on a biaxial servo-hydraulic material testing machine (MTS 858 Mini Bionix II, MTS Systems, Eden Prairie, MN, USA) equipped with a 5 kN load cell (MCS10-005, HBM, Darmstadt, Germany). The test setup was in accordance with the ASTM F2077-18 test standards for intervertebral body fusion devices [27] and adopted from Calek et al. [16]. The vertebra was mounted and secured on a tiltable device connected to the machine base, which allowed translation. Vertical load applied by the machine actuator with the interconnected load cell was transmitted via a flat cylindrical indenter, measuring 30 mm in diameter. The loading protocol for quasi-static testing comprised a single uniaxial quasi-static ramp in compression at a rate of 2 mm/min, starting at a manually set preload of 20 N. Cyclic testing was performed at 3 Hz applying a physiological loading profile. Whereas the valley load was held constant at 50 N throughout each test, the peak load, starting at 200 N, was monotonically increased cycle by cycle at a rate of 0.1 N/cycle, until the test stop criterion of 10 mm machine displacement with respect to the test begin was reached. A triggered lateral x-ray was shot every 200th cycle. The test ended when the test stop criterion was fulfilled, i.e., when the actuator displacement reached 10 mm compared to the initial preloaded position.

2.4. Data Acquisition and Analysis

Machine data of axial displacement and axial load were continuously recorded from the machine transducers at 2 Hz throughout the quasi-static and cyclic tests. The load-to-failure tests were recorded using continuous lateral fluoroscopy. Based on these, and with respect to the actuator position at the test start, machine displacements of 3 mm and 5 mm were considered as failure criteria for cage subsidence from quasi-static loading. The peak load reached before the corresponding criterion had been fulfilled was calculated. With regard to cyclic loading, the failure criteria were set to 3 mm, 5 mm, and 10 mm, and the corresponding numbers of cycles until fulfillment of these criteria were calculated together with the corresponding peak loads. To document subsidence macroscopically, a video camera recorded all tests.

Statistical analysis among the parameters of interest was performed using the Prism software package (Prism; IBM Corp., Armonk, NY, USA). Anderson–Darling and Shapiro–Wilk tests were used to assess the normality of the data distribution for each group separately. Based on a Gaussian distribution for all data sets, unpaired, two-tailed *t*-tests were used to compare the loads and number of cycles until fulfillment of the failure criteria between the two groups. F-test was used to compare variances of the mean failure loads to determine whether the groups' variability can be assumed equal. The association between BMD and failure load was analyzed using linear regression. The level of significance was set at $p \leq 0.05$, and a trend towards significance was defined as $0.05 < p < 0.075$.

3. Results

3.1. Quasi-Static Testing

The failure load for 3 mm subsidence did not differ significantly between the groups (887.5 ± 577.0 N in augmenting screw group, 891.4 ± 441.2 N in control group, $p = 0.956$, Figure 2, Table 1). The variances of the failure loads for 3 mm subsidence did not differ significantly between the groups ($p = 0.392$). The means of the failure loads reached during severe subsidence differed significantly between the groups (1426.0 ± 863.6 N in augmenting screw group, 682.2 ± 174.5 N in control group, $p = 0.032$). There was no significant association between BMD and failure load in linear regression analysis in either group. The variances of the failure loads for 5 mm subsidence were significantly different between the groups ($p < 0.001$).

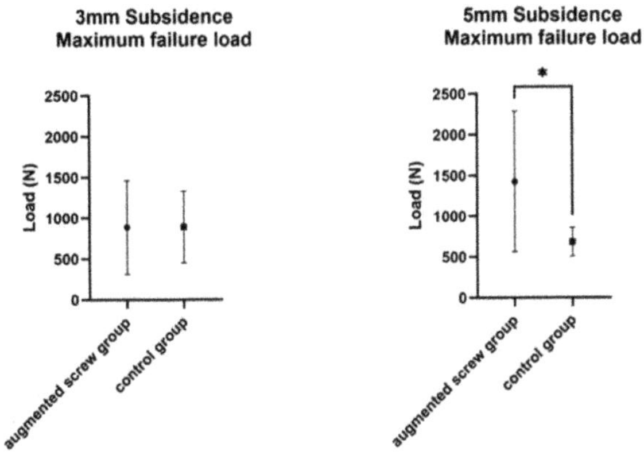

Figure 2. Failure load reached during moderate subsidence (3 mm displacement, **left**) and severe subsidence (5 mm displacement, **right**) of the augmenting screw group and the control group. There was higher data scattering for the augmenting screw technique. Asterisk indicates significance *: $p < 0.05$.

Table 1. Results for failure loads.

	Augmenting Screw Group (Mean ± SD) n = 10	Control Group (Mean ± SD) n = 10	*p*-Value	*p*-Value Variance
Failure load 3 mm subsidence	887.5 ± 577.0	891.4 ± 441.2	0.956	0.392
Failure load 5 mm subsidence	1426.0 ± 863.6	682.2 ± 174.5	<0.05 *	<0.001 ***

* Asterisk indicates significance *: $p < 0.05$, ***: $p < 0.001$.

3.2. Cyclic Testing

In the second testing protocol, the number of cycles to reach 3 mm subsidence in the augmenting screw group was not significantly different from the control group. The number of cycles to reach 5 mm subsidence was higher in the augmenting screw group compared to the control group and showed a statistical trend but did not reach significance ($p = 0.068$). The number of cycles to reach 10 mm subsidence was significantly higher in the augmenting screw group than in the control group ($p = 0.015$; Figure 3, Table 2).

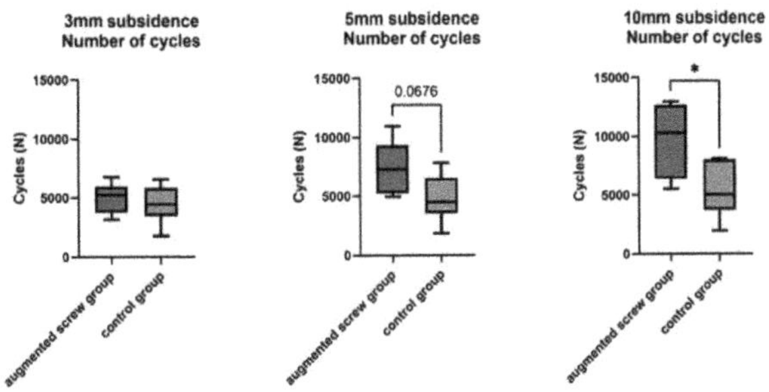

Figure 3. Number of cycles to reach 3 mm subsidence (**left**), 5 mm subsidence (**middle**), and 10 mm subsidence (**right**) of the augmenting screw group and the control group. Asterisk indicates significance *: $p < 0.05$.

Table 2. Results cyclic loading protocol.

Variable	Augmenting Screw Group (Mean ± SD) n = 10	Control Group (Mean ± SD) n = 10	p-Value
Number of cycles to 3 mm subsidence	4960 ± 1369	4466 ± 1639	0.567
Number of cycles to 5 mm subsidence	7437 ± 2388	4812 ± 2035	0.068
Number of cycles to 10 mm subsidence	9645 ± 3050	5395 ± 2340	<0.05 *

* Asterisk indicates significance *: $p < 0.05$.

3.3. Failure Modes

The failure modes observed during the quasi-static load-to-failure tests were similar to those seen during the cyclic loading protocol and differed between the two groups (Figure 4). In the control group, the failure mode was an endplate fracture at the site of the cage and continuous sinking of the cage into the vertebra reaching moderate and finally severe subsidence. In the augmenting screw group, the cage first settled into the endplate, then the two subcortical screws were bent slightly directly under the cage in the direction of force action with preserved screw anchorage in the cortices. During this phase, the cage was pushed downwards into the cancellous bone. With test progression, a clear cutting-through of the screws not only through the trabecular bone but also at the cortices of the vertebra was observed, allowing further sinking of the cage into the vertebra, reaching severe subsidence. Cut-out was more evident in the vertebrae that underwent quasi-static load-to-failure test than those tested cyclically.

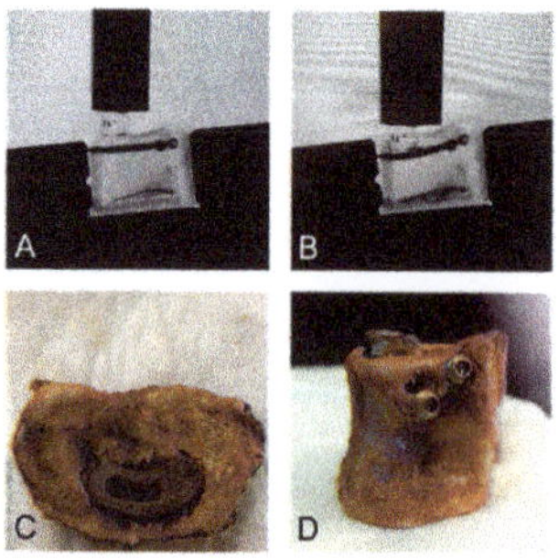

Figure 4. Failure modes. Lateral x-ray of a lumbar vertebra TLIF cage with augmenting screws (**A**) before and (**B**) during quasi-static load-to-failure testing. In the augmenting screw group, bending of the screws was observed radiologically. (**C**) Macroscopically, endplate fracture and cage subsidence were confirmed after axial compression loading, and (**D**) cortical cut-out of the screws was observed.

4. Discussion

The main findings of the current study are that (1) the failure load to reach severe TLIF cage subsidence can be altered by subcortically placed augmenting screws, (2) the number of cycles to failure can be increased by augmenting screws, and (3) the failure modes are different between the two techniques. These findings indicate that the augmenting screw technique potentially prevents severe subsidence and may help in selected cases where conventional solutions for subsidence prevention are not feasible.

While the effects of mild subsidence may not necessarily be clinically significant, the clinical relevance of severe subsidence is undoubtedly significant from a biomechanical perspective. Mild subsidence occurs frequently and is not necessarily associated with worse outcomes [10,28]. When severe subsidence occurs, biomechanically, the correction is lost, and the cage loses its ability to support the anterior column, and, clinically, severe subsidence has been described to predict TLIF pseudarthrosis and screw loosening [12].

The investigated augmentation technique may help prevent severe subsidence, but not mild or moderate subsidence.

The augmenting screw technique investigated in the present study mitigated severe subsidence, but not mild and moderate subsidence up to 3 mm. A possible explanation is that the screws were placed 4 mm subcortically. This distance was necessary due to the differing endplate concavity depth of the specimens. Osteoporotic, degenerated vertebrae were selected to better reflect the anatomy of patients undergoing TLIF surgery, but degeneration is associated with a change in endplate anatomy with a more variable surface geometry [29]. To allow for standardized screw placement in all specimens without endplate perforation, the entry point was required to suit the vertebra with the greatest concavity, which was, in the present case, set as 4 mm below the endplate, measured at the lateral cortex. Clinically, the prevention of subsidence appears to be particularly relevant for vertebrae with a deep endplate concavity, as previous studies indicate that shallow endplates are associated with a lower subsidence rate [30–32]. It remains uncertain whether screws placed closer to the endplate could engage earlier to diminish the moderate subsidence of TLIF cages. Overall, particularly the prevention of severe subsidence seems clinically important. Specifically, for patients with osteoporotic vertebrae and deep endplate concavity, a proposed solution is the placement of screws 4 mm subcortically.

To prevent subsidence, pharmacological osteoporosis therapy is a widely used option in clinics but may be limited by patient-related contraindications and requires time to improve bone density. It can be costly, and there is a high geographical variability regarding the criteria for reimbursability, as shown by the example of teriparatide [33]. Apart from systemic therapy and reducing patient-based risk factors, biomedical strategies include advancements in cage design and biomaterials, as well as cement augmentation [34]. Wilke et al. were among the first to demonstrate that cement augmentation can restore strength and stiffness in osteoporotic vertebral bodies, helping to prevent cage subsidence [23,26,35]. However, preventive vertebroplasty remains controversial due to the rare but potentially severe complications of cement extravasation and embolism [36–41]; additionally, an increased fracture risk of the vertebrae adjacent to the augmented ones has been described [14,42,43].

Wang et al. investigated a screw augmentation technique to prevent subsidence in oblique lumbar interbody fusion combined with anterolateral fixation (OLIF). The authors demonstrated that the trajectory, specifically, the position and angle of the vertebral screws was associated with cage subsidence. The authors concluded that inserting screws as close to the endplate as possible and parallel to each other while keeping the cage inside the range of the superior and inferior screws is an optimal implantation strategy [31]. Differences to the present study are the surgical technique and associated cage geometry, with OLIF cages having a larger footprint compared to TLIF T-pal cages. Wang et al. used one screw above and one below, while the present study investigates the effect of two screws augmenting the cage. Although there are distinct differences between the studies, the main findings of the present study are broadly in line with the findings of Wang et al., suggesting screw augmentation as an alternative to mitigate subsidence.

In the present study, the failure modes observed differed between the groups. The two subcortical screws were first bent slightly under the cage, allowing moderate but no severe subsidence. With test progression, a clear cutting through of the screws, not only through the trabecular bone but also at the cortices of the vertebra, was observed, with further sinking of the cage into the vertebra resulting in delayed severe subsidence. A possible explanation may be that the subcortical screws partially transmitted the axial load to the cortical bone accounting for a higher failure load requiring more cycles to reach severe subsidence. In the present study, fully threaded screws were used. In other anatomic areas, the use of partially threaded screws has been shown to decrease the risk of cut-out and could pose an alternative for vertebral use [44]. There was higher data scattering for the augmenting screw technique compared with the control group. This effect may be attributed to the use of osteoporotic specimens exhibiting degenerative changes including subchondral sclerosis. This increased density close to the endplates could provide improved screw fixation resulting in better screw purchase of the investigated technique. Overall, new failure modes need to be explored further as they may lead to a different kind of complications.

This biomechanical study inherits several limitations. First, isolated axial compression force was applied through the cage to the endplate and vertebra, which does not represent the physiological forces acting on the human spine in vivo. The primary biomechanical role of an intervertebral implant is to support the anterior column and transmit compressive forces, as 80% of all compressive forces are reported to be transmitted through the anterior column. Therefore, we considered this simplified model as appropriate. Second, the setup ensured a maximum contact area by carefully preparing the endplate and precise alignment of the upper surface of the cage upon the isolated vertebra, which could be tilted and translated as needed parallel to the stamp. However, in surgical settings, cage placement is far more complex and can be restricted. Peak loads at sites of smaller contact areas between the cage and vertebra may occur, for instance, when a cage is slightly tilted. In clinics, the threshold of endplate fracture may be lower. Third, screw placement to prevent cage subsidence in the present study was only tested for ideal placement conditions. In vitro conditions allowed unlimited access to an isolated vertebra lateral cage for optimal subcortical screw placement. The feasibility, failure rates, and modes of the investigated technique in vivo need to be assessed.

The concept is unique for TLIF and may prove valuable for selected cases with compromised bone quality, contributing to the existing body of literature. A relatively large sample size of forty fresh-frozen human cadaveric lumbar specimens was utilized in the current work to enhance the statistical power. Biomechanical studies with combined motion patterns and extensive cyclic testing should be performed to determine whether migration or loosening of the subcortical screws can occur. In addition, the biomechanical effects of different entry points and screw trajectories need to be investigated.

5. Conclusions

Augmenting screws can help prevent severe cage subsidence in vertebrae with low bone quality when conventional solutions are not feasible. However, this technique should be applied with caution as its failure modes differ and require further exploration to understand potential complications.

Author Contributions: Conceptualization, A.J., T.F.F., M.L., and D.H.; methodology, A.J., T.F.F., I.Z., J.C., and B.G.; software, A.J., A.F., and I.Z.; formal analysis, A.J., I.Z., and A.F.; investigation, A.J.; resources, T.F.F., B.G., M.L., C.R.W., and G.R.; writing—original draft preparation, A.J.; writing—review and editing, A.F., I.Z., A.J., D.H., and T.F.F.; visualization, A.J.; supervision, I.Z., T.F.F., C.R.W., G.R., B.G., M.L., and D.H.; project administration, B.G.. All authors have read and agreed to the published version of the manuscript.

Funding: The APC was funded by the University of Ulm, the department of Neurosurgery of the University of Ulm, and the department of Neurosurgery, Schulthess Clinic, Zurich.

Institutional Review Board Statement: All donors gave their informed consent inherent within the donation of the anatomical gift statement during their lifetime. In addition, the institutional review boards of the donor supplier company (ScienceCare, Phoenix, AZ, USA), as well as of the institution at which this study was conducted, approved this project, approval number BMD.MRF.23.5.1, approval date 10 January 2023.

Informed Consent Statement: Informed consent was obtained from all subjects involved in the study.

Data Availability Statement: Data available on request due to restrictions eg privacy or ethical.

Conflicts of Interest: The authors declare no conflicts of interest.

Abbreviations

The following abbreviations are used in this manuscript:

BMD	Bone mineral density;
OLIF	Oblique lumbar interbody fusion;
PEEK	Polyether ether ketone;
TLIF	Transforaminal lumbar interbody fusion.

References

1. Vos, T.; Lim, S.S.; Abbafati, C.; Abbas, K.M.; Abbasi, M.; Abbasifard, M.; Abbasi-Kangevari, M.; Abbastabar, H.; Abd-Allah, F.; Abdelalim, A.; et al. Global burden of 369 diseases and injuries in 204 countries and territories, 1990–2019: A systematic analysis for the Global Burden of Disease Study 2019. *Lancet* **2020**, *396*, 1204–1222. [CrossRef]
2. Reisener, M.-J.; Pumberger, M.; Shue, J.; Girardi, F.P.; Hughes, A.P. Trends in lumbar spinal fusion-a literature review. *J. Spine Surg.* **2020**, *6*, 752–761. [CrossRef]
3. Ravindra, V.M.; Senglaub, S.S.; Rattani, A.; Dewan, M.C.; Härtl, R.; Bisson, E.; Park, K.B.; Shrime, M.G. Degenerative Lumbar Spine Disease: Estimating Global Incidence and Worldwide Volume. *Glob. Spine J.* **2018**, *8*, 784–794. [CrossRef]
4. O'Lynnger, T.M.; Zuckerman, S.L.; Morone, P.J.; Dewan, M.C.; Vasquez-Castellanos, R.A.; Cheng, J.S. Trends for Spine Surgery for the Elderly: Implications for Access to Healthcare in North America. *Neurosurgery* **2015**, *77* (Suppl. 4), S136–S141. [CrossRef] [PubMed]
5. Martin, B.I.; Mirza, S.K.; Spina, N.; Spiker, W.R.; Lawrence, B.; Brodke, D.S. Trends in Lumbar Fusion Procedure Rates and Associated Hospital Costs for Degenerative Spinal Diseases in the United States, 2004 to 2015. *Spine* **2019**, *44*, 369–376. [CrossRef]
6. McAfee, P.C.; DeVine, J.G.; Chaput, C.D.; Prybis, B.G.; Fedder, I.L.; Cunningham, B.W.; Farrell, D.J.; Hess, S.J.; Vigna, F.E. The indications for interbody fusion cages in the treatment of spondylolisthesis: Analysis of 120 cases. *Spine* **2005**, *30*, S60–S65. [CrossRef] [PubMed]
7. Blumenthal, S.L.; Ohnmeiss, D.D. Intervertebral cages for degenerative spinal diseases. *Spine J.* **2003**, *3*, 301–309. [CrossRef]
8. Harms, J.G.; Jeszenszky, D. Die posteriore, lumbale, interkorporelle Fusion in unilateraler transforaminaler Technik. *Oper. Orthop. Traumatol.* **1998**, *10*, 90–102. [CrossRef]
9. Rathbone, J.; Rackham, M.; Nielsen, D.; Lee, S.M.; Hing, W.; Riar, S.; Scott-Young, M. A systematic review of anterior lumbar interbody fusion (ALIF) versus posterior lumbar interbody fusion (PLIF), transforaminal lumbar interbody fusion (TLIF), posterolateral lumbar fusion (PLF). *Eur. Spine J.* **2023**, *32*, 1911–1926. [CrossRef]
10. Parisien, A.; Wai, E.K.; ElSayed, M.S.A.; Frei, H. Subsidence of Spinal Fusion Cages: A Systematic Review. *Int. J. Spine Surg.* **2022**, *16*, 1103–1118. [CrossRef]
11. Liu, Y.; Li, N.-H. Factors associated with intervertebral cage subsidence in posterior lumbar fusion. *J. Orthop. Surg. Res.* **2024**, *19*, 7. [CrossRef] [PubMed]
12. Levy, H.A.; Pinter, Z.W.; Reed, R.; Harmer, J.R.; Raftery, K.; Nathani, K.R.; Katsos, K.; Bydon, M.; Fogelson, J.L.; Elder, B.D.; et al. Transforaminal lumbar interbody fusion subsidence: Computed tomography analysis of incidence, associated risk factors, and impact on outcomes. *J. Neurosurg. Spine* **2024**, *1*, 463–472. [CrossRef]
13. Filley, A.; Baldwin, A.; Ben-Natan, A.R.; Hansen, K.; Arora, A.; Xiao, A.; Hammond, D.; Chen, C.; Tweedt, I.; Rohde, J.; et al. The influence of osteoporosis on mechanical complications in lumbar fusion surgery: A systematic review. *N. Am. Spine Soc. J.* **2024**, *18*, 100327. [CrossRef]
14. Choi, S.S.; Kim, H.; Choung, Y.J.; Jeong, S.J.; Lee, C.H. Risk factors for new vertebral compression fracture after kyphoplasty and efficacy of osteoporosis treatment: A STROBE-compliant retrospective study. *Medicine* **2022**, *101*, e32018. [CrossRef] [PubMed]
15. Okano, I.; Jones, C.; Salzmann, S.N.; Reisener, M.-J.; Sax, O.C.; Rentenberger, C.; Shue, J.; Carrino, J.A.; Sama, A.A.; Cammisa, F.P.; et al. Endplate volumetric bone mineral density measured by quantitative computed tomography as a novel predictive measure of severe cage subsidence after standalone lateral lumbar fusion. *Eur. Spine J.* **2020**, *29*, 1131–1140. [CrossRef]
16. Calek, A.-K.; Cornaz, F.; Suter, M.; Fasser, M.-R.; Farshad, M.; Widmer, J. Endplate weakening during cage bed preparation significantly reduces endplate load capacity. *Eur. Spine J.* **2024**, *33*, 2621–2629. [CrossRef] [PubMed]

17. Amorim-Barbosa, T.; Pereira, C.; Catelas, D.; Rodrigues, C.; Costa, P.; Rodrigues-Pinto, R.; Neves, P. Risk factors for cage subsidence and clinical outcomes after transforaminal and posterior lumbar interbody fusion. *Eur. J. Orthop. Surg. Traumatol.* **2022**, *32*, 1291–1299. [CrossRef]
18. Yu, Y.; Robinson, D.L.; Ackland, D.C.; Yang, Y.; Lee, P.V.S. Influence of the geometric and material properties of lumbar endplate on lumbar interbody fusion failure: A systematic review. *J. Orthop. Surg. Res.* **2022**, *17*, 224. [CrossRef]
19. Zhang, Y.; Nüesch, C.; Mündermann, A.; Halbeisen, F.; Schären, S.; Netzer, C. Is Age a Risk Factor for Early Postoperative Cage Subsidence After Transforaminal Lumbar Interbody Fusion? A Retrospective Study in 170 Patients. *Glob. Spine J.* **2023**, *15*. [CrossRef]
20. Sultana, T.; Hossain, M.; Jeong, J.H.; Im, S. Comparative Analysis of Radiologic Outcomes Between Polyetheretherketone and Three-Dimensional-Printed Titanium Cages After Transforaminal Lumbar Interbody Fusion. *World Neurosurg.* **2023**, *179*, e241–e255. [CrossRef]
21. Galbusera, F.; Schmidt, H.; Wilke, H.-J. Lumbar interbody fusion: A parametric investigation of a novel cage design with and without posterior instrumentation. *Eur. Spine J.* **2012**, *21*, 455–462. [CrossRef] [PubMed]
22. Wilke, H.-J.; Volkheimer, D.; Robie, B.; Christensen, F.B. Two-piece ALIF cage optimizes the bone-implant interface in a 360° setting. *Eur. Spine J.* **2017**, *26*, 2747–2753. [CrossRef] [PubMed]
23. Wilke, H.J.; Kettler, A.; Claes, L. Primary stabilizing effect of interbody fusion devices for the cervical spine: An in vitro comparison between three different cage types and bone cement. *Eur. Spine J.* **2000**, *9*, 410–416. [CrossRef]
24. Lowe, T.G.; Hashim, S.; Wilson, L.A.; O'Brien, M.F.; Smith, D.A.B.; Diekmann, M.J.; Trommeter, J. A biomechanical study of regional endplate strength and cage morphology as it relates to structural interbody support. *Spine* **2004**, *29*, 2389–2394. [CrossRef]
25. Teng, I.; Han, J.; Phan, K.; Mobbs, R. A meta-analysis comparing ALIF, PLIF, TLIF and LLIF. *J. Clin. Neurosci.* **2017**, *44*, 11–17. [CrossRef] [PubMed]
26. Wilke, H.J.; Kettler, A.; Claes, L. Stabilizing effect and sintering tendency of 3 different cages and bone cement for fusion of cervical vertebrae segments. *Orthopade* **2002**, *31*, 472–480. [CrossRef]
27. F04 Committee. *Test Methods For Intervertebral Body Fusion Devices*; ASTM International: West Conshohocken, PA, USA, 2022.
28. Fushimi, K.; Miyagawa, T.; Iwai, C.; Nozawa, S.; Iinuma, N.; Tanaka, R.; Shirai, G.; Tanahashi, H.; Yokoi, T.; Akiyama, H. Transforaminal Lumbar Interbody Fusion with Double Banana Cages: Clinical Evaluations and Finite Element Model Analysis. *Glob. Spine J.* **2023**, *14*, 2031–2038. [CrossRef]
29. He, X.; Liang, A.; Gao, W.; Peng, Y.; Zhang, L.; Liang, G.; Huang, D. The relationship between concave angle of vertebral endplate and lumbar intervertebral disc degeneration. *Spine* **2012**, *37*, E1068–E1073. [CrossRef]
30. Hu, Z.; He, D.; Gao, J.; Zeng, Z.; Jiang, C.; Ni, W.; Yik, J.H.N.; Zhao, X.; Fan, S. The Influence of Endplate Morphology on Cage Subsidence in Patients With Stand-Alone Oblique Lateral Lumbar Interbody Fusion (OLIF). *Glob. Spine J.* **2023**, *13*, 97–103. [CrossRef]
31. Wang, K.; Wang, X.; Li, Z.; Xie, T.; Wang, L.; Luo, C.; Huang, S.; Zeng, J. The Influence of Screw Positioning on Cage Subsidence in Patients with Oblique Lumbar Interbody Fusion Combined with Anterolateral Fixation. *Orthop. Surg.* **2023**, *15*, 3263–3271. [CrossRef]
32. Xie, T.; Pu, L.; Zhao, L.; Lu, Y.; Yang, Z.; Wang, X.; Song, Y.; Zeng, J. Influence of coronal-morphology of endplate and intervertebral space to cage subsidence and fusion following oblique lumbar interbody fusion. *BMC Musculoskelet. Disord.* **2022**, *23*, 633. [CrossRef]
33. Minisola, S.; Cipriani, C.; Della Grotta, G.; Colangelo, L.; Occhiuto, M.; Biondi, P.; Sonato, C.; Vigna, E.; Cilli, M.; Pepe, J. Update on the safety and efficacy of teriparatide in the treatment of osteoporosis. *Ther. Adv. Musculoskelet. Dis.* **2019**, *11*, 1759720X19877994. [CrossRef]
34. Oshtory, R.; Harris, J.A.; Patel, P.D.; Mirabile, B.A.; Bucklen, B.S. Lumbar Intervertebral Spacer With Cement Augmentation of Endplates and Integrated Screws as a Fixation Device in an Osteoporotic Model: An In Vitro Kinematic and Load-to-Failure Study. *Int. J. Spine Surg.* **2021**, *15*, 324–333. [CrossRef]
35. Hung, S.-F.; Tsai, T.-T.; Wang, S.-F.; Hsieh, M.-K.; Kao, F.-C. Additional cement augmentation reduces cage subsidence and improves clinical outcomes in oblique lumbar interbody fusion combined with anterolateral screw fixation: A retrospective cohort study. *Curr. Probl. Surg.* **2024**, *61*, 101441. [CrossRef]
36. Hatzantonis, C.; Czyz, M.; Pyzik, R.; Boszczyk, B.M. Intracardiac bone cement embolism as a complication of vertebroplasty: Management strategy. *Eur. Spine J.* **2017**, *26*, 3199–3205. [CrossRef]
37. Malik, M.K.; Wroblewski, I.; Darki, A. Pulmonary Cement Embolism After Vertebroplasty. *Cureus* **2023**, *15*, e39194. [CrossRef]
38. Rothermich, M.A.; Buchowski, J.M.; Bumpass, D.B.; Patterson, G.A. Pulmonary cement embolization after vertebroplasty requiring pulmonary wedge resection. *Clin. Orthop. Relat. Res.* **2014**, *472*, 1652–1657. [CrossRef]
39. Waler, A.R.; Sanchez, K.J.; Parikh, A.A.; Okorie, O.N. A Case of Pulmonary Cement Embolism Managed through Symptomatic Treatment. *Case Rep. Crit. Care* **2020**, *2020*, 2425973. [CrossRef]

40. Gomez, F.A.; Herrera, O.M.; Gaona, J.L.V.; Reyes, C.A.F.; Gutiérrez, M.L.C.; Saenz, L.C.M. Pulmonary cement embolism following transpedicular screws placement for thoracolumbar fractures. *Surg. Neurol. Int.* **2021**, *12*, 495. [CrossRef]
41. Song, Y.; Huang, X.; Wu, L. Removal of intracardiac bone cement embolism after percutaneous kyphoplasty: A case report. *Medicine* **2020**, *99*, e19354. [CrossRef] [PubMed]
42. Hulme, P.A.; Krebs, J.; Ferguson, S.J.; Berlemann, U. Vertebroplasty and kyphoplasty: A systematic review of 69 clinical studies. *Spine* **2006**, *31*, 1983–2001. [CrossRef]
43. Adida, S.; Taori, S.; Wong, V.R.; Tang, A.; Sefcik, R.K.; Zhang, X.; Gerszten, P.C. Analysis of injected cement volume and clinical outcomes following balloon-assisted kyphoplasty in a series of 368 patients. *Clin. Neurol. Neurosurg.* **2024**, *243*, 108367. [CrossRef]
44. Wang, Y.; Han, N.; Zhang, D.; Zhang, P.; Jiang, B. Comparison between headless cannulated screws and partially threaded screws in femoral neck fracture treatment: A retrospective cohort study. *Sci. Rep.* **2022**, *12*, 1743. [CrossRef] [PubMed]

bioengineering

MDPI

Article

Using Surface Topography to Visualize Spinal Motion During Gait—Examples of Possible Applications and All Tools for Open Science

Jürgen Konradi [1,*], Ulrich Betz [1], Janine Huthwelker [1], Claudia Wolf [1], Irene Schmidtmann [2], Ruben Westphal [2], Meghan Cerpa [3], Lawrence G. Lenke [3] and Philipp Drees [4]

[1] Institute of Physical Therapy, Prevention and Rehabilitation, University Medical Center of the Johannes Gutenberg University Mainz, Langenbeckstrasse 1, D-55131 Mainz, Germany; ulrich.betz@unimedizin-mainz.de (U.B.); janine.huthwelker@unimedizin-mainz.de (J.H.); claudia.wolf@unimedizin-mainz.de (C.W.)

[2] Institute of Medical Biostatistics, Epidemiology and Informatics, University Medical Center of the Johannes Gutenberg University Mainz, Rhabanusstraße 3/Tower A, D-55118 Mainz, Germany; irene.schmidtmann@unimedizin-mainz.de (I.S.); rubenswestphal@gmail.com (R.W.)

[3] Department of Orthopedic Surgery, Columbia University Medical Center, 5141 Broadway, New York, NY 10032, USA; meghancerpa@yahoo.com (M.C.); ll2989@cumc.columbia.edu (L.G.L.)

[4] Department of Orthopedics and Trauma Surgery, University Medical Center of the Johannes Gutenberg University Mainz, Langenbeckstrasse 1, D-55131 Mainz, Germany; philipp.drees@unimedizin-mainz.de

* Correspondence: juergen.konradi@unimedizin-mainz.de

Abstract: Precise segmental spinal analysis during gait has various implications for clinical use and basic research. Here, we report the use of Surface Topography (ST) to analyze three-dimensional spinal segment movements, in combination with foot pressure measuring, to describe individual vertebral bodies' motion relative to specific phases of gait. Using Statistical Analysis System (SAS) scripts, single files were merged into one raw data table and were used to generate a standardized gait cycle (SGC) for each measurement, including all measured gait cycles for each individual patient, with a spline function to obtain smooth curve progressions. Graph templates from Statistical Package for the Social Sciences create detailed visualizations of the SGCs. Previously obtained measurements from healthy participants were used to demonstrate possible applications of our method. An impressive inter-individual variability as well as intra-individual consistency of spinal motion is shown. The transformation into an SGC facilitates intra- and inter-individual comparisons for qualitative and quantitative analyses. In future studies, we want to use this method to distinguish between physiologic and pathologic spinal motion. Artificial intelligence-based analysis can facilitate this process. All tools and visualizations used are freely available in repositories to enable the replication and validation of our findings.

Keywords: spine biomechanics; graph-based representation; motion analysis; surface topography; rasterstereography

1. Introduction

Precise segmental spinal analysis during gait would have various implications for basic research and clinical use, exemplarily in the context of low back pain (LBP). LBP in the German population has a point prevalence of 25–40%, a 12-month prevalence of 60–70% [1], and a lifetime prevalence in the American population of up to 85% [2] in adults. According to the German Medical Association [3], LBP is currently ranked as the most frequent musculoskeletal disorder, with an annual cost of 3.6 billion Euros. Thus, companies and stakeholders have a strong interest in reducing such often work-related diseases [4]. As many as 90% of LBP complaints have no anatomic structure abnormalities that can be identified as the source of the patient's pathologies [5]. Most of these occur in motion [6]. Hence, static and structure-orientated diagnostic approaches like X-rays or MRI cannot detect the etiology of pathology in the majority of LBP patients, meanwhile incurring unnecessary costs. Therefore, systems for multidimensional motion analysis are becoming instrumental in the diagnosis of unspecific musculoskeletal problems, as they are able to provide additional dynamic and functional information for individualized diagnoses [7], even though they do not provide ground-truth imaging, such as fluoroscopy or dynamic X-ray imaging.

Even though this three-dimensional approach is popular for musculoskeletal problems of the pelvic–leg region, the spine and trunk are often neglected due to metrological limitations [8,9]. For example, the assessment of each functional spinal unit requires the application of three non-collinear markers per segment [9]. Due to the close anatomical vicinity of adjacent vertebrae, unintended marker contact can cause significant measuring artifacts. Furthermore, due to the variety of spinal segments, a complex preparation is required, which is immensely prone to palpation bias and can result in measurement error [8]. Even though current research recognizes the spine's active role in balance and locomotion dynamics [10], truncal measurements in instrumented gait analysis mostly regard it as a rigid body [11], usually called the "passenger unit", which is transported by the "locomotor", with no relevant contribution to ambulation [12]. Based on the described limitations, there is little literature regarding three-dimensional segment-related spinal movement during gait. Additionally, the results of the few existing reports are not comparable because of methodological differences, ranging from three-dimensional motion analyses of isolated spinal segments to invasive measurement procedures [13–23].

The most valid method [19] uses markers to capture three-dimensional motion by inserting bone pins under local anesthesia into the spinal processes of each lumbar vertebra under the control of an image converter. This procedure can theoretically be considered a gold standard [24]. However, due to the surgical invasiveness and the resulting radiation exposure, this approach is inappropriate for use as a routine assessment in clinical as well as in scientific practice. Furthermore, only very low amplitudes of spinal motion have been observed, which may be either influenced by pain-induced inhibition of habitual movement or by residual effects of local anesthesia.

Hence, the ability to measure the spinal motion of each single segment during gait without extensive preparation or the usage of invasive or radiation-based measurement approaches, however, is a valuable tool for clinical practice as well as basic research. It can expand on our knowledge of the spine's role in maintaining balance and upright posture during gait, as well as provide further understanding of the underlying biomechanical mechanisms behind unspecific musculoskeletal conditions, such as LBP [24]. Rasterstereography (RS), more recently called Surface Topography (ST) [25], is a non-invasive and reliable [26] alternative high-precision technique to analyze the shapes of surfaces [27], even in 360° [28]. Resting upon back shape data and orientated on visible anatomical landmarks, the Turner–Smith model [29] combined with other models [30–33] can be used for the estimation of the segmental spine posture. Originally, it was used for static or quasi-dynamic measurements during stance [25], specifically within the context of scoliosis [34–37]. However, its use in the setting of degenerative disk disease has been questioned [38]. More recently, dynamic measurements have been introduced [39]. Using DIERS formetric's standard software during gait analysis, its reliability to detect certain measuring points (e.g., max of T4) at some point during the gait cycle has been demonstrated (Gipsman et al., 2014) [40]. What former ST approaches lacked were the precise determination of spinal movement in direct relation to phases of gait during the gait cycle, and subsequent standardization of data from various gait cycles to make data intra- and inter-individually comparable, regardless of aspects confounded by walking speed or stride length. As already demonstrated [41,42], we successfully further developed this method in this direction.

In our approach, we utilized DIERS formetric as a means for gathering dynamic ST measurements. The system generates 3D images of the surface, estimates corresponding 3D movements of the spine for each individual segment starting at vertebra prominens [30] and ending at the pelvis [31], and features a treadmill with an embedded foot pressure measuring plate to analyze ground reaction forces. This can be used to identify certain moments in gait, as gait follows certain identifiable determinants [43]. According to the common model [12], a gait cycle can be divided into two periods (60% stance and 40% swing) and consists of eight total phases. The most relevant phases pertaining to this study are Initial Contact (IC), which divides gait cycles, and Initial Swing (IS), which departs the stance from the swing phase.

In this methodologic contribution, we aim to fully visualize and describe spinal movement in direct relation to gait phases, thereby concentrating on spinal rotation. We therefore used previously obtained measurements from healthy participants. First, we describe the use of foot pressure measuring data to encode for the step and swing phases as well as for complete gait cycles. Secondly, there had to be a modification of the system's export functions in order to synchronize spinal motion data with foot pressure measuring data and combine them into the same raw data export. Since all exports are separate for each measurement, the third task was to merge single export files to create a complete raw data table. Finally, we were able to then standardize the combined raw data set of three or more gait cycles by interpolating splines to make spinal motion analysis relative to gait cycles intra- and inter-individually comparable. Together, this enables us to create oscillographs of spinal movement for and across each gait cycle, resulting in interpretable depths and precision for analyses. This analysis will address the described methods separately and provide the developed solutions for all spinal movement oscillographs of individual gait patterns as well as their standardized counterparts in several repositories [44–50].

2. Materials and Methods

In this methodologic contribution, we use previously obtained dynamic ST measurements from 201 healthy participants (132 females, 69 males, aged 18 to 70) that have already been used as a normative reference data pool. All of them had passed several functional tests and very strict inclusion criteria [41]. They were taken with a 4D spine and posture analysis device (DIERS Formetric III 4D™, DICAM v3.7Beta, Wiesbaden, Germany). It projects structured light onto the textile-free back of the individual person. A camera unit in defined positions records the movement with a frequency of 60 Hz. The software analyzes all three dimensions of each individual measuring point (up to 150,000, depending on body size) and generates a 3D image of the surface. The system then calculates [31] the corresponding 3D movements for each spinal segment, starting at the spinous process of C7 and ending at the pelvis. Due to the low correlation accuracy between surface structure and spinal position in the lower back area, L5 is not measured. Additionally, it is supplemented by one rear axis and two lateral cameras that record a video signal, enabling a subsequent visual inspection for multiple purposes. The measurement setup can be seen in Figure 1.

An integrated Zebris™ foot pressure measuring plate (5376 sensors, scanning frequency 120 Hz, sensitivity 1 N/cm^2, accuracy 5% FS) enables the analysis of ground reaction forces.

When beginning the analysis, all measuring devices start simultaneously, but due to the different measuring frequencies this results in an unequal number of observation times. This first needs to be reconciled in order to enable analysis of spinal motion directly related to gait.

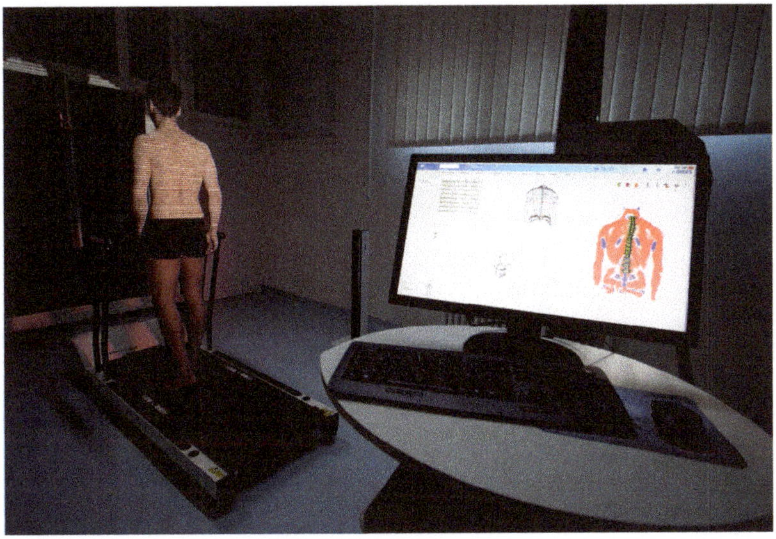

Figure 1. Dynamic Surface Topography measuring setup. Participant walking on a treadmill with an integrated foot pressure measuring plate while structured light is projected on the textile-free back. System generates a 3D image of the surface and then calculates the corresponding 3D movements for each spinal segment.

Therefore, the necessary data were obtained from an already existing normative referent data pool [41]. The study was approved by the responsible ethics committee of the medical chamber Rhineland-Palatinate (837.194.16) and is registered with WHO (INT: DRKS00010834). All participants gave informed consent. For further information on the ST measuring device, experimental procedures, or characteristics of participants, refer to Huthwelker et al. (2022) [41].

Here, we provide detailed descriptions of the four central processes.

2.1. Encoding of Step and Swing Phases and Complete Gait Cycles into the Spinal Model

In our approach, we started to measure a gait cycle with the first full IC of the right foot, which is also the start of encoding for the stance phase right. Hence, the start of the next gait cycle is the next IC of the right foot and so forth. Both leg swing phases are also marked in the data export. Along their assigned time stamp, the respective gait phases were encoded and synchronized into the raw data of the spinal model. We introduced the notion to the manufacturer DIERS. They implemented the concept in the software (DICAM v3.7Beta). We validated the updated software by taking a random sample of 20 participants' video recordings of the back and lateral axis cameras in which single frames were checked for face-valid results compared to the automated detection. In all recordings, the visually identified moment in time of IC was in the range of ± 3 Frames/3/60 Hz compared to the frame number in the exported raw data. For a detailed explanation of the notion, compare the related repository [46]. Since all observed video recordings revealed valid results, the next data step could be addressed.

2.2. Assembly of Individual Measurement Export Files and Creation of Rotation Graphs

DICAM evaluates over one hundred global (e.g., lordosis and kyphosis angles) and local (e.g., 3D position data of each vertebra) parameters of the spine that can be exported as raw data in .CSV format. The exported data are separate for each measurement that was chosen in the menu. Depending on the parameter choice, different columns are generated. Based on the parameter settings, specific column distribution results are produced. A Statistical Analysis System (SAS v9.4) syntax script assembles all single exports into one complete raw data table [44]. Since not all characters in the DICAM export are compatible with SAS, the script defines its own column headings and labels them accordingly. Consequently, the parameter settings of DICAM are crucial; otherwise, wrong data would be read in. Columns containing the subject's code and the speed in km/h are imported based on the filename of the former DICAM export. Now, all kinds of statistical analyses can be applied to a full set of data.

The resulting complete raw data table can also be imported by Statistical Package for the Social Sciences (SPSS v23). We used this to check data for plausibility and potential outlying values and to generate oscillation graphs based on individual movement data with a direct relation to the phases of gait. For the graphs, we used the position data of the vertebra and the pelvis in the transversal plane. For the indication of gait cycles, we used foot pressure measuring data indicating the phases of gait. A graph template, as well as the respective SPSS script for execution, is openly provided [49].

2.3. Standardization Combining Raw Data of Three or More Gait Cycles for Interpolating Splines and Creation of Rotational Graphs

We used an openly available SAS (v9.4) syntax script [44] to generate a standardized gait cycle (SGC) for each measurement. By computing a standardized gait cycle on a scale of 0–100%, measurements all differing in the number of observations are thus made comparable. One SGC per measurement was generated out of three (all available) gait cycles by using a spline function as part of the syntax. The resulting raw data table was analyzed using SPSS to create rotational graphs of the spine now within an SGC. Therefore, another graph template and the respective executing SPSS script were created and made openly available [48]. It can be downloaded and used on its own measurement data from a similar ST device after files have been exported.

3. Results

After visual inspection of all spinal rotation graphs of our 201 participants, we present a selection of interesting cases and related motion analytical considerations in the following subsection. The purpose is to provide an overview of typical findings and emerging patterns as well as highlight rare cases. Each of these graphs (Figures 2–5) represents the spinal motion of one of our participants across three gait cycles and the same subject again (Figures 6–9) after computing an SGC. Therefore, we analyzed average walking speeds of approximately 82–84 m per minute/5 km/h [12] as this speed provides data for most habitual movement patterns. In addition, we limited visualizations of rotational curves in the transverse plane as a first approach to make spinal motion visible. All anonymized single graphs containing only the graph number, but not the subject code, and an all-encompassing visualization are openly available [50].

3.1. Raw Data Visualizations of Rotational Curves Directly Related to Phases Gait

The pelvis and the lumbar segments show an opposite progression in comparison to the thoracic segments. The rotational direction changes gradually through each of the segments. As expected, periodic near sinusoidal oscillations are seen, revealing a phase shift between the pelvis and the upper thoracic segments with their maxima facing each other, meaning a direct equalization of the pelvic rotation by the thoracic spine (Figure 2). At IC right, the pelvis and L4 are maximally rotated to the left, reaching the zero point in the middle of the stance phase. L4 constantly follows the rotation of the pelvis. T12 has a fairly small amplitude, being close to the zero point where the intersection of movement directions takes place. Maximum antagonistic rotation is displayed by T7 and T8. The movement direction then changes again, with T4 also rotating in the opposite direction of the pelvis but to a lesser extent. Along the gait cycle, the amplitude and the period are intra-individually constant. Even though we see expected movement patterns, there are also broad variations with specific manifestations for each individual (Figures 3–5).

After visual analysis of all individual cases, we made the following observations: Graphs of rotation patterns are usually characterized by sinusoidal curve progression (Figure 2). Rhythmic movements can superimpose these typical curves (Figure 3). They can be assumed as individually characterizing features and not attributed to measuring artifacts as they appear consistently throughout the whole spine and across all gait cycles. In particular cases, potentially non-sinusoidal but still periodic curves, mostly with steep rises, occur (Figure 3). Rarely, for instance, in the pelvis (Figure 3), very little movement (<5°) or even no systematic curve course could be detected. The measured values of individual segments oscillate around a "stable level" but not necessarily around the zero point. Frequently, this "symmetry line" (SL) of oscillation is shifted several degrees into one of the two directions of movement (Figure 5); this "level shift" (LS) might depend on the alignment during stance. During gait cycles, the amplitude and the period are intra-individually constant (Figures 2–5) and behave relative to the stance phase for all participants but in different ways, varying for each individual. The direction of movement of the pelvis and the lower lumbar spine is usually opposite to that of the upper lumbar and thoracic spine. However, in some cases, the pelvis and the majority of all segments rotate in the same phase (Figure 4). Usually, the graphs of neighboring vertebral bodies rotate nearly in phase but with differently prominent maximum and minimum segmental motions. Maximum antagonistic rotation to the pelvis is mostly displayed by T8 (Figure 2). The movement excursions of these two regions can be equally large, or they can differ significantly. The "point of intersection" (PoI), the height of the segment where the two directions of movement exchange, varies depending on the subject and can be between the middle lumbar and lower thoracic spine. For the resulting phase shifts between these "counterparts", we found multiple patterns reaching from exact antagonistic 100% (180°, sine-to-sine), over 50% (90°, sine-to-cosine) to 0% (0°, oscillating in phase). Relative to the stance phase of the right leg, the rotational maximum of the pelvis to the left predominately occurs between IC and mid stance.

In order to make the described spinal motion analysis intra- and inter-individually comparable, we had to standardize combined raw data of all gait cycles for interpolating splines.

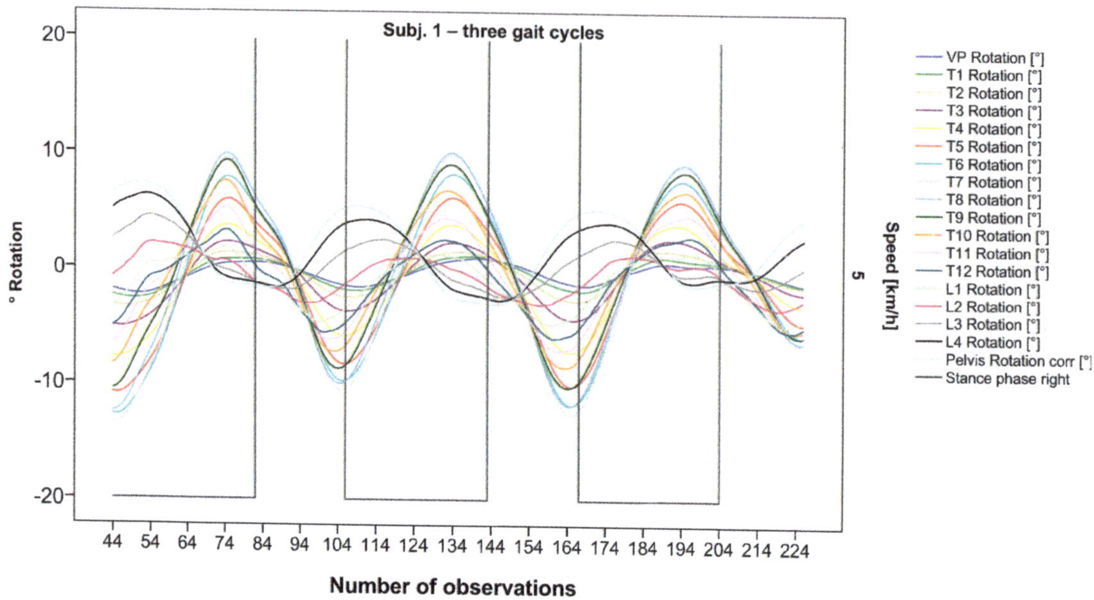

Figure 2. Visual representation for all segments of graph Nr.641 [50]. Positive values show rotation to the left, and negative values show rotation to the right. Observation number is displayed on the abscissa, always starting with Initial Contact of the right foot. Durations of right stance phases are delineated with a vertical black line.

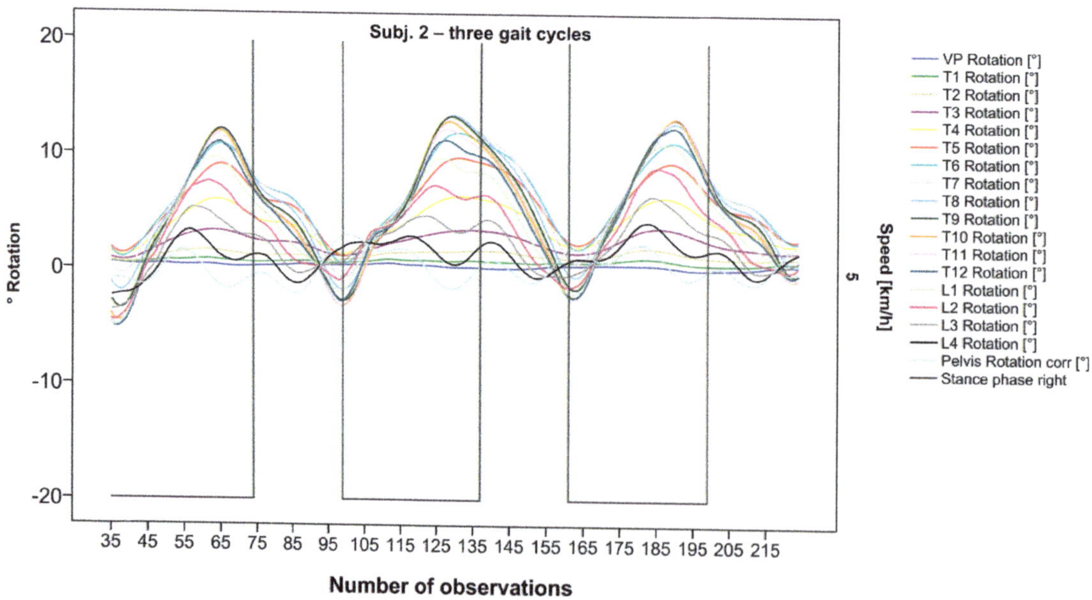

Figure 3. Illustration of all segments of graph Nr.622 [50]. Beginning at the pelvis, rhythmic movements superimpose the curve progressions of all vertebral bodies upward. Durations of right stance phases are delineated with a vertical black line.

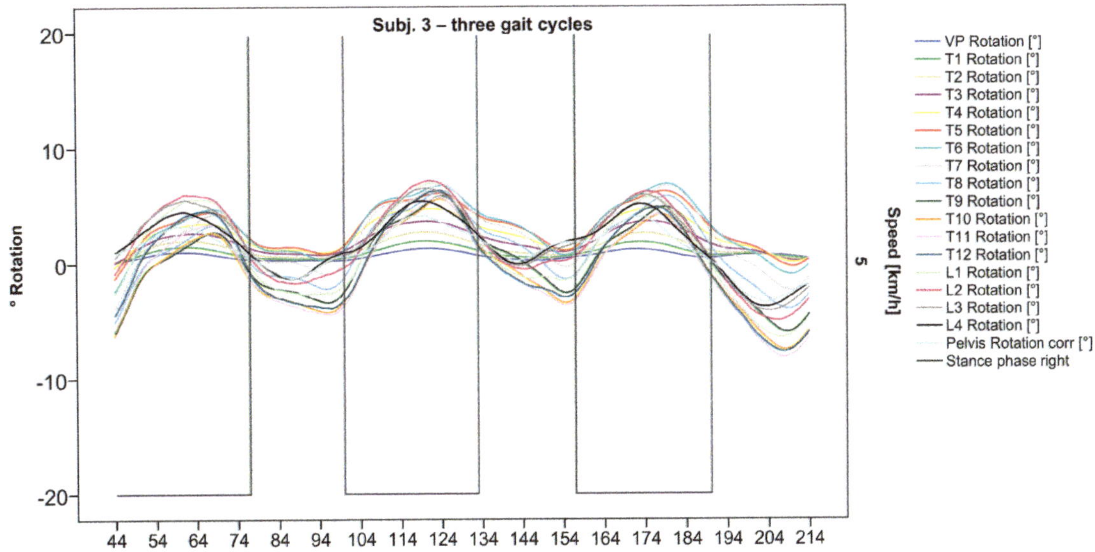

Figure 4. Visual representation of graph Nr.615 [50] depicting a rarely seen reverse pattern. Most parts of the spine are rotating nearly in phase. Durations of right stance phases are delineated with a vertical black line.

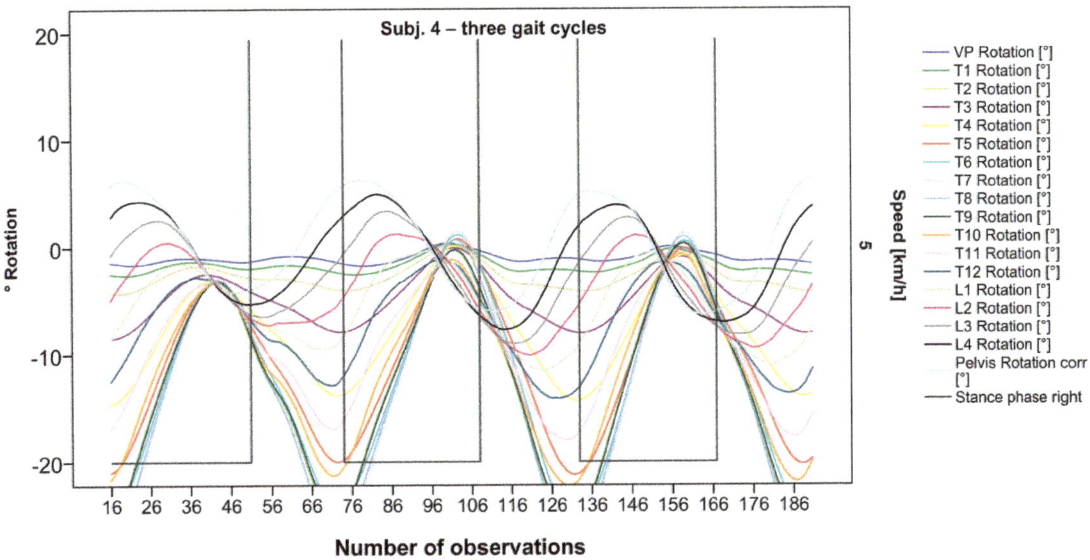

Figure 5. Visual representation of graph Nr.608 [50] depicting a shift of rotation to the right for the entire thoracic spine. Durations of right stance phases are delineated with a vertical black line.

3.2. Standardization of Combining Raw Data of Three or More Gait Cycles for Interpolating Splines and Creation of Rotational Graphs

After standardization, graphs combining raw data of all gait cycles can be created. In order to demonstrate the effects of standardization and interpolation, subsequent figures take up their counterparts from the previous section. Comparing Figures 2 and 6, vertebral bodies of the middle and lower thoracic spine during the first and third gait cycle, previously presenting an asymmetrical curve progression, especially toward the left (Figure 2), now show a symmetrical curve progression leading to a much more precise identification of maxima (Figure 6). In given contexts, where rhythmic movements superimpose curve progressions, thereby constituting individually characterizing features (Figure 3), the standardization, nevertheless, preserves them, meanwhile improving maxima identification (Figure 7). The standardization not only enables comparability but can also clarify individual features. At first, Figure 4 reveals that the pelvis, lumbar spine, and all thoracic segments are rotating nearly in phase, but after transformation in SGC, we see that this must be subdivided so that the lumbar and middle thoracic spine rotate nearly in phase while the upper thoracic spine displays hardly any movement (Figure 8). A similar clarification of an individual feature occurs when comparing the LS to the right of the SL before (Figure 5) and after standardization (Figure 9), when the isolated LS of the thoracic spine and T12, being the PoI, is much easier to recognize. All single graphs within an SGC and an all-encompassing video are openly available [47].

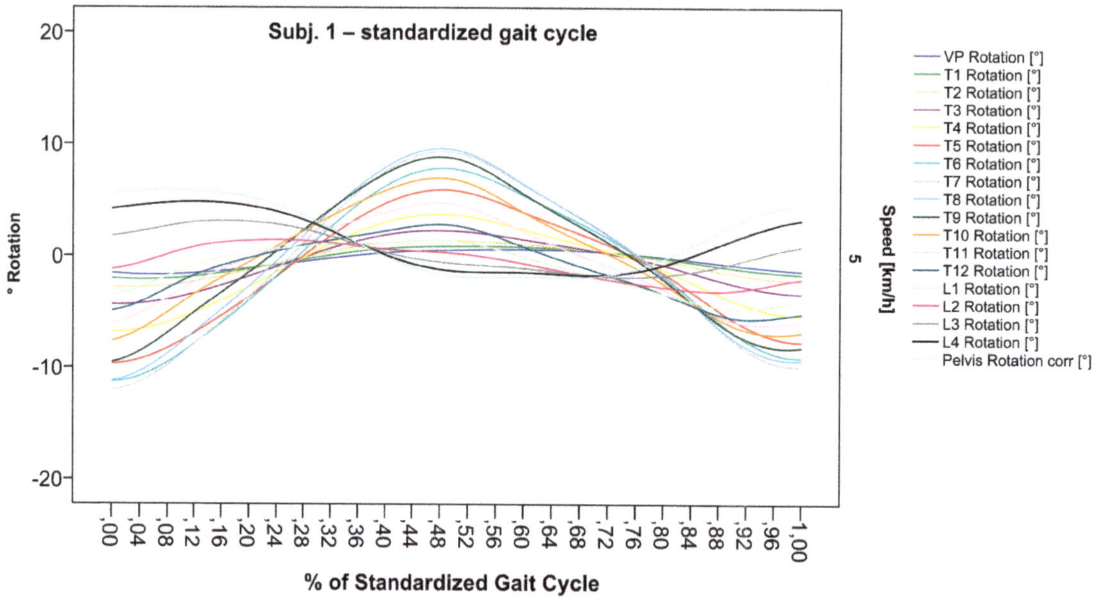

Figure 6. Illustration of all segments of graph Nr.641 SGC [47]. Near sinusoidal wave form, a further specified identification of maxima (rotation to the left) for all vertebral bodies occurred.

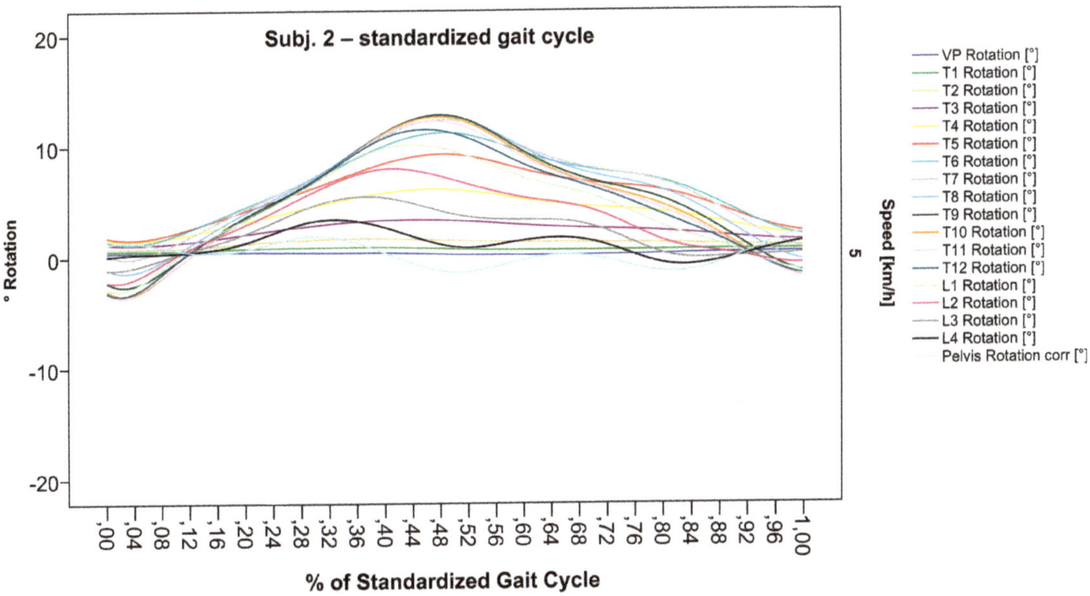

Figure 7. Illustration of all segments of graph Nr.622 SGC [47]. Superimposed oscillation as an individually characteristic feature is still visible; maxima identification nevertheless much more precise.

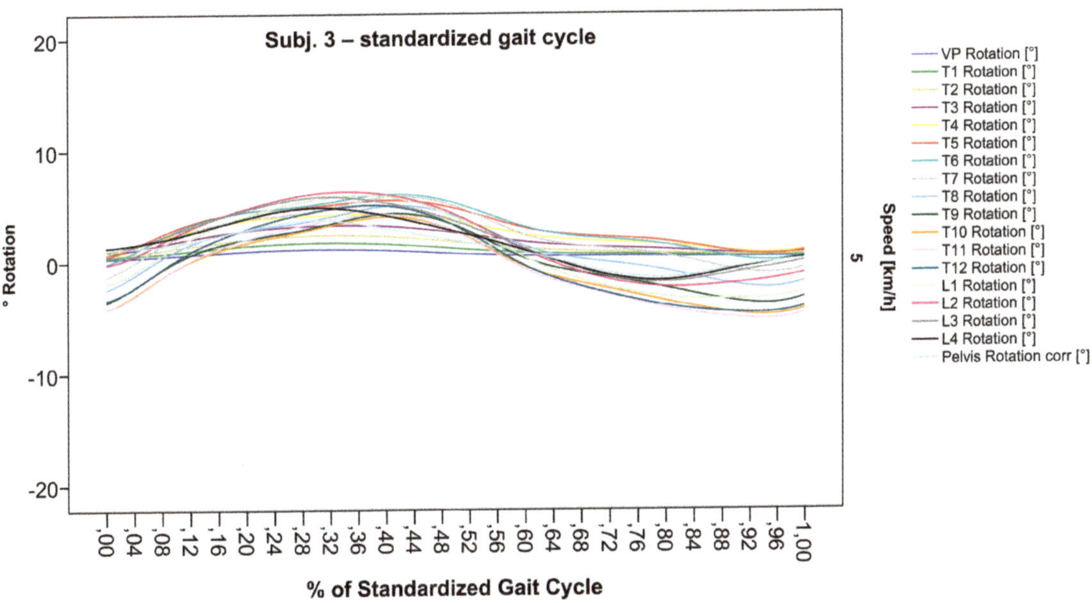

Figure 8. Illustration of all segments of graph Nr.613 SGC [47]. Isolated rotation in the phase of the lumbar and middle thoracic spine, as the individual feature becomes more apparent after transformation in SGC.

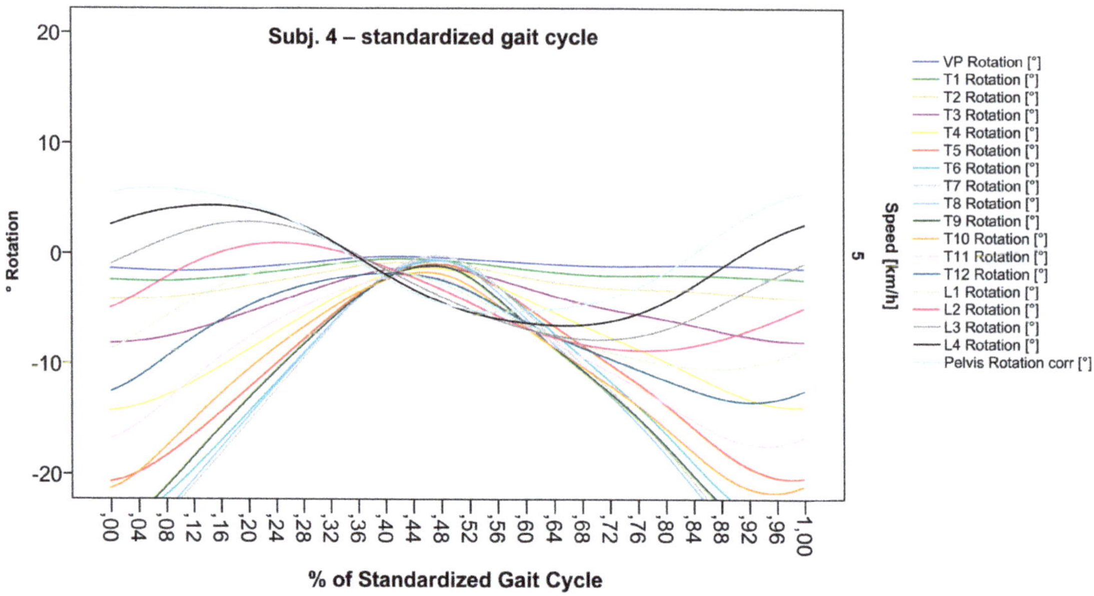

Figure 9. Illustration of all segments of graph Nr.608 SGC [47]. The isolated thoracic rotation shift to the right, with T12 being the point of intersection, is now easier to recognize.

4. Discussion

With the described methodologic advancements, ST can be used to visualize spinal motion as it directly relates to phases of gait and, after standardization, to compare these results intra- and inter-individually. As demonstrated, interpolating spline functions work for average walking speed measurements, leading to a more precise determination of relevant and characteristic points (e.g., maxima, phase shifts, lumbar and thoracic movement behavior), thereby aiding in the clarification of individual features. Additionally, this constitutes the basis for calculating phase shifts between different vertebral bodies as a future parameter to describe patterns of spinal motion in gait.

Using this method, we observed high intra-individual consistency of movement patterns as well as extensive inter-individual variation. When applying AI-based analysis, e.g., a Siamese neural network architecture, we were already able to identify 100% of individuals, thus constituting a spinal fingerprint [51]. This individuality of motion patterns is similar to previous work [17,19]. What contradicts these former findings is that our results detected regions (T6–T8) of many volunteers where they actually contained the largest movement excursions [42], whereas previous work detected the least spinal motion. Especially in regards to phase shift patterns of spinal segments, we detected subjects where the majority of all segments rotated in the same phase. One would expect this phenomenon while walking in amble. In this unusual pattern for humans, the ipsilateral instead of the contralateral arm is advanced by the leg. This type of ambulation is normally restricted to quadruped mammals, such as the elephant, but can also be examined by primate species [52], although this finding requires further examination. Taken together, our approach can directly relate segmental data to specific phases of gait and moments in time for an SGC.

Regarding the interpretation of normative reference data of asymptomatic healthy controls [42], the inter-individual variation due to differences in gait types, the alignment during stance [41,53], and other confounding factors must be determined. It is still unclear to which extent this would help to discern between physiological and pathological movement patterns. Thus, a relevant goal of future research will be to identify movement parameters and resulting characteristic patterns in which Artificial Intelligence-based analysis could be very helpful, as has been demonstrated for gait patterns [54]. As an example of dynamic ST, longitudinal changes of similarity in subjects might be of interest for the detection of early change, thereby supporting experts with an objective metric that might be helpful when approaching pathologies [51]. In order to test these assumptions, we will look for differences in spinal motion between the healthy controls and back pain patients, as well as other musculoskeletal disorders, in separate projects.

Limitations arise from the methodology, as reliability, reproducibility, and intra-individual consistency for the use in dynamic gait analysis have been shown [40,51], though validity is only for dynamic stance measures [55]. Although the transfer from validated stance measurements to dynamic gait analysis is feasible, there has not been validation of the spinal model in dynamic ST measurements so far. Hence, dynamic ST represents 3D spinal position data estimated from the back surface, which can be affected by contraction of the back muscles or movements and deformations of the soft tissue. Therefore, we were only able to apply internal validity measures to our results, e.g., the slightly displaced but still parallel course of the pelvis and L4. Furthermore, we only measured three gait cycles, and we need to investigate whether the inclusion of more than three gait cycles would alter the reported results. Alternatively, local Fourier transformations could also serve the same purpose in a superior way. Further research is needed to identify the most appropriate method of gait analysis necessary for adequate assessments of slower speeds in the setting of patients with back pain or hip/knee arthrosis where pathology inhibits walking (reducing speed).

Comprehension of these relationships will facilitate future research to understand the nature of pathologies, for example, back pain, arthrosis, scoliosis, and the effect of orthopedic surgery on spinal motion for comparison between physiological and pathological variations.

5. Conclusions

As the result of our methodologic advancement, ST now enables a face-valid and reproducible description of estimated spinal motion in direct relation to gait without extensive preparation procedures, significant radiation exposure, or other forms of invasive strategies. The transformation into an SGC facilitates intra- and inter-individual comparisons while preserving individual characteristic features. Hence, we conclude that this novel form of gait-related spinal motion analysis appears to have several advantages over existing methodology and holds much promise for future research in this field.

Regarding necessary future work and in order to address the lack of validity for dynamic ST, we propose two approaches: As a direct validation of dynamic ST, when compared to a gold standard, we suggest the application of 3D X-ray while walking on a treadmill (e.g., EOS). As a clinical challenge for dynamic ST, in order to provide even more face validity, a research project could evaluate the potential of ST to detect spinal fusions.

Nevertheless, before dynamic ST has been validated with a gold-standard measurement, crucial clinical decisions should not solely rely on its results.

Author Contributions: Conceptualization, J.K., M.C., L.G.L. and U.B.; methodology, I.S., R.W., M.C. and J.K.; software, I.S., R.W., J.K. and C.W.; validation, J.K., U.B. and J.H.; formal analysis, J.K., U.B., J.H. and C.W.; investigation, J.H.; resources, U.B. and P.D.; data curation, J.K., U.B., J.H. and C.W.; writing—original draft preparation, J.K. and M.C.; writing—review and editing, J.K., U.B., J.H., C.W., I.S., R.W., M.C., L.G.L. and P.D.; visualization, J.K. and C.W.; supervision, U.B., L.G.L. and P.D.;

project administration, J.K., U.B. and P.D. All authors have read and agreed to the published version of the manuscript.

Funding: This research received no external funding.

Institutional Review Board Statement: Data used in this report were collected according to the guidelines of the Declaration of Helsinki, were approved by the responsible ethics committee of the State Chamber of Physicians of Rhineland-Palatinate (Nr.: 837.194.16), and are registered with WHO (INT: DRKS00026822).

Informed Consent Statement: Informed consent was obtained from all subjects involved in the original study.

Data Availability Statement: To enable our own studies and for the replication of our findings, all developed tools are provided in several publicly available repositories that can be found in the references [41–47]. This includes the aggregated data for the oscillographs. The underlying participant data could not be provided since the original informed consent only enables sharing with one research partner. Furthermore, public accessibility is not covered by the given ethics vote.

Acknowledgments: We would like to thank all the participants of the original study. The colleagues of the Institute for Physical Therapy, Prevention and Rehabilitation, University Medical Center of the Johannes Gutenberg University Mainz, Germany, are acknowledged for their support in the recruitment of participants. We express our gratitude to DIERS Company, in particular to Kjell Heitmann and Amira Basic, for statistical and technological support during the implementation of our notion into the Formetric III 4D™ analyzing system. Finally, special thanks to Theresa Dersch for visual inspection and corrections of the captured data and Henriette Rönsch for her help in creating the final graph versions of this manuscript.

Conflicts of Interest: The authors declare no conflicts of interest.

References

1. Schmidt, C.O.; Raspe, H.; Pfingsten, M.; Hasenbring, M.; Basler, H.D.; Eich, W.; Kohlmann, T. Back pain in the German adult population: Prevalence, severity, and sociodemographic correlates in a multiregional survey. *Spine* **2007**, *32*, 2005–2011. [CrossRef] [PubMed]
2. Urits, I.; Burshtein, A.; Sharma, M.; Testa, L.; Gold, P.A.; Orhurhu, V.; Viswanath, O.; Jones, M.R.; Sidransky, M.A.; Spektor, B.; et al. Low Back Pain, a Comprehensive Review: Pathophysiology, Diagnosis, and Treatment. *Curr. Pain Headache Rep.* **2019**, *23*, 23. [CrossRef] [PubMed]
3. Fachgesellschaften, Arbeitsgemeinschaft der Wissenschaftlichen Medizinischen; Bundesärztekammer (BÄK) Kassenärztliche Bundesvereinigung (KBV). Programm für Nationale VersorgungsLeitlinien. *AZQ* **2017**, *2*. [CrossRef]
4. Brambilla, C.; Lavit Nicora, M.; Storm, F.; Reni, G.; Malosio, M.; Scano, A. Biomechanical Assessments of the Upper Limb for Determining Fatigue, Strain and Effort from the Laboratory to the Industrial Working Place: A Systematic Review. *Bioengineering* **2023**, *10*, 445. [CrossRef]
5. Koes, B.W.; van Tulder, M.W.; Thomas, S. Diagnosis and treatment of low back pain. *BMJ* **2006**, *332*, 1430–1434. [CrossRef]
6. Jöllenbeck, T. Bewegungsanalyse–Wesentliches Element moderner sportmedizinischer Diagnostik. *Dtsch. Z. Sportmed.* **2012**, *63*, 2.
7. Horst, F.; Mildner, M.; Schollhorn, W.I. One-year persistence of individual gait patterns identified in a follow-up study—A call for individualised diagnose and therapy. *Gait Posture* **2017**, *58*, 476–480. [CrossRef]
8. Chockalingam, N.; Dangerfield, P.H.; Giakas, G.; Cochrane, T. Study of marker placements in the back for opto-electronic motion analysis. *Stud. Health Technol. Inform.* **2002**, *88*, 105–109.
9. Konz, R.J.; Fatone, S.; Stine, R.L.; Ganju, A.; Gard, S.A.; Ondra, S.L. A kinematic model to assess spinal motion during walking. *Spine* **2006**, *31*, E898–E906. [CrossRef]
10. Alshehre, Y.M.; Alkhathami, K.; Brizzolara, K.; Weber, M.; Wang-Price, S. Effectiveness of Spinal Stabilization Exercises on Dynamic Balance in Adults with Chronic Low Back Pain. *Int. J. Sports Phys. Ther.* **2023**, *18*, 173–187. [CrossRef]
11. Yang, Y.T.; Yoshida, Y.; Hortobagyi, T.; Suzuki, S. Interaction between thorax, lumbar, and pelvis movements in the transverse plane during gait at three velocities. *J. Appl. Biomech.* **2013**, *29*, 261–269. [CrossRef] [PubMed]
12. Perry, J.; Burnfield, J.M. *Gait Analysis: Normal and Pathological Function*; SLACK Incorporated: Thorofare, NJ, USA, 1992.
13. Bruijn, S.M.; Meijer, O.G.; van Dieen, J.H.; Kingma, I.; Lamoth, C.J. Coordination of leg swing, thorax rotations, and pelvis rotations during gait: The organisation of total body angular momentum. *Gait Posture* **2008**, *27*, 455–462. [CrossRef] [PubMed]

14. Ceccato, J.C.; de Seze, M.; Azevedo, C.; Cazalets, J.R. Comparison of trunk activity during gait initiation and walking in humans. *PLoS ONE* **2009**, *4*, e8193. [CrossRef] [PubMed]
15. Crosbie, J.; de Faria Negrao Filho, R.; Nascimento, D.P.; Ferreira, P. Coordination of spinal motion in the transverse and frontal planes during walking in people with and without recurrent low back pain. *Spine* **2013**, *38*, E286–E292. [CrossRef]
16. Feipel, V.; De Mesmaeker, T.; Klein, P.; Rooze, M. Three-dimensional kinematics of the lumbar spine during treadmill walking at different speeds. *Eur. Spine J.* **2001**, *10*, 16–22. [CrossRef]
17. Gregersen, G.G.; Lucas, D.B. An in vivo study of the axial rotation of the human thoracolumbar spine. *J. Bone Jt. Surg. Am.* **1967**, *49*, 247–262. [CrossRef]
18. Leardini, A.; Berti, L.; Begon, M.; Allard, P. Effect of trunk sagittal attitude on shoulder, thorax and pelvis three-dimensional kinematics in able-bodied subjects during gait. *PLoS ONE* **2013**, *8*, e77168. [CrossRef]
19. MacWilliams, B.A.; Rozumalski, A.; Swanson, A.N.; Wervey, R.A.; Dykes, D.C.; Novacheck, T.F.; Schwartz, M.H. Assessment of three-dimensional lumbar spine vertebral motion during gait with use of indwelling bone pins. *J. Bone Jt. Surg. Am.* **2013**, *95*, e1841–e1848. [CrossRef]
20. Needham, R.; Naemi, R.; Healy, A.; Chockalingam, N. Multi-segment kinematic model to assess three-dimensional movement of the spine and back during gait. *Prosthet. Orthot. Int.* **2016**, *40*, 624–635. [CrossRef]
21. Schmid, S.; Bruhin, B.; Ignasiak, D.; Romkes, J.; Taylor, W.R.; Ferguson, S.J.; Brunner, R.; Lorenzetti, S. Spinal kinematics during gait in healthy individuals across different age groups. *Human. Mov. Sci.* **2017**, *54*, 73–81. [CrossRef]
22. Stokes, V.P.; Andersson, C.; Forssberg, H. Rotational and translational movement features of the pelvis and thorax during adult human locomotion. *J. Biomech.* **1989**, *22*, 43–50. [CrossRef] [PubMed]
23. Thurston, A.J.; Harris, J.D. Normal kinematics of the lumbar spine and pelvis. *Spine* **1983**, *8*, 199–205. [CrossRef] [PubMed]
24. Needham, R.; Stebbins, J.; Chockalingam, N. Three-dimensional kinematics of the lumbar spine during gait using marker-based systems: A systematic review. *J. Med. Eng. Technol.* **2016**, *40*, 172–185. [CrossRef]
25. Betsch, M.; Wild, M.; Johnstone, B.; Jungbluth, P.; Hakimi, M.; Kuhlmann, B.; Rapp, W. Evaluation of a novel spine and surface topography system for dynamic spinal curvature analysis during gait. *PLoS ONE* **2013**, *8*, e70581. [CrossRef]
26. Betsch, M.; Wild, M.; Jungbluth, P.; Hakimi, M.; Windolf, J.; Haex, B.; Horstmann, T.; Rapp, W. Reliability and validity of 4D rasterstereography under dynamic conditions. *Comput. Biol. Med.* **2011**, *41*, 308–312. [CrossRef]
27. Knott, P.; Sturm, P.; Lonner, B.; Cahill, P.; Betsch, M.; McCarthy, R.; Kelly, M.; Lenke, L.; Betz, R. Multicenter Comparison of 3D Spinal Measurements Using Surface Topography With Those From Conventional Radiography. *Spine Deform.* **2016**, *4*, 98–103. [CrossRef]
28. Michalik, R.; Siebers, H.; Eschweiler, J.; Quack, V.; Gatz, M.; Dirrichs, T.; Betsch, M. Development of a new 360-degree surface topography application. *Gait Posture* **2019**, *73*, 39–44. [CrossRef]
29. Turner-Smith, A.R. A television/computer three-dimensional surface shape measurement system. *J. Biomech.* **1988**, *21*, 515–529. [CrossRef]
30. Drerup, B.; Hierholzer, E. Objective determination of anatomical landmarks on the body surface: Measurement of the vertebra prominens from surface curvature. *J. Biomech.* **1985**, *18*, 467–474. [CrossRef]
31. Drerup, B.; Hierholzer, E. Automatic localization of anatomical landmarks on the back surface and construction of a body-fixed coordinate system. *J. Biomech.* **1987**, *20*, 961–970. [CrossRef]
32. Frobin, W.; Hierholzer, E. Automatic measurement of body surfaces using rasterstereography. Part I: Image scan and control point measurement. *Photogramm. Eng. Remote Sens.* **1983**, *49*, 377–384.
33. Frobin, W.; Hierholzer, E. Automatic measurement of body surfaces using rasterstereography. Part II: Analysis of the rasterstereographic line pattern and three-dimensional surface reconstruction. *Photogramm. Eng. Remote Sens.* **1983**, *49*, 1443–1452.
34. Drerup, B.; Ellger, B.; Meyer zu Bentrup, F.M.; Hierholzer, E. Functional rasterstereographic images. A new method for biomechanical analysis of skeletal geometry. *Orthopade* **2001**, *30*, 242–250. [CrossRef] [PubMed]
35. Hackenberg, L.; Hierholzer, E. 3-D back surface analysis of severe idiopathic scoliosis by rasterstereography: Comparison of rasterstereographic and digitized radiometric data. *Stud. Health Technol. Inform.* **2002**, *88*, 86–89. [CrossRef]
36. Liljenqvist, U.; Halm, H.; Hierholzer, E.; Drerup, B.; Weiland, M. Three-dimensional surface measurement of spinal deformities using video rasterstereography. *Z. Fur Orthop. Und Ihre Grenzgeb.* **1998**, *136*, 57–64. [CrossRef]
37. Tabard-Fougere, A.; Bonnefoy-Mazure, A.; Dhouib, A.; Valaikaite, R.; Armand, S.; Dayer, R. Radiation-free measurement tools to evaluate sagittal parameters in AIS patients: A reliability and validity study. *Eur. Spine J.* **2019**, *28*, 536–543. [CrossRef]
38. Wanke-Jellinek, L.; Heese, O.; Krenauer, A.; Wurtinger, C.; Siepe, C.J.; Wiechert, K.; Mehren, C. Is there any use? Validity of 4D rasterstereography compared to EOS 3D X-ray imaging in patients with degenerative disk disease. *Eur. Spine J.* **2019**, *28*, 2162–2168. [CrossRef]
39. Michalik, R.; Hamm, J.; Quack, V.; Eschweiler, J.; Gatz, M.; Betsch, M. Dynamic spinal posture and pelvic position analysis using a rasterstereographic device. *J. Orthop. Surg. Res.* **2020**, *15*, 389. [CrossRef]

40. Gipsman, A.; Rauschert, L.; Daneshvar, M.; Knott, P. Evaluating the Reproducibility of Motion Analysis Scanning of the Spine during Walking. *Adv. Med.* **2014**, *2014*, 721829. [CrossRef]
41. Huthwelker, J.; Konradi, J.; Wolf, C.; Westphal, R.; Schmidtmann, I.; Drees, P.; Betz, U. Reference Values for 3D Spinal Posture Based on Videorasterstereographic Analyses of Healthy Adults. *Bioengineering* **2022**, *9*, 809. [CrossRef]
42. Huthwelker, J.; Konradi, J.; Wolf, C.; Westphal, R.; Schmidtmann, I.; Schubert, P.; Drees, P.; Betz, U. Reference values and functional descriptions of transverse plane spinal dynamics during gait based on surface topography. *Human. Mov. Sci.* **2023**, *88*, 103054. [CrossRef] [PubMed]
43. Saunders, J.B.d.M.; Inman, V.T.; Eberhart, H.D. The major determinants in normal and pathological gait. *JBJS* **1953**, *35*, 543–558. [CrossRef]
44. Westphal, R.; Konradi, J. SAS syntax script for creation of a Standardized Gait Cycle. *Mendeley Data* **2022**, *V1*. [CrossRef]
45. Schmidtmann, I.; Konradi, J. SAS syntax script for merging export files. *Mendeley Data* **2022**, *V1*. [CrossRef]
46. Konradi, J.; Betz, U. Validation of automatic detection of gait phases. *Mendeley Data* **2022**, *V1*. [CrossRef]
47. Konradi, J. Visualizations of rotational curves within a Standardized Gait Cycle. *Mendeley Data* **2022**, *V1*. [CrossRef]
48. Konradi, J. SPSS syntax script to create graphs of spinal motion for a Standardized Gait Cycle. *Mendeley Data* **2022**, *V1*. [CrossRef]
49. Konradi, J. SPSS syntax script to create graphs of spinal motion relative to phases of gait. *Mendeley Data* **2022**, *V1*. [CrossRef]
50. Konradi, J. Visualizations of rotational curves directly related to gait phases. *Mendeley Data* **2022**, *V1*. [CrossRef]
51. Dindorf, C.; Konradi, J.; Wolf, C.; Taetz, B.; Bleser, G.; Huthwelker, J.; Werthmann, F.; Drees, P.; Fröhlich, M.; Betz, U. Machine learning techniques demonstrating individual movement patterns of the vertebral column: The fingerprint of spinal motion. *Comput. Methods Biomech. Biomed. Eng.* **2021**, *25*, 821–831. [CrossRef]
52. Schmitt, D.; Cartmill, M.; Griffin, T.M.; Hanna, J.B.; Lemelin, P. Adaptive value of ambling gaits in primates and other mammals. *J. Exp. Biol.* **2006**, *209*, 2042–2049. [CrossRef] [PubMed]
53. Wolf, C.; Betz, U.; Huthwelker, J.; Konradi, J.; Westphal, R.; Cerpa, M.; Lenke, L.; Drees, P. Evaluation of 3D Vertebral and Pelvic Position by Surface Topography in Asymptomatic Females: Presentation of Normative Reference Data. *J. Orthop. Surg. Res.* **2021**, *16*, 1–10. [CrossRef]
54. Horst, F.; Lapuschkin, S.; Samek, W.; Muller, K.R.; Schollhorn, W.I. Explaining the unique nature of individual gait patterns with deep learning. *Sci. Rep.* **2019**, *9*, 2391. [CrossRef]
55. Mohokum, M.; Schülein, S.; Skwara, A. The validity of rasterstereography: A systematic review. *Orthop. Rev.* **2015**, *7*, 68–73. [CrossRef]

bioengineering

MDPI

Article

Upper Limb Neural Tension Test and Spinal Biomechanics: Insights from a Longitudinal Pilot Study

Massimo Rossi [1], Marianna Signorini [2], Ali Baram [3], Mario De Robertis [3,4], Gabriele Capo [3], Marco Riva [3,4], Maurizio Fornari [3], Federico Pessina [3,4] and Carlo Brembilla [3,*]

1 Scuolaecomskbo, Istituto Ortopedico Rizzoli, Via Giulio Cesare Pupilli 1, 40100 Bologna, Italy; massimo.rossi.md@gmail.com
2 Dipartimento di Ingegneria Matematica, Politecnico di Milano, Via A.M. Ampere 2, 20133 Milan, Italy; marianna.signorini@gmail.com
3 Department of Neurosurgery, IRCCS Humanitas Research Hospital, Via Alessandro Manzoni 56, 20089 Rozzano, Italy; ali.baram@humanitas.it (A.B.); mario.derobertis@humanitas.it (M.D.R.); gabriele.capo@humanitas.it (G.C.); marco.riva@hunimed.eu (M.R.); maurizio.fornari@humanitas.it (M.F.); federico.pessina@hunimed.eu (F.P.)
4 Department of Biomedical Sciences, Humanitas University, Via Rita Levi Montalcini 4, 20090 Pieve Emanuele, Italy
* Correspondence: carlo.brembilla@humanitas.it

Abstract: Background: The Upper Limb Neural Tension Test (ULNTT) is a common assessment for neurodynamic function, yet the relationship between ULNTT findings and specific spinal biomechanical patterns remains poorly understood, particularly in the context of cervicobrachial neuralgia. This study aimed to investigate the association between ULNTT asymmetry and cervicothoracic spine biomechanics using advanced motion capture analysis. **Methods:** A longitudinal experimental study was conducted on two groups of asymptomatic participants: one with ULNTT asymmetry > 10° (AS group, n = 12) and another with symmetrical ULNTT (S group, n = 11). Neurodynamic testing and 3D motion capture of spinal kinematics during head lateral bending were performed at baseline. The AS group then underwent manual medicine intervention targeting spinal mobility impairments, followed by post-intervention reassessment. Spine biomechanics data, focusing on the C5-T4 region, were analyzed using the least squares approximation method to derive parameters describing upper thoracic (T1-T4_VERT) and lower cervical (C5-T1_CONC) lateral bending, and their interrelationship (ANGLE_TANG). **Results:** At baseline, the AS group showed significant differences between sides in neurodynamic parameters and T1-T4_VERT, with limited upper thoracic lateral bending contralateral to the side of the restricted ULNTT. Significant intergroup differences were also observed for these parameters. Following intervention in the AS group, significant improvements were noted in neurodynamic parameters and T1-T4_VERT, with no significant between-side differences post-intervention. **Conclusions:** These are preliminary results and preliminary conclusions based on the first study on a small group of patients. Given the limitations, this study provides evidence for a relationship between ULNTT asymmetry and upper thoracic spine biomechanics, specifically a contralateral limitation in lateral bending. These findings suggest a functional link between brachial plexus neurodynamics and upper thoracic spine mobility, offering potential insights into the pathophysiology of cervicobrachial conditions and highlighting the potential role of manual therapy in addressing both neurodynamic and biomechanical impairments. The developed motion capture analysis method offers a novel approach to quantify fine spinal motion patterns.

Keywords: Upper Limb Neural Tension Test; spine biomechanics; motion capture; least squares approximation; cervicobrachial neuralgia; manual therapy

1. Introduction

The Upper Limb Neural Tension Test (ULNTT) is a widely used clinical tool for assessing the mobility and displacement of cervical nerve roots and the brachial plexus within their surrounding structures [1–3]. Characterized by good to excellent intra-tester reliability [4–6], it plays a crucial role in identifying the contribution of the nervous system to arm pain [7]. Alterations in neurodynamics, as detected by a positive ULNTT, are often implicated in the development of cervicobrachial neuralgia, a condition characterized by neuropathic pain in the neck and upper limbs [8,9].

In functional approaches to spine-related problems, such as osteopathy and chiropractic, the analysis of fine spinal motion patterns—subtle movements reflecting spinal function—is of paramount importance. However, a significant gap exists in the understanding of the relationship between specific spinal motion patterns and clinical conditions, particularly in the context of cervicobrachial neuralgia. Therefore, investigating the connection between ULNTT positivity and dysfunctional cervicothoracic spinal biomechanics is essential for elucidating the underlying pathophysiology and identifying potential therapeutic targets.

Traditional methods for studying spine biomechanics often rely on motion capture technology, which tracks the positional variations of marker clusters placed on anatomical landmarks [10–14]. While effective for capturing gross movements, these methods struggle to accurately quantify fine spinal motion patterns. Furthermore, as highlighted in recent studies, capturing spinal movements presents challenges due to the difficulty in identifying spinal joint centers with skin-mounted sensors, and the low amplitude of intervertebral motions [15,16]. To address these limitations, this study employs the least squares approximation method, a new advanced mathematical tool that enables a more precise and reliable analysis of motion capture data.

This longitudinal experimental study aims to explore and describe any potential relationship between ULNTT positivity and dysfunctional cervicothoracic spine biomechanics using motion capture technology enhanced by the least squares approximation method. By identifying and characterizing any potential relationship, we seek to provide new insights into the pathophysiology of cervicobrachial neuralgia and pave the way for more targeted therapeutic interventions. We hypothesize that ULNTT positivity will be associated with specific patterns of dysfunctional biomechanics in the cervicothoracic spine.

2. Materials and Methods

2.1. Study Design

This was a longitudinal pilot study.

2.2. Participants

Participants were recruited from students enrolled in the M.D. program at the University of Milano-Bicocca and patients attending the clinics of physicians specializing in musculoskeletal conditions who collaborated with the authors. Those who agreed to participate in the study underwent an eligibility assessment with the lead investigator. To ensure maximum reliability in testing, the same investigator conducted the ULNTT for all participants, considering the good to excellent intra-tester reliability of the test.

The study involved two groups of participants: one group with asymmetry observed during the ULNTT (AS group) and another group with symmetry during the same test (S group). Inclusion criteria for the AS group required a greater than 10° asymmetry on the ULNTT, while the S group included participants with symmetrical results on the same test. The threshold of 10° asymmetry was chosen based on previous studies of the ULNTT in asymptomatic individuals with arm pain, where asymmetry greater than this was considered significant [17–20]. Sample size calculations for the AS group were based on achieving 80% power (alpha = 5%) with a smallest detectable difference of 7° and a standard deviation of 10%, which is four times the typical value reported in the literature [21]. This resulted in a required sample size of 10, but we opted to include 12 participants for greater confidence in the results. For the S group, the sample size was also determined to achieve 80% power (alpha = 5%) based on preliminary data comparing differences between the two sides in parameters related to upper thoracic spine side bending. Given the novel motion capture method used, we selected a smallest detectable difference of 5°, although smaller differences were detectable when comparing clinical and instrumental measurements. With a standard deviation of 5°, the required sample size was 10, which was increased to 11 participants.

To focus on the biomechanical aspect of the study, only asymptomatic participants were included to exclude pain-related confounding factors in the ULNTT analysis. Participants were recruited for the study following an assessment of their eligibility through medical history and a targeted physical examination. Individuals were excluded if they had a history of trauma or surgery to the shoulder girdle, upper limb, or spine. Further exclusion criteria included the presence of peripheral neuropathies or central nervous system (CNS) diseases, a history of mental health disorders, and known allergies to plastic materials, latex, or adhesives. To ensure participant safety and the integrity of the study, individuals with a medical history or physical examination findings indicative of contraindications to spinal manipulation were also excluded, such as collagenopathies, elastopathies, suspected fracture, clinical signs of occipito-atlanto-axial instability identified through clinical tests, and bone metabolism disorders. Inclusion was also precluded for subjects with limited range of motion in any joint of the upper limb (shoulder, elbow, or wrist) and for individuals with a Body Mass Index (BMI) greater than 25, the latter criterion being motivated by the increased difficulty in accurately identifying the anatomical landmarks necessary for marker placement during motion capture acquisition. Participants were also excluded if they reported any pain in the upper quadrant of the body within the three months prior to enrollment, or in the presence of a diagnosis of rheumatoid arthritis or diabetes.

The AS group consisted of 4 men and 8 women, with a median age of 30 years and a median BMI of 20.3. The S group consisted of 6 men and 5 women, with a median age of 27 years and a median BMI of 22.5. Participants who met the inclusion criteria and provided consent were evaluated at the Motion Capture Laboratory of the University of Milano-Bicocca, located at the Istituti Clinici Zucchi in Carate Brianza, where both neurodynamics and spine biomechanics were assessed using the ULNTT and motion capture techniques (Figure 1).

2.3. Neurodynamic Testing

The ULNTT was performed according to the universally accepted procedure described by Butler [22] (Figure 2). The electrogoniometer used in the study was the Biometrics SG110 (Biometrics Ltd., Newport, UK). Two parameters were recorded during the test: resistance to elbow extension, referred to as R2 (operator-dependent), and the onset of severe pain, referred to as P2 (patient-dependent). To ensure accurate timing of P2, a light with a remote control was placed behind the electrogoniometer display. The patient was instructed to press a button on the remote control with their non-testing hand when severe pain was first perceived. This allowed the operator to continue testing without interruption and enabled the second operator to record the P2 value displayed. The P2 value was later reviewed by examining the recorded videos of the test. This method was believed to improve the intra-tester reliability for the R2 parameter, as it reduced potential interruptions during the test. To maintain the highest reliability in the ULNTT, all tests were conducted by the same operator, who remained blinded to the values displayed on the electrogoniometer. A second operator noted the readings. The test was performed four times per side, with a one-minute interval between repetitions to avoid increasing the range of motion (ROM), as reported in previous literature [4]. For the AS group, the first side tested was the one with the greater ROM in clinical evaluation to prevent conditioning that could affect the onset of symptoms.

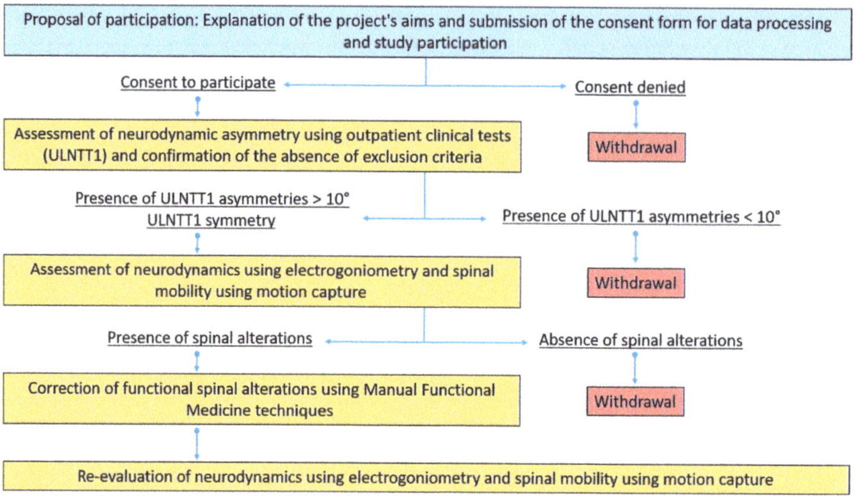

Figure 1. Study design flowchart.

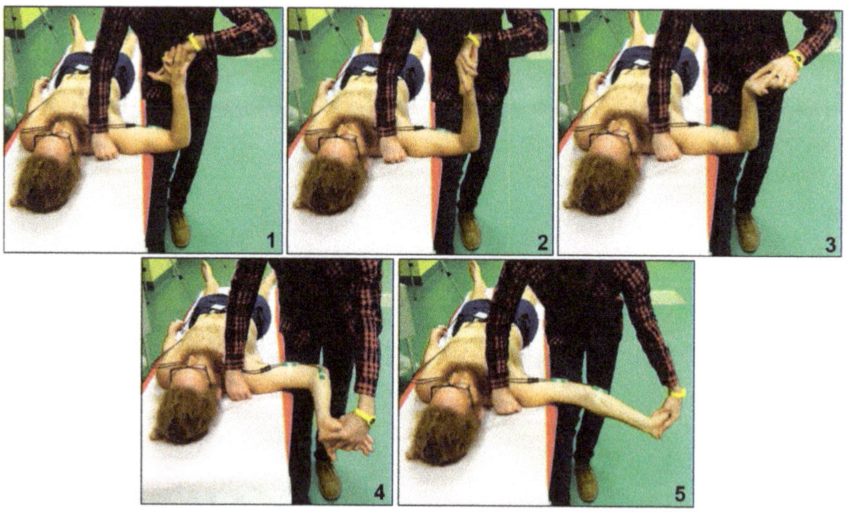

Figure 2. Performance of the ULNTT1, highlighting the 5 phases of the test. (**1**) Shoulder abduction to 90° and cranial stabilization of the scapular girdle with elbow flexion to 90°. (**2**) Forearm supination. (**3**) Wrist and first three fingers extension. (**4**) Shoulder lateral rotation to 90°. (**5**) Elbow extension, maintaining stabilization of the scapular girdle, distal arm, and wrist extension.

2.4. Spine Mobility Testing

Following the neurodynamic testing procedures, spine biomechanics were assessed using motion capture technology. Reflective markers were affixed to specific anatomical landmarks with double-sided adhesive tape, including C2, C5, C7, all spinous processes from T1 to L5, the acromions, the jugular notch, the xiphoid process, and the posterior and anterior superior iliac spines (Figure 3A). Additionally, a cap with four markers, two positioned on the sagittal plane and two on the frontal plane, was worn by each patient (Figure 3B). Movements were recorded by seven cameras (Qualisys Pro Reflex, Qualisys, Gothenburg, Sweden), four positioned posteriorly and three anteriorly relative to the patient.

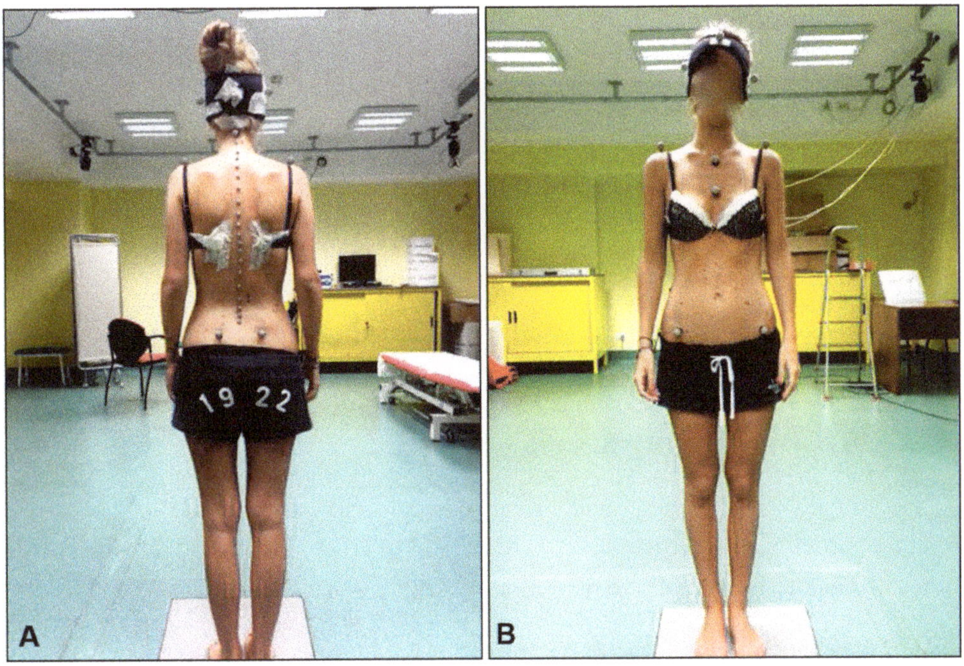

Figure 3. Marker placement for motion capture acquisition: (**A**) rear view, (**B**) front view.

The movement analyzed in this study was lateral bending of the head. Patients were positioned in front of a reference-less wall and instructed to maintain a forward gaze throughout the movement, continuing until they reached the end of the motion. Four repetitions were recorded for each side, alternating between the left and right sides, with the starting side chosen randomly.

2.5. Spine Intervention

At the conclusion of spine biomechanics assessments, patients with fine mobility impairments, defined as asymmetry in side bending during clinical testing, underwent manual medicine interventions aimed at correcting these impairments. The correction approach followed a functional perspective, treating the musculoskeletal system as an integrated unit. Due to the variability in patients' specific biomechanical characteristics, no standardized treatment protocol was established.

The manipulative procedures performed on each patient are detailed in Table 1. These procedures included high-velocity, low-amplitude spinal manipulations and muscle energy techniques. The intervention lasted only a few minutes, and no spine marker except for C2 was removed during the intervention.

Table 1. Treatments applied to the patients in group PZ; hvla: high-velocity low-amplitude spinal manipulations; mit: Mitchell techniques. Letters N/F/E, S, R following each vertebra: primary direction of motion of the vertebra in the sagittal plane (N: no primary direction; F: increased range of motion in flexion; E: increased range of motion in extension), frontal plane (S left: increased side bending to the left; S right: increased side bending to the right), and horizontal plane (R left: increased rotation to the left; R right: increased rotation to the right).

PZ1	hvla L3 N S left R left mit T9-L1 ERS right	PZ8	hvla L3 N S left R left mit T9-L1 ERS right mit T9-L1 ERS right
PZ3	hvla L4 N S left R left mit T8-T12 ERS right	PZ9	hvla L3 N S left R left mit T8-T12 ERS right hvla C5 N S left R left
PZ4	hvla L3 N S left R left mit T9-L1 ERS right hvla C5 N S left R left	PZ10	hvla L4 N S left R left mit T9-L1 ERS right hvla C6 N S left R left mit T7-T9 ERS right
PZ5	hvla L3 N S left R left mit T10-L1 ESR left hvla C6 N S right R right	PZ11	hvla L3 N S left R left mit T9-L1 ERS left hvla C6 N S left R left
PZ6	hvla L3 N S left R left mit T9-T12 ERS right hvla C5 N S left R left	PZ12	hvla L3 N S right R right mit T10-L1 ERS left hvla C5 N S left R left
PZ7	hvla L4 N S left R left mit T9-L1 ERS right hvla C6 N S left R left	PZ13	hvla L4 N S right R right mit T9-L1 ERS right hvla C6 N S left R right

Immediately following the intervention, both spine biomechanics and neurodynamics were re-evaluated using the same methods as before.

2.6. Spine Biomechanics Data Analysis

The spine mobility data obtained using motion capture technology were analyzed using the least squares (LS) approximation method. This mathematical approach provides an analytical function (Equation (1)) that approximates discrete datasets without forcing the function to pass through the data points, as interpolation methods would. The decision to use approximation rather than interpolation was based on the nature of the dataset, which includes experimental measurements that may contain errors. Interpolation techniques could amplify these errors, leading to inconsistent results, while approximation avoids this issue.

To approximate the shape of the spine, focusing on the regions from C5 to T4, a custom routine was developed in Matlab® (version R2014, The MathWorks Inc., Natick, MA, USA). The spine was approximated in two segments: the first function approximated the coordinates of C5, C7, and T1, and the second approximated T1, T2, T3, and T4. This division was based on the distinct biomechanical patterns of the lower cervical and upper thoracic spine. The routine automatically identifies the maximum side bending of the spine. Figure 4A illustrates the results for patient number 1.

$$\sum_{i=0}^{n} [y_i - \tilde{f}(x_i)]^2 \leq \sum_{i=0}^{n} [y_i - p_m(x_i)]^2 \quad (1)$$

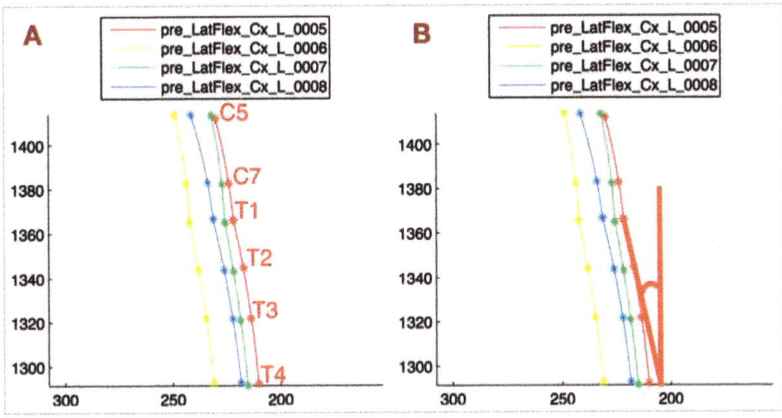

Figure 4. (**A**) Least squares approximation applied to markers on C5-T4 spinous processes during left head lateral flexion. Axes in millimeters. Four curves represent four movement repetitions. (**B**) Angle between T1-T4 curve tangent at T1 and vertical. Represents thoracic spine lateral movement and its contribution to cervical lateral flexion.

These approximated functions were used to derive objective parameters that describe the biomechanics of the spine. The key parameters are as follows:

a. T1-T4_VERT: The angle between the tangent at T1 to the T1-T4 curve and the vertical line, representing the side bending of the upper thoracic spine (Figure 4B).
b. C5-T1_CONC: The C5-T1 concavity angle, which defines the side bending of C5-C7, independent of the upper thoracic side bending (Figure 5A).
c. ANGLE_TANG: The angle between the tangent at T1 to the C5-T1 curve and the tangent at T1 to the T1-T4 curve, describing the relationship between the lower cervical and upper thoracic spine (Figure 5B).
d. Head rotation (HEAD_ROT): The angular variation of the segment connecting the lateral markers on the head projected onto the horizontal plane, used to monitor head rotation. This parameter helps avoid bias in measuring spine lateral bending due to head rotation.

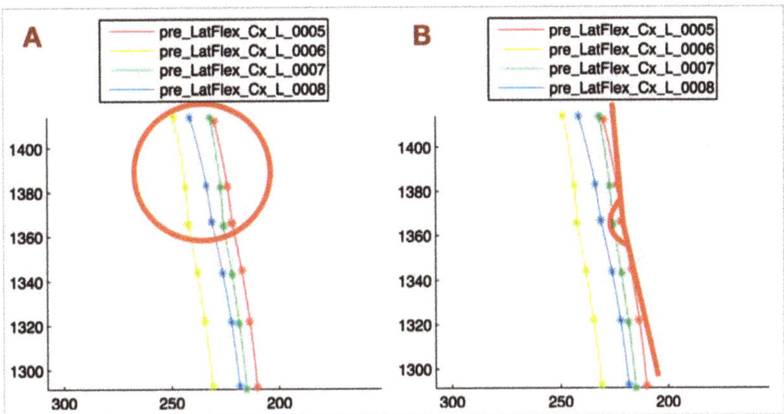

Figure 5. (**A**) C5-T1 curve concavity. Represents cervical spine lateral flexion magnitude. (**B**) Angle between T1-T4 and C5-T1 curve tangents at T1. Represents the relationship between cervical and thoracic spine curves.

2.7. Statistical Analysis

All statistical analyses were conducted using IBM SPSS Statistics software (Version 22, IBM Corporation, Somers, NY, USA). The Wilcoxon Mann–Whitney test was used for comparison between groups. The level of significance for all statistical tests, including comparisons between pre- and post-intervention measures and between the two groups, was set at $p < 0.05$.

3. Results

The results are presented as medians in Tables 2–4. Angular measurements were used for all parameters, except C5-T1_CONC (1/m). P2 and R2 are reported in degrees, representing the difference in elbow extension between these points and full extension (0°); elbow hyperextension is expressed as a negative value.

Table 2. Intragroup parameters comparison at baseline.

Null Hypothesis	Group	*p* Value	Median (Min/Max) Greater ROM Side	Median (Min/Max) Smaller ROM Side
Median of following parameter differences equals zero				
R2 greater ROM side/R2 smaller ROM side	S	0.003 *	−2.25 (−7.25/3.75)	0.25 (−7.00/5.75)
	AS	0.002 *	0.12 (−7.25/6.25)	13.87 (7.25/23.25)
P2 greater ROM side/P2 smaller ROM side	S	0.123	0.00 (−7.25/12.00)	0.5 (−7.00/12.00)
	AS	0.002 *	3.12 (−5/7.75)	17.62 (9.26/28.00)
T1-T4_VERT controlat ULNTT greater ROM side/T1-T4_VERT controlat ULNTT smaller ROM side	S	0.306	11.72 (2.68/24.37)	12.73 (5.38/22.22)
	AS	0.003 *	12.13 (4.70/19.83)	−1.17 (−3.90/7.63)
C5-T1_CONC controlat ULNTT greater ROM side/ C5-T1_CONC controlat ULNTT smaller ROM side	S	0.824	0.0029 (−0.0025/0.0086)	0.0024 (0.0006/0.0056)
	AS	0.195	0.0022 (0.0003/0.112)	0.0026 (−0.001/0.0083)
ANGLE_TANG controlat ULNTT greater ROM side /ANGLE_TANG controlat ULNTT smaller ROM side	S	0.722	4.34 (0.05/9.78)	4.6 (1.55/6.83)
	AS	0.077	5.73 (1.50/15.51)	3.26 (0.13/18.02)
ROT_CERV controlat ULNTT greater ROM side/ROT_CERV controlat ULNTT smaller ROM side	S	1	12.45 (2.81/46.42)	17.05 (7.44/24.64)
	AS	0.099	12.74 (3.97/31.48)	6.51 (2.18/18.42)

* means that that value has statistical significance ($p < 0.05$).

Table 3. Intergroup parameters comparison at baseline.

Null Hypothesis	*p* Value	Median (Min/Max) S	Median (Min/Max) AS
Parameter distribution is the same in S and AS groups			
R2 greater ROM side	0.104	−2.25 (−7.25/3.75)	0.12 (−7.25/6.25)
R2 smaller ROM side	<0.001 *	0.25 (−7.00/5.75)	13.87 (7.25/23.25)
P2 greater ROM side	0.535	0.00 (−7.25/12.00)	3.12 (−5/7.75)
P2 smaller ROM side	<0.001 *	0.5 (−7.00/12.00)	17.62 (9.26/28.00)
T1-T4_VERT controlat ULNTT greater ROM side	1	11.72 (2.68/24.37)	12.13 (4.70/19.83)
T1-T4_VERT controlat ULNTT smaller ROM side	<0.001 *	12.73 (5.38/22.22)	−1.17 (−3.90/7.63)
C5-T1_CONC controlat ULNTT greater ROM side	0.695	0.0029 (−0.0025/0.0086)	0.0022 (0.0003/0.112)
C5-T1_CONC controlat ULNTT smaller ROM side	0.695	0.0024 (0.0006/0.0056)	0.0026 (−0.001/0.0083)

Table 3. *Cont.*

Null Hypothesis	p Value	Median (Min/Max) S	Median (Min/Max) AS
ANGLE_TANG controlat ULNTT greater ROM side	0.487	4.34 (0.05/9.78)	5.73 (1.50/15.51)
ANGLE_TANG controlat ULNTT smaller ROM side	0.651	4.6 (1.55/6.83)	3.26 (0.13/18.02)
ROT_CERV controlat ULNTT greater ROM side	0.976	12.45 (2.81/46.42)	17.05 (7.44/24.64)
ROT_CERV controlat ULNTT smaller ROM side	0.007 *	12.74 (3.97/31.48)	6.51 (2.18/18.42)

* means that that value has statistical significance ($p < 0.05$).

Table 4. AS pre/post-intervention parameters comparison.

Null Hypothesis	p Value	Median (Min/Max) Pre	Median (Min/Max) Post
Median of following parameter differences equals zero			
R2 greater ROM side pre/R2 greater ROM side post	0.182	0.12 (−7.25/6.25)	−1.62 (−8.00/4.00)
R2 smaller ROM side pre/R2 smaller ROM side post	0.002 *	13.87 (7.25/23.25)	−0.12 (−8.25/5.75)
P2 greater ROM side pre/P2 greater ROM side post	0.109	3.12 (−5/7.75)	0.62 (−8.00/6.75)
P2 smaller ROM side pre/P2 smaller ROM side post	0.002 *	17.62 (9.26/28.00)	1.87 (−8.25/5.75)
T1-T4_VERT controlat ULNTT greater ROM side pre/ T1-T4_VERT controlat ULNTT greater ROM side post	0.638	12.13 (4.70/19.83)	9.84 (6.52/20.68)
T1-T4_VERT controlat ULNTT smaller ROM side pre/ T1-T4_VERT controlat ULNTT smaller ROM side post	0.003 *	−1.17 (−3.90/7.63)	7.33 (3.11/14.58)
C1-C5_CONC controlat ULNTT greater ROM side pre/ C1-C5_CONC controlat ULNTT greater ROM side post	0.41	0.0022 (0.0003/0.112)	0.0023 (0.0003/0.0010)
C1-C5_CONC controlat ULNTT smaller ROM side pre/ C1-C5_CONC controlat ULNTT smaller ROM side post	0.034 *	0.0026 (−0.001/0.0083)	0.0043 (0.0005/0.0010)
ANGLE_TANG controlat ULNTT greater ROM side pre/ANGLE_TANG controlat ULNTT greater ROM side post	0.754	5.73 (1.50/15.51)	4.71 (2.51/14.80)
ANGLE_TANG controlat ULNTT smaller ROM side pre/ANGLE_TANG controlat ULNTT smaller ROM side post	0.754	3.26 (0.13/18.02)	4.01 (1.96/10.87)
CONC_CERV controlat ULNTT greater ROM side pre/CONC_CERV controlat ULNTT greater ROM side post	0.117	17.05 (7.44/24.64)	15.82 (1.73/38.73)
CONC_CERV controlat ULNTT smaller ROM side pre/CONC_CERV controlat ULNTT smaller ROM side post	0.308	6.51 (2.18/18.42)	6.69 (2.66/25.50)

* means that that value has statistical significance ($p < 0.05$).

3.1. Intragroup Comparisons at Baseline (Table 2)

a. S group:
   Only R2 showed a statistically significant difference (2.5°), which is likely clinically insignificant and consistent with the literature on ULNTT asymmetry in asymptomatic individuals [17–20,23,24].
   No significant differences were observed in spine biomechanics parameters [25–27].

b. AS group:
   ULNTT confirmed the inclusion criterion of >10° asymmetry (R2 median difference: 13.5°).
   P2 also showed a significant difference (14.5°), reinforcing R2 findings.
   T1-T4_VERT showed a significant difference, with 12.13° contralateral to the greater ULNTT ROM side and −1.17° contralateral to the smaller ROM side, indicating opposite T1-T4 movement to the cervical spine.

No other spine biomechanics parameters, including HEAD_ROT, showed significant differences.

3.2. Intergroup Comparisons at Baseline (Table 3)

a. Significant differences between S and AS groups were observed in R2 and P2 at the smaller ULNTT ROM side, and in T1-T4_VERT contralateral to the smaller ROM side.
b. HEAD_ROT also differed significantly, but its clinical relevance is uncertain [28].
c. R2 and P2 at the greater ULNTT ROM side, and T1-T4_VERT contralateral to the greater ROM side, were similar between groups.
d. These findings suggest a relationship between upper limb neurodynamics and upper thoracic spine biomechanics.

3.3. Pre- to Post-Intervention Comparisons (AS Group—Table 4)

a. Significant changes were observed in R2, P2, T1-T4_VERT, and C5-T1_CONC at the smaller ULNTT ROM side.
b. These parameters were the same parameters that showed differences between groups at baseline.
c. The clinical relevance of C5-T1_CONC changes is uncertain.

3.4. Post-Intervention Intragroup Comparisons (AS Group—Table 5)

No significant differences were observed between sides in any parameters post-intervention.

Table 5. AS group post-intervention parameters comparison.

Null Hypothesis	p Value	Median (Min/Max) Greater ROM Side	Median (Min/Max) Smaller ROM Side
Median of following parameter differences equals zero			
R2 greater ROM side/R2 smaller ROM side	0.789	−1.62 (−8.00/4.00)	−0.12 (−8.25/5.75)
P2 greater ROM side/P2 smaller ROM side	0.755	0.62 (−8.00/6.75)	1.87 (−8.25/5.75)
T1-T4_VERT controlat ULNTT greater ROM side/T1-T4_VERT controlat ULNTT smaller ROM side	0.158	9.84 (6.52/20.68)	7.33 (3.11/14.58)
C5-T1_CONC controlat ULNTT greater ROM side/C5-T1_CONC controlat ULNTT smaller ROM side	0.13	0.0023 (0.0003/0.0010)	0.0043 (0.0005/0.0010)
ANGLE_TANG controlat ULNTT greater ROM side /ANGLE_TANG controlat ULNTT smaller ROM side	0.99	4.71 (2.51/14.80)	4.01 (1.96/10.87)
ROT_CERV controlat ULNTT greater ROM side/ROT_CERV controlat ULNTT smaller ROM side	0.209	15.82 (1.73/38.73)	6.69 (2.66/25.50)

3.5. Data Interpretation

The results indicate a potential association between ULNTT findings and cervicothoracic spine biomechanics. Particularly, the T1-T4_VERT parameter consistently showed significant differences related to ULNTT asymmetry and intervention effects. The observed changes in neurodynamic parameters (R2, P2) alongside T1-T4_VERT support a functional link between upper limb neurodynamics and upper thoracic spine mobility. The lack of significant differences in other spine biomechanics parameters suggests that the relationship may be specific to T1-T4 lateral bending. The elimination of significant differences between sides post-intervention in the AS group highlights the potential impact of manual therapy on both neurodynamic and biomechanical parameters.

4. Discussion

This research addresses a gap in the current understanding of brachial plexus pathophysiology, specifically concerning the altered interface with surrounding structures. There is a notable absence of data linking clinical functional spinal alterations to specific organic diagnoses, particularly brachial plexopathy, and a lack of methods to objectively quantify these functional changes [19,21,24,27]. Moreover, effective therapeutic strategies for this condition are limited.

Our study's findings reveal that neurodynamic parameters and the angle between the T1-T4 tangent at T1 and the vertical line exhibit statistically significant differences within the AS group, but not within the S group. These parameters, reflecting both neurodynamics and upper thoracic spine biomechanics, show significant intergroup differences at the side of restricted neurodynamics and contralateral to this side for spinal motion. Furthermore, these parameters improve significantly following functional correction and show no significant differences post-intervention.

Based on these observations, it is plausible to suggest, within the limitations of our study, that a relationship between brachial plexus neurodynamics and thoracic spine mobility may exist. Specifically, this relationship manifests as a limitation in thoracic spine lateral bending contralateral to the brachial plexus neurodynamic restriction.

The limitation in upper thoracic spine lateral flexion may reduce the contribution of the thoracic spine to the overall lateral bending movement, potentially leading to an increased range of motion demand on the cervical spine and subsequent tension on contralateral paravertebral cervical musculature, particularly the anterior and middle scalene muscles. Given their insertion on the first rib, which is mechanically linked to T1, increased tension in these muscles could alter their relationship with the brachial plexus, which courses between them, thereby limiting its mobility.

Nerve structure sensitization due to increased mechanical stress on the brachial plexus during daily head movements, mediated by the scalene muscles, may predispose patients to cervicobrachialgia in conjunction with other factors like trauma or adverse environmental conditions such as poor workplace ergonomics or repetitive upper limb strain [14,15].

Our study's sample size limits the assessment of the prevalence of thoracic spine lateral flexion limitations contralateral to neurodynamic restrictions. Although the observed limitation in 11 out of 12 AS group patients suggests a potential association, larger studies are needed to contextualize its role in the pathophysiological process. Nevertheless, our findings establish a relationship between this limitation and neurodynamics. This connection aligns with the conceptual premise of functional diagnosis, emphasizing the importance of biomechanical assessment in neuromusculoskeletal symptom management.

Traditional approaches to spine biomechanics analysis, often employing motion capture systems that track skin-mounted markers [10–14], are known to have limitations in accurately quantifying subtle spinal movements. As highlighted by the recent literature [15,16], the challenges arise from difficulties in precisely locating spinal joint centers with external sensors and the inherently small amplitude of intervertebral motion. In contrast, our study utilized the least squares approximation method, an advanced mathematical tool allowing for a more precise and reliable analysis of the motion capture data we acquired. These methods offer the advantage of analyzing the contribution of individual vertebrae to spinal mechanics, though they are limited to single-plane analysis and require specialized mathematical expertise. Consequently, their application to the upper and mid-cervical spine, characterized by complex coupled movements, is restricted. For lumbar spine analysis, vertical planes through the posterior superior iliac spines may mitigate pelvic girdle rotation effects [15,16]. Multiplanar motion analysis is feasible with these methods but requires separate assessments. Finally, the time required for data acquisition and analysis and the need for specialized personnel may limit their broader research application.

Several limitations inherent in this preliminary study warrant careful consideration when interpreting these findings. The study's limitations included a small sample size, which inherently restricts the generalizability of our results to a larger population. The lack of randomization in participant selection may have introduced potential biases that could influence the observed associations. The analysis of spinal motion was limited to a single plane, specifically lateral bending, which may not fully capture the complex interplay between ULNTT asymmetry and overall cervicothoracic biomechanics. Additionally, the manual therapy interventions were not standardized in this study; each participant received individually selected techniques, which limits the comparability of the results and their reproducibility in other studies. Further clinical studies in patients with cervicobrachialgia due to altered brachial plexus interface are necessary to evaluate the therapeutic potential of correcting functional spinal alterations, particularly in those with upper thoracic spine lateral flexion limitations contralateral to the plexopathy. Future research should aim to address these limitations by employing larger, randomized controlled trials and by investigating spinal motion in multiple planes, alongside standardized therapeutic interventions, to provide more robust and generalizable conclusions.

5. Conclusions

In conclusion, our findings support the hypothesis that a relationship exists between the Upper Limb Neural Tension Test (ULNTT) and spine biomechanics. Specifically, this relationship is characterized by a limitation in upper thoracic spine lateral bending on the side contralateral to the restricted ULNTT in patients exhibiting asymmetry greater than 10 degrees. This observation underscores the biomechanical interplay between spinal mobility and neurodynamics, contributing insights for future research and the possible development of targeted therapeutic approaches. By highlighting this connection, we open avenues to explore more effective interventions for patients presenting ULNTT asymmetry and related spinal biomechanical alterations. It should be emphasized that these are preliminary results, based on a study of a small group of patients. Further research with a larger patient population is needed to confirm these preliminary results, including symptomatic individuals.

Author Contributions: Conceptualization, M.R. (Massimo Rossi) and M.S.; Data curation, M.S., A.B., M.D.R. and G.C.; Investigation, M.S.; Methodology, M.R. (Massimo Rossi) and M.S.; Resources, G.C.; Software, A.B., M.D.R. and G.C.; Supervision, M.R. (Marco Riva), M.F., F.P. and C.B.; Validation, F.P. and C.B.; Visualization, M.R. (Marco Riva) and M.F.; Writing—original draft, M.R. (Massimo Rossi); Writing—review and editing, C.B. All authors have read and agreed to the published version of the manuscript.

Funding: This research received no external funding.

Institutional Review Board Statement: This study was approved by the Università degli Studi di Milano Bicocca Ethics Committee (Approval code: 0035126/15; Approval date: 6 July 2015).

Informed Consent Statement: Informed consent was obtained from the subjects involved in the study.

Data Availability Statement: The corresponding author will share the data upon request due to legal and ethical reasons.

Acknowledgments: The authors would like to thank the reviewers for their comments and suggestions that allowed us to improve the manuscript.

Conflicts of Interest: The authors declare that the research was conducted in the absence of any commercial or financial relationships that could be construed as a potential conflict of interest.

References

1. Lohman, C.; Gilbert, K.; Sobczak, S.; Brismée, J.-M.; James, R.C.; Day, M.; Smith, M.P.; Taylor, L.; Dugailly, P.-M.; Pendergrass, T.; et al. Cervical Nerve Root Displacement and Strain During Upper Limb Neural Tension Testing. *Spine* **2015**, *40*, 793–800. [CrossRef]
2. Kleinrensink, G.; Stoeckart, R.; Mulder, P.G.H.; Hoek, G.V.D.; Broek, T.H.; Vleeming, A.; Snijders, C.J. Upper limb tension tests as tools in the diagnosis of nerve and plexus lesions. Anatomical and biomechanical aspects. *Clin. Biomech.* **2000**, *15*, 9–14. [CrossRef]
3. Selvaratnam, P.J. Brachial Plexus Tension Test in Patients And cadavers. Ph.D. Thesis, Department of Anatomy, Faculty of Medicine, Monash University, Clayton VIC, Australia, 1991.
4. Vanti, C.; Conteddu, L.; Guccione, A.; Filomena, M.; Sergio, P.; Carlotta, V.; Paolo, P. The Upper Limb Neurodynamic Test 1: Intra- and intertester reliability and the effect of several repetitions on pain and resistance. *J. Manip. Physiol. Ther.* **2010**, *33*, 292–299.
5. Van der Heide, B.; Allison, G.T.; Zusman, M. Pain and muscular responses to a neural tissue provocation test in the upper limb. *Man. Ther.* **2001**, *6*, 154–162. [CrossRef]
6. Coppieters, M.; Stappaerts, K.; Janssens, K.; Jull, G. Reliability of detecting 'onset of pain' and 'submaximal pain' during neural provocation testing for the upper quadrant. *Physiother. Res. Int.* **2002**, *7*, 146–156. [CrossRef] [PubMed]
7. Selvaratnam, P.J. Noninvasive Discrimination of Brachial Plexus Involvement in Upper Limb Pain. *Spine* **1994**, *19*, 26–33. [CrossRef] [PubMed]
8. Elvey, R.L. Brachial plexus tension test and the pathoanatomical origin of arm pain. In *Aspects of Manipulative Therapy*, 2nd ed.; Glasgow, E.F., Twomey Scull, E.R., Kleynhans, A.M., Eds.; Churchill Livingstone: Melbourne, VIC, Australia, 1985; pp. 116–122.
9. Elvey, R.L. The investigation of arm pain: Signs of adverse responses to the physical examination of the brachial plexus and related tissues. In *Grieve's Modern Manual Therapy*, 2nd ed.; Boyling, J.D., Palastanga, N., Eds.; Churchill Livingstone: New York, NY, USA, 1994; pp. 577–585.
10. Ferrario, V.; Sforza, C. Active range of motion of the head and cervical spine: A three dimensional investigation in healthy young adults. *J. Orthop. Res.* **2002**, *20*, 122–129. [CrossRef]
11. Sidorkewicz, N.; McGill, S. Male Spine Motion During Coitus. Implications for the Low Back Pain Patient. *Spine* **2014**, *39*, 1633–1639. [CrossRef]
12. Grip, H.; Gerdle, G.; Karlsson, S. Cervical helical axis characteristics and its center of rotation during active head and upper arm movements—Comparisons of whiplash-associated disorders, non-specific neck pain and asymptomatic individuals. *J. Biomech.* **2008**, *41*, 2799–2805. [CrossRef]
13. Duc, C.; Salvia, P.; Lubansu, A.; Feipel, V.; Aminian, K. A wearable inertial system to assess the cervical spine mobility: Comparison with an optoelectronic-based motion capture evaluation. *Med. Eng. Phys.* **2014**, *36*, 49–56. [CrossRef]
14. Cobian, D.; Daehn, N.; Anderson, P.; Heiderscheit, B.C. Active Cervical and Lumbar Range of Motion During Performance of Activities of Daily Living in Healthy Young Adults. *Spine* **2013**, *38*, 1754–1763. [CrossRef]
15. Harsh, S.; Supreeth, S.K.; Ayushi, K.A.; Vemuri, K. The role of individual physical body measurements and activity on spine kinematics during flexion, lateral bending and twist tasks in healthy young adults–Comparing marker(less) data. *Biomed. Signal Process. Control.* **2023**, *82*, 104517. [CrossRef]
16. Digo, E.; Pierro, G.; Pastorelli, S.; Gastaldi, L. Tilt-Twist Method Using Inertial Sensors to Assess Spinal Posture During Gait. In *Advances in Service and Industrial Robotics. RAAD 2019*; Advances in Intelligent Systems and Computing; Berns, K., Görges, D., Eds.; Springer: Cham, Switzerland, 2020; Volume 980. [CrossRef]
17. Lohkamp, M.; Small, K. Normal response to Upper Limb Neurodynamic Test 1 and 2A. *Man. Ther.* **2011**, *16*, 125–130. [CrossRef]
18. Van Hoof, T.; Vangestel, C.; Shacklock, M.; Kerckaert, I.; D'Herde, K. Asymmetry of the ULNT1 elbow extension range-of-motion in a healthy population: Conse-quences for clinical practice and research. *Phys. Ther. Sport* **2012**, *13*, 141–149. [CrossRef]
19. Stalioraitis, V.; Robinson, K.; Hall, T. Side-to-side range of movement variability in variants of the median and radial neurodynamic test sequences in asymptomatic people. *Man. Ther.* **2014**, *19*, 338–342. [CrossRef] [PubMed]
20. Owen, T.J.; Brew, J. A single Blind Investigation into the potential differences in Passive range of movement at the elbow, between dominant and non-dominant arm when using the upper limb tension test 1. *Physiotherapy* **2000**, *86*, 40. [CrossRef]
21. Oliver, S.G.; Rushton, A. A study to explore the reliability and precision of intra and inter-rater measures of ULNT1 on an asymptomatic population. *Man. Ther.* **2011**, *16*, 203–206. [CrossRef] [PubMed]
22. Butler, D.S. *The Sensitive Nervous System*; Uley, Noigroup Publication: Adelaide, SA, Australia, 2000.
23. Quarteroni, A.; Sacco, R.; Saleri, F. Approximation of a Function in the Least Square sense. In *Numerical Mathematics*; Quarteroni, A., Sacco, R., Saleri, F., Eds.; Springer: Berlin/Heidelberg, Germany, 2000; pp. 431–433.
24. Covill, L.; Petersen, S. Upper extremity neurodynamic tests: Range of motion asymmetry may not indicate impairment. *Physiother. Theory Pract.* **2012**, *28*, 535–541. [CrossRef]
25. Block, C.A.; Cantrall, C.E.; Threlkeld, A.J. Responses of asymptomatic subjects to the upper limb tension test. *Phys. Ther.* **1998**, *78*.

26. Hines, T.; Noakes, R.; Manners, B. The upper limb tension test: Inter-tester reliability for assessing the onset of passive resistance R1. *J. Man. Manip. Ther.* **1993**, *1*, 95–98. [CrossRef]
27. Pullos, J. The upper limb tension test. *Aust. J. Physiother.* **1986**, *32*, 258e9.
28. White, A.; Panjabi, M. The occipital-atlanto-axial complex. In *Clinical Biomechanics of the Spine*, 2nd ed.; White, A., Panjabi, M., Eds.; Lippincot Williams and Wilkins: Philadelphia, PA, USA, 1990; pp. 97–102.

bioengineering MDPI

Article

Finite Element Analysis of Biomechanical Assessment: Traditional Bilateral Pedicle Screw System vs. Novel Reverse Transdiscal Screw System for Lumbar Degenerative Disc Disease

Utpal K. Dhar [1], Kamran Aghayev [2], Hadi Sultan [3], Saahas Rajendran [3], Chi-Tay Tsai [1,*] and Frank D. Vrionis [4,*]

1 Department of Ocean and Mechanical Engineering, Florida Atlantic University, Boca Raton, FL 33431, USA; udhar2020@fau.edu
2 BHT Clinic, 34307 Istanbul, Türkiye
3 Charles E. Schmidt College of Medicine, Florida Atlantic University, Boca Raton, FL 33431, USA
4 Department of Neurosurgery, Marcus Neuroscience Institute, Boca Raton Regional Hospital, Boca Raton, FL 33486, USA
* Correspondence: tsaict@fau.edu (C.-T.T.); fvrionis@baptisthealth.net (F.D.V.); Tel.: +86-156-1297-2824 (C.-T.T.); +86-156-1955-4600 (F.D.V.)

Abstract: The traditional bilateral pedicle screw system has been used for the treatment of various lumbar spine conditions including advanced degenerative disc disease. However, there is an ongoing need to develop more effective and less invasive techniques. The purpose of this study was to compare the traditional bilateral pedicle screw system (BPSS) with the novel reverse transdiscal screw system (RTSS) for lumbar disc degenerative disease. A 3D solid lumbar L1–L5 spine model was developed and validated based on a human CT scan. Fusions were simulated at L3–L4. The first scenario comprised a transforaminal lumbar interbody cage in combination with the bilateral pedicle screw-rod system (BPSS-TLIF). In the second scenario, the same TLIF cage was combined with reverse L3–L4 transdiscal screws (RTSS-TLIF). Testing parameters included range of motion (ROM) in three orthogonal axes, hardware (cage and screw) stress, and shear load resistance. The ROM of the surgical model was reduced by approximately 90% compared to the intact model at the fused level. The RTSS model demonstrated less ROM compared to the BPSS model at the fused level for all loading conditions. Overall, the RTSS model exhibited lower stress on both screws and cage compared with the BPSS model in all biomechanical testing conditions. The RTSS model also exhibited higher anterior and posterior shear load resistance than the BPSS model. In conclusion, the RTSS model proved superior to the BPSS model in all respects. These findings indicate that the RTSS could serve as a feasible option for patients undergoing lumbar fusion, especially for adjacent segment disease, potentially enhancing surgical outcomes for disc degeneration.

Keywords: disc degeneration; finite element study; range of motion; von Mises stress; lumbar spine

1. Introduction

Interbody fusion is a well-established surgical option for a variety of lumbar spine diseases including degenerative, infectious, and traumatic conditions. Standard surgical technique comprises discectomy, insertion of interbody cage, and bone graft into the disc space with a supplementary bilateral pedicle screw–rod stabilization system (BPSS) [1]. It is known that the combination of an interbody cage and a BPSS yields the best fusion rates and, thus, favorable clinical outcomes. Yet the BPSS requires placing two pedicle screws per vertebra and connecting them with rods and is an indirect stabilization technique which relies on moving parts fastened to each other. For a single level stabilization, the required parts are four screws, four tulips, two rods, and four nuts total, making 14 parts [2]. Inherently, such a system is more prone to mechanical failure under physiological multiaxial load. Clinical failure may present as hardware breakage and/or screw loosening. Such long-term complications may lead to pseudoarthrosis and poor clinical outcomes and significantly influence patients' life quality. There is much need for simpler, stronger, and more resilient hardware options that can yield higher fusion rates, less failures, and better clinical outcomes.

Transdiscal screw (TSS) stabilization is an alternative technique which was developed, assessed, and clinically implemented [3,4]. It comprises only two screws inserted from one vertebra to another through the disc space [5] and is a direct stabilization technique with no moving parts and therefore more resistant to failure. Furthermore, the inherent simplicity of TSS makes it superior to the BPSS in terms of soft tissue damage during implantation. The inferior to superior S1-L5 TSS trajectory was previously developed and implemented [3,4]. This was possible due to the unique anatomical features of the lumbo-sacral segment. Such an infero–superior trajectory is unfeasible for other lumbar segments in the absence of spondylolisthesis due to anatomical restrictions, yet a superior to inferior trajectory was deemed possible. Thus, the model used in this study features a "reverse" transdiscal screw starting at the L3 pedicle and oriented caudally, passing through L3–4-disc space and terminating at the inferior border of the L4 vertebral body. This technique may circumvent the need for exposing and revising existing instrumentation in cases of superior upper adjacent segment disease that typically requires extension of the fusion and reconnection to existing hardware.

Finite element analysis is a biomechanical evaluation tool that provides several advantages in spinal matters, including evaluations of physiological and various pathological and surgical scenarios. It is an alternative method to measure hardware stress which is difficult to obtain from cadaveric studies [6–12]. In addition, finite element analysis provides reliable simulations of molecular, pathological, and biomechanical characteristics of degenerated discs without the invasiveness and replicability issues associated with other research methods [13–19]. Studies on the biomechanical efficacy of transdiscal screw approaches for lumbar disc degeneration are still limited. No finite element studies were conducted to compare the traditional BPSS with the RTSS for lumbar disc degeneration. The purpose of this study was to compare biomechanical features of the reverse transdiscal screw system with the standard bilateral pedicle screw–rod system using finite element analysis.

2. Materials and Methods

2.1. Model Creation

The L1–L5 model was generated using the computed tomography (CT) imaging data of a healthy patient (64 years old, female) in Mimics 25.0 (Materialize, Leuven, Belgium). The volunteer was informed; consent forms authorizing the use of her images for educational and research purposes were secured. All methods were carried out in accordance with relevant guidelines and regulations. All experimental protocols were approved by Boca Raton Regional Hospital and Florida Atlantic University. Figure 1 shows how a typical CT scan is converted into a 3D model through Mimics software. Digital Imaging and Medical Communication (DICOM) imaging data of the patient's spine are transferred into Mimics. Based on these images, generation of vertebral bodies was achieved through threshold segmentation, yielding surface mesh elements that are exported through STL format into 3-Matic (Materialize, Leuven, Belgium) for further operations. Careful post-processing in

3-Matic was then carried out for geometry mesh preparation. Following this, the cortical and cancellous bones were derived from the processed vertebral body with an offset of 1 mm [20]. The endplates were generated with a thickness of 0.5 mm on the top and bottom surfaces of the cortical bone, with generation of the disc and its constituent components, comprising the annulus fibrosis and nucleus pulposus, connecting contiguous endplates within the same software. Meshing of cortical and cancellous bodies was then finally completed using tetrahedral elements.

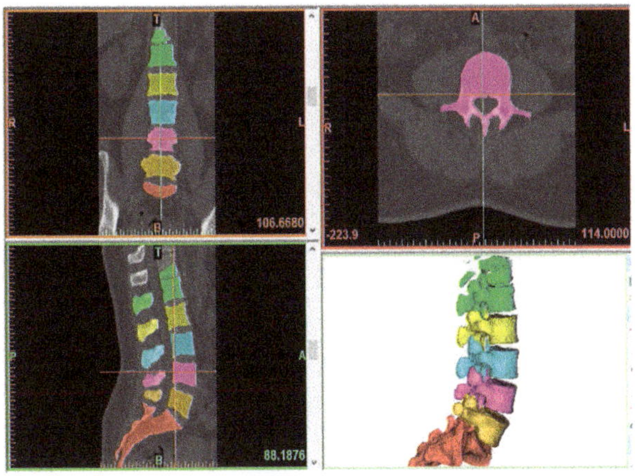

Figure 1. A typical diagram for converting a 2D CT scan to a 3D model.

A Unilateral L3–L4 facetectomy was performed. After excision of the posterior longitudinal ligament and annulus fibrosus, the nucleus pulposus was removed. An appropriately sized TLIF cage was designed with CAD software (SolidWorks 2023, Dassault Systèmes Inc., Vélizy-Villacoubla, France) based on the L3–L4 gap and was placed. In the BPSS model, four pedicle screws and two rods were also crafted with the same CAD software and employed to fix L3 and L4. Figure 2A–C depict the BPSS model from sagittal, anterior, and posterior perspectives, respectively. In the RTSS model, a similar TLIF cage was inserted between L3 and L4. Two transdiscal screws passed through the L3 pedicle inferiorly and terminated at the inferior border of the L4 vertebral body. The RTSS model is displayed in sagittal, anterior, and posterior views in Figure 2D–F, respectively. In both the BPSS and RTSS models, the diameter of the screw size was chosen to be 5.5 mm. The length of the transdiscal screw was 68 mm, while the pedicle screw length was 40 mm.

2.2. Finite Element Model

The completed components were then exported. In Ansys Workbench (Ansys Workbench 2023 R1, Ansys Inc. Canonsburg, PA, USA), the annulus geometry prepared in 3-Matic was meshed using a hexahedral mesh with embedded tension-only fiber bar elements for simulation of the ground substance and fibrous components, respectively. Fibers were constructed in a crisscross pattern with angles of approximately 30° to the horizontal and Young moduli of 550 MPa in the outside band to 360 MPa in the inner band. The intermediate layers varied linearly in their Young moduli. Detailed material properties are listed in Table 1 for the constructed model with the corresponding literature. Ligaments were subsequently added to the model in Ansys Workbench and programmed as tension-only spring elements. Based on prior studies, nonlinear force–displacement relationships, which describe the response of each ligament to varying vertebral loads, were also incorporated. Similarly, ligament material properties were referenced in Table 1. The included ligaments in this model consisted of the anterior longitudinal ligament, posterior longitudinal ligament, transverse ligaments, spinous and supraspinous ligaments, and capsular ligaments.

The mesh convergence was performed for the intact model. Three distinct meshes such as Mesh1, Mesh2, and Mesh3 were created consecutively for different parts of the bone. Mesh1 has the lowest number of elements and nodes, whereas Mesh3 has the highest among the three meshes. For the endplate and nucleus pulposus, a relatively fine mesh was chosen. The difference between Mesh1 and Mesh2 showed more than 10%, whereas it was less than 5% between Mesh2 and Mesh3 when accounting for equivalent von Mises stresses for the identical loading conditions. The mesh was deemed to converge when the variance between the predicted von Mises stresses of various components from two consecutive mesh refinements was below 5% under a torque of 7.5 Nm. Hence, Mesh2 was chosen for the entire FE analysis. The final model consisted of 435,808 elements and 787,522 nodes, comprising five vertebrae with their corresponding interconnected discs, ligaments, and connections.

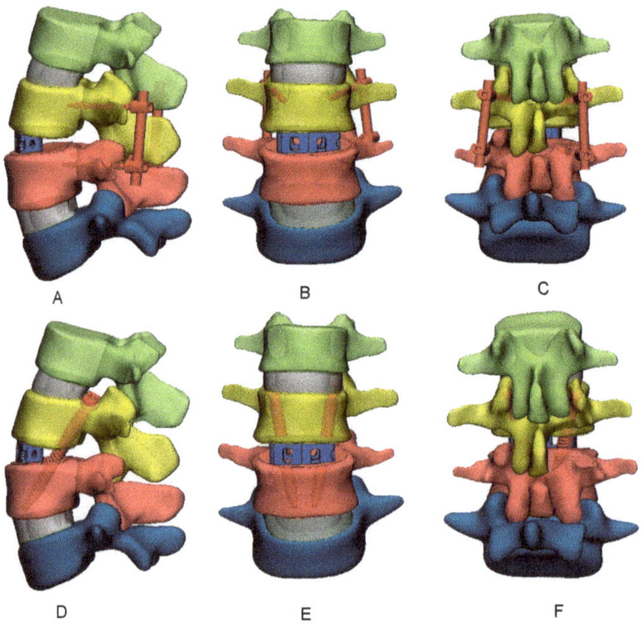

Figure 2. Traditional bilateral pedicle screw system: sagittal (**A**), anterior (**B**), and posterior view (**C**); reverse trandiscal screw system: sagittal (**D**), anterior (**E**), and posterior view (**F**).

2.3. Contact, Loading, and Boundary Conditions

Model loading and boundary conditions were applied in accordance with cadaveric experiments previously described [21,22]. For the simulations, the inferior surface of the L5 vertebral body was fixed and a 7.5 N.m. moment was applied to the upper endplate of L1. This moment was varied in direction to simulate each of the loading conditions. Furthermore, in each load case, 500 N of compressive force was applied and evenly distributed across the upper endplate of L1.

Table 1. Element type and material properties used in the model.

Component	Young Modulus (Mpa)	Poisson Ratio	Element Type
Cortical [23]	12,000	0.3	Tetrahedral element
Cancellous [23,24]	100	0.2	Tetrahedral element
Endplate [25]	500	0.45	Tetrahedral element
Annulus ground [23]	4.2	0.45	Hexahedral element
Annulus fiber [17,26–28]	360–450	Cross-sectional area (0.15 mm^2)	Link element
Ligament [23,28–30]	Calibrated force–deflection curve		Spring element
Nucleus [29]	1	0.4	Tetrahedral element
Titanium [30]	110,000	0.3	Tetrahedral element

Lastly, anterior and posterior shear loads of 150 N were applied parallel to the upper endplate of the L3 by fixing the bottom surface of L4. A representative boundary condition for the intact model is illustrated in Figure 3A for axial rotation and Figure 3B for anterior shear loading. Because of the body-dependent mesh sizing, which would lead to highly distorted elements for shared nodes between bodies, bonded contacts were utilized to connect the vertebral discs to the endplates. Bonded contacts were also utilized for connection of the endplates to the vertebral bodies. All implants were similarly connected to the existing vertebral bodies using bonded connections, with Boolean operations where appropriate to achieve a cohesive model.

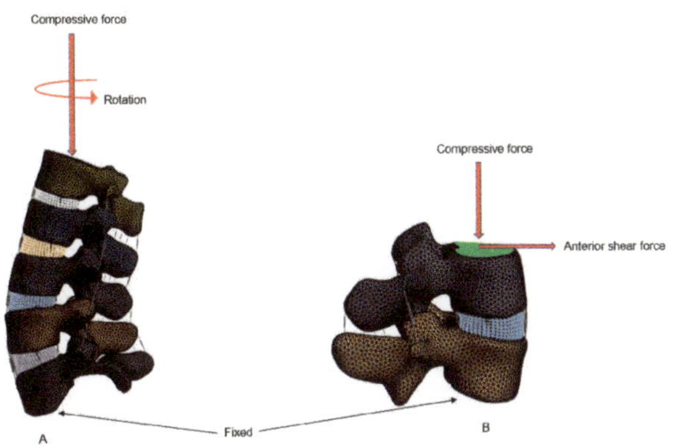

Figure 3. A typical boundary condition for left-axial rotation (**A**) and anterior shear loading (**B**) for the intact model.

3. Results

3.1. FE Model Validation

After subjecting similar boundary and loading conditions, we compared our range of motion (ROM) with those derived from a cadaveric investigation conducted by previous in vitro experiment and other FEA study [31,32]. Figure 4 illustrates the ROM of the current study compared with the experimental and FEA results and congruence with the previously documented data.

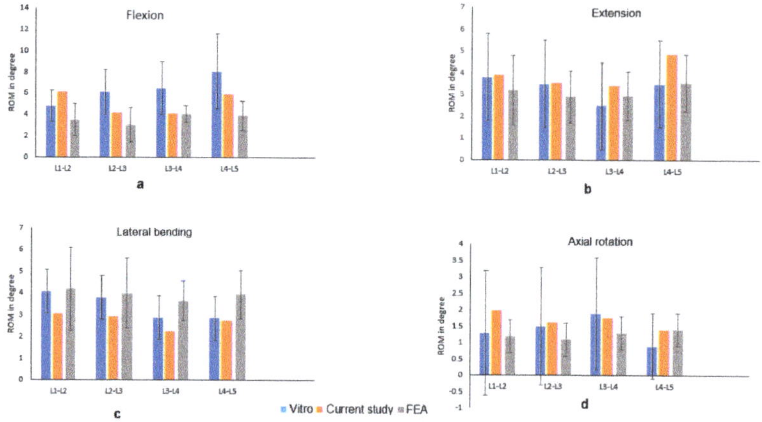

Figure 4. Comparison of ROM with the in vitro and FEA study for flexion (**a**), extension (**b**), lateral bending (**c**), axial rotation (**d**).

Consequently, the intact model in the present study was successfully established and was deemed ready for further analysis.

3.2. ROM

In Figure 5, the ROM for flexion, extension, left bending, right bending, left rotation, and right rotation is compared between the BPSS and RTSS models for the index level. The ROM of the BPSS and the RTSS was compared with the intact lumbar model. Both treatment models showed significant reduction in ROM across all loading scenarios when compared to the intact model. Additionally, the RTSS model demonstrated a further reduction in ROM compared to the BPSS model under all loading conditions. The greatest difference between the two models was observed in extension, where the RTSS model reduced ROM by 58% compared to the BPSS model. The smallest difference was seen in right rotation, with a 13% reduction.

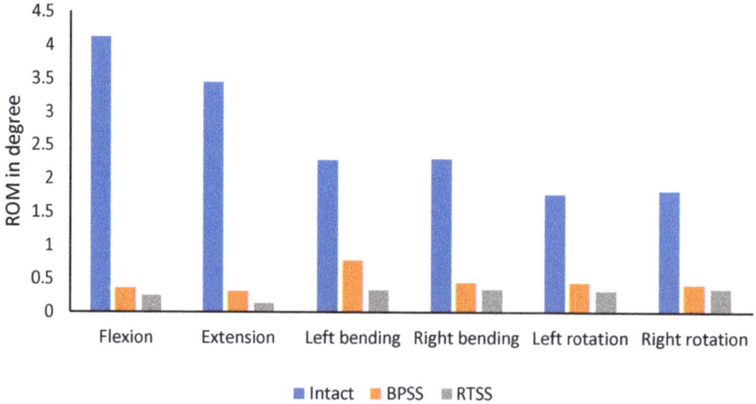

Figure 5. Comparing range of motion between the intact, BPSS, and RTSS models.

3.3. Cage von Mises Stress

The cage stress under various loading scenarios was calculated for both the BPSS and the RTSS models and is shown in Figure 6b, while Figure 6a is the typical contour plot for cage von Mises stress for the RTSS model in flexion condition. In both flexion and extension, the cage stress in the BPSS model was over 30% higher compared to the RTSS model. For rotation and right bending, there was no significant difference in cage stress between the two models. In left bending, the BPSS model exhibited a cage stress of 86.5 MPa, while the RTSS model showed a significantly lower stress of 49.6 MPa. The highest stress for the BPSS system was observed in flexion, reaching 88.7 MPa, whereas for the RTSS system, the highest stress occurred in left rotation, measuring 67.3 MPa.

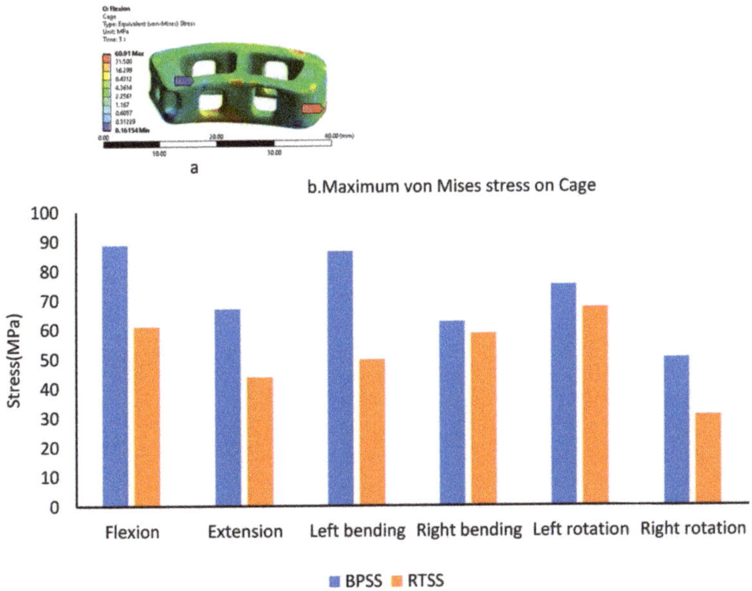

Figure 6. A typical von Mises stress contour on RTSS cage under flexion condition (**a**). Comparing maximum von Mises stress on the cage between the BPSS and the RTSS (**b**).

3.4. Screw von Mises Stress

A typical von Mises stress for the BPSS and RTSS models is shown in Figure 7a,b, respectively, for the flexion condition. Figure 7c illustrates the screw stress for the BPSS and the RTSS models under various physiological loading conditions. In the BPSS model, higher stress was observed during axial rotation compared to flexion–extension and lateral bending scenarios. The maximum screw stress in the BPSS model was recorded at 418 MPa during right rotation, while the minimum stress was 208 MPa during right bending. In contrast, the RTSS model exhibited the highest screw stress (239 MPa) during extension and the lowest stress (108 MPa) during left bending. In lateral bending, the RTSS model showed a 40% lower screw stress compared to the BPSS model. For extension, the difference in screw stress between the RTSS and BPSS models was minimal, with both recording a value of 239 MPa, a smaller variation compared to other loading conditions.

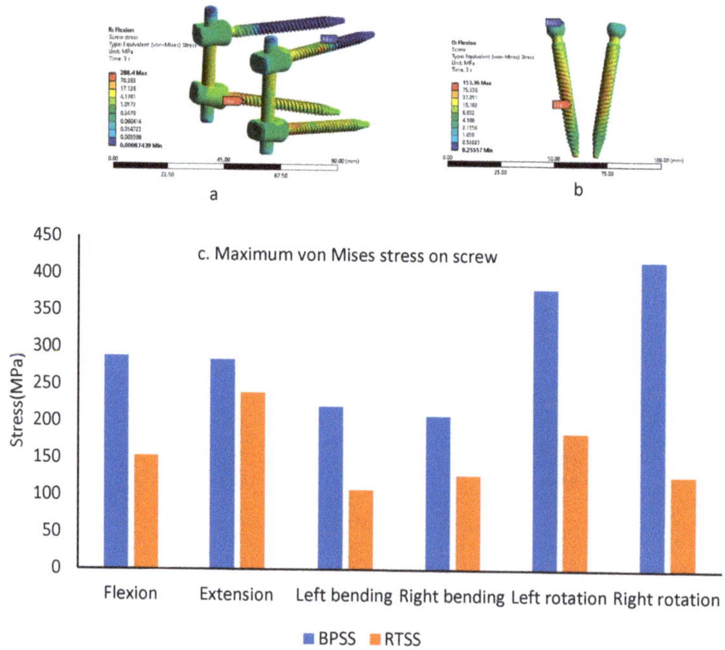

Figure 7. A typical von Mises stress contour on internal fixation under flexion condition (**a**). BPSS (**b**). RTSS (**c**). Comparing maximum von Mises stress on the screw between the BPSS and the RTSS.

3.5. Resistance to Shear Load

Regarding anterior and posterior shear loads, a shear load of magnitude 150 N was applied at the center of the L3 vertebra in a horizontal direction parallel to the base of L4. These loads replicated the loading scenarios in the mechanical assessment of cadaveric motion segments [33]. The displacement comparison due to shear load is depicted in Figure 8. The RTSS model illustrated anterior and posterior displacement of 3.43 mm and 3.44 mm, respectively, whereas the BPSS model showed 3.70 mm and 4.48 mm. No notable difference was experienced in these two models due to the anterior shear load, but the posterior shear load RTSS model exhibited 23% lower displacement than the BPSS model. Overall, the RTSS model showed better shear load resistance than the BPSS model for the same loading condition.

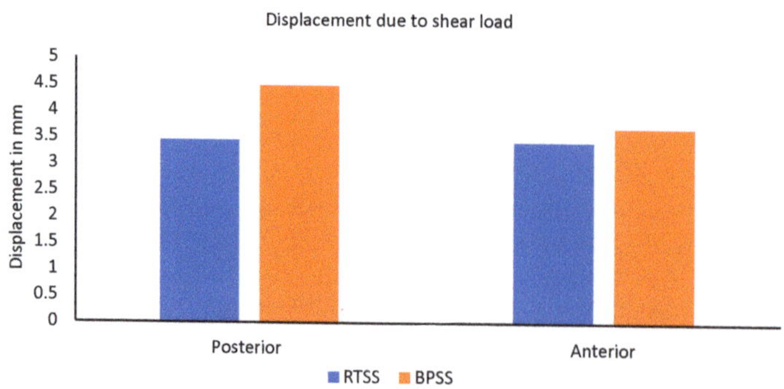

Figure 8. Comparing shear load resistance between the RTSS and BPSS models due to anterior and posterior shear load.

4. Discussion

In the present study, we compared the RTSS and BPSS models to evaluate their mechanical properties. This study focuses on a novel RTSS designed to address some of the biomechanical shortcomings associated with the traditional BPSS, particularly reducing the incidence of non-union and improving post-surgical outcomes. In line with the previous studies [3,4], the RTSS demonstrated higher stiffness across all loading conditions, particularly in extension, where a 58% reduction in ROM was observed compared to the BPSS model. These findings suggest that the RTSS provides enhanced rigidity, which could potentially facilitate fusion.

The stress analysis of both the cage and screws shows the advantages of the RTSS. The cage stress was notably lower in the RTSS model under flexion and lateral bending conditions, reduced by over 30% compared to the BPSS. These results indicate that the RTSS may distribute mechanical loads more evenly, avoiding high pressure points, and can potentially reduce the risk of cage-related complications such as subsidence and migration [34]. Similarly, screw stress was significantly lower in the RTSS model during lateral bending and axial rotation, with a 40% reduction compared to the BPSS model. This decrease in stress is crucial for long-term stability, as reduced screw stress may lower the risk of hardware breakage or loosening—potential complications in spinal fusion surgeries [35]. The screw stress ranged from 200 MPa to 700 MPa, while the cage stress was approximately 100 MPa when this type of construction is inserted in the lumbar spine. In this study, we found that the von Mises stresses were significantly lower than the yield strength of titanium, indicating that this construction can be safely used in real scenarios.

The vertebral slippage was investigated after applying the anterior and posterior shear force. Most biomechanical studies have neglected this parameter in their analysis. This information is necessary for daily activities that produce such loads, including bending and lifting [36]. Overall, the magnitude of displacement by posterior shear load was lower than the anterior shear force condition for both the BPSS and RTSS models. Lee et al. [33] also reported decreases in displacement of the upper vertebra due to posterior shear loading relative to anterior shear loading. In this study, the RTSS model exhibited higher resistance to both anterior and posterior shear loading compared to the BPSS model, further highlighting the advantage of the RTSS. We attribute this effect to the geometry of the screw in relation to the axis of the spine. The RTSS axis is almost parallel to the spine, whereas the BPSS screws are orthogonal to the spine's axis. Thus, the RTSS is much more resistant to orthogonal forces (shear force) in comparison to the BPSS. Overall increased stiffness in combination with lesser hardware stress theoretically makes the RTSS a much more effective alternative to the BPSS. Moreover, it is potentially useful in adjacent segment disease where typically the surgeon needs to expose and connect with the previous instrumentation. But reverse TSS technique bypasses the need. The RTSS can be less invasive than the BPSS even with modern paramedian muscle splitting minimally invasive techniques. As the RTSS does not require a tulip for the screw head, it can be placed entirely percutaneously through a 1 cm incision on each side. Biplanar fluoroscopy or an image-guided technique is clearly needed for such placement.

In the mechanical perspective view, the RTSS has larger bone–interface contact and a longer lever arm. This scenario will enhance the pull-out strength and resistance to shear load. Another advantage of the RTSS is its insertion angle of approximately 45 degrees, compared to the BPSS, where screws are typically inserted parallel to the horizontal plane. Due to this oblique trajectory, the resulting compressive force acting on the RTSS is reduced to approximately two-thirds of that experienced by the BPSS. That could be another reason the RTSS screws experience lower stresses than the BPSS. That could potentially lower the risk of implant failure under axial loading. In the RTSS, the screws align more directly with the load-bearing axis of the spinal column. This allows for more axial load sharing between the bone and implant. On the other hand, in the BPSS, screws are inserted to the posterior side and somewhat off-axis, which is responsible for higher bending moments at the screw–rod junction. In the osteoporotic bone, the RTSS may offer greater initial fixation and long-term stability because of the load distribution along the transdiscal path.

While these biomechanical advantages highlight the potential benefits of the RTSS, it is important to consider the limitations of FEA. Only clinical studies can fully unlock the potential of new stabilization techniques. FEA studies of spinal implants rely on a single spine model and may not be able to fully account for clinical pathologies including disc degeneration, limiting the clinical application of the results to a wide population. Additionally, some models assume linear material properties and ignore the contributions of paravertebral muscles which may not be fully accurate when considering vivo loading conditions.

To further improve the RTSS design and optimization, it is necessary to analyze the multiple insertion angle by using FE analyses. Screws could be designed as double thread to improve the anchorage in both cortical and cancellous bone. Diameter and length could be customized to reduce the damage of the endplate. Moreover, interbacterial coating on the screws could be useful for reducing post-operative infection. Hydroxyapatite or titanium plasma spray could be used for promoting osteointegration.

In finite element modeling, the material assignment for different components is very important, as the simulation influences outcome significantly. Therefore, careful selection of appropriate material properties is essential for obtaining meaningful and reliable results. In this study, we also employed an appropriate boundary condition as they play a critical role in accurately replicating physiological loading scenarios. Moreover, mesh sensitivity was performed to check the mesh quality. It is essential to validate the current model by comparing it with existing experimental and other finite element studies to ensure the results are accurate and meaningful. Common mistakes should be avoided such as misaligning the anatomical landmark, over constraining the model that may affect the outcome significantly. This insight will assist other engineers not only with this type of disc degeneration model but with common biomechanical design projects.

5. Conclusions

These findings suggest that the RTSS may offer a viable alternative for patients undergoing lumbar fusion, potentially improving overall surgical outcomes for disc degeneration. However, further research, including in vivo studies and long-term clinical trials, is necessary to fully assess the clinical impact of the RTSS and its potential role in the treatment of lumbar degenerative disc disease.

6. Limitations

This BPSS compared with the RTSS for lumbar disc degeneration is grounded in FE techniques and presents specific limitations. First, we utilized skeletal data from only one individual's FEA simulation, without considering individual differences. Secondly, applying a compressive force of 500 N combined with a moment of 7.5 Nm does not precisely simulate the stress experienced by the cage and screws in patients following spinal fusion surgery. Another limitation of this study is that the RTSS may be technically demanding to be inserted in such a position, and it would not be suitable for the multi-level fixation. Finally, in the present model, ligaments are modeled as nonlinear spring elements that are affected only by tensile forces, while muscles are excluded from the simulation. This may influence the movement and stress variations in the lumbar spine after post-surgery.

Author Contributions: U.K.D., Conception and design, Literature search, Analysis and interpretation of data, Drafting the article, Reviewed submitted version of manuscript; K.A., Conception and design, Drafting the article, Reviewed submitted version of manuscript, H.S., Drafting the article, Critically revising the article, Design; S.R., Drafting the article, Literature search; C.-T.T., Study supervision, Conception and design, Administrative/technical/material support, Reviewed submitted version of manuscript; F.D.V., Drafting, Designing, Funding, Critically revising the article, Supervision, and Approved the final version of the manuscript on behalf of all authors. The 1st and 2nd authors equally contributed in this paper. All authors have read and agreed to the published version of the manuscript.

Funding: This research was supported by the Boca Raton Regional Hospital Foundation and the Helene and Stephen Weicholz Foundation. (Award # SP 19-579).

Institutional Review Board Statement: Not applicable.

Informed Consent Statement: Not applicable.

Data Availability Statement: The datasets used and/or analyzed during the current study are available from the corresponding author on reasonable request.

Acknowledgments: We thank the reviewers for their valuable suggestions, especially for helping improve the figures in the manuscript.

Conflicts of Interest: The authors declare no conflicts of interest.

References

1. Lee, Y.C.; Zotti, M.G.T.; Osti, O.L. Operative Management of Lumbar Degenerative Disc Disease. *Asian Spine J.* **2016**, *10*, 801–819. [CrossRef] [PubMed]
2. Mok, J.M.; Forsthoefel, C.; Diaz, R.L.; Lin, Y.; Amirouche, F. Biomechanical Comparison of Unilateral and Bilateral Pedicle Screw Fixation after Multilevel Lumbar Lateral Interbody Fusion. *Glob. Spine J.* **2022**, *14*, 1524–1531. [CrossRef]
3. Dhar, U.K.; Sultan, H.; Aghayev, K.; Tsai, C.-T.; Vrionis, F.D. Biomechanical assessment of anterior plate system, bilateral pedicle screw and transdiscal screw system for high-grade spondylolisthesis: A finite element study. *Front. Bioeng. Biotechnol.* **2024**, *12*, 1491420. [CrossRef]
4. Aghayev, K.; Dhar, U.K.; Tsai, C.-T.; Ahmedov, M.; Vrionis, F.D. Sacrolumbar Interbody Fusion (SLIF): Feasibility, Technical Nuances, Biomechanical Assessment, and Clinical Outcomes. *Spine Surg. Relat. Res.* **2024**, *8*, 448–457. [CrossRef]
5. Pietrantonio, A.; Trungu, S.; Famà, I.; Forcato, S.; Miscusi, M.; Raco, A. Long-term clinical outcomes after bilateral laminotomy or total laminectomy for lumbar spinal stenosis: A single-institution experience. *Neurosurg. Focus* **2019**, *46*, E2. [CrossRef] [PubMed]
6. Natarajan, R.N.; Garretson, R.B.; Biyani, A.; Lim, T.H.; Andersson, G.B.J.; An, H.S. Effects of slip severity and loading directions on the stability of isthmic spondylolisthesis: A finite element model study. *Spine* **2003**, *28*, 1103–1112. [CrossRef] [PubMed]
7. Zhang, R.; Mo, Z.; Li, D.; Yang, B.; Huang, S.; Tang, S. Biomechanical Comparison of Lumbar Fixed-Point Oblique Pulling Manipulation and Traditional Oblique Pulling Manipulation in Treating Lumbar Intervertebral Disk Protrusion. *J. Manip. Physiol. Ther.* **2020**, *43*, 446–456. [CrossRef]
8. Obeidallah, M.; Hamad, M.K.; Holland, R.; Mammis, A. The Use of a Transforaminal Lumbar Interbody Fusion (TLIF) Cage Following Inadequate Access to the Disc Space During an Anterior Lumbar Interbody Fusion Procedure. *Cureus* **2022**, *14*, e22792. [CrossRef]
9. Gum, J.L.; Reddy, D.; Glassman, S. Transforaminal Lumbar Interbody Fusion (TLIF). *JBJS Essent. Surg. Tech.* **2016**, *6*, e22. [CrossRef]
10. Cai, X.; Sun, M.; Huang, Y.; Liu, Z.; Liu, C.; Du, C.; Yang, Q. Biomechanical Effect of L_4–L_5 Intervertebral Disc Degeneration on the Lower Lumbar Spine: A Finite Element Study. *Orthop. Surg.* **2020**, *12*, 917–930. [CrossRef]
11. Li, Q.Y.; Kim, H.-J.; Son, J.; Kang, K.-T.; Chang, B.-S.; Lee, C.-K.; Seok, H.S.; Yeom, J.S. Biomechanical analysis of lumbar decompression surgery in relation to degenerative changes in the lumbar spine—Validated finite element analysis. *Comput. Biol. Med.* **2017**, *89*, 512–519. [CrossRef] [PubMed]
12. Matsumoto, K.; Shah, A.; Kelkar, A.; Parajuli, D.; Sudershan, S.; Goel, V.K.; Sairyo, K. Biomechanical evaluation of a novel decompression surgery: Transforaminal full-endoscopic lateral recess decompression (TE-LRD). *N. Am. Spine Soc. J.* **2021**, *5*, 100045. [CrossRef]

13. Kang, S.; Park, C.-H.; Jung, H.; Lee, S.; Min, Y.-S.; Kim, C.-H.; Cho, M.; Jung, G.-H.; Kim, D.-H.; Kim, K.-T.; et al. Analysis of the physiological load on lumbar vertebrae in patients with osteoporosis: A finite-element study. *Sci. Rep.* **2022**, *12*, 11001. [CrossRef] [PubMed]
14. Zhu, R.; Niu, W.-X.; Zeng, Z.-L.; Tong, J.-H.; Zhen, Z.-W.; Zhou, S.; Yu, Y.; Cheng, L.-M. The effects of muscle weakness on degenerative spondylolisthesis: A finite element study. *Clin. Biomech.* **2017**, *41*, 34–38. [CrossRef]
15. Lin, M.; Paul, R.; Dhar, U.K.; Doulgeris, J.; O'cOnnor, T.E.; Tsai, C.-T.; Vrionis, F.D. A Review of Finite Element Modeling for Anterior Cervical Discectomy and Fusion. *Asian Spine J.* **2023**, *17*, 949–963. [CrossRef]
16. Xu, Z.; Zheng, Q.; Zhang, L.; Chen, R.; Li, Z.; Xu, W. Biomechanical evaluation of different oblique lumbar interbody fusion constructs: A finite element analysis. *BMC Musculoskelet. Disord.* **2024**, *25*, 97. [CrossRef] [PubMed]
17. Shin, D.S.; Lee, K.; Kim, D. Biomechanical study of lumbar spine with dynamic stabilization device using finite element method. *Comput. Des.* **2007**, *39*, 559–567. [CrossRef]
18. Spina, N.T.; Moreno, G.S.; Brodke, D.S.; Finley, S.M.; Ellis, B.J. Biomechanical effects of laminectomies in the human lumbar spine: A finite element study. *Spine J.* **2021**, *21*, 150–159. [CrossRef]
19. Ahuja, S.; Moideen, A.N.; Dudhniwala, A.G.; Karatsis, E.; Papadakis, L.; Varitis, E. Lumbar stability following graded unilateral and bilateral facetectomy: A finite element model study. *Clin. Biomech.* **2020**, *75*, 105011. [CrossRef]
20. Edwards, W.T.; Zheng, Y.; Ferrara, L.A.; Yuan, H.A. Structural features and thickness of the vertebral cortex in the thoracolumbar spine. *Spine* **2001**, *26*, 218–225. [CrossRef]
21. Nachemson, A.L.; Schultz, A.B.; Berkson, M.H. Mechanical properties of human lumbar spine motion segments. Influence of age, sex, disc level, and degeneration. *Spine* **1979**, *4*, 1–8. [CrossRef] [PubMed]
22. Panjabi, M.M.; Krag, M.H.; Chung, T.Q. Effects of Disc Injury on Mechanical Behavior of the Human Spine. *Spine* **1984**, *9*, 707–713. [CrossRef] [PubMed]
23. Shirazi-Adl, S.A.; Shrivastava, S.C.; Ahmed, A.M. Stress analysis of the lumbar disc-body unit in compression. A three-dimensional nonlinear finite element study. *Spine* **1984**, *9*, 120–134. [CrossRef] [PubMed]
24. Carter, D.R.; Caler, W.E.; Spengler, D.M.; Frankel, V.H. Fatigue behavior of adult cortical bone: The influence of mean strain and strain range. *Acta Orthop.* **1981**, *52*, 481–490. [CrossRef]
25. Panjabi, M.M.; Crisco, J.J.; Vasavada, A.; Oda, T.; Cholewicki, J.; Nibu, K.; Shin, E. Mechanical properties of the human cervical spine as shown by three-dimensional load–displacement curves. *Spine* **2001**, *26*, 2692–2700. [CrossRef]
26. Bereczki, F.; Turbucz, M.; Kiss, R.; Eltes, P.E.; Lazary, A. Stability Evaluation of Different Oblique Lumbar Interbody Fusion Constructs in Normal and Osteoporotic Condition–A Finite Element Based Study. *Front. Bioeng. Biotechnol.* **2021**, *9*, 749914. [CrossRef] [PubMed]
27. Eberlein, R.; Holzapfel, G.A.; Fröhlich, M. Multi-segment FEA of the human lumbar spine including the heterogeneity of the annulus fibrosus. *Comput. Mech.* **2004**, *34*, 147–163. [CrossRef]
28. Wang, Q.-D.; Guo, L.-X. Biomechanical role of cement augmentation in the vibration characteristics of the osteoporotic lumbar spine after lumbar interbody fusion. *J. Mater. Sci. Mater. Med.* **2022**, *33*, 52. [CrossRef]
29. Zhang, S.; Liu, Z.; Lu, C.; Zhao, L.; Feng, C.; Wang, Y.; Zhang, Y. Oblique lateral interbody fusion combined with different internal fixations for the treatment of degenerative lumbar spine disease: A finite element analysis. *BMC Musculoskelet. Disord.* **2022**, *23*, 206. [CrossRef]
30. Lv, Q.-B.; Gao, X.; Pan, X.-X.; Jin, H.-M.; Lou, X.-T.; Li, S.-M.; Yan, Y.-Z.; Wu, C.-C.; Lin, Y.; Ni, W.-F.; et al. Biomechanical properties of novel transpedicular transdiscal screw fixation with interbody arthrodesis technique in lumbar spine: A finite element study. *J. Orthop. Transl.* **2018**, *15*, 50–58. [CrossRef]
31. Dreischarf, M.; Zander, T.; Shirazi-Adl, A.; Puttlitz, C.; Adam, C.; Chen, C.; Goel, V.; Kiapour, A.; Kim, Y.; Labus, K.; et al. Comparison of eight published static finite element models of the intact lumbar spine: Predictive power of models improves when combined together. *J. Biomech.* **2014**, *47*, 1757–1766. [CrossRef] [PubMed]
32. Panjabi, M.M.; Oxland, T.R.; Yamamoto, I.; Crisco, J.J. Mechanical behavior of the human lumbar and lumbosacral spine as shown by three-dimensional load-displacement curves. *J. Bone Jt. Surg.* **1994**, *76*, 413–424. [CrossRef] [PubMed]
33. Lee, K.; Teo, E. Effects of laminectomy and facetectomy on the stability of the lumbar motion segment. *Med. Eng. Phys.* **2004**, *26*, 183–192. [CrossRef]
34. Dhar, U.K.; Menzer, E.L.; Lin, M.; Hagerty, V.; O'cOnnor, T.; Tsai, C.-T.; Vrionis, F.D. Factors influencing cage subsidence in anterior cervical corpectomy and discectomy: A systematic review. *Eur. Spine J.* **2023**, *32*, 957–968. [CrossRef] [PubMed]

35. Lonstein, J.E.; Denis, F.; Perra, J.H.; Pinto, M.R.; Smith, M.D.; Winter, R.B. Complications associated with pedicle screws. *J. Bone Jt. Surg.* **1999**, *81*, 1519–1528. [CrossRef]
36. Gallagher, S.; Marras, W.S. Tolerance of the lumbar spine to shear: A review and recommended exposure limits. *Clin. Biomech.* **2012**, *27*, 973–978. [CrossRef]

bioengineering

MDPI

Article

AI-Assisted Image Recognition of Cervical Spine Vertebrae in Dynamic X-Ray Recordings

Esther van Santbrink [1,2,3,*], Valérie Schuermans [1,2,3], Esmée Cerfonteijn [4], Marcel Breeuwer [4,5], Anouk Smeets [1,2], Henk van Santbrink [1,2,3] and Toon Boselie [1,2,3]

1 Department of Neurosurgery, Maastricht University Medical Centre, 6229 HX Maastricht, The Netherlands; valerie.schuermans@mumc.nl (V.S.); anouk.smeets@mumc.nl (A.S.); h.van.santbrink@mumc.nl (H.v.S.); t.boselie@mumc.nl (T.B.)

2 Department of Neurosurgery, Zuyderland Medical Centre, Henri Dunantstraat 5, 6419 PC Heerlen, The Netherlands

3 CAPHRI Institute for Public Health and Primary Care, Maastricht University, 6200 MD Maastricht, The Netherlands

4 Department of Biomedical Engineering, Eindhoven University of Technology, 5600 MB Eindhoven, The Netherlands; e.e.j.cerfonteijn@student.tue.nl (E.C.); m.breeuwer@tue.nl (M.B.)

5 Department of Electrical Engineering, Eindhoven University of Technology, 5600 MB Eindhoven, The Netherlands

* Correspondence: esther.van.santbrink@mumc.nl

Abstract

Background: Qualitative motion analysis revealed that the cervical spine moves according to a consistent pattern. This analysis calculates the relative rotation between vertebral segments to determine the sequence in which they contribute to extension, demonstrating a mean sensitivity of 90% and specificity of 85%. However, the extensive time that is required limits its applicability. This study investigated the feasibility of implementing a deep-learning model to analyze qualitative cervical motion. *Methods:* A U-Net architecture was implemented as 2D and 2D+t models. Dice similarity coefficient (DSC) and Intersection over Union (IoU) were used to assess the performance of the models. Intraclass Correlation Coefficient (ICC) was used to compare the relative rotation of individual vertebrae and segments to the ground truth. *Results:* IoU ranged from 0.37 to 0.74 and DSC ranged from 0.53 to 0.80. The ICC scores for relative rotation ranged from 0.62 to 0.96 for individual vertebrae and from 0.28 to 0.72 for vertebral segments. For segments, 2D+t models presented higher ICC scores compared to 2D models. *Conclusions:* This study showed the feasibility of implementing deep-learning models to analyze qualitative cervical motion in dynamic X-ray recordings. Future research should focus on improving model segmentation by enhancing recording contrast and applying post-processing methods. Improved segmentation accuracy will enable routine use of the analysis of motion patterns in clinical research. The absence or presence of a motion pattern, or identification of new patterns has the potential to aid in clinical decision-making.

Keywords: cervical spine; qualitative motion analysis; image segmentation; U-Net; X-ray recording; deep-learning model; motion pattern

1. Introduction

A clear definition of physiological cervical spine motion is lacking. Segmental range of motion (sROM) has been proposed as a metric to evaluate cervical spine motion. However, sROM is limited by high intra- and interindividual variability and restricted to static endpoints [1,2]. The sequence of segmental contribution (SSC) is an alternative metric that is used to analyze qualitative motion of the cervical spine [3]. To determine SSCs, relative rotations of the segments are tracked in dynamic X-ray recordings. The SSC is a consistent metric and has a mean sensitivity of 90% and specificity of 85% to differentiate motion patterns between asymptomatic individuals and patients with a radicular syndrome [4]. In a cross-validation study, two individual methods to analyze cervical spine motion in dynamic X-ray recordings demonstrated excellent agreement with a median difference below 0.2 degrees [5]. Motion pattern analysis revealed that various segments contribute at different moments during the extension movement of the cervical spine.

In young asymptomatic individuals, a consistent motion pattern of vertebrae C4 to C7 is observed during extension [4]. The pattern is defined as the order of maximum contribution of each motion segment to the extension movement, i.e., a peak in the relative rotation with respect to the cumulative amount of rotation of vertebrae C4 to C7. An RCT comparing anterior cervical discectomy with arthroplasty (ACDA) to anterior cervical discectomy (ACD) in patients with radicular syndrome demonstrated the impact of surgery on observed motion patterns. Before surgery, the consistent motion pattern that was previously defined in asymptomatic individuals was present in 37.5% of patients. In the ACDA group, in 80% of patients this same motion pattern was restored one year after surgery, compared to 20% in the ACD group [6]. The difference in motion patterns between the groups can be explained by the presence of fusion in the ACD group, and maintenance of movement in the ACDA group. However, in a subsequent study, we observed that the motion pattern was absent in both treatment groups after a mean follow-up of 11 years. The range of motion (ROM) was however preserved in the ACDA group compared to the ACDF group in this long-term follow-up [7]. It has already been shown that ROM decreases with age due to the natural aging process. Qualitative analysis of cervical motion in elderly showed that the consistent motion pattern that was found in young asymptomatic individuals was absent [8]. The presence or absence of a consistent motion pattern among different groups raises questions and requires further investigation of cervical spine motion. In clinical practice, the presence of a consistent motion pattern might play a role in clinical decision-making such as choosing surgical strategies.

Qualitative motion analysis has not been widely applied in research or clinical practice due to its time-consuming nature and the need for trained and experienced individuals. To increase feasibility, artificial intelligence (AI) algorithms are implemented to design deep-learning models for the segmentation of cervical vertebrae. Promising results have been reported in studies that use these models to recognize cervical injuries on lateral X-rays, with accuracy for segmentation of cervical vertebrae ranging from 92 to 99% [9,10]. One study focused on calculating ROM of cervical vertebrae in static lateral flexion and extension radiographs and showed good accuracy [11].

In contrast to static X-rays, dynamic X-ray recordings can incorporate a temporal dimension, enabling a 2D + time (2D+t) model. Implementation of both a U-Net and capsule network (CapsNet) for 2D, 2D+t and 3D model approaches to segment brain structures in MRI data showed comparable performance in the segmentation of the third ventricle [12]. The 3D U-Net achieved the highest segmentation accuracy of 96% Dice score, a metric that measures the overlap between the predicted segmentation and the ground truth. The segmentation accuracy of the 2D+t U-Net was similar to the results of the 2D U-Net with a Dice score of 91% and 90%, respectively. Similar findings were observed when comparing a U-Net with a SegNet architecture for the 2D, 2D+t and 3D model approaches [13]. Five medical datasets were investigated, both CT and MRI data, and the segmentation performance of the 2D+t approach did not improve compared to both 2D and 3D. The effectiveness of the U-Net architecture for segmentation using a

2D+t approach has been demonstrated, but conclusions on the efficacy of 2D+t versus 2D methods varied.

To enable qualitative analysis of the cervical spine, this study aimed to develop a 2D and 2D+t deep-learning model, trained on previously annotated dynamic X-ray recordings, to automatically calculate the relative rotation of each vertebra.

2. Materials and Methods

2.1. Population

Previous studies by our group investigated motion patterns in healthy, asymptomatic individuals in different age groups. Detailed information can be found in the published full-text articles [4,6,7,14,15]. Manually annotated data from these studies were combined for the training of this algorithm. Dynamic X-ray recordings were all made following the same protocol. Participants were seated on a crutch, adjustable in height, with their neck, shoulders, and head free. The subjects were asked to move their heads from maximal extension to maximal flexion, and vice versa, as fluent as possible and without moving the upper part of his body in about 10 s, using a metronome. While making the recordings the participants shoulders are kept as low as possible to ensure that all the cervical vertebrae are visible. The recordings were either made with the Siemens Arcadis Avantic VC10A (Siemens AG, Munich, Germany), Toshiba Infinix Vi (Ōtawara-shi, Tochigi-ken, Japan), or Philips Allura Xper FD20 X-ray system (Best, The Netherlands), capturing frames of 1024 × 1024 pixels at 15 frames per second. Images were not compressed.

2.2. Manual Annotation

Template areas containing each vertebral body and the skull base were manually drawn by two trained individuals (TB and VS) in the median frame of the recording, labeled C0–C7. Two-way mixed ICCs were calculated to assess reproducibility and consistency. The bottom part of the sella turcica and clivus are projected in the X-ray image as a stable and structure rich midline area of the skull. This is therefore an ideal template to track the movement of C0 (the skull base). Image recognition software, based on matching normalized gradient field images on a best-fit principle, was then used to locate the vertebrae in the other frames of the recordings, as described by Reinartz et al. [16]. The contours of the vertebra were manually checked and if necessary adjusted to fit the corresponding vertebrae in every frame of the recording (Figure A1, Appendix A). These image recognition-assisted manual annotations were considered the ground truth.

2.3. Development of the Model

Two image dimensions were evaluated to determine the best approach for analyzing motion patterns in clinical research. To choose a uniform image size for both image dimensions, the maximum width and height of the manual annotations of vertebrae C1–C7 were determined in each frame across all recordings. The maximum Region of Interest (RoI) that centered around the vertebrae in a single frame was identified to be 608 × 505 pixels. To capture the entire spine movement in a single view, the absolute minimum and maximum x- and y-coordinates of the vertebrae in each recording were used to determine the RoI of 826 × 543 pixels. To ensure a margin around the vertebrae and optimal visibility of vertebra C0, the two image dimensions were set to 640 × 640 pixels and 832 × 576 pixels, respectively.

2D models were compared to 2D+t models. The 2D+t models integrated temporal information with the hypothesis that the predicted vertebral shape will become more consistent over time, leading to more accurate relative rotation values. The neural network that was used for the segmentation task of this study was a five layered U-Net, widely used in medical segmentation studies [17,18]. For the 2D model, each frame was presented as input to the model. For the 2D+t model, multiple frames were presented as input to the model. Consecutive frames were given as input to the model, with the segmentation only generated on the middle frame. All models have nine output channels, one for each of the eight vertebrae and one for the background class. To determine the ideal number of frames for the 2D+t models, an ablation study was performed. As the application was not needed in real-time, it was possible to use the frames acquired before and after frame t, e.g., $t-1$ and $t+1$ in case of three input frames. As pre-processing step, both image sizes were normalized with the mean and standard deviation of pixel values of the training subset. On the training dataset, data augmentation was applied to artificially increase the diversity of the training data by applying various transformations. Specifically, Gaussian noise was introduced, and the contrast was adjusted. The hyperparameters settings for the model training were the Adam optimizer with a learning rate of 0.0001, the weighted Dice loss function (Equation (1)), a batch size of eight images, and for a maximum of 300 epochs. The weighted loss function is a strategy to increase the significance of the vertebrae, by determining the ratio of foreground pixels (g) to the total amount of pixels (N) to use as weights (Equation (2)).

$$L_{Dice} = 1 - \frac{2 \cdot \sum_i^N p_i \cdot gt_i}{\sum_i^N p_i^2 + \sum_i^N gt_i^2} \tag{1}$$

where p_i represents the predicted probability of pixel i and gt_i represents the ground truth of pixel i.

$$W = \frac{1}{\left(\sum_{i=1}^N g_i\right)^2}, \tag{2}$$

For both 2D and 2D+t models, the two dimensions were analyzed, resulting in four different models, models A to D. An overview of models A to D can be found in Table 1.

Table 1. Overview of the tested models A to D.

Model	Dimension	
A	640 × 640	2D
B	832 × 576	2D
C	640 × 640	2D + time
D	832 × 576	2D + time

The output of the model, given in probability values, was transformed into a binary mask by applying an optimized threshold between 0.1 and 0.9 (Figure A2, Appendix A). The optimal threshold for each vertebra was determined via the Precision-Recall (PR) curve, as it addresses the class imbalance by focusing on foreground pixel performance. For thresholds ranging from 0.1 to 0.9 in steps of 0.1, the PR-curve was generated, and the optimal threshold was selected based on the highest F1-score, calculated from the harmonic mean of precision and recall derived from the PR-curve. The threshold varied between vertebrae per model (Table A1, Appendix A). With the optimal threshold-vertebra combinations, the Intersection over Union (IoU) and Dice score were computed for each vertebra in every frame for each trained model.

2.4. Dataset

Dynamic X-ray recordings with annotated and corrected contours were available for 39 healthy controls. Extension recordings were made at two time points, resulting in 78 extension recordings. In eleven cases, flexion recordings were also analyzed, resulting in a total of 89 recordings in these 39 subjects. Each recording consisted of ±52 individual frames that were analyzed. Altogether this resulted in a total of 4671 frames. The recordings were split into subsets for training, validation, and testing, with a split of approximately 55–20–25%, respectively (Table 2). These percentages are commonly used as a standard distribution for deep-learning models, resulting in 2709 frames for training, 964 frames for validation, and 998 frames for testing. Recordings were allocated per individual into one of the subsets.

Table 2. Overview of the included data divided into subsets, displayed for number of individuals and number of recordings.

Data Subset	Individuals (*n* =)	Recordings (*n* =)
Training (55%)	21	52
Validation (20%)	8	18
Testing (25%)	10	19

2.5. Generation of Mean Shape

The model predictions were made per individual frame of a recording, resulting in a slightly different shape in each frame, which could be caused by overprojection of the 3D structures in the 2D plane or due to coupled motion. However, vertebrae are rigid bodies, and thus their actual shape does not vary throughout the movement. To include this physical property, a mean shape was created for each vertebra per recording to calculate the relative rotation.

To quantify the overlap between the mean shape and the ground truth, the Intersection over Union (IoU) was calculated. Similarly, the IoU was calculated between the mean shape and the model predictions.

2.6. Outcome Measures

To assess the performance of each model, the IoU and the Dice similarity coefficient (DSC) were calculated.

The IoU measures the overlap between the predicted segmentation and the ground truth by calculating the ratio of the intersection area to the union area (Equation (3)). Typically, IoU values above 0.7 are deemed good, those between 0.5 and 0.7 indicate moderate overlap, and values below 0.5 suggest poor overlap:

$$IoU\ (A,B) = (|A \cap B|)/(|A \cup B|), \tag{3}$$

The DSC is also a measure of overlap between the predicted segmentation and the ground truth, by dividing twice the size of the intersection area with the sum of their individual sizes (Equation (4)) [19]. A DSC is considered good when above 0.7, a DSC between 0.5 and 0.7 is considered moderate, and below 0.5 is considered poor [20]. The main difference between IoU and DSC is the fact that under- and over-segmentation is penalized more by the IoU.

$$DSC\ (A,B) = (2 \times |A \cap B|)/(|A| + |B|), \tag{4}$$

The primary outcome will be the intraclass correlation coefficient (ICC) of the relative rotation of individual vertebrae over time. The relative rotation values will be calculated using a mean shape of each vertebra per recording. The mean shape was positioned on the centroid of the model prediction on each frame and subsequently rotated iteratively until maximal overlap was achieved. The fitted mean shape was used to calculate the rotation of each vertebra in the consecutive frames of the entire recording ($d\theta_C$), as shown in Equation (5), where θ denotes an angle in frame t.

$$d\theta_C = \theta_{C(t)} - \theta_{C(t-1)}, \tag{5}$$

On the first frame of the recording $d\theta_C = 0$ as there was no preceding frame $(t-1)$. For the 2D+t models, $d\theta_C = 0$ for frames that were not considered as a middle frame, such as the first and last frames of the recording in case of three input frames. C indicates one of the eight vertebrae.

The secondary outcome measure was the ICC of the relative rotation of vertebral segments over time. Relative rotation for segments was also calculated using a mean shape. With the relative rotation values of each vertebra, the relative rotation between two vertebrae ($d\theta R$) was calculated via Equation (6).

$$d\theta_R = d\theta_{Ck} - d\theta_{Cl}, \tag{6}$$

with Ck and Cl each indicating another vertebra.

The Wilcoxon Signed Rank test was used to assess significant differences in outcome performance between the different models. A p-value ≤ 0.05 was considered statistically significant.

3. Results

3.1. Ablation Study 2D+T

The ablation study of the 2D+t models was conducted using three, five, seven, and nine input frames. It was observed that the evaluation scores did not further improve when comparing nine input frames to seven. For both model configurations, seven input frames most frequently resulted in the significantly highest score. Results of the ablation study for Model C and Model D can be found in Appendix A, Tables A2 and A3.

3.2. Model Performance

The mean IoU score ranged from 0.37 to 0.72 for model A, from 0.51 to 0.72 for model B, from 0.45 to 0.74 for model C, and from 0.49 to 0.72 for model D. The mean DSC ranged from 0.53 to 0.82 for model A, from 0.66 to 0.82 for model B, from 0.61 to 0.83 for model C, and from 0.64 to 0.82 for model D. The mean IoU and DSC for each vertebra across all models are presented in Table 3, with the highest score indicated in bold. An asterisk denotes a significant difference between the highest score and the scores from the other three models.

Table 3. Performance of model options A–D on the test set.

	IoU				DSC			
Model	**A**	**B**	**C**	**D**	**A**	**B**	**C**	**D**
C0	0.37	**0.51** *	0.45	0.49	0.53	**0.66** *	0.61	0.64
C1	0.71	**0.72** *	**0.72**	0.7	0.81	**0.82** *	0.82	0.81
C2	**0.72**	0.71	**0.72**	**0.72**	**0.82**	0.81	**0.82**	**0.82** *
C3	0.7	0.72	**0.74** *	0.72	0.8	0.82	**0.83** *	0.82
C4	0.6	**0.64** *	0.63	0.63	0.72	**0.76** *	0.74	0.75
C5	0.51	0.56	0.52	**0.58** *	0.64	0.69	0.64	**0.71** *
C6	0.51	0.55	0.57	**0.59** *	0.65	0.69	0.7	**0.73** *
C7	0.51	0.52	**0.55** *	0.54	0.65	0.66	**0.68** *	0.67

IoU = Intersection over Union, DSC = Disc Similarity Coefficient. The highest IoU and DSC values for each vertebra are indicated in bold. Significant differences compared to all other options are marked with an asterisk.

Figure 1 illustrates the segmentation of each vertebra created by model C comparing the first, median, and final frames of the recording to the ground truth. From top to bottom, good, average, and poor model segmentation results are displayed compared to the ground truth. Equal results were observed for models A, B, and D, as seen in Figures A3–A5, Appendix A.

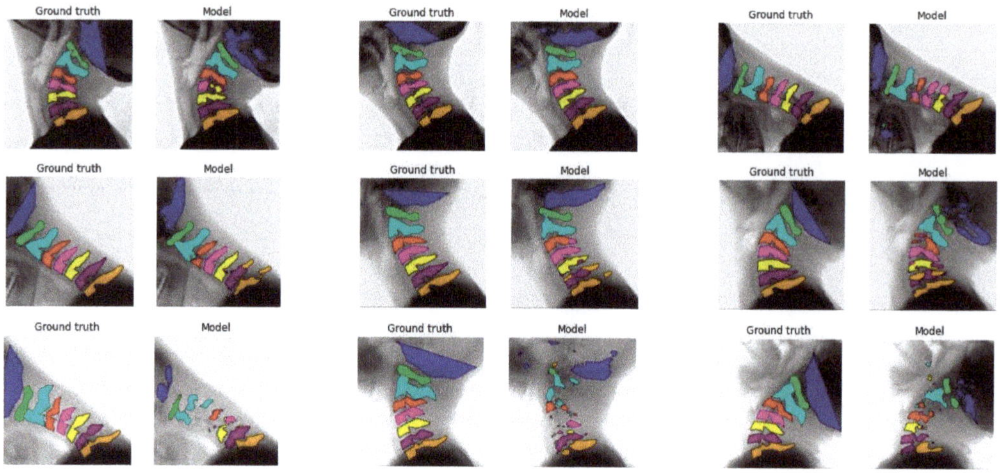

Figure 1. Comparison between the ground truth and the segmentation made by model C. Each vertebra is indicated with another color. Each row depicts a recording of a different test subject. In the first row, the recording with the best visual results is shown; in the second row, the median visual accuracy is displayed, and the last row shows the worst visual result. The first column shows the first frame of the recording, the middle column shows the median frame of the recording, and the last column shows the last frame of the recording. The first row depicts a flexion recording, while the second and third rows show extension.

3.3. Mean Shape

The mean shape was generated for vertebrae C1–C7 for each model option, excluding the skull base C0 because of its variable shape. For several recordings per model, it was not possible to create a mean shape due to deviations in the model segmentation. Figure 2 displays the mean shape of C5 compared to the ground truth and the model segmentation. The mean shape that was generated for the other vertebrae can be seen in Figures A6 and A7, Appendix A.

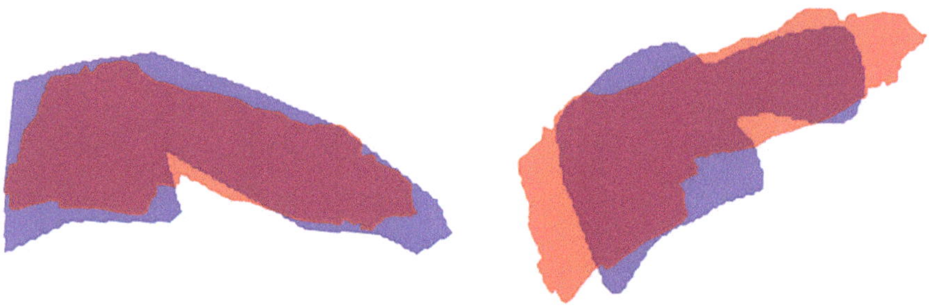

(a) (b)

Figure 2. Visualization of the mean shape of C5. Both (**a**) and (**b**) are calculated on the median frame. (**a**) The ground truth is blue, the mean shape is red, and overlap is displayed in purple; (**b**) The model segmentation is blue, the mean shape is red, and the overlap is displayed in purple.

To compare the accuracy of the mean shape generated by the model to the ground truth, the IoU was calculated for each vertebra and model, presented in Table 4. The IoU score ranged from 0.56 to 0.80 for model A, from 0.56 to 0.84 for model B, from 0.61 to 0.84 for model C, and from 0.61 to 0.84 for model D.

Table 4. The mean IoU score calculated between the ground truth and mean shape of each vertebra, per model A–D.

	A	**B**	**C**	**D**
C1	0.76	0.76	0.78	0.75
C2	0.80	0.79	0.78	0.76
C3	0.79	0.84	0.84	0.84
C4	0.69	0.81	0.78	0.75
C5	0.56	0.62	0.61	0.61
C6	0.60	0.56	0.63	0.66
C7	0.63	0.63	0.62	0.56

IoU = Intersection over Union.

3.4. *ICC of Relative Rotation of Individual Vertebrae*

The ICC score for individual vertebrae ranged from 0.819 to 0.962 for model A, from 0.798 to 0.948 for model B, from 0.683 to 0.907 for model C, and from 0.620 to 0.878 for model D, presented in Table 5. The highest mean score between the models is indicated in a red bold font. The number of recordings on which the ICC is based is also presented in Table 5. Figure 3a illustrates the trajectory of C1 Model A with a high ICC (0.990), and Figure 3b illustrates the trajectory of C3 Model D with a low ICC (0.298).

Table 5. ICC scores individual vertebrae calculated per model using the mean shape.

Vertebra	**Model A ICC [min–max]**	*n*	**Model B ICC [min–max]**	*n*	**Model C ICC [min–max]**	*n*	**Model D ICC [min–max]**	*n*
C1	**0.962** [0.916–0.993]	7	0.948 [0.834–0.996]	13	0.888 [0.471–0.997]	12	0.843 [0.479–0.982]	12
C2	**0.904** [0.699–0.996]	10	0.882 [0.449–0.978]	12	0.868 [0.413–0.988]	12	0.796 [0.400–0.985]	12
C3	0.871 [0.422–0.993]	7	**0.917** [0.826–0.976]	9	0.741 [0.132–0.979]	7	0.620 [0.298–0.909]	6
C4	0.880 [0.814–0.960]	3	0.812 [0.601–0.927]	7	**0.907** [0.899–0.923]	3	0.636 [0.343–0.820]	3
C5	**0.904** [n/a]	1	0.798 [0.650–0.945]	2	0.683 [0.658–0.680]	2	0.775 [0.707–0.864]	3
C6	**0.982** [n/a]	1	0.830 [0.665–0.995]	2	0.769 [0.471–0.979]	4	0.878 [0.639–0.966]	8
C7	0.819 [0.732–0.905]	2	0.869 [0.650–0.954]	5	**0.879** [0.785–0.974]	5	0.863 [0.697–0.942]	4

ICC = Intraclass Correlation Coefficient. Per vertebra the highest mean score is indicated in red bold.

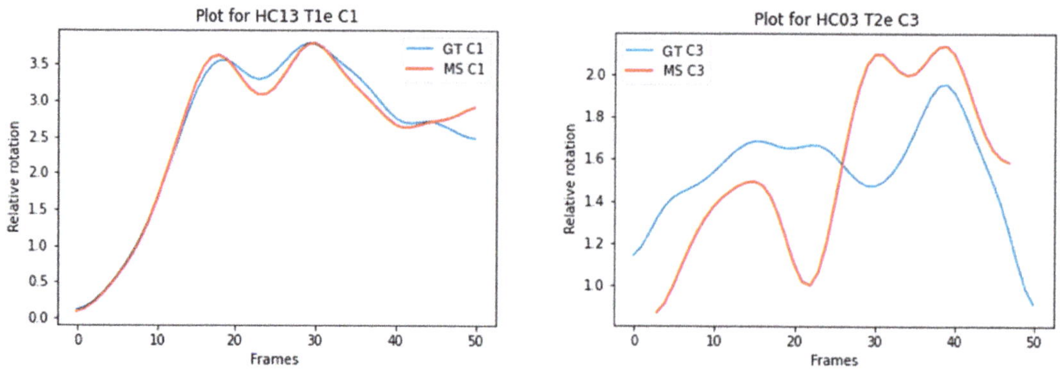

(a) (b)

Figure 3. (**a**) Graphical representation of the C1 trajectory in model A (ICC = 0.990), comparing the ground truth (blue line) with the mean shape (red line). The *x*-axis shows the vertebral rotation in degrees, and the *y*-axis indicates the number of frames; (**b**) Graphical representation of the C3 trajectory in model D (ICC = 0.298), comparing the ground truth (blue line) with the mean shape (red line).

3.5. ICC of Relative Rotation of Vertebral Segments

The ICC score for the vertebral segments ranged from 0.511 to 0.685 for model A, from 0.408 to 0.627 for model B, from 0.382 to 0.770 for model C, and from 0.281 to 0.724 for model D, presented in Table 6. The highest mean score between the models is indicated in a red bold font. The number of recordings on which the ICC is based is also presented in Table 6. Figure 4a illustrates the trajectory of C1–C2 Model D with a high ICC (0.890), and Figure 4b illustrates the trajectory of C5–C6 Model B with a low ICC (0.000).

Table 6. ICC scores for vertebral segments calculated per model using the mean shape.

Segment	Model A ICC [min–max]	*n*	Model B ICC [min–max]	*n*	Model C ICC [min–max]	*n*	Model D ICC [min–max]	*n*
C1–C2	0.685 [0.481–0.988]	5	0.627 [0.136–0.938]	5	0.713 [0.283–0.937]	7	**0.724** [0.559–0.890]	4
C2–C3	0.512 [0.181–0.934]	4	0.408 [0.025–0.661]	4	**0.500** [0.321–0.615]	6	0.340 [0.006–0.647]	4
C3–C4	0.511 [n/a]	1	0.412 [0.025–0.831]	5	0.382 [0.355–0.409]	2	**0.645** [n/a]	1
C4–C5	[n/a]	0	0.489 [0.464–0.514]	2	**0.578** [0.492–0.663]	2	0.281 [n/a]	1
C5–C6	[n/a]	0	0.605 [0.505–0.705]	2	0.535 [n/a]	1	**0.542** [0.314–0.772]	3
C6–C7	0.674 [n/a]	1	[n/a]	0	**0.770** [n/a]	1	0.685 [n/a]	1

ICC = Intraclass Correlation Coefficient. Per vertebra the highest mean score is indicated in red bold.

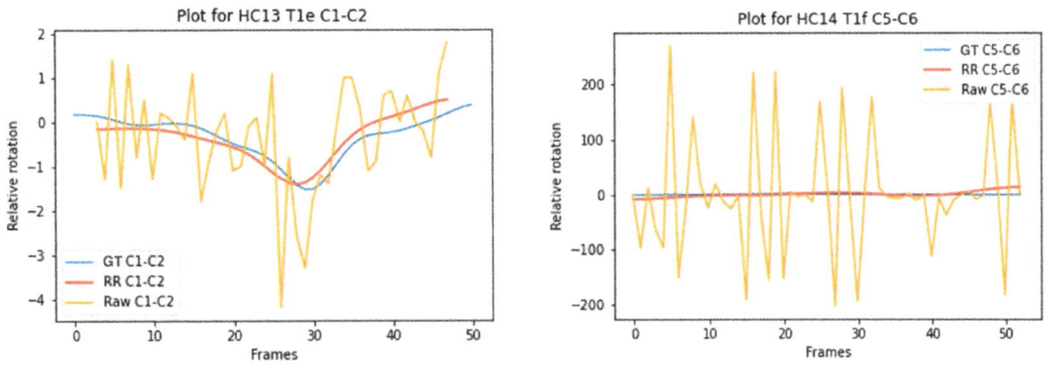

(a) (b)

Figure 4. (**a**) Graphical representation of the C1–C2 trajectory in model D (ICC = 0.890), comparing the ground truth (blue line) with the mean shape (red line) and the unfiltered mean shape trajectory (yellow line). The *x*-axis shows the relative rotation of the vertebral segment in degrees, and the *y*-axis indicates the number of frames; (**b**) Graphical representation of the C5–C6 trajectory in model B (ICC = 0.000), comparing the ground truth (blue line) with the mean shape (red line) and the unfiltered mean shape trajectory (yellow line).

4. Discussion

The results of this study show the feasibility of implementing a 2D and 2D+t deep-learning model to recognize cervical vertebrae in dynamic X-ray recordings. Based on acceptable ICC scores for the relative rotation of individual vertebrae, the investigated models can accurately track vertebrae over time. To calculate the segmental relative rotation, the models need to be improved further, as the current results are still poor to moderate.

In the ablation study for the 2D+t models, an improvement in the IoU score was observed when seven frames were compared to three or five frames. In the current literature, three or five frames are often used in static imaging [12,21,22]. As cervical spine motion can be observed between frames, a model that analyzes the temporal dimension by comparing consecutive frames can benefit from additional information. However, using nine frames did not lead to further improvement in the IoU score, indicating that too much information away from the median frame leads to a poorer outcome.

The IoU results of the different models are better overall for C1 to C4 compared to C0 and C5 to C7. Figures 1 and A3–A5 also display that the lower vertebrae are more fragmented. This can be explained by the different features of the cervical vertebrae and the imaging technique. Vertebrae C1 and C2 each have a unique shape that facilitates easier prediction by the model. In contrast, vertebrae C3–C7 exhibit more similar features. In particular, vertebrae C5 and C6 pose a challenge for the model, as they are frequently misclassified due to their strong resemblance [23]. The segmentation accuracy of the skull base C0 is compromised due to the absence of a visual feature at the position of the straight line in the ground truth, especially when C0 is partially out of view in certain frames of the recording. Due to the imaging technique, C6 and C7 are often obstructed by the shoulders, which makes it challenging for the model to accurately predict these vertebrae. These challenges may reduce model performance in cases where precise segmentation is essential for angle determination.

Inaccuracies in model segmentation can to some extent be explained by the quality of recordings. In multiple recordings, the contrast between the vertebrae and background is less pronounced, which leads to worse outcomes. In contrast, recordings with distinct edges of the vertebrae and enhanced contrast lead to improved outcomes. In some included recordings, increased tube voltage was necessary to enhance visualization of vertebrae C6 and C7 located behind the shoulders, albeit at the expense of overexposure to the other structures, resulting in light grey vertebrae with minimal contrast against the background. These findings provide valuable knowledge for the acquisition of future X-ray recordings. Examples of recordings with low and good results are presented in Figures A8 and A9, Appendix A.

Even though the shape of vertebral bodies is rigid, model segmentation shows different shapes of the vertebral bodies over time. This variance in segmentation is caused by overprojection of structures on the recording. The generated mean shape for all vertebrae had an acceptable IoU compared to the ground truth, presented in Table 4. Similarly to the model segmentation, C5 to C7 demonstrated lower IoU scores. This emphasizes the complexity for the model to accurately segment the lower vertebrae. The proposed solution for the different shapes, creating a mean shape, led to better results for the ICC scores, as presented in Tables 6 and A4.

The ICC scores for the relative rotation of individual vertebrae are similar between the 2D and 2D+t models. The mean ICC scores ranged from good to excellent, indicating that all models can accurately track individual vertebrae over time. The ICC scores for relative rotation of vertebral segments are lower. This can be explained by comparing the overall rotation of an individual vertebra to a segment. Boselie et al. calculated that between frames a segment can rotate as little as 0.3 degrees, less than the measurement error [4]. Rotation of individual vertebrae between frames ranges from 2 to 3.5 degrees, illustrated in Figure 3a,b. Therefore, rotation over time of individual vertebrae is less sensitive to measurement errors or segmentation inaccuracies. Additionally, inaccuracies in segmentation of vertebrae above or below do not affect the measured relative rotation of the individual vertebra.

When the 2D and 2D+t models are compared to track vertebral segments over time, the 2D+t results in higher ICC values, confirming that incorporating a time dimension will lead to a more consistent segmentation shape than the 2D models. We also analyzed two image sizes to measure if a more centralized frame led to more accurate model segmentation. Between the image sizes, we did not observe a relevant difference in our outcomes, indicating that both can be used for qualitative cervical motion analysis.

A limitation of this study is the angular shape of the ground truth segmentation, as the natural shape of a vertebra is smoother. The ground truth is created with the algorithm of Reinartz et al., which is already very time-consuming [16]. Adding a greater number of points to achieve a smoother and more realistic shape would further extend the processing time. The difference in shape can lead to lower IoU or DSC scores. It is favorable that the model segmentation already creates a more natural looking vertebra.

Another limitation is that the 2D recordings capture only the sagittal rotation. Other motion planes, e.g., axial rotation or lateral flexion, have not been analyzed. To address these limitations, future work could explore the integration of multi-modal imaging data to improve vertebral segmentation, especially in complex cases. Incorporating axial rotation information by combining biplanar X-ray with CT or MRI could enhance the model's ability to capture three-dimensional motion patterns, which are clinically relevant for assessing cervical spine kinematics.

Lastly, the number of recordings included in ICC calculation is limited, and we only reported the minimum and maximum values. Some ICC values are based on one or two recordings. Therefore, these values should be interpreted with caution. For each level, the recording was excluded if the mean ICC was negative or if the recording contained at least one outlier. A negative value occurs when the ground truth segmentation moves opposite to the model segmentation. An outlier occurs when the threshold, based on the minimum and maximum raw values of the gold standard, is exceeded. The most probable cause of an outlier is the model segmentation used to fit either the mean shape or the model segmentation in the subsequent frame. Deviating shapes and differences in shape between consecutive frames can result in sub-optimal fits during rotation. Also, most outliers were detected for C5 and C6. The shape of these vertebrae is very similar, making it more complex for the model to differentiate between C5 and C6. We decided to exclude these recordings because the outliers originate from model segmentation issues.

For future use, model segmentation can be improved by several methods. One method is improving the quality of the recordings by enhancing the contrast. This can be achieved by histogram equalization or histogram matching [24]. Another method is training the model to recognize the vertebra in a specific sequence. Incorporating a sequence will prevent the model from interchanging vertebrae. Also, post-processing techniques can be applied when the model segmentation is fragmented. Currently, only the largest segment is selected for further analysis in the mean shape and relative rotation calculations. By incorporating post-processing methods, fragmented segments can be merged into a single segmentation using techniques like morphological operations, e.g., dilation and closing, and majority voting for consistent foreground pixels across multiple frames [25]. Another method is to recognize frames that cause outliers and replace these frames with the preceding or following frame. This is especially relevant for recordings with a limited number of outliers. Improving the model segmentation in each frame with post-processing techniques will result in a more accurate mean shape and more precise relative rotation values, ultimately improving the ICC scores. With improved segmentation, the model can be used for measuring other metrics, such as the cervical lordosis and T1 slope. It has already been shown to be feasible to measure this in static X-rays [26]. Analyzing these metrics in dynamic recordings will provide further insights into the normal movement of the cervical spine.

This model can accurately calculate the relative rotation of an individual vertebra. However, the relative rotation analysis of vertebral segments needs to be further improved to enable motion pattern analysis. Such motion patterns have previously been shown to allow differentiation between asymptomatic individuals and patients with cervical radiculopathy. More clinically relevant applications are likely, particularly in evaluating spinal instability, degenerative changes over time, and post-surgical biomechanics. For instance, in case segmental motion is preserved preoperatively, a motion-preserving device may be considered over fusion in anterior cervical discectomy procedures. Conversely, the absence of motion may justify a fusion approach. Additionally, routinely analyzing changes in motion patterns before and after surgery could yield insight into the development of complications such as adjacent segment disease.

By leveraging big data in combination with these motion analyzes, previously unrecognized motion patterns that might influence clinical outcome and possible future treatment strategies could be identified. These applications underscore the potential of AI-based motion tracking not only as a diagnostic tool but also as a future individualized decision-support system in cervical spine surgery.

5. Conclusions

Implementing a 2D or 2D+t deep-learning model to analyze the relative rotation of individual vertebrae leads to accurate results compared to the ground truth. For the calculation of the relative rotations of vertebral segments, the 2D+t models, incorporating a temporal dimension, seem to do better than 2D models. However, these results remain poor and require further improvement. Future research should focus on improving the model segmentation by enhancing contrast and applying post-processing methods. This will enable motion pattern analysis in large groups. Additionally, integrating multi-dimensional or multi-modal data and applying the model to clinical datasets could help to increase the robustness and clinical relevance of the approach. Ultimately, these advancements could enable the analysis of motion patterns in larger and more diverse patient groups, forecasting long term clinical outcome and, for the far future, serving as a support for more personalized patient care.

Author Contributions: Conceptualization, E.v.S., V.S., H.v.S. and T.B.; methodology, E.v.S., V.S., E.C. and T.B.; software, V.S., E.C. and T.B.; validation, E.C.; formal analysis, E.v.S., V.S., E.C. and T.B.; investigation, E.v.S. and E.C.; resources, V.S. and T.B.; data curation, E.v.S., V.S. and E.C.; writing—original draft preparation, E.v.S., V.S. and E.C.; writing—review and editing, M.B., A.S., H.v.S. and T.B.; visualization, E.v.S., V.S., E.C. and T.B.; supervision, M.B., A.S., H.v.S. and T.B.; project administration, E.v.S. and E.C. All authors have read and agreed to the published version of the manuscript.

Funding: This research received no external funding.

Institutional Review Board Statement: The clinical studies of which the data have been obtained have all been approved by the Medical Ethical Committee of Zuyderland Medical Center and Maastricht University Medical Center + (Z2019087, METC 06-1-098, Z2020101) They were all registered before the start of the study (NCT04222777, NCT00868335, NCT04545983) The studies were conducted according to principles enshrined in the Declaration of Helsinki and in accordance with the Medical Research Involving Human Subjects Act (WMO).

Informed Consent Statement: Informed consent was obtained from all subjects involved in the studies.

Data Availability Statement: The data presented in this study are available on request from the corresponding author due to privacy restrictions.

Acknowledgments: We would like to thank the subjects who participated in the studies and the reviewers for their constructive feedback.

Conflicts of Interest: The authors declare no conflicts of interest.

Abbreviations

The following abbreviations are used in this manuscript:

2D	Two dimensional
2D+t	Two dimensional + time
AI	Artificial intelligence
DSC	Dice similarity coefficient
ICC	Intraclass Correlation Coefficient
IoU	Intersection over Union
ROM	Range of motion
sROM	Segmental range of motion
SSC	Sequence of segmental contribution

Appendix A

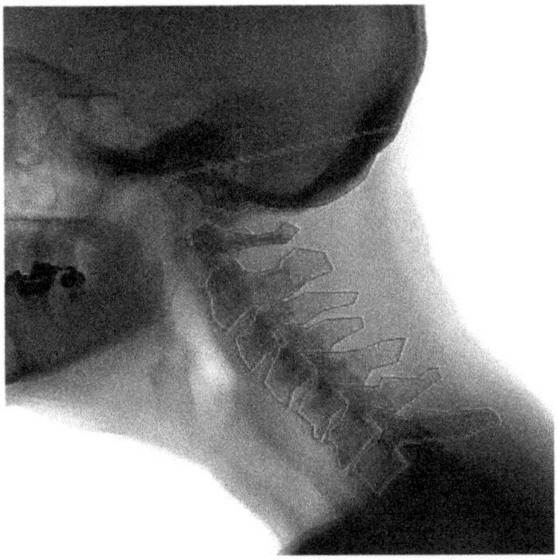

Figure A1. Manual annotations on a median frame (blue lines).

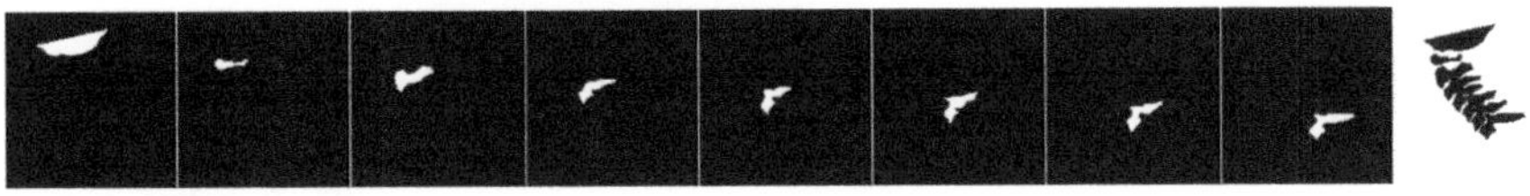

Figure A2. Binary mask.

Table A1. Optimal threshold per vertebra of models A to D.

Channel	A	B	C	D
C0	0.7	0.8	0.9	0.4
C1	0.1	0.1	0.1	0.6
C2	0.9	0.9	0.5	0.9
C3	0.1	0.1	0.7	0.7
C4	0.1	0.1	0.3	0.1
C5	0.1	0.1	0.1	0.1
C6	0.1	0.1	0.1	0.1
C7	0.1	0.1	0.1	0.1
Background	0.1	0.9	0.1	0.1

Table A2. Ablation study model C: IoU and DSC per vertebra for the varying number of input frames.

	IoU				DSC			
Frames	3	5	7	9	3	5	7	9
C0	0.45	**0.50** *	0.45	0.44	0.61	**0.65** *	0.61	0.61
C1	0.68	0.69	**0.72** *	0.69	0.79	0.8	**0.82** *	0.8
C2	0.49	0.72	0.72	**0.73** *	0.65	**0.82**	**0.82**	**0.82** *
C3	0.72	0.73	**0.74** *	0.7	0.82	0.82	**0.83**	0.8
C4	0.59	0.62	**0.63**	0.6	0.72	**0.74**	**0.74**	0.71
C5	0.47	0.5	**0.52** *	0.49	0.61	**0.64**	**0.64** *	0.62
C6	0.5	0.52	**0.57** *	0.54	0.63	0.65	**0.7** *	0.67
C7	0.49	0.5	**0.55** *	0.51	0.64	0.64	**0.68** *	0.65

IoU = Intersection over Union, DSC = Disc Similarity Coefficient. The highest IoU and DSC values for each vertebra are indicated in bold. Significant differences compared to all other options are marked with an asterisk.

Table A3. Ablation study model D: IoU and DSC per vertebra for the varying number of input frames.

	IoU				DSC			
Frames	**3**	**5**	**7**	**9**	**3**	**5**	**7**	**9**
C0	0.48	0.48	**0.49**	0.46	0.63	0.63	**0.64**	0.62
C1	0.68	**0.71** *	0.7	0.69	0.78	**0.81**	**0.81**	0.79
C2	0.71	0.71	**0.72**	**0.72** *	0.81	0.81	**0.82**	**0.82** *
C3	0.71	0.69	**0.72** *	0.69	0.81	0.8	**0.82** *	0.8
C4	0.62	0.57	**0.63**	0.6	0.74	0.71	**0.75**	0.72
C5	0.57	0.49	**0.58**	0.5	0.7	0.64	**0.71** *	0.64
C6	**0.59**	0.52	**0.59**	0.53	0.72	0.66	**0.73**	0.66
C7	**0.55** *	0.52	0.54	0.46	**0.68** *	0.66	0.67	0.6

IoU = Intersection over Union, DSC = Disc Similarity Coefficient. The highest IoU and DSC values for each vertebra are indicated in bold. Significant differences compared to all other options are marked with an asterisk.

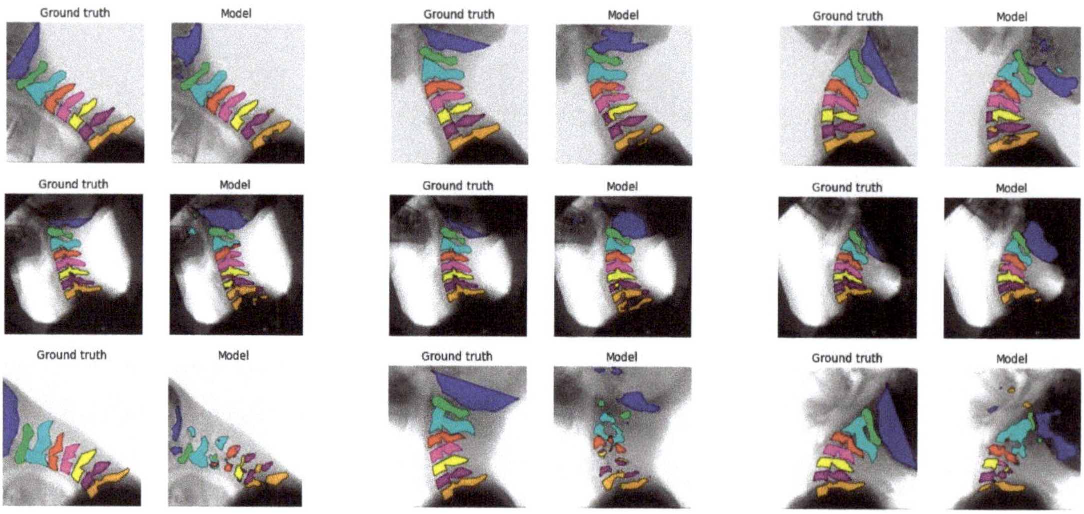

Figure A3. Comparison between the ground truth and the segmentation made by model A. Each vertebra is indicated with another color. From left to right: first-median-last frame of the recording. From top to bottom, the comparison with the best visual comparison to the worst visual comparison.

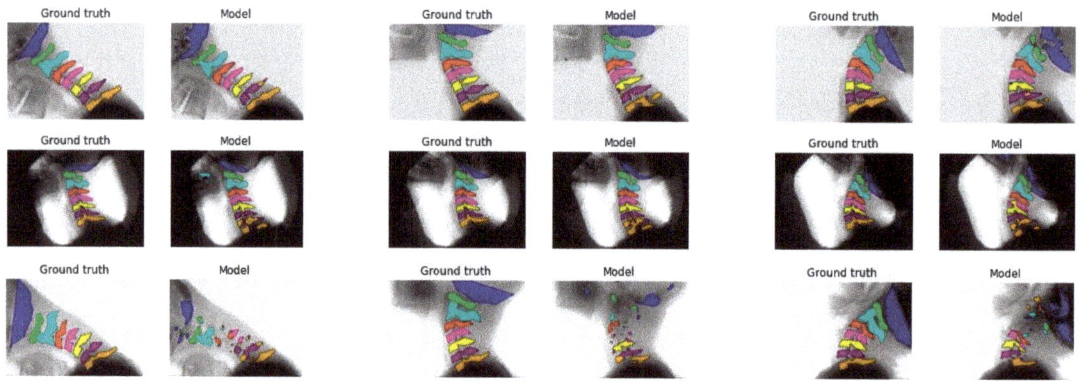

Figure A4. Comparison between the ground truth and the segmentation made by model B. Each vertebra is indicated with another color. From left to right: first-median-last frame of the recording. From top to bottom, the comparison with the best visual comparison to the worst visual comparison.

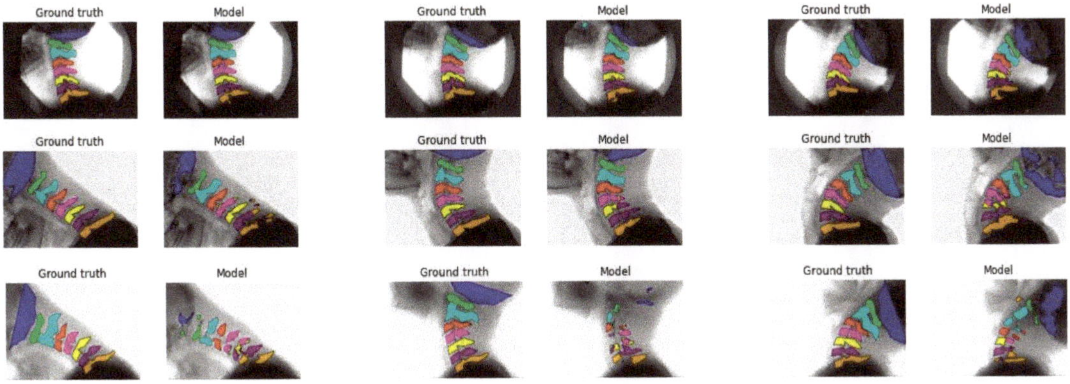

Figure A5. Comparison between the ground truth and the segmentation made by model D. Each vertebra is indicated with another color. From left to right: first-median-last frame of the recording. From top to bottom, the comparison with the best visual comparison to the worst visual comparison.

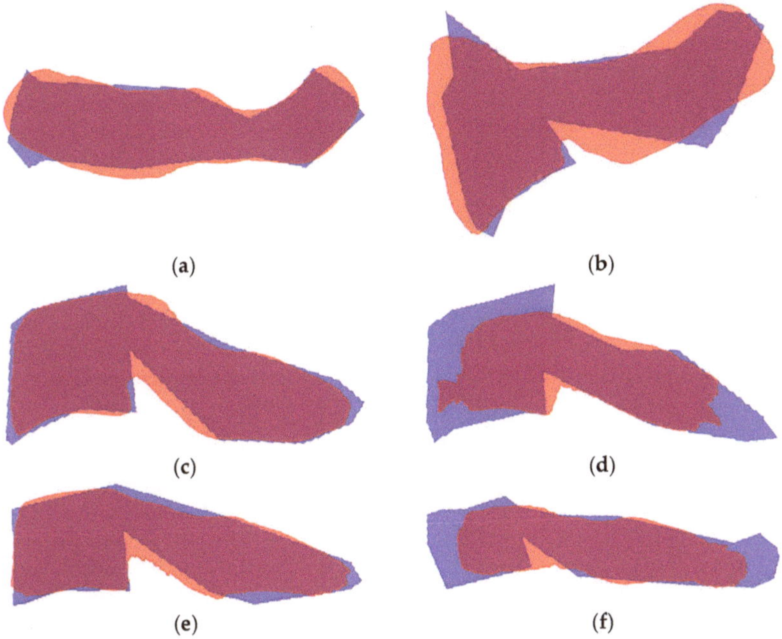

Figure A6. Mean shape compared to the ground truth of C1–C7. Ground truth is indicated in blue, mean shape is given in red. Overlapping areas appear as purple: (**a**) Vertebra C1; (**b**) Vertebra C2; (**c**) Vertebra C3; (**d**) Vertebra C4 (**e**) Vertebra C6; (**f**) Vertebra C7.

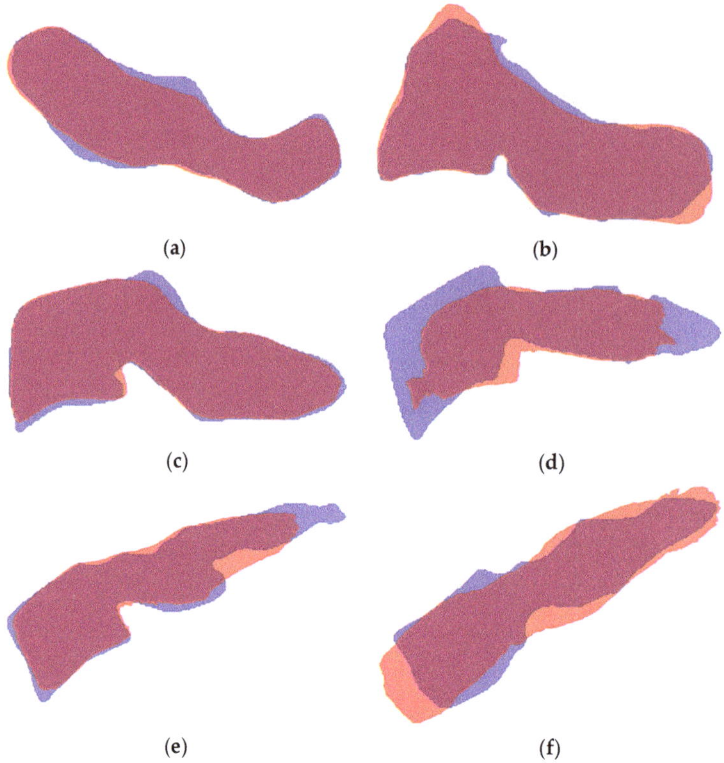

Figure A7. Mean shape compared to the model segmentation of C1–C7. Model segmentation is indicated in blue, mean shape is given in red. Overlapping areas appear as purple: (**a**) Vertebra C1; (**b**) Vertebra C2; (**c**) Vertebra C3; (**d**) Vertebra C4 (**e**) Vertebra C6; (**f**) Vertebra C7.

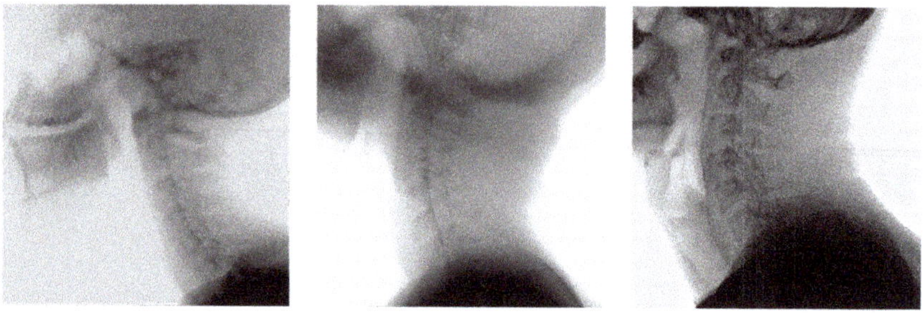

Figure A8. Median frame of recordings with low performance scores.

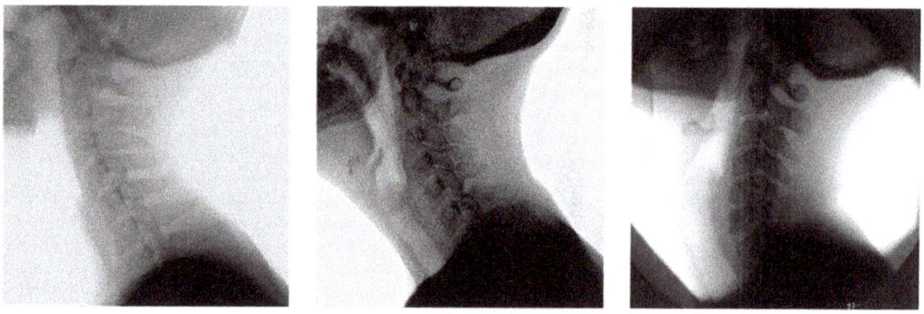

Figure A9. Median frame of recordings with high performance scores.

Table A4. ICC scores for vertebral segments calculated per model using model segmentation.

Segment	Model A ICC [min–max]	*n*	Model B ICC [min–max]	*n*	Model C ICC [min–max]	*n*	Model D ICC [min–max]	*n*
C1–C2	0.143 [0.056–0.218]	4	0.146 [0.081–0.251]	3	0.205 [0.106–0.346]	3	0.082 [0.052–0.112]	2
C2–C3	0.258 [0.139–0.325]	3	0.238 [0.221–0.254]	2	0.178 [0.063–0.344]	3	0.078 [0.022–0.133]	2
C3–C4	0.017 [n/a]	1	0.112 [0.028–0.245]	3	0.018 [0.007–0.063]	2	[n/a]	0
C4–C5	[n/a]	0	[n/a]	0	[n/a]	0	0.1 [0.031–0.069]	2
C5–C6	[n/a]	0	0.003 [n/a]	1	[n/a]	0	0.041 [n/a]	1
C6–C7	[n/a]	0	[n/a]	0	[n/a]	0	0.043 [0.0–0.086]	2

ICC = Intraclass Correlation Coefficient.

References

1. Bogduk, N.; Mercer, S. Biomechanics of the cervical spine. I: Normal kinematics. *Clin. Biomech.* **2000**, *15*, 633–648. [CrossRef] [PubMed]
2. van Mameren, H. Motion Patterns in the Cervical Spine. Ph.D. Thesis, Maastricht University, Maastricht, The Netherlands, 1988.
3. Van Mameren, H.; Drukker, J.; Sanches, H.; Beursgens, J. Cervical spine motion in the sagittal plane (I) range of motion of actually performed movements, an X-ray cinematographic study. *Eur. J. Morphol.* **1990**, *28*, 47–68. [PubMed]
4. Boselie, T.F.M.; van Santbrink, H.; de Bie, R.A.; van Mameren, H. Pilot Study of Sequence of Segmental Contributions in the Lower Cervical Spine During Active Extension and Flexion: Healthy Controls Versus Cervical Degenerative Disc Disease Patients. *Spine* **2017**, *42*, E642–E647. [CrossRef]
5. Schuermans, V.N.E.; Breen, A.; Branney, J.; Smeets, A.; van Santbrink, H.; Boselie, T.F.M. Cross-Validation of two independent methods to analyze the sequence of segmental contributions in the cervical spine in extension cineradiographic recordings. In *Outcomes of Anterior Cervical Spine Surgery*; Maastricht University: Maastricht, The Netherlands, 2024.
6. Boselie, T.F.; van Mameren, H.; de Bie, R.A.; van Santbrink, H. Cervical spine kinematics after anterior cervical discectomy with or without implantation of a mobile cervical disc prosthesis; an RCT. *BMC Musculoskelet. Disord.* **2015**, *16*, 34. [CrossRef] [PubMed]
7. Schuermans, V.N.E.; Smeets, A.; Curfs, I.; van Santbrink, H.; Boselie, T.F.M. A randomized controlled trial with extended long-term follow-up: Quality of cervical spine motion after anterior cervical discectomy (ACD) or anterior cervical discectomy with arthroplasty (ACDA). *Brain Spine* **2024**, *4*, 102726. [CrossRef]
8. Schuermans, V.N.E.; Smeets, A.; Breen, A.; Branney, J.; Curfs, I.; van Santbrink, H.; Boselie, T.F.M. An observational study of quality of motion in the aging cervical spine: Sequence of segmental contributions in dynamic fluoroscopy recordings. *BMC Musculoskelet. Disord.* **2024**, *25*, 330. [CrossRef]
9. Al Arif, S.; Knapp, K.; Slabaugh, G. Fully automatic cervical vertebrae segmentation framework for X-ray images. *Comput. Methods Programs Biomed.* **2018**, *157*, 95–111. [CrossRef]
10. Shim, J.H.; Kim, W.S.; Kim, K.G.; Yee, G.T.; Kim, Y.J.; Jeong, T.S. Evaluation of U-Net models in automated cervical spine and cranial bone segmentation using X-ray images for traumatic atlanto-occipital dislocation diagnosis. *Sci. Rep.* **2022**, *12*, 21438. [CrossRef]
11. Fujita, K.; Matsuo, K.; Koyama, T.; Utagawa, K.; Morishita, S.; Sugiura, Y. Development and testing of a new application for measuring motion at the cervical spine. *BMC Med. Imaging* **2022**, *22*, 193. [CrossRef]
12. Avesta, A.; Hossain, S.; Lin, M.; Aboian, M.; Krumholz, H.M.; Aneja, S. Comparing 3D, 2.5D, and 2D Approaches to Brain Image Auto-Segmentation. *Bioengineering* **2023**, *10*, 181. [CrossRef]
13. Vu, M.H.; Grimbergen, G.; Nyholm, T.; Lofstedt, T. Evaluation of multislice inputs to convolutional neural networks for medical image segmentation. *Med. Phys.* **2020**, *47*, 6216–6231. [CrossRef]
14. Branney, J. An Observational Study of Changes in Cervical Inter-Vertebral Motion and the Relationship with Patient-Reported Outcomes in Patients Undergoing Spinal Manipulative Therapy for Neck Pain. Ph.D. Thesis, Bournemouth University, Poole, UK, 2014.

15. Branney, J.; Breen, A.C. Does inter-vertebral range of motion increase after spinal manipulation? A prospective cohort study. *Chiropr. Man. Therap.* **2014**, *22*, 24. [CrossRef]
16. Reinartz, R.; Platel, B.; Boselie, T.; van Mameren, H.; van Santbrink, H.; Romeny, B. Cervical vertebrae tracking in video-fluoroscopy using the normalized gradient field. In Proceedings of the Medical Image Computing and Computer-Assisted Intervention—MICCAI 2009: 12th International Conference, London, UK, 20–24 September 2009; Volume 12, pp. 524–531.
17. Ronneberger, O.; Fischer, P.; Brox, T. U-Net: Convolutional Networks for Biomedical Image Segmentation. In Proceedings of the Medical Image Computing and Computer-Assisted Intervention—MICCAI 2015, Munich, Germany, 5–9 October 2015; Springer International Publishing: Cham, Switzerland, 2015; pp. 234–241.
18. Siddique, N.; Paheding, S.; Elkin, C.P.; Devabhaktuni, V. U-Net and Its Variants for Medical Image Segmentation: A Review of Theory and Applications. *IEEE Access* **2021**, *9*, 82031–82057. [CrossRef]
19. Dice, L.R. Measures of the Amount of Ecologic Association Between Species. *Ecology* **1945**, *26*, 297–302. [CrossRef]
20. Zou, K.H.; Warfield, S.K.; Bharatha, A.; Tempany, C.M.; Kaus, M.R.; Haker, S.J.; Wells, W.M., 3rd; Jolesz, F.A.; Kikinis, R. Statistical validation of image segmentation quality based on a spatial overlap index. *Acad. Radiol.* **2004**, *11*, 178–189. [CrossRef]
21. Kittipongdaja, P.; Siriborvornratanakul, T. Automatic kidney segmentation using 2.5D ResUNet and 2.5D DenseUNet for malignant potential analysis in complex renal cyst based on CT images. *EURASIP J. Image Video Process.* **2022**, *2022*, 5. [CrossRef] [PubMed]
22. Li, J.; Liao, G.; Sun, W.; Sun, J.; Sheng, T.; Zhu, K.; von Deneen, K.M.; Zhang, Y. A 2.5D semantic segmentation of the pancreas using attention guided dual context embedded U-Net. *Neurocomputing* **2022**, *480*, 14–26. [CrossRef]
23. Gilad, I.; Nissan, M. A study of vertebra and disc geometric relations of the human cervical and lumbar spine. *Spine* **1986**, *11*, 154–157. [CrossRef]
24. Choukali, M.A.; Valizadeh, M.; Amirani, M.C.; Mirbolouk, S. A desired histogram estimation accompanied with an exact histogram matching method for image contrast enhancement. *Multimed. Tools Appl.* **2023**, *82*, 28345–28365. [CrossRef]
25. Salvi, M.; Acharya, U.R.; Molinari, F.; Meiburger, K.M. The impact of pre- and post-image processing techniques on deep learning frameworks: A comprehensive review for digital pathology image analysis. *Comput. Biol. Med.* **2021**, *128*, 104129. [CrossRef]
26. Vogt, S.; Scholl, C.; Grover, P.; Marks, J.; Dreischarf, M.; Braumann, U.D.; Strube, P.; Holzl, A.; Bohle, S. Novel AI-Based Algorithm for the Automated Measurement of Cervical Sagittal Balance Parameters. A Validation Study on Pre- and Postoperative Radiographs of 129 Patients. *Glob. Spine J.* **2024**, *15*, 1155–1165. [CrossRef] [PubMed]

bioengineering

MDPI

Article

Does Malpositioning of Pedicle Screws Affect Biomechanical Stability in a Novel Quasistatic Test Setup?

Stefan Schleifenbaum [1,2,*,†], Florian Metzner [1,†], Janine Schultze [1], Sascha Kurz [1], Christoph-Eckhard Heyde [1,2] and Philipp Pieroh [2]

1 ZESBO—Center for Research on Musculoskeletal Systems, Semmelweisstraße 14, 04103 Leipzig, Germany; florian.metzner@medizin.uni-leipzig.de (F.M.); sascha.kurz@medizin.uni-leipzig.de (S.K.)

2 Department of Orthopedic, Trauma, and Plastic Surgery, University of Leipzig, Liebigstraße 20, 04103 Leipzig, Germany; philipp.pieroh@medizin.uni-leipzig.de

* Correspondence: stefan.schleifenbaum@medizin.uni-leipzig.de

† These authors contributed equally to this work.

Abstract

Pedicle screw fixation is a common spinal surgery technique, but concerns remain about stability when screws are malpositioned. Traditional in vitro pull-out tests assess anchorage but lack physiological accuracy. This study examined the stability of correctly placed and intentionally malpositioned pedicle screws on forty vertebrae from five cadavers. Optimal screw paths were planned via CT scans and applied using 3D-printed guides. Four malposition types—medial, lateral, superior, and superior-lateral—were created by shifting the original trajectory. A custom setup applied three consecutive cycles of tensile and compressive load from 50 N to 200 N. Screw inclination under load was measured with a 3D optical system. The results showed increasing screw inclination with higher forces, reaching about 1° at 50 N and 2° at 100 N, similar in both load directions. Significant differences in inclination were only found at 100 N tensile load, where malpositioned screws showed a lower inclination. Overall, malpositioning had no major effect on screw loosening. These findings suggest that minor deviations in screw placement do not significantly compromise mechanical stability. Clinically, the main concern with malpositioning lies in the potential for injury to nearby structures rather than reduced screw fixation strength.

Keywords: pedicle screw; malposition; thoracolumbar; biomechanics; stability; screw loosening

1. Introduction

Appropriate pedicle screw placement is essential for the stability of posterior fixation constructs of the spine [1,2]. Laterally, medially, or cranially misplaced screws might impair the implant's biomechanical properties, affecting the stability and long-term safety of the treatment [3,4]. Gertzbein and Robbins [5] provided a systematic classification of malpositioning based on the extent to which the screw breaches the pedicle cortex, focusing on possible neurological deficits [3,5,6]. In particular, laterally or medially misplaced screws might become anchor increasing pull-out strength and reducding the risk of implant failure [4,7]. For example, it has been shown that medial and lateral malpositioning significantly reduce the tensile and torsional strength of the screws, increasing the risk of implant failure [4].

Recent donor-based biomechanical studies have shown that stability differs between medial and lateral malpositioning. Studies have demonstrated a general reduction in screw load-bearing capacity, and the reduction was less pronounced in medial than in lateral malpositioning [7–9]. Furthermore, the repositioning of the screws after initial malpositioning (e.g., using the previous pilot hole) did not significantly reduce the fixation strength [10,11]). The pull-out strength after the reinsertion of the screws due to lateral or endplate-related breakthrough is also an important parameter that underlines the clinical relevance of the screw position. It has been shown that the pull-out force has not changed and therefore the screw can be reinserted [10–12].

This study investigated the biomechanical stability of correctly placed and misplaced pedicle screws. In contrast to uniaxial pull-out tests [7–9], we analyzed the cycling cranio-caudal stability by means of a biomechanical test setup presented by Javers et al. [13]. This should be performed by measuring the tilt angle between the faulty and optimum screw. Can the comparison of different malpositionings with the optimal screw position provide insights that emphasize the importance of precise placement and provide information for clinical practice?

2. Materials and Methods

2.1. Ethical Statement

All body donors gave their informed and written consent to the donation of their bodies for teaching and research purposes while alive. Being part of the body donor program regulated by the Saxonian Death and Funeral Act of 1994 (third section, paragraph 18 item 8), institutional approval for the use of the post-mortem tissues of human body donors was obtained from the Institute of Anatomy (University of Leipzig) by the Ethics Committee of the University of Leipzig Medical Center (ethical approval No. 129/21-ck). The authors declare that all experiments were conducted according to the principles of the Declaration of Helsinki (as revised in 2013).

2.2. Specimen and Tissue Preparation

Ten vertebrae (T8-L5) from five body donors (mean age 77.2 ± 13.2 years; n = 4 male, n = 1 female) with different bone densities (each one non-osteoporotic and osteopenic, n = 3 osteoporosis) were taken fresh frozen. The specimens were stored at −80 °C and thawed to room temperature (20 °C) 24 h before testing. On testing day, all vertebrae were separated, and all soft tissue was removed. Osteoporosis was classified by determining the volumetric bone mineral content (vBMD) using quantitative CT scans in which a calcium hydroxyapatite phantom was inserted. The mean Hounsfield unit (HU) of each vertebral cancellous bone was determined using Mimics Innovation Suite 24 (Materialise, Leuven, Belgium) [14], the corresponding vBMD was calculated by linear regression [15] for each lumbar vertebra, and a mean value for each donor was calculated.

The samples were then embedded in aluminum sleeves using rapid casting resin consisting of the three components, RenCast FC 52/53 isocyanate, FC 53 polyol (Huntsman Advanced Materials, Basel, Switzerland) and aluminum hydroxide (filler DT 082, Gößl + Pfaff GmbH, Brautlach/Karlskron, Germany), at a ratio of 1:1:3 equivalent as published previously [13,16].

2.3. Planning and Screw Positioning

The traditional transpedicular trajectory was defined as the correct screw position (c) for the experimental design. They were individually planned using CAD software (Rhinoceros Version 7 SR37, Robert McNeel & Associates, Seattle, WA, USA), taking all spatial directions into account. Starting from the correct trajectory, four malpositionings were defined as follows: medial (m), lateral (l), superior (s), and superior-lateral (sl). Malpositioning was generated by defined parallel shifting from the correct trajectory

(Figure 1) without angular correction. The displacement value corresponds to the distance by which the screw breaks out of the pedicle wall and is based on previous reports [5] (Figure 1), All malpositioning is classified as Grade C, following the classification of Gertzbein and Robbins [5].

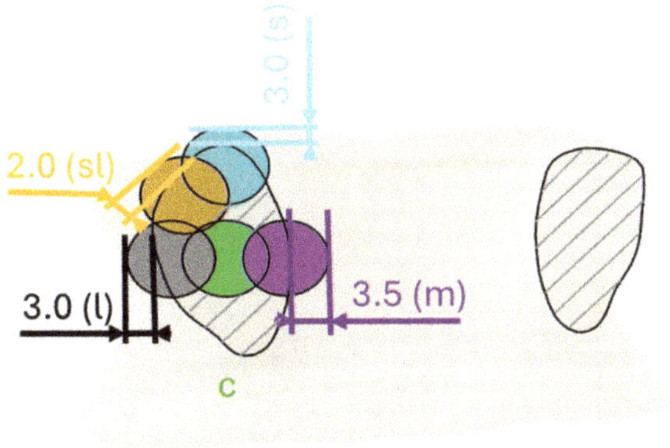

Figure 1. Planned screw trajectories within the pedicle: transpedicular coronal cross-section. Correct (c), lateral (l), superior-lateral (sl), superior (s), and medial (m) screw trajectories with indicated amount of cortical breach (mm).

The screw trajectories were distributed evenly across the donors and randomly across the pedicles (left/right) to avoid possible side dependencies (Table 1).

Table 1. Each trajectory (medial—m; lateral—l; superior—s; superior-lateral—sl; correct—c) occurred twice per donor, and the side selection (left or right) was randomized.

Donor	1		2		3		4		5	
Level	**Left**	**Right**	**Left**	**Right**	**Left**	**Right**	**Left**	**Right**	**Left**	**Right**
T8	m	c	c	sl	s	c	c	c	l	c
T9	l	c	m	c	c	sl	c	s	c	c
T10	c	c	c	l	m	c	c	sl	s	c
T11	c	s	c	c	c	l	c	m	sl	c
T12	c	sl	s	c	c	c	c	l	c	m
L1	c	m	c	sl	s	c	c	c	l	c
L2	l	c	m	c	c	sl	c	s	c	c
L3	c	c	l	c	c	m	sl	c	c	s
L4	c	s	c	c	l	c	m	c	c	sl
L5	sl	c	c	s	c	c	c	l	c	m

Individual drilling templates for the respective vertebrae were then designed using the above-mentioned CAD software and 3D printed using the Polyjet printing process (Stratasys J850 DAP, Stratasys Inc., Minnetonka, MN, USA) from VeroPureWhite material. After determining the entry points, the traditional trajectory along the pedicle was defined (see Figure 2A), and the appropriate screw diameter was selected, which is represented as a cylinder in the software (see Figure 2B). This was followed by the aforementioned parallel displacement of the correct screw position in the desired direction, using the predefined offset (1). Connecting elements and supports were then designed onto the drilling templates to ensure secure placement on the vertebra (Figure 2C). The templates were printed.

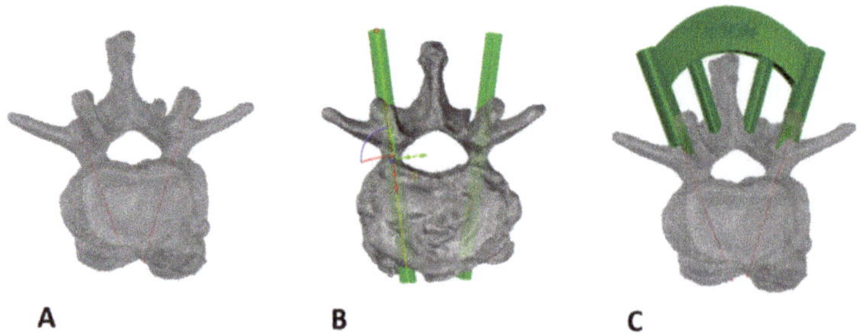

Figure 2. The procedure for designing drilling templates. First, the trajectory was set as a curve (**A**), followed by creating cylinders as a preview of the screw diameter (**B**). Finally, a predefined design was applied to the curves, and the vertebral geometry was subtracted (**C**).

M.U.S.T. pedicle screws from Medacta International (M.U.S.T., Medacta International, Castel San Pietro, Switzerland) were utilized in the experimental setting, along with the corresponding original instruments for inserting and fixing the screw–rod system in the test bench. The fixation of the rod was executed in accordance with the manufacturer's specifications.

2.4. Biomechanical Experiments

An optimized test setup in accordance with ASTM F1717-15 [17] was used for the biomechanical tests, similar to the dynamic tests used by Javers et al. and Schleifenbaum et al. [13,16]. The setup is designed to enable standard positioning of the pedicle screw relative to the lower axis of rotation (Figure 3). Testing was then performed in a servo-electric testing machine (RetroLine Z10, Zwick/Roell GmbH & Co. KG, Ulm, Germany) equipped with a 2.5 kN force sensor. The movement of the screw head relative to the bone was recorded during the test using an optical 3D measurement system (ZEISS ARAMIS 3D Camera, Carl Zeiss GOM Metrology GmbH, Braunschweig, Germany). This measurement can output the movement of the screw in degrees and thus describes the movement of the screw that can lead to cut-outs from the vertebra, among other things.

The load application was executed as a quasi-static test consisting of compressive and tensile loads in the craniocaudal direction. These loads were applied with 5 N/s to the pedicle screw via the implant rod in three cycles. The target force values increased gradually, reaching ±50 N (first cycle), ±100 N (second cycle), and ±200 N (third cycle) (Figure 3). In the event of significant loosening of the screws during compression loading, a risk of collision was identified due to the compact design of the test setup. This was prevented by premature manual shutdown. The following exclusion criteria were defined for data evaluation:

1. Test not possible (no pedicle screw instrumented due to fracture, for example);
2. Embedding error (e.g., embedded too deeply, screw in embedding compound);
3. Error during instrumentation (e.g., drill guide slipped, screw position out of plane)
4. Incorrect data recording;
5. Side comparison not possible (because data evaluation of the comparison pedicle was excluded due to the above criteria).

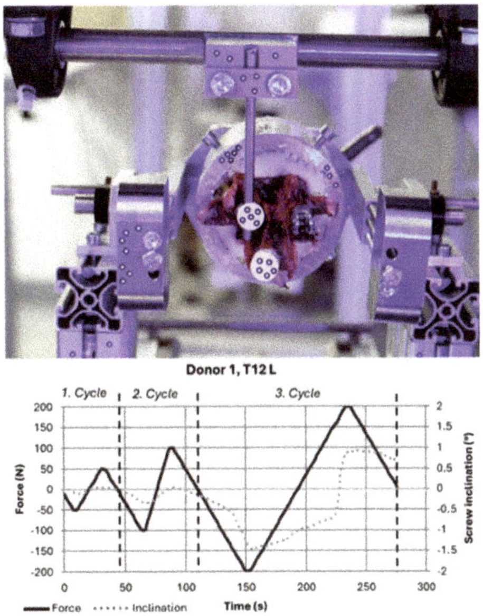

Figure 3. A mounted specimen with optical markers on the screw head and the spinal process (**top**). Axial force and screw inclination as a function of time (**bottom**).

2.5. Statistical Analysis

The data were processed using MS Excel (MS Office 2016, Microsoft Corporation, Redmond, WA, USA) and statistically evaluated using SPSS software (IBM SPSS Statistics 24.0, IBM Corporation, Armonk, NY, USA). The study investigated whether the pedicle screw inclination depends on correct or incorrect screw positioning. A paired Wilcoxon test was used to account for the differences between misplaced and correctly implanted screws, after a test of normal distribution (Kolmogorov–Smirnov-Test). The significance level α was set to 0.05.

3. Results

During the preparation and execution of the mechanical tests, a total of 17 vertebrae were excluded from the 40 vertebrae planned, with a total of 80 screws. The previously defined exclusion criteria were applied. The excluded screws were distributed as follows: six times criterion 1, eight times criterion 2, five times criterion 3, and seven times criterion 4. For a comparative evaluation, vertebrae with usable data recording were required that were instrumented as planned on both sides (criterion 5). From this, twentythree vertebrae with 46 screws were used for the final evaluation. Twenty-three vertebrae instrumented with a misplaced pedicle screw on one side could be used for evaluation, nine of which had medial (m), six lateral (l), three superior (s), and five superolateral (sl) malpositionings, as shown in Figure 4. Due to the low number of individual misalignments (l, m, s, sl), these were grouped together as malpositioning (F).

A total of eight pedicle screws completed the load test: four with correct positioning, two with lateral malpositioning, and two with medial malpositioning. The remaining tests resulted in screw loosening or severe tilting during the third cycle of the test. For this reason, only data from loads between $\pm$50 N and $\pm$100 N of all 46 screws were evaluated.

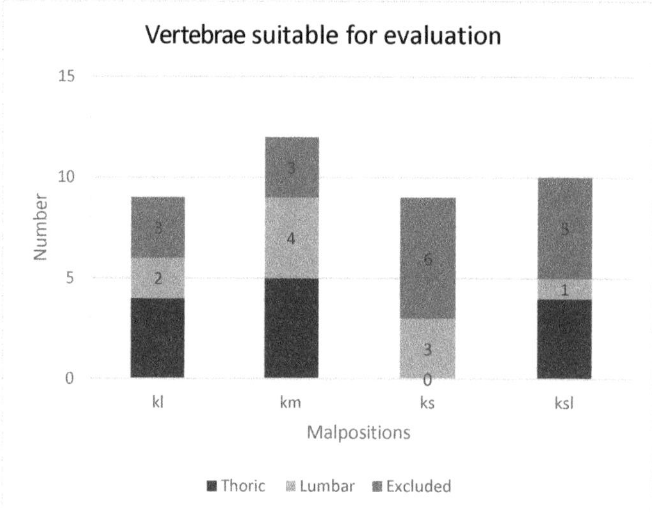

Figure 4. Distribution of suitable vertebrae for data evaluation grouped into each malpositioning (lateral—l; medial—m; superior—s; superior-lateral—sl).

Comparison of Malpositioning

The Wilcoxon test showed that there were only significant differences between correctly and incorrectly positioned screws at a tensile load of 100 N (Table 2). The said difference in inclination at 100 N is approximately 2° (Figure 5). The box plots indicate increasing dispersion of the inclination with increasing loads (Figure 5), but no differences between the groups are apparent.

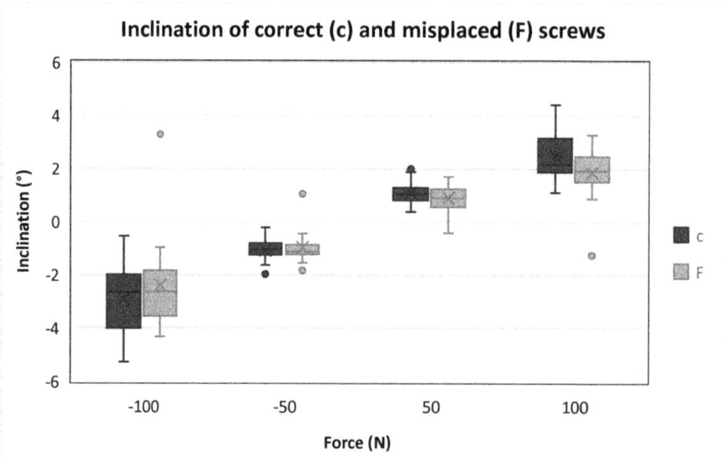

Figure 5. Boxplots showing the inclination of correct and misplaced pedicle screws at each maximum or minimum force.

Table 2. Comparison of median inclination at maximum and minimum forces during the first two cycles for correctly and misplaced screws using the Wilcoxon test.

Cycle	Force (N)	Correct Inclination (°)	Misplacement Inclination (°)	*p*-Value
1	−50	−1.0	−1.1	0.615
	50	1.0	0.9	0.277
2	−100	−2.7	−2.6	0.355
	100	2.2	1.9	0.026

4. Discussion

The study shows no biomechanical difference between the incorrectly placed screws and the correctly placed screws. There is a slight tendency for the misplaced screws to have a slightly lower tilt. This is supported by the significant differences at 100 N and the visualization of the box plots. However, the high data variability reduces the statistical power, limiting the interpretability of these trends.

From a biomechanical perspective, misplaced screws may remain in the vertebral body, provided there are no clinical concerns or symptoms. Previous studies have shown that even suboptimal placed screws can provide sufficient primary stability, unless there is a perforation of cortical structures or a risk to neurovascular elements [3,6,9,18]. This is further supported by pull-out tests, which assess the maximum extraction force of screws and, in some cases, show no significant difference between correctly and slightly misplaced screws [7,9].

Most recently, a biomechanical study revealed a higher pull-out force for medially misplaced lumbar pedicle screws compared to correctly implanted screws, whereas laterally misplaced showed a reduced pull-out force [4]. Therefore, medially misplaced screw should be reinserted to prevent neurological damage, whereas laterally misplaced screws should be reinserted to improve construct stability [4]. Clinically, in a 12-month follow-up, lateral breaches were found to have no impact on functional outcome and fusion but affected life quality and the postoperative improvement of leg and back pain [19].

However, the transferability of pedicle screw pull-out to clinical scenarios is limited. As demonstrated by Jarvers et al. [13], the results from biomechanical test environments can only be applied to in vivo conditions to a limited extent. A key advantage of the test setup used in this study lies in its time efficiency while still capturing dynamic loading scenarios. In contrast to traditional pull-out tests, which only simulate static maximum loads, the test method employed here allows for the simulation of physiologically relevant cyclic loads. This corresponds more closely to the actual mechanical load on the spine in everyday life [20]. Song et al. [21] showed in an FE study that the cranial–caudal movement corresponds to failure more than the pull-out, This supports the thesis of our test-setup as a practical addition to established test procedures and expands the biomechanical assessment of screw stability under quasi in vivo conditions.

Historically, a malpositioning of >4 mm was defined as a risk for neurovascular injury [5]. But there is low agreement between surgeons on which screws should be revised. A survey revealed that in asymptomatic patients, only 25% of surgeons tended to remove medially or inferiorly misplaced screws with >6 mm breach [22]: neurological symptoms were a strong indication for screw correction [22].

The rate of cortical breach or malpositioning of pedicle screws ranged from 2 to 50% depending on the definition [23]. Screw malpositioning accounts for the top four reasons leading to reoperation during the first 30 days following spine surgery [24]. In pediatric spine deformity surgery, screw malpositioning was the most frequently reported complication and reason for reoperation [25].

5. Limitations

A limitation of this study is that the investigations were conducted exclusively on human donor specimens, which significantly restricted the availability and number of samples. Due to the limited sample size, the statistical power of the results is constrained, potentially affecting the generalizability of the findings. To ensure a broad coverage of different bone density qualities, the specimens were carefully selected based on their bone quality and filtered through strict exclusion criteria. This approach helped to ensure data quality and improve comparability of the results. Another aspect is that, due to the limited number of samples, a paired comparison was not always possible. Consequently, some specimens had to be excluded from the analysis. This limitation is common in biomechanical studies using human specimens, as the availability of suitable samples poses a significant challenge [26].

6. Conclusions

Our results show that pedicle screw malpositioning of this extent has no significant influence on screw loosening. The biomechanical test results show that the malpositioning of pedicle screws does not necessarily lead to reduced stability of the implant in the bone. From a clinical point of view, malpositionings seem to have a greater impact on the potential injury to surrounding structures. The test procedure also reflects the pull-out results but is more physiological in terms of load application and therefore more transferable to clinical contexts in terms of perspective.

Author Contributions: Conceptualization, all; methodology, S.S., S.K. and F.M.; validation, J.S., S.S. and F.M.; investigation, F.M. and J.S.; data curation, S.S., F.M. and J.S.; writing—original draft preparation, S.S., F.M. and P.P.; writing—review and editing, all; visualization, S.S., F.M. and J.S.; supervision, C.-E.H. and P.P.; project administration, S.S. All authors have read and agreed to the published version of the manuscript.

Funding: This research received no external funding.

Institutional Review Board Statement: Being part of the body donor program regulated by the Saxonian Death and Funeral Act of 1994 (third section, paragraph 18, item 8), institutional approval for the use of the post-mortem tissues of human body donors was obtained. In addition, an institutional IRB approval in favor of the use of body donor material was obtained by the Ethics Committee of the Medical Faculty of the University of Leipzig (129/21-ek).

Data Availability Statement: The data that support the findings of this study are available on reasonable request from the corresponding author.

Acknowledgments: We acknowledge the support of the Open Access Publication Fund of Leipzig University, which enabled the open access publication of this article.

Conflicts of Interest: The authors declare no conflicts of interest.

Abbreviations

The following abbreviations are used in this manuscript:

vBMD	Bolumetric bone mineral content ()
CT	Computer tomography
HU	Hounsfield unit

References

1. Burval, D.J.; McLain, R.F.; Milks, R.; Inceoglu, S. Primary pedicle screw augmentation in osteoporotic lumbar vertebrae: Biomechanical analysis of pedicle fixation strength. *Spine* **2007**, *32*, 1077–1083. [CrossRef]
2. Cho, W.; Cho, S.K.; Wu, C. The biomechanics of pedicle screw-based instrumentation. *J. Bone Jt. Surg. Br.* **2010**, *92*, 1061–1065. [CrossRef]
3. Nevzati, E.; Marbacher, S.; Soleman, J.; Perrig, W.N.; Diepers, M.; Khamis, A.; Fandino, J. Accuracy of pedicle screw placement in the thoracic and lumbosacral spine using a conventional intraoperative fluoroscopy-guided technique: A national neurosurgical education and training center analysis of 1236 consecutive screws. *World Neurosurg.* **2014**, *82*, 866–871.e1-2. [CrossRef] [PubMed]
4. Tsagkaris, C.; Fasser, M.-R.; Farshad, M.; Passaplan, C.; Cornaz, F.; Widmer, J.; Spirig, J.M. Stability of medially and laterally malpositioned screws: A biomechanical study on cadavers. *Spine J.* **2025**, *25*, 380–388. [CrossRef] [PubMed]
5. Gertzbein, S.D.; Robbins, S.E. Accuracy of pedicular screw placement in vivo. *Spine* **1990**, *15*, 11–14. [CrossRef]
6. Floccari, L.V.; Larson, A.N.; Crawford, C.H.; Ledonio, C.G.; Polly, D.W.; Carreon, L.Y.; Blakemore, L. Which Malpositioned Pedicle Screws Should Be Revised? *J. Pediatr. Orthop.* **2018**, *38*, 110–115. [CrossRef]
7. Maeda, T.; Higashino, K.; Manabe, H.; Yamashita, K.; Hayashi, F.; Goda, Y.; Tsuruo, Y.; Sairyo, K. Pullout Strength of Pedicle Screws Following Redirection After Lateral or Medial Wall Breach. *Spine* **2018**, *43*, E983–E989. [CrossRef]
8. Ge, D.-W.; Chen, H.-T.; Qian, Z.-Y.; Zhang, S.; Zhuang, Y.; Yang, L.; Cao, X.-J.; Sui, T. Biomechanical strength impact of lateral wall breach on spinal pedicle screw fixation. *Eur. Rev. Med. Pharmacol. Sci.* **2018**, *22*, 63–68. [CrossRef]
9. Goda, Y.; Higashino, K.; Toki, S.; Suzuki, D.; Kobayashi, T.; Matsuura, T.; Fujimiya, M.; Hutton, W.C.; Fukui, Y.; Sairyo, K. The Pullout Strength of Pedicle Screws Following Redirection After Lateral Wall Breach or End-plate Breach. *Spine* **2016**, *41*, 1218–1223. [CrossRef]
10. Huang, X.; Huang, Z.; Xu, L.; Liang, D.; Zhang, M.; Zhang, H. Pullout strength of reinserted pedicle screws using the previous entry point and trajectory. *J. Orthop. Surg. Res.* **2019**, *14*, 205. [CrossRef]
11. Kang, D.G.; Lehman, R.A.; Wagner, S.C.; Bevevino, A.J.; Bernstock, J.D.; Gaume, R.E.; Dmitriev, A.E. Pedicle screw reinsertion using previous pilot hole and trajectory does not reduce fixation strength. *Spine* **2014**, *39*, 1640–1647. [CrossRef]
12. Kalemci, O.; Kizmazoglu, C.; Ozyoruk, S.; Uzunoglu, I.; Husemoglu, B.; Atar, M.; Aydin, H.E.; Karakasli, A. What is the Effect of Pedicle Screw Reinsertion Through the Same Trajectory on Pullout Strength? *Turk. Neurosurg.* **2022**, *32*, 635–640. [CrossRef] [PubMed]
13. Jarvers, J.-S.; Schleifenbaum, S.; Pfeifle, C.; Oefner, C.; Edel, M.; von der Höh, N.; Heyde, C.-E. Comparison of three different screw trajectories in osteoporotic vertebrae: A biomechanical investigation. *BMC Musculoskelet. Disord.* **2021**, *22*, 418. [CrossRef]
14. Metzner, F.; Reise, R.; Heyde, C.-E.; von der Höh, N.H.; Schleifenbaum, S. Side specific differences of Hounsfield-Units in the osteoporotic lumbar spine. *J. Spine Surg.* **2024**, *10*, 232–243. [CrossRef]
15. Brett, A.D.; Brown, J.K. Quantitative computed tomography and opportunistic bone density screening by dual use of computed tomography scans. *J. Orthop. Translat.* **2015**, *3*, 178–184. [CrossRef]
16. Schleifenbaum, S.; Vogl, A.-C.; Heilmann, R.; von der Hoeh, N.H.; Heyde, C.-E.; Jarvers, J.-S. Biomechanical comparative study of midline cortical vs. traditional pedicle screw trajectory in osteoporotic bone. *BMC Musculoskelet. Disord.* **2023**, *24*, 395. [CrossRef]
17. *ASTM F1717-15*; Standard Test Method for Spinal Implant Constructs in a Vertebrectomy Model. ASTM: West Conshohocken, PA, USA, 2015.
18. Reichle, E.; Morlock, M.; Sellenschloh, K.; Eggers, C. Zur Definition der Pedikelfehllage. Primärstabilität und Lockerungsverhalten von Pedikelschrauben in Abhängigkeit von ihrer Lage: Spongiöse Verankerung, kortikale Verankerung, Perforation und Fehllage. *Orthopade* **2002**, *31*, 402–405. [CrossRef]
19. LeRoy, T.E.; Smith, I.C.; Kim, D.H.; Golenbock, S.W.; Baker, K.C.; Arnold, P.M.; Sasso, R.C.; Park, D.K.; Fischgrund, J.S.; Zaidi, Q.H.; et al. Clinical Significance of Lateral Pedicle Screw Malposition in Lumbar Spine Fusion. *Clin. Spine Surg.* **2023**, *36*, E258–E262. [CrossRef]
20. Rohlmann, A.; Bergmann, G.; Graichen, F. Loads on internal spinal fixators measured in different body positions. *Eur. Spine J.* **1999**, *8*, 354–359. [CrossRef]
21. Song, F.; Liu, Y.; Fu, R.; Gao, X.; Iqbal, K.; Yang, D.; Liu, Y.; Yang, H. Craniocaudal toggling increases the risk of screw loosening in osteoporotic vertebrae. *Comput. Methods Programs Biomed.* **2023**, *238*, 107625. [CrossRef]
22. Aoude, A.; Ghadakzadeh, S.; Alhamzah, H.; Fortin, M.; Jarzem, P.; Ouellet, J.A.; Weber, M.H. Postoperative Assessment of Pedicle Screws and Management of Breaches: A Survey among Canadian Spine Surgeons and a New Scoring System. *Asian Spine J.* **2018**, *12*, 37–46. [CrossRef]
23. Woo, E.J.; DiCuccio, M.N. Clinically significant pedicle screw malposition is an underestimated cause of radiculopathy. *Spine J.* **2018**, *18*, 1166–1171. [CrossRef]
24. Ouyang, H.; Hu, Y.; Hu, W.; Zhang, H.; Sun, Z.; Tang, Y.; Jiang, Y.; Chen, J.; Dong, S.; Li, W.; et al. Incidences, causes and risk factors of unplanned reoperations within 30 days of spine surgery: A single-center study based on 35,246 patients. *Spine J.* **2022**, *22*, 1811–1819. [CrossRef]

25. Demura, S.; Ohara, T.; Tauchi, R.; Takimura, K.; Watanabe, K.; Suzuki, S.; Uno, K.; Suzuki, T.; Yanagida, H.; Yamaguchi, T.; et al. Incidence and causes of instrument-related complications after primary definitive fusion for pediatric spine deformity. *J. Neurosurg. Spine* **2023**, *38*, 192–198. [CrossRef]
26. Schmoelz, W.; Kienle, A. Mechanical and Biomechanical Testing of Spinal Implants. In *Manual of Spine Surgery*; Vieweg, U., Grochulla, F., Eds.; Springer: Berlin/Heidelberg, Germany, 2023; pp. 23–27, ISBN 978-3-662-64062-3.

bioengineering

MDPI

Article

Comprehensive Analysis of Chronic Low Back Pain: Morphological and Functional Impairments, Physical Activity Patterns, and Epidemiology in a German Population-Based Cross-Sectional Study

Bernhard Ulrich Hoehl [1,2,*], Nima Taheri [1,2], Lukas Schönnagel [1,2], Luis Alexander Becker [1,2], Lukas Mödl [3], Sandra Reitmaier [2], Matthias Pumberger [1,†] and Hendrik Schmidt [2,†]

1 Center for Musculoskeletal Surgery, Charité—Universitätsmedizin Berlin, 10117 Berlin, Germany
2 Julius Wolff Institute, Berlin Institute of Health, Charité—Universitätsmedizin Berlin, 10117 Berlin, Germany
3 Institute of Biometry and Clinical Epidemiology, Charité—Universitätsmedizin Berlin, 10117 Berlin, Germany
* Correspondence: bernhard.hoehl@charite.de; Tel.: +49-30-450-652474
† These authors contributed equally to this work.

Abstract

Low back pain (LBP) is the leading cause of disability worldwide. While studies often focus on the relationship between magnetic resonance imaging (MRI) findings and symptoms or the link between pain and disability, comprehensive assessments that incorporate both structural and functional impairments are lacking. This study prospectively includes standardized questionnaires, medical histories, clinical exams, and lumbar–pelvic MRI. Participants were grouped by pain status, physical activity, structural impairments (e.g., Pfirrmann, Krämer, Fujiwara, Meyerding), and posture/mobility deviations. Data were analyzed using the Kruskal–Wallis test. Of the 1262 participants, 392 (31%) reported chronic low back pain (cLBP), 226 (18%) had intermittent low back pain (iLBP), and 335 (27%) were pain-free. Significant differences were observed in high physical activity levels based on WHO criteria (cLBP: 79%, iLBP: 78%, no-BP(2): 86%, $p = 0.020$, $\eta^2 = 0.008$). Morphological impairments were more prevalent in cLBP (75%) and iLBP (76%) compared to no-BP(2) (55%) ($p = 0.000$, $\eta^2 = 0.043$). Functional impairments showed similar patterns (cLBP: 42%, iLBP: 51%, no-BP(2): 38%, $p = 0.014$, $\eta^2 = 0.010$). Participants with functional impairments tended to be younger. Consequently, the current classification system for diagnostics needs to incorporate alternative categories to more accurately differentiate types of back pain, which could enhance therapeutic outcomes.

Keywords: chronic; intermittent; low back pain; LBP; morphologic; functional; impairment

1. Introduction

Low back pain (LBP) is a leading contributor to age-standardized years lived with disability (YLD) globally, affecting both men and women [1]. The lifetime prevalence of LBP ranges between 49% and 70% [2], with recurrent episodes occurring in 20% to 44% of individuals within a single year, particularly within the working population [3]. Chronic LBP (cLBP), defined as pain persisting beyond 12 weeks, affects 5% to 10% of those initially experiencing LBP [4]. The Global Burden of Disease Study 2021 shows a significant increase in individuals suffering from LBP since 1990. The age-standardized prevalence of LBP is the highest in central and eastern Europe and Australasia and the lowest in East Asia,

Andean Latin America, and Southeast Asia. The impact of age on case numbers increases with advancing age, and peaking occurred within the 50–54 age group [5].

Management strategies for LBP encompass up to 50 clinical practice guidelines [6], with over 5000 trials evaluating various therapies and their modifications. However, many of these therapies have failed to replicate claimed effects, leading to ongoing modifications [7]. This may be attributed to inadequate application of evidence-based strategies or insufficient evidence supporting the effectiveness of commonly employed treatments [8,9]. Many clinicians argue that subgrouping LBP is crucial for improving individualized therapy [10]. The multidimensional nature of the condition presents a clinical challenge, requiring that each contributing factor be individually assessed and appropriately prioritized [11]. Effective daily management can be resource-intensive and may necessitate improved infrastructure and the integration of modern technologies [12]. Therefore, there is a pressing need for more precise and objective diagnostic assessments to tailor treatment plans to individual biological factors [13].

In 90% of LBP cases, no specific pathophysiological cause can be identified [2]. Potential sources of pain include lumbar structures such as intervertebral discs, facet joints, or spondylolisthesis. However, clinical tests often fail to consistently pinpoint these structures as the source of pain [14]. Magnetic resonance imaging (MRI) has become increasingly prevalent in recent years, revealing more morphological abnormalities in lumbar structures among LBP patients [15,16]. Yet, many imaging findings are also common in asymptomatic individuals, raising questions about their diagnostic relevance [17]. Additionally, while there is debate over the relationship between LBP and morphological changes, a correlation with restricted movement appears plausible, as a limited range of motion (ROM) is rarely observed in LBP patients [18–20]. Recent reviews suggest an inverse relationship between LBP and physical activity, though definitions of physical activity vary, including total activity, leisure time activities, and intensity [21,22]. The overall health benefits of physical activity are well-established, with the World Health Organization (WHO) recommending 75–150 min of physical activity per week, depending on intensity [23].

Despite extensive research on the relationships between LBP and MRI findings, mobility, and physical activity, comprehensive analyses that simultaneously examine all three factors are lacking. This study aims to provide a detailed analysis of morphological and functional impairments in relation to physical activity levels to better understand the multifactorial nature of LBP and identify key factors contributing to its persistence. The specific objectives of the study are as follows:

1. To characterize structural and functional impairments of the spine/back and levels of functional activity in individuals with and without LBP.
2. To analyze demographic factors (age, sex, Body Mass Index (BMI)) and clinical characteristics (pain intensity, pain duration).

2. Materials and Methods

2.1. Study Design

The ongoing "Berlin Back Study" is a prospective cross-sectional investigation registered with the German Clinical Trial Register (DRKS-ID: DRKS00027907), scheduled from 1 January 2022 to 31 December 2025. Recruitment of participants is conducted through multiple channels, including local promotion at Charité-Universitätsmedizin Berlin (via mailed flyers, notice boards, online platforms, and social media), public outreach (including newspapers, magazines, podcasts, and television), collaborations with local businesses and administrative bodies, and word-of-mouth referrals.

The study protocol adheres to the ethical principles outlined in the Helsinki Declaration [24] and has received approval from the Ethics Committee of Charité-

Universitätsmedizin Berlin (approval numbers: EA1/058/21). Written informed consent has been obtained from all participants. The research follows the Strengthening the Reporting of Observational Studies in Epidemiology (STROBE) guidelines [25]. Data collection commenced on 1 January 2022, with the analysis scheduled to conclude on 24 April 2024, at a university hospital research center.

2.2. Study Participants

To be eligible for participation in the study, individuals were required to meet the following inclusion criteria: provision of written informed consent; being either asymptomatic (with no history of back, pelvis, or hip pain, and no prior spinal surgery) or symptomatic with cLBP; and being aged 18–72 years with a minimum of 12 weeks of daily pain.

Exclusion criteria included the following: professional, competitive, or elite athletes; individuals with acute infections; substance abuse; pregnancy; a BMI exceeding 28 kg/m^2; central or peripheral neurological impairments (such as spinal cord injury, radicular symptoms, or sensory deficits); irritated, inflamed, or infected tissues in the back measurement areas; spinal fractures; osteoporosis; tumors or bone metastases; previous spinal surgery; current use of strong medications (e.g., antiepileptics, long-acting antihistamines, systemic glucocorticoids, immunosuppressive drugs); rheumatic diseases; active systemic diseases (including tuberculosis, collagenosis, multiple sclerosis, autoimmune diseases, acquired immune deficiency syndrome); internal conditions that could pose risks during measurements (such as coronary heart disease, heart failure, severe hypertension, chronic obstructive pulmonary disease); and malformations or anomalies of the lower extremities (e.g., knee or hip arthroplasty, arthrodesis).

2.3. Quantitative Variables and Data Collection

The study coordinator provided participants with detailed instructions regarding the study protocol and facilitated their progression through the following stages: (1) completion of questionnaires; (2) a clinical physical examination; and (3) assessment of back morphology and motion. These evaluations were conducted on the same day, with a total duration averaging 90 min. Additionally, participants had the option to undergo MRI and gait analysis within a 14-day period following the initial assessments.

2.4. Questionnaires

Participants primarily completed the questionnaires digitally using a survey program specifically developed for the study. The data collection was conducted under consistent conditions, including the same room and computer for all subjects at the study center. The standardized approach ensured uniformity in the completion of the questionnaires, which included measures of pain intensity and duration, pain-related disability (based on Von Korff et al., 1992 [26]), and physical activity levels assessed using the International Physical Activity Questionnaire (IPAQ) [27].

2.5. Physical Examination

The physical examination, conducted by a skilled orthopedic consultant, encompassed an assessment of organ function; overall patient impression; and vital signs including respiratory rate, heart rate, and blood pressure. Neurological status was evaluated through tests of coordination, reflexes, sensitivity, and motor function.

Functional parameters of the lumbar spine and pelvis, such as posture, shape, orientation, and movement, were assessed using established clinical tests (e.g., Ott and Schober test, 3-step hyperextension test, passive lumbar extension test) and participant self-assessment. Measurements, recorded with precision (e.g., distances in centimeters, angles in degrees), focused on pain provocation. Self-reported functional limitations were rated on a scale from 1 (least severe) to 10 (most severe).

Pain characteristics, including location, type, course, radiation, intensity, quality, and duration, were detailed, along with exacerbating or relieving factors, triggers, and patient-reported causes. The medical history covered previous illnesses of the spine, surgeries of the lower extremity, pain history, treatments, treatment effects, and medication use. Family and social history included occupational status, family medical history of back pain, and stressful life circumstances, as well as past or current use of addictive substances such as alcohol and nicotine. Demographic data, including age, sex, body height, weight, hip diameter, and waist diameter, were also recorded.

2.6. Back Morphology and Motion Analysis

All participants underwent back shape measurements in both the sagittal and frontal planes while standing and sitting, using the Idiag M360® (MediMouse, Idiag AG, Fehraltorf, Switzerland). This device measures segmental angles of the thoracic and lumbar spine. Participants were assessed in various postures—upright, flexed, extended, and left and right lateral bending—both in standing and sitting positions, with each position held for approximately 10 s and repeated three times. For standing measurements, maximum upper body flexion, extension, and lateral bending were performed with knees extended and arms crossed during extension to avoid interfering with the measurement. The sequence of tasks was randomized. Measurements were conducted by trained medical students, and the validity and reliability of the SpinalMouse have been confirmed in previous studies [28–33].

2.7. Spino-Pelvic MRI

MRI scans were performed using a 1.5 Tesla MRI scanner (Philips, Hamburg, Germany) with the following sequences: (1) Sagittal T1 (4 mm slices); (2) Sagittal T2 (4 mm slices); (3) Coronal STIR-T2 (4 mm slices); and (4) Axial T2 (3 mm slices). These scans were analyzed for various spinal pathologies, including intervertebral disc degeneration (Pfirrmann classification; [34,35]), disc herniation (Kramer classification; [35,36]), facet joint arthrosis (Fujiwara classification; [37]), spondylolisthesis (Meyerding classification; [38]), osteochondrosis intervertebralis (Modic classification; [39]), and spinal canal stenosis (Schizas classification; [40]) at each lumbar spine level.

The MRI assessments were conducted independently and blindly by three experienced investigators. In cases of discrepancies, the median classification was used.

2.8. Definition of Pain Status

Participants were categorized into three groups: chronic low back pain (cLBP), intermittent low back pain (iLBP), and no back pain (no-BP(2)). To determine these classifications, participants reported their current and past back pain experiences via questionnaires. During the clinical examination, a skilled orthopedic consultant verified the self-reported pain status, adhering to the definition of daily back pain lasting at least 12 weeks. Pain localization was documented through a detailed examination and classified into primary and secondary areas. Participants with daily lower back pain lasting 12 weeks or more were classified into the cLBP group. In contrast, participants who had experienced low back pain for less than 12 weeks and/or in intermittent episodes were classified as having intermittent low back pain (iLBP). Individuals without any history of back pain were assigned to the no-BP(2) group.

2.9. Characterization of Pain Status by Subgroups

Participants were systematically assigned to 24 subgroups based on morphological and functional parameters, physical activity levels, and pain status (Table 1). A participant was classified as morphologically or functionally impaired if any associated parameter was classified as impaired. For example, to be assigned to groups i, v, ix, xiii, xvii, or xxi, both morphological and functional assessments must be classified as non-impaired (Table 1).

Table 1. Assignment of study participants into the subgroups based on their pain status, physical activity, and functional or morphologic impairment: ✓: not impaired, ✗: impaired, no-BP(2) = no back pain, iLBP = intermittent low back pain, cLBP = chronic low back pain.

Physical Activity	Pain Status					
Low Physical Activity	no-BP(2)		**Group i**	**Group ii**	**Group iii**	**Group iv**
		Morphology	✓	✗	✓	✗
		Function	✓	✓	✗	✗
	iLBP		**Group v**	**Group vi**	**Group vii**	**Group viii**
		Morphology	✓	✗	✓	✗
		Function	✓	✓	✗	✗
	cLBP		**Group ix**	**Group x**	**Group xi**	**Group xii**
		Morphology	✓	✗	✓	✗
		Function	✓	✓	✗	✗
High Physical Activity	no-BP(2)		**Group xiii**	**Group xiv**	**Group xv**	**Group xvi**
		Morphology	✓	✗	✓	✗
		Function	✓	✓	✗	✗
	iLBP		**Group xvii**	**Group xviii**	**Group xix**	**Group xx**
		Morphology	✓	✗	✓	✗
		Function	✓	✓	✗	✗
	cLBP		**Group xxi**	**Group xxii**	**Group xxiii**	**Group xxiv**
		Morphology	✓	✗	✓	✗
		Function	✓	✓	✗	✗

2.9.1. Morphological Impairments

- MRI Analysis: Intervertebral discs and facet joints were graded using established degeneration scores. According to Pfirrmann et al., discs were categorized into five grades: grades 1 and 2 indicated no to minor changes and were considered non-impaired, while grades 3 to 5, reflecting moderate to severe degeneration, were deemed impaired [34,35]. Facet joints were evaluated based on Krämer et al.'s classification, with grades 0 to 2 indicating no to minor changes (non-impaired) and grades 3 to 5 indicating significant degeneration (impaired) [35,41]. Facet joint degeneration was also classified into four grades by Fujiwara et al., where grades 1 and 2 were non-impaired and grades 3 and 4 indicated impairment [37]. Spondylolisthesis was assessed using Meyerding's classification, with grade 1 (spondylolisthesis $\leq$25%) considered non-impaired and grades 2 to 5 (spondylolisthesis >25%) considered impaired [38].
- Back Morphology Measurements: Segmental angles of the lumbar spine in an upright standing position were compared to a reference database of healthy individuals. Participants exhibiting deviations beyond two standard deviations from the norm were classified as impaired.

2.9.2. Functional Impairments

Segmental angles of the lumbar spine during flexion, extension, and lateral bending (left and right) in a standing position were compared with a reference database of healthy individuals. Participants with deviations exceeding two standard deviations from the norm in any posture were considered functionally impaired.

2.9.3. Level of Physical Activity

Participants' physical activity levels were retrospectively categorized according to the WHO guidelines, which recommend a minimum of 150 min of moderate-intensity or 75 min of vigorous-intensity physical activity per week, or an equivalent combination. Meeting this criterion was classified as high physical activity. Vigorous physical activity was defined as activities such as aerobics, running, fast cycling, or fast swimming performed for at least 10 consecutive minutes. Moderate physical activity included activities such as carrying light loads, cycling at a regular pace, or swimming at a regular pace, also for a minimum of 10 uninterrupted minutes. Walking was not classified as either moderate or vigorous activity [23]. This assessment was conducted using the IPAQ.

2.10. Analysis of Demographic and Clinical Characteristics

To gain deeper insights into back pain, demographic data (age, sex, BMI), and clinical characteristics were visualized using boxplots, violin plots, and bar plots. Clinical characteristics included chronic pain grade and characteristic pain intensity, which was derived from current pain, the worst pain in the past three months, and average pain over the past three months, along with pain duration. Group differences were assessed using the non-parametric Kruskal–Wallis test to account for potential non-normal distributions.

2.11. Statistical Analysis

Descriptive analyses were conducted using boxplots, violin plots, bar, and column plots. Group differences were assessed using the non-parametric Kruskal–Wallis test. Effect sizes (Kruskal–Wallis η^2) were interpreted based on standard thresholds: 0.01 to <0.06 for small effects, 0.06 to <0.14 for moderate effects, and $\geq$0.14 for large effects [42,43]. The significance level was set at $p = 0.05$. All statistical analyses were performed using R software [44] and RStudio software Version 2023.12.1.402 [45].

3. Results

3.1. Study Population

From January 2022 to April 2024, a total of 1262 participants were examined. The presented analysis on LBP enrolled 335 participants without back pain (male = 140, female = 195), 226 participants with iLBP (male = 104, female = 122), and 392 participants with cLBP (male = 172, female = 220). The remaining 309 participants were not included in this analysis due to back pain without involvement of the lower back. A significant but small age difference was observed between the groups ($p = 0.001$, Kruskal–Wallis $\eta^2 = 0.014$), with median ages of 39 years (IQR: 29–52) for the no-BP(2) group, 40 years (IQR: 32–52) for the iLBP group, and 44 years (IQR: 34–53) for the cLBP group. No significant differences were found between groups in terms of sex ($p = 0.609$) or BMI ($p = 0.805$). Within the cLBP group, 40 (10.2%) participants had a history of lower extremity surgery, compared to 25 (11.1%) participants in the iLBP group and 23 (6.9%) participants in the no-BP(2) group ($p = 0.183$, $\eta^2 = 0.004$).

3.2. Characterization of Pain Status to Subgroups

3.2.1. Morphological Impairment

The classification of lumbar spine morphological changes was based on the Pfirrmann, Krämer, Fujiwara, and Meyerding grading systems. Significant differences between the pain groups were observed in all segments, except for the Meyerding classification and in the L1/2 and L2/3 segments of the Fujiwara classification (Figure 1). Five participants presented with six lumbar vertebrae. Lumbar spine alignment in the standing position showed significant but small differences between the pain status groups ($p = 0.001$, $\eta^2 = 0.015$). Median lumbar spine alignment was −25° (IQR: −31° to −19°) for the cLBP group, −24° (IQR: −28° to −19°) for the iLBP group, and −26° (IQR: −31° to −21°) for the no-BP(2) group. Overall, significant differences in morphological impairments were observed between the groups ($p < 0.001$, $\eta^2 = 0.043$), with 75% of cLBP, 76% of iLBP, and 55% of no-BP(2) participants showing impairments.

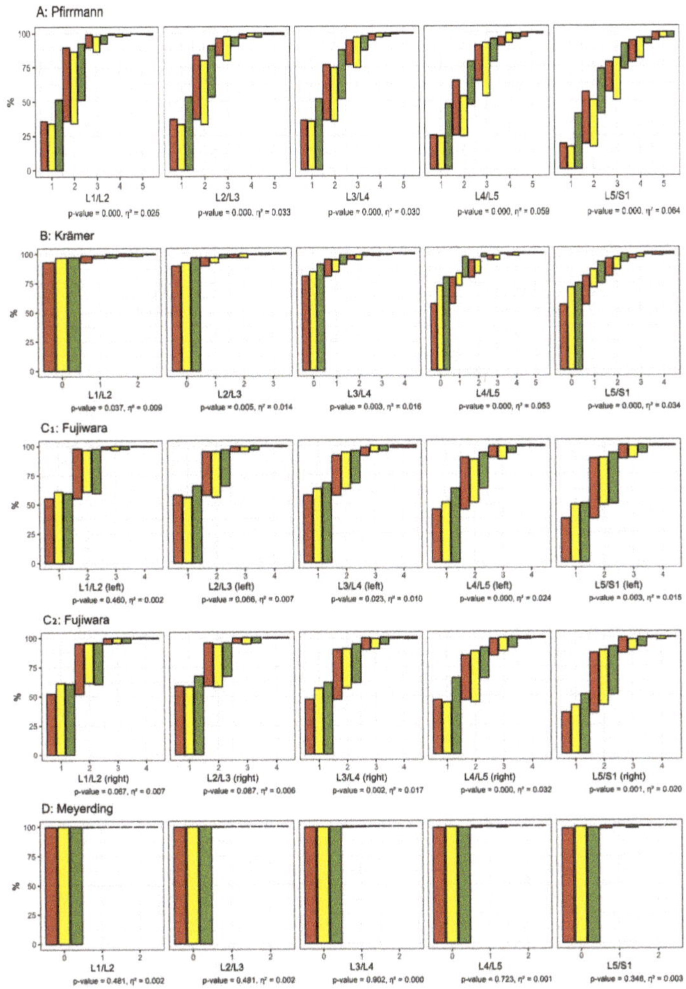

Figure 1. Comparison of the morphologic changes of the lumbar spine classified by Pfirrmann (disc degeneration, (**A**); Krämer (disc herniation, (**B**); Fujiwara (facet joint degeneration, left: (**C_1**); right: (**C_2**) and Meyerding (spondylolisthesis, (**D**) of the group with chronic low back pain (cLBP, red) and intermittent low back pain (iLBP, yellow) and without back pain (no-BP(2), green).

3.2.2. Functional Impairment

In basic functional tests assessing lumbar spine mobility while standing, significant differences were observed between participants without back pain and those with back pain. Mobility in extension ($\eta^2 = 0.022$, $p < 0.001$), right lateral bending ($\eta^2 = 0.027$, $p < 0.001$), and left lateral bending ($\eta^2 = 0.011$, $p = 0.007$) were notably different in the no-BP(2) group in comparison to the groups of cLBP and iLBP. The mobility cut-off values for 90% of no-BP(2) participants were as follows: flexion (8.9° to 36°), extension (−48° to −20°), right lateral bending (−29° to −1°), and left lateral bending (9° to 31°) (Figure 2). Overall, functional impairments differed significantly between pain groups ($p = 0.014$, $\eta^2 = 0.010$), with 42% of cLBP, 51% of iLBP, and 38% of no-BP(2) participants exhibiting functional impairments.

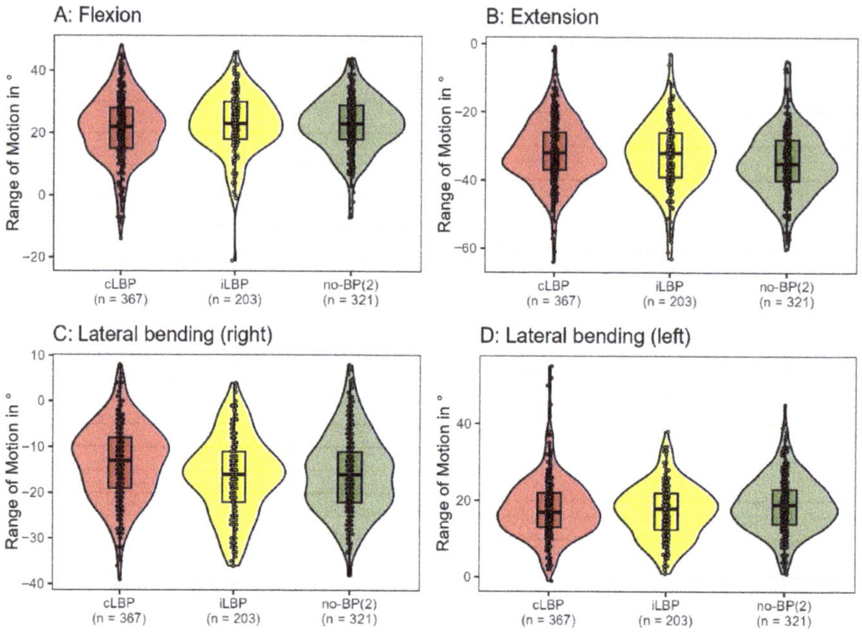

Figure 2. Lumbar spine mobility assessed using the Idiag M360®(MediMouse, Idiag AG, Fehraltorf, Switzerland), across three primary groups: chronic low back pain (cLBP), intermittent low back pain (iLBP), and pain-free individuals (no-BP(2)). Significant intergroup differences were found in extension ($\eta^2 = 0.022$, $p < 0.001$), right lateral bending ($\eta^2 = 0.027$, $p < 0.001$), and left lateral bending ($\eta^2 = 0.011$, $p = 0.007$), indicating reduced mobility in specific movements among those with LBP.

3.2.3. Physical Activity

The metabolic equivalent of task (MET) minutes, as assessed by the IPAQ, did not reveal significant differences between the pain groups (cLBP: median = 2837, IQR = 1555–4513; iLBP: median = 3099, IQR = 1673–5088; no-BP(2): median = 3092, IQR = 1710–4971 [all in MET minutes]; $\eta^2 = 0.006$, $p = 0.052$). Based on the WHO classification, a significantly higher proportion of participants without back pain demonstrated high levels of physical activity (cLBP: 79%, iLBP: 78%, no-BP(2): 86%; $\eta^2 = 0.008$, $p = 0.020$) (Figure 3).

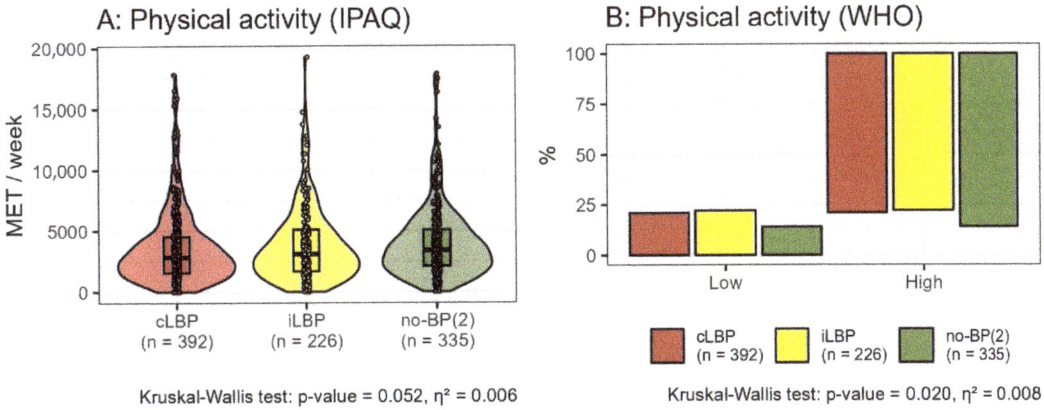

Figure 3. Level of physical activity, measured in metabolic equivalent (MET) minutes per week as assessed by the International Physical Activity Questionnaire (IPAQ) (**A**). IPAQ shows no significant difference (p = 0.052). However, according to World Health Organization (WHO) classifications (**B**), there are significant differences in physical activity levels among participants with chronic low back pain (cLBP), intermittent low back pain (iLBP), and those without back pain (no-BP(2)). Participants are classified as having a high level of physical activity if they engage in either 150 min of moderate-intensity physical activity or 75 min of vigorous-intensity physical activity per week.

3.2.4. Assigning Participants to Subgroups

A total of 10% of study participants with either low or high physical activity levels are pain-free and exhibit no morphological or functional impairments. Among cLBP patients, 8.7% (low activity) and 5.9% (high activity) show no morphological or functional impairments. Interestingly, the pain-free groups exhibit both morphological and functional impairments in 7.2% (low activity) and 7.9% (high activity) of cases, which is only slightly different from the iLBP groups but significantly and clinically lower compared to the cLBP groups, with 13.8% and 11% respectively (Table 2).

Table 2. Assignment of study participants with no back pain (no-BP(2)), intermittent low back pain (iLBP), and chronic low back pain (cLBP) into the subgroups (i)–(xxiv): Each table cell indicates the number of participants (n), number of sexes specified in brackets (males, females), the proportion (%) relative to the corresponding activity group, and the proportion (%) relative to of the whole group (excluding the group with missing data).

<table>
<tr><td rowspan="6">Low Physical Activity</td><td rowspan="2">no-BP(2) n = 47 (17, 30) (4.9%)</td><td>Group i</td><td>Group ii</td><td>Group iii</td><td>Group iv</td><td>Missing Data</td></tr>
<tr><td>n = 14 (6, 8)
10%/1.9%</td><td>n = 10 (4, 6)
7.2%/1.3%</td><td>n = 4 (1, 3)
2.9%/0.5%</td><td>n = 10 (3, 7)
7.2%/1.3%</td><td>n = 9 (3, 6)</td></tr>
<tr><td rowspan="2">iLBP n = 50 (18, 32) (5.2%)</td><td>Group v</td><td>Group vi</td><td>Group vii</td><td>Group viii</td><td rowspan="2">n = 14 (6, 8)</td></tr>
<tr><td>n = 6 (2, 4)
4.3% / 0.8%</td><td>n = 18 (4, 14)
13% / 2.4%</td><td>n = 3 (0, 3)
2.1% / 0.4%</td><td>n = 9 (6, 3)
6.5%/1.2%</td></tr>
<tr><td rowspan="2">cLBP n = 82 (38, 44) (8.6%)</td><td>Group ix</td><td>Group x</td><td>Group xi</td><td>Group xii</td><td rowspan="2">n = 18 (6, 12)</td></tr>
<tr><td>n = 12 (7, 5)
8.7%/1.6%</td><td>n = 26 (12, 14)
19%/13.5%</td><td>n = 7 (1, 6)
5.1%/0.9%</td><td>n = 19 (12, 7)
13.8%/2.6%</td></tr>
<tr><td colspan="7">n = 138 (19% of the entire group) + 41 with missing data (MRI or SpinalMouse)</td></tr>
</table>

Table 2. *Cont.*

High Physical Activity	**no-BP(2)** *n* = 288 (123, 165) (30%)	**Group xiii**	**Group xiv**	**Group xv**	**Group xvi**	**Missing Data**
		n = 74 (40, 34) 12%/10%	*n* = 65 (22, 43) 11%/8.7%	*n* = 30 (15, 15) 4.9%/4%	*n* = 59 (22, 37) 9.7%/7.9%	*n* = 60 (24, 36)
	iLBP *n* = 176 (86, 90) (18%)	**Group xvii**	**Group xviii**	**Group xix**	**Group xx**	
		n = 13 (8, 5) 2.1%/1.7%	*n* = 45 (18, 27) 7.4%/6%	*n* = 21 (10, 11) 3.5% / 2.8%	*n* = 58 (34, 24) 9.6%/7.8%	*n* = 39 (16, 23)
	cLBP *n* = 310 (134, 176) (33%)	**Group xxi**	**Group xxii**	**Group xxiii**	**Group xxiv**	
		n = 44 (20, 24) 7.2%/5.9%	*n* = 97 (39, 58) 16%/13%	*n* = 18 (7, 11) 3%/2.4%	*n* = 83 (36, 47) 14%/11%	*n* = 68 (32, 36)
n = 607 (81% of the entire group) + 167 with missing data (MRI or SpinalMouse)						

3.3. *Impact of Demographic and Clinical Characteristics on Subgroup Differentiation*

There were no significant differences between sexes in subgroups with low physical activity (no-BP(2): $\eta^2 = 0.018$, $p = 0.877$; iLBP: $\eta^2 = 0.194$, $p = 0.078$; cLBP: $\eta^2 = 0.084$, $p = 0.152$) or high physical activity (no-BP(2): $\eta^2 = 0.032$, $p = 0.065$; iLBP: $\eta^2 = 0.031$, $p = 0.244$; cLBP: $\eta^2 = 0.002$, $p = 0.924$). Participants with a high level of physical activity and functional impairment only (groups xv, xix, xxiii) were significantly younger (no-BP(2): $\eta^2 = 0.108$, $p < 0.001$; iLBP: $\eta^2 = 0.144$, $p < 0.001$; cLBP: $\eta^2 = 0.059$, $p = 0.924$). Body Mass Index (BMI) displayed significant differences among groups i–iv ($\eta^2 = 0.268$, $p = 0.022$) (Figure 4).

3.4. *Impact of Physical Activity on Subgroup Differentiation*

For participants with low levels of physical activity, there were no significant differences in chronic pain grade ($\eta^2 = 0.061$, $p = 0.551$), characteristic pain intensity ($\eta^2 = 0.138$, $p = 0.064$), or duration of pain ($\eta^2 = 0.120$, $p = 0.264$) between those with iLBP and cLBP. Within each pain status group, no significant differences were observed between participants with no morphological or functional impairment (Table 2, groups v, ix), morphological impairment only (Table 2, groups vi, x), functional impairment only (Table 2, groups vii, xi), or both morphological and functional impairment (Table 2, groups viii, xii).

For participants with high levels of physical activity, there were significant differences in chronic pain grade ($\eta^2 = 0.058$, $p = 0.003$) and characteristic pain intensity ($\eta^2 = 0.146$, $p < 0.001$) between those with iLBP and cLBP. However, there was no significant difference in pain duration ($\eta^2 = 0.015$, $p = 0.763$) between these groups. Within each pain status group, no significant differences were observed between participants with no morphological or functional impairment (Table 2, groups xvii, xxi), morphological impairment only (groups xviii, xxii), functional impairment only (Table 2, groups xix, xxiii), or both morphological and functional impairment (Table 2, groups xx, xxiv) (Figure 5).

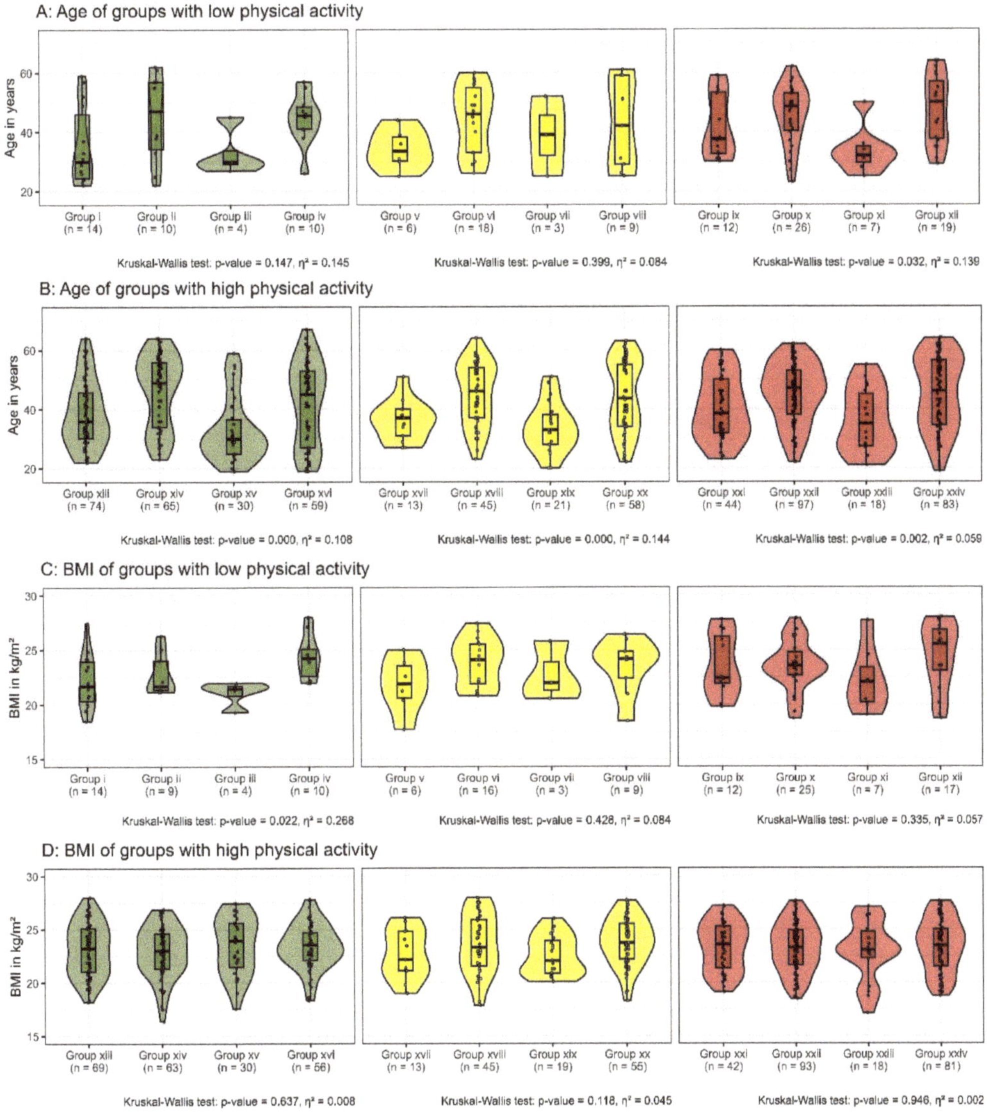

Figure 4. Age (**A**,**B**) and Body Mass Index (BMI) (**C**,**D**) are presented for participants categorized by their pain status: without back pain (green), with intermittent low back pain (yellow), and with chronic low back pain (red). Within each color-coded group, the four subsets represent no morphological or functional impairment (first), morphological impairment only (second), functional impairment only (third), and both morphological and functional impairment (fourth). Participants are classified as having a high level of physical activity (**B**,**D**) if they engage in either 150 min of moderate-intensity physical activity or 75 min of vigorous-intensity physical activity per week. Differences within each individual graph are evaluated using the Kruskal–Wallis test.

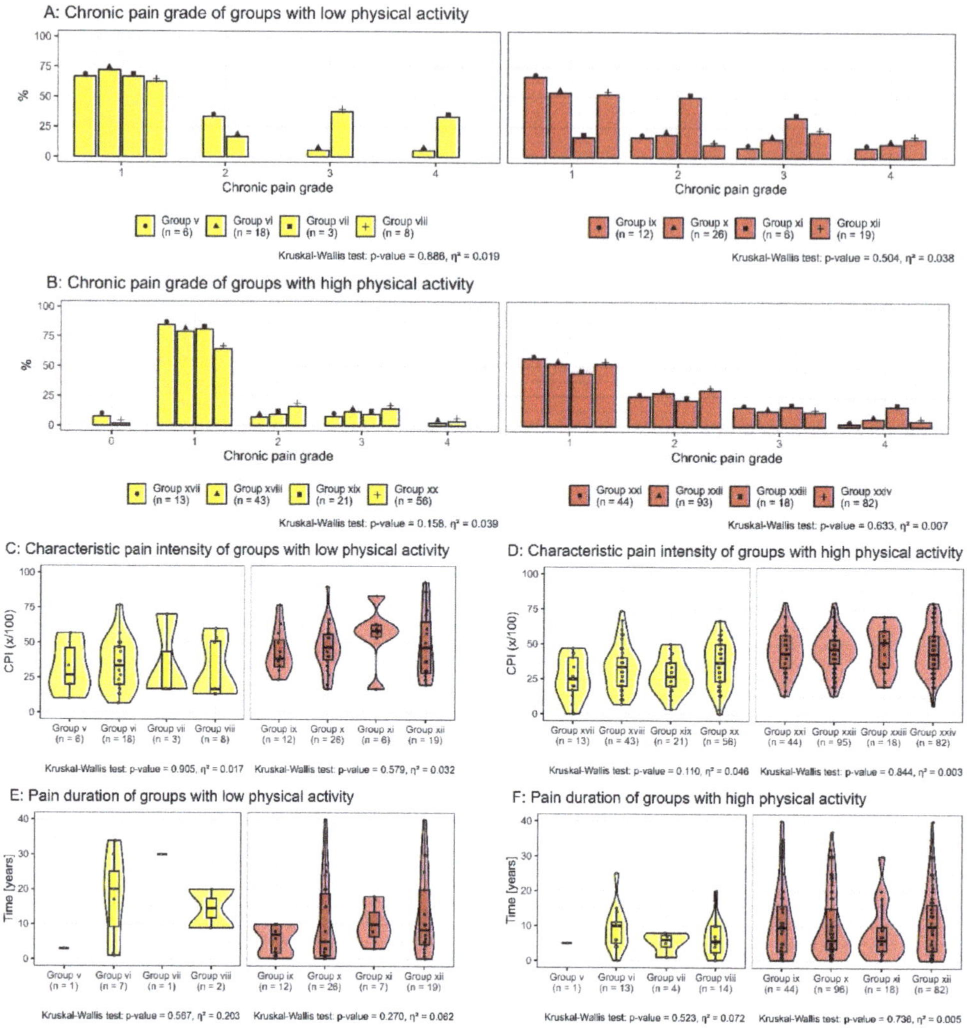

Figure 5. The chronic pain grade (**A**,**B**) and characteristic pain intensity (**C**,**D**) according to Korff et al. among pain duration (**E**,**F**) are displayed for the subgroups of intermittent low back pain (yellow) and chronic low back pain (red). In the four-pack of each color, the first group had neither morphologic nor functional impairment, the second one had morphologic impairment only, the third one had functional impairment only, and the fourth one had both morphologic and functional impairment. The participants are considered to have a high level of physical activity (**B**,**D**,**F**) when they spend either 150 min of moderate physical activity or 75 min of vigorous physical activity per week. The Kruskal–Wallis test checks for differences within each individual graph.

4. Discussion

This study examines the influence of physical activity levels, morphological, and functional impairments on the subclassification of LBP into cLBP, iLBP, and no-BP(2) categories. Our findings reveal that 10% of participants across both low and high activity levels were pain-free and displayed no morphological or functional impairments. Notably, cLBP patients showed a lower percentage of individuals without impairments (8.7% for low activity and 5.9% for high activity) compared to pain-free participants, who exhibited a slightly higher incidence of both morphological and functional impairments.

For high physical activity participants, significant differences were observed in chronic pain grade ($\eta^2 = 0.058$, $p = 0.003$) and characteristic pain intensity ($\eta^2 = 0.146$, $p < 0.001$) between iLBP and cLBP groups, although no significant difference in pain duration was found. This suggests that high physical activity influences pain severity and intensity but not necessarily the duration of pain.

Our results further indicate that a significant number of participants with high levels of physical activity were pain-free and exhibited fewer impairments overall. According to the WHO classification, there were significantly more participants with high physical activity among those without back pain (cLBP: 79%; iLBP: 78%; no-BP(2): 86%; $p = 0.020$, $\eta^2 = 0.008$). The percentage of morphological impairments was significantly different between pain statuses, with the highest prevalence in cLBP (75%) and iLBP (76%) compared to no-BP(2) (55%; $\eta^2 = 0.043$; $p < 0.001$). Functional impairments also varied significantly, being more prevalent in cLBP (42%) and iLBP (51%) than in no-BP(2) (38%; $\eta^2 = 0.010$; $p = 0.014$).

The proposed subclassification revealed significant age differences, with participants having functional impairments being the youngest in all pain status groups (cLBP: $\eta^2 = 0.072$, $p < 0.001$; iLBP: $\eta^2 = 0.126$, $p < 0.001$; no-BP(2): $\eta^2 = 0.109$, $p < 0.001$). However, within each pain status, there was no significant correlation between functional or morphological impairments and chronic pain grade, pain intensity, or pain duration. This suggests that while age and physical activity levels influence impairment and pain status, the specific impact of these factors on pain severity remains complex.

Our study's results are consistent with other large-scale studies regarding the physical activity levels of participants. Notably, our participants reported a median of approximately 3000 MET minutes per week, which is higher than the 1000 MET minutes reported in other studies [21]. This suggests that our voluntary participant group is particularly active.

Interestingly, our findings on BMI and age align with the existing literature. We observed that morphologic impairments tended to increase with age and were marginally associated with higher BMI, although no significant differences in BMI were found across pain statuses ($p = 0.805$). This consistency with the literature [17,46] reinforces the importance of considering age and BMI in the context of LBP.

The MRI findings of our study indicated a tendency for morphologic impairments to increase towards the lower segments, particularly involving vertebral disc degeneration (Pfirrmann and Krämer classification in Figure 1), aligning with the body of literature [47]. In our study, in the majority of lumbar segments, the grades of degeneration were associated with pain status. Existing studies show inconsistent correlations between disc degeneration and LBP [48,49], which may be attributed to different characteristics of the study populations [50]. The extent to which subclassification can be refined by integrating further clinical parameters remains unclear. Recent studies have begun to investigate genetic predisposition as a means to deepen our understanding of low back pain [51].

In our study population, functional impairment showed a statistically significant ($p = 0.014$) but weak correlation with pain status ($\eta^2 = 0.010$). Despite the statistical significance, the clinical relevance remains questionable, particularly given the absence of an established minimal clinically important difference for spinal ROM. Based on the distribution of spinal ROM presented in Figure 2 (Section 3.2.2. Functional Impairment), it appears unlikely that a clinician could reliably assign a patient to a specific pain category based solely on movement amplitude. In our study population, the clinical impact on ROM may be further attenuated due to the relatively high levels of physical activity and low BMI among participants. This finding is consistent with the previous literature; the systematic review and meta-analysis by Errabity et al. demonstrated significant reductions in lumbar ROM across sagittal, frontal, and transverse planes, although these findings were not uniform across all studies [18].

Several limitations must be considered. The cross-sectional design limits our ability to infer causality and assess changes over time. Additionally, our voluntary participant sample may not be fully representative, and the exclusion of individuals with a BMI over 28 kg/m^2 or previous surgeries could introduce bias towards a healthier population. The subjective nature of pain also introduces variability in the data. In addition, comorbidities (e.g., diabetes, high blood pressure) possibly influencing the results were assessed.

Future research should focus on prospective interventional studies to explore the effectiveness of individualized treatments based on physical activity levels and impairment types. Improved subclassification of LBP can potentially enhance therapeutic strategies and outcomes.

Our study contributes to the understanding of LBP by incorporating individual characteristics, such as physical activity and impairments, into the diagnosis. Although age and BMI showed significant differences between subgroups, a clinically meaningful subclassification based solely on these or other individual parameters remains challenging.

Musculoskeletal health professionals should remain aware of the multifactorial nature of LBP. A thorough, interdisciplinary assessment that includes lifestyle, spinal morphology, and functional capacity is essential to address patient-specific needs effectively. Future research is needed to assess the impact of these characteristics on treatment efficacy and to refine LBP management strategies.

Author Contributions: M.P. and H.S. contributed equally to the work. B.U.H. drafted the manuscript and finalized the study design. L.M. performed the statistical analysis. N.T., L.S. and L.A.B. collected and curated the data. S.R., M.P. and H.S. contributed to the study design and coordination. M.P. and H.S. conceived the study and equally supervised its design and coordination. All authors have read and agreed to the published version of the manuscript.

Funding: This study was financially supported by the German Research Foundation (DFG): RE 4292/3-1, SCHM 2572/11-1, SCHM 2572/13-1, PU 762/1-1.

Institutional Review Board Statement: The study was conducted in accordance with the Declaration of Helsinki and approved by the Ethics Committee of Charité-Universitätsmedizin Berlin (approval numbers: EA1/058/21, date of approval: 26 March 2021).

Informed Consent Statement: Informed consent was obtained from all subjects involved in the study.

Data Availability Statement: The data that support the findings of this study are available from the corresponding author, B.U.H., upon reasonable request.

Acknowledgments: The authors thank Barbara Schiller and Theresa Ahnert for their efficient administrative support throughout the project. Special thanks are extended to our clinical and research colleagues for their constructive input and to the reviewers for their thoughtful comments.

Conflicts of Interest: The authors declare no conflicts of interest.

Abbreviations

The following abbreviations are used in this manuscript:

BMI	Body Mass Index
cLBP	chronic Low Back Pain
DRKS	German Clinical Trial Register
iLBP	intermittent Low Back Pain
IPAQ	International Physical Activity Questionnaire
IQR	Interquartile Range
LBP	Low Back Pain
MET	Metabolic Equivalent of Task
MRI	Magnetic Resonance Imaging

no-BP(2)	pain-free
ROM	Range of Motion
STROBE	Strengthening the Reporting of Observational Studies in Epidemiology
WHO	World Health Organization
YLD	Years Lived with Disability

References

1. Vos, T.; Abajobir, A.; Abbafati, C.; Abbas, K.; Abate, K.; Abd-Allah, F. Global, regional, and national incidence, prevalence, and years lived with disability for 328 diseases and injuries for 195 countries, 1990-2016: A systematic analysis for the Global Burden of Disease Study 2016. *Lancet* **2017**, *390*, 1211–1259. [CrossRef]
2. Koes, B.W.; van Tulder, M.W.; Thomas, S. Diagnosis and treatment of low back pain. *BMJ* **2006**, *332*, 1430–1434. [CrossRef] [PubMed]
3. Andersson, G.B. Epidemiological features of chronic low-back pain. *Lancet* **1999**, *354*, 581–585. [CrossRef] [PubMed]
4. Meucci, R.D.; Fassa, A.G.; Faria, N.M. Prevalence of chronic low back pain: Systematic review. *Rev. Saude Publica* **2015**, *49*, 1. [CrossRef] [PubMed]
5. Yang, N.; Di, J.; Wang, W.; Feng, H. Global burden of low back pain from 1990 to 2021: A comprehensive analysis of risk factors and trends using the Global Burden of Disease Study 2021. *BMC Public Health* **2025**, *25*, 1886. [CrossRef]
6. Jenkins, H.J.; Ferreira, G.; Downie, A.; Maher, C.; Buchbinder, R.; Hancock, M.J. The available evidence on the effectiveness of 10 common approaches to the management of non-specific low back pain: An evidence map. *Eur. J. Pain* **2022**, *26*, 1399–1411. [CrossRef]
7. Maher, C.; Underwood, M.; Buchbinder, R. Non-specific low back pain. *Lancet* **2017**, *389*, 736–747. [CrossRef]
8. Buchbinder, R.; van Tulder, M.; Öberg, B.; Costa, L.M.; Woolf, A.; Schoene, M.; Croft, P. Low back pain: A call for action. *Lancet* **2018**, *391*, 2384–2388. [CrossRef]
9. Foster, N.E.; Anema, J.R.; Cherkin, D.; Chou, R.; Cohen, S.P.; Gross, D.P.; Ferreira, P.H.; Fritz, J.M.; Koes, B.W.; Peul, W.; et al. Prevention and treatment of low back pain: Evidence, challenges, and promising directions. *Lancet* **2018**, *391*, 2368–2383. [CrossRef]
10. Bishop, F.L.; Dima, A.L.; Ngui, J.; Little, P.; Moss-Morris, R.; Foster, N.E.; Lewith, G.T. "Lovely Pie in the Sky Plans": A Qualitative Study of Clinicians' Perspectives on Guidelines for Managing Low Back Pain in Primary Care in England. *Spine* **2015**, *40*, 1842–1850. [CrossRef]
11. Wirth, B.; Schweinhardt, P. Personalized assessment and management of non-specific low back pain. *Eur. J. Pain* **2024**, *28*, 181–198. [CrossRef] [PubMed]
12. Anselmo, A.; Pagano, M.; Cappadona, I.; Cardile, D.; Calabrò, R.S.; Corallo, F. Psychological Support for Chronic Low Back Pain: A Systematic Review on the Validity of a Growing Remote Approach. *Curr. Pain Headache Rep.* **2025**, *29*, 51. [CrossRef] [PubMed]
13. Knezevic, N.N.; Candido, K.D.; Vlaeyen, J.W.S.; Van Zundert, J.; Cohen, S.P. Low back pain. *Lancet* **2021**, *398*, 78–92. [CrossRef]
14. Hancock, M.J.; Maher, C.G.; Latimer, J.; Spindler, M.F.; McAuley, J.H.; Laslett, M.; Bogduk, N. Systematic review of tests to identify the disc, SIJ or facet joint as the source of low back pain. *Eur. Spine J.* **2007**, *16*, 1539–1550. [CrossRef]
15. Deyo, R.A.; Weinstein, J.N. Low back pain. *N. Engl. J. Med.* **2001**, *344*, 363–370. [CrossRef]
16. Hartvigsen, J.; Hancock, M.J.; Kongsted, A.; Louw, Q.; Ferreira, M.L.; Genevay, S.; Hoy, D.; Karppinen, J.; Pransky, G.; Sieper, J.; et al. What low back pain is and why we need to pay attention. *Lancet* **2018**, *391*, 2356–2367. [CrossRef]
17. Brinjikji, W.; Luetmer, P.H.; Comstock, B.; Bresnahan, B.W.; Chen, L.E.; Deyo, R.A.; Halabi, S.; Turner, J.A.; Avins, A.L.; James, K.; et al. Systematic literature review of imaging features of spinal degeneration in asymptomatic populations. *AJNR Am. J. Neuroradiol.* **2015**, *36*, 811–816. [CrossRef]
18. Errabity, A.; Calmels, P.; Han, W.S.; Bonnaire, R.; Pannetier, R.; Convert, R.; Molimard, J. The effect of low back pain on spine kinematics: A systematic review and meta-analysis. *Clin. Biomech.* **2023**, *108*, 106070. [CrossRef]
19. Steiger, F.; Wirth, B.; de Bruin, E.D.; Mannion, A.F. Is a positive clinical outcome after exercise therapy for chronic non-specific low back pain contingent upon a corresponding improvement in the targeted aspect(s) of performance? A systematic review. *Eur. Spine J.* **2012**, *21*, 575–598. [CrossRef]
20. Wernli, K.; Tan, J.S.; O'Sullivan, P.; Smith, A.; Campbell, A.; Kent, P. Does Movement Change When Low Back Pain Changes? A Systematic Review. *J. Orthop. Sports Phys. Ther.* **2020**, *50*, 664–670. [CrossRef] [PubMed]
21. Alzahrani, H.; Mackey, M.; Stamatakis, E.; Zadro, J.R.; Shirley, D. The association between physical activity and low back pain: A systematic review and meta-analysis of observational studies. *Sci. Rep.* **2019**, *9*, 8244. [CrossRef]
22. Shiri, R.; Falah-Hassani, K. Does leisure time physical activity protect against low back pain? Systematic review and meta-analysis of 36 prospective cohort studies. *Br. J. Sports Med.* **2017**, *51*, 1410–1418. [CrossRef]

23. Bull, F.C.; Al-Ansari, S.S.; Biddle, S.; Borodulin, K.; Buman, M.P.; Cardon, G.; Carty, C.; Chaput, J.P.; Chastin, S.; Chou, R.; et al. World Health Organization 2020 guidelines on physical activity and sedentary behaviour. *Br. J. Sports Med.* **2020**, *54*, 1451–1462. [CrossRef]
24. World Medical Association. World Medical Association Declaration of Helsinki: Ethical principles for medical research involving human subjects. *JAMA* **2013**, *310*, 2191–2194. [CrossRef] [PubMed]
25. von Elm, E.; Altman, D.G.; Egger, M.; Pocock, S.J.; Gøtzsche, P.C.; Vandenbroucke, J.P. The Strengthening the Reporting of Observational Studies in Epidemiology (STROBE) statement: Guidelines for reporting observational studies. *Lancet* **2007**, *370*, 1453–1457. [CrossRef] [PubMed]
26. Von Korff, M.; Ormel, J.; Keefe, F.J.; Dworkin, S.F. Grading the severity of chronic pain. *Pain.* **1992**, *50*, 133–149. [CrossRef]
27. Lee, P.H.; Macfarlane, D.J.; Lam, T.H.; Stewart, S.M. Validity of the International Physical Activity Questionnaire Short Form (IPAQ-SF): A systematic review. *Int. J. Behav. Nutr. Phys. Act.* **2011**, *8*, 115. [CrossRef]
28. Guermazi, M.; Ghroubi, S.; Kassis, M.; Jaziri, O.; Keskes, H.; Kessomtini, W.; Ben Hammouda, I.; Elleuch, M.H. Validity and reliability of Spinal Mouse to assess lumbar flexion. *Ann. Readapt. Med. Phys.* **2006**, *49*, 172–177. [CrossRef]
29. Post, R.B.; Leferink, V.J. Spinal mobility: Sagittal range of motion measured with the SpinalMouse, a new non-invasive device. *Arch. Orthop. Trauma. Surg.* **2004**, *124*, 187–192. [CrossRef] [PubMed]
30. Mannion, A.F.; Knecht, K.; Balaban, G.; Dvorak, J.; Grob, D. A new skin-surface device for measuring the curvature and global and segmental ranges of motion of the spine: Reliability of measurements and comparison with data reviewed from the literature. *Eur. Spine J.* **2004**, *13*, 122–136. [CrossRef]
31. Topalidou, A.; Tzagarakis, G.; Souvatzis, X.; Kontakis, G.; Katonis, P. Evaluation of the reliability of a new non-invasive method for assessing the functionality and mobility of the spine. *Acta Bioeng. Biomech.* **2014**, *16*, 117–124.
32. Barrett, E.; McCreesh, K.; Lewis, J. Reliability and validity of non-radiographic methods of thoracic kyphosis measurement: A systematic review. *Man. Ther.* **2014**, *19*, 10–17. [CrossRef]
33. Dreischarf, B.; Koch, E.; Dreischarf, M.; Schmidt, H.; Pumberger, M.; Becker, L. Comparison of three validated systems to analyse spinal shape and motion. *Sci. Rep.* **2022**, *12*, 10222. [CrossRef] [PubMed]
34. Pfirrmann, C.W.; Metzdorf, A.; Zanetti, M.; Hodler, J.; Boos, N. Magnetic resonance classification of lumbar intervertebral disc degeneration. *Spine* **2001**, *26*, 1873–1878. [CrossRef] [PubMed]
35. Han, C.S.; Hancock, M.J.; Sharma, S.; Sharma, S.; Harris, I.A.; Cohen, S.P.; Magnussen, J.; Maher, C.G.; Traeger, A.C. Low back pain of disc, sacroiliac joint, or facet joint origin: A diagnostic accuracy systematic review. *eClinicalMedicine* **2023**, *59*, 101960. [CrossRef]
36. Kraemer, R.; Wild, A.; Haak, H.; Herdmann, J.; Krauspe, R.; Kraemer, J. Classification and management of early complications in open lumbar microdiscectomy. *Eur. Spine J.* **2003**, *12*, 239–246. [CrossRef]
37. Fujiwara, A.; Tamai, K.; Yamato, M.; An, H.S.; Yoshida, H.; Saotome, K.; Kurihashi, A. The relationship between facet joint osteoarthritis and disc degeneration of the lumbar spine: An MRI study. *Eur. Spine J.* **1999**, *8*, 396–401. [CrossRef]
38. Meyerding, H. Spondylolisthesis. *Surg. Gynecol. Obs.* **1932**, *54*, 371–377.
39. Modic, M.T.; Steinberg, P.M.; Ross, J.S.; Masaryk, T.J.; Carter, J.R. Degenerative disk disease: Assessment of changes in vertebral body marrow with MR imaging. *Radiology* **1988**, *166*, 193–199. [CrossRef]
40. Schizas, C.; Theumann, N.; Burn, A.; Tansey, R.; Wardlaw, D.; Smith, F.W.; Kulik, G. Qualitative Grading of Severity of Lumbar Spinal Stenosis Based on the Morphology of the Dural Sac on Magnetic Resonance Images. *Spine* **2010**, *35*, 1919–1924. [CrossRef] [PubMed]
41. Krämer, J.; Ludwig, J. Surgical treatment of lumbar disc herniation. Indication and methods. *Orthopade* **1999**, *28*, 579–584. [CrossRef]
42. Tomczak, M.; Tomczak-Łukaszewska, E. The need to report effect size estimates revisited. An overview of some recommended measures of effect size. *Sport Sci.* **2014**, *21*, 19–25.
43. Cohen, J.; Cohen, J.W. *Statistical Power Analysis for the Behavioral Sciences*, 2nd ed.; Erlbaum: Hillsdale, NJ, USA, 1988.
44. R Core Team. R: A Language and Environment for Statistical Computing. 2023. Available online: https://www.R-project.org/ (accessed on 3 June 2024).
45. Posit Team. RStudio: Integrated Development Environment for R. 2024. Available online: http://www.posit.co/ (accessed on 3 June 2024).
46. Liuke, M.; Solovieva, S.; Lamminen, A.; Luoma, K.; Leino-Arjas, P.; Luukkonen, R.; Riihimäki, H. Disc degeneration of the lumbar spine in relation to overweight. *Int. J. Obes.* **2005**, *29*, 903–908. [CrossRef]
47. Chen, X.-l.; Li, X.-y.; Wang, Y.; Lu, S.-b. Relation of lumbar intervertebral disc height and severity of disc degeneration based on Pfirrmann scores. *Heliyon* **2023**, *9*, e20764. [CrossRef] [PubMed]
48. Brinjikji, W.; Diehn, F.E.; Jarvik, J.G.; Carr, C.M.; Kallmes, D.F.; Murad, M.H.; Luetmer, P.H. MRI Findings of Disc Degeneration are More Prevalent in Adults with Low Back Pain than in Asymptomatic Controls: A Systematic Review and Meta-Analysis. *AJNR Am. J. Neuroradiol.* **2015**, *36*, 2394–2399. [CrossRef] [PubMed]

49. Watanabe, T.; Otani, K.; Sekiguchi, M.; Konno, S.I. Relationship between lumbar disc degeneration on MRI and low back pain: A cross-sectional community study. *Fukushima J. Med. Sci.* **2022**, *68*, 97–107. [CrossRef] [PubMed]
50. Rahyussalim, A.J.; Zufar, M.L.L.; Kurniawati, T. Significance of the Association between Disc Degeneration Changes on Imaging and Low Back Pain: A Review Article. *Asian Spine J.* **2020**, *14*, 245–257. [CrossRef] [PubMed]
51. Suri, P.; Naeini, M.K.; Heagerty, P.J.; Freidin, M.B.; Smith, I.G.; Elgaeva, E.E.; Compte, R.; Tsepilov, Y.A.; Williams, F.M.K. The association of lumbar intervertebral disc degeneration with low back pain is modified by underlying genetic propensity to pain. *Spine J.* **2025**, *25*, 8–17. [CrossRef]

bioengineering

MDPI

Article

Association of Spinopelvic Anatomy with the Level of Lumbar Disc Herniation

Jannis Löchel [1], Moritz Hanisch [2,*], Justus Bürger [2], Kirsten Labbus [2] and Robert Zahn [2,3]

1. Meoclinic Berlin, MEOSPINE, Friedrichstraße 71, 10117 Berlin, Germany
2. Center for Musculoskeletal Surgery, Charité-Universitätsmedizin Berlin, Corporate Member of Freie Universität Berlin, Humboldt-Universität zu Berlin, and Berlin Institute of Health, 13353 Berlin, Germany; justus.buerger@icloud.com (J.B.); kirsten.labbus@charite.de (K.L.)
3. Clinic for Orthopaedics, Ulm University Hospital, Campus Nord Oberer Eselsberg, Oberer Eselsberg 45, 89081 Ulm, Germany

* Correspondence: moritz.hanisch@charite.de

Abstract

Aim: The aim of this study was to investigate the association between the level of lumbar disc herniation (LDH) and individual spinopelvic anatomy. Material and methods: Spinopelvic parameters were retrospectively evaluated in 57 patients with symptomatic LDH at L4/5 and L5/S1 undergoing minimal invasive sequestrectomy at our institution. LDH was diagnosed in 23 patients at L5/S1 and in 34 patients at L4/5. Patients with further segment degeneration at the index level were excluded from the study. Results: Spinopelvic parameters between the two groups were significantly different. Patients with LDH at L5/S1 had statistically significant lower Pelvic Incidence (PI), Pelvic Tilt (PT), Relative Lumbar Lordosis (RLL) and PI-LL than patients with LDH at L4/5. C7 Sagittal Vertical Axis (C7SVA) was statistically significant lower in patients with LDH at L5/S1. Both groups had no sagittal imbalance. Patients with LDH at L5/S1 were significantly younger than patients with LDH at L4/5. There was a significant positive correlation between age and PT. We observed no significant differences for preoperative values of Lumar Lordosis (LL) and Sacral Slope between the two groups. Conclusions: This is the first study to reveal individual spinopelvic anatomy and, in particular, PI to be associated with the distinct level of LDH. These findings substantiate the biomechanical influence of the sagittal profile on the pathogenesis of LDH. Individual spinopelvic compensatory mechanisms were available independently of the patient's age. Minimal invasive sequestrectomy is a reliable treatment for symptomatic LDH without further segment degeneration.

Keywords: spinopelvic alignment; spinopelvic anatomy; lumbar disc herniation; level

1. Introduction

Lumbar spinal degeneration is a common condition in the aging population. Degenerative disc disease (DDD) and lumbar disc herniation (LDH) represent early manifestations of segmental degeneration. The progression of spinal degeneration can lead to more complex pathologies such as osteochondrosis (OC), spondylarthrosis, spinal stenosis, spondylolisthesis, and adult spinal deformities [1].

Despite significant advancements in understanding the genesis of several spinal pathologies, the specific impact of individual spinopelvic anatomy and resulting biomechanical factors on the development at the most affected levels of LDH has not been investigated in detail until now.

The influence of spinopelvic anatomy and biomechanics is increasingly recognized as a determinant in the development and progression of degenerative spinal conditions [2,3]. Among the parameters of the spinal sagittal profile, Pelvic Incidence (PI) stands out as the invariable anatomical parameter, influencing the overall spinal shape and load distribution [4,5]. Prior research has demonstrated the influence of PI, along with other parameters such as Pelvic Tilt (PT), Sacral Slope, Lumbar Lordosis (LL), and C7 Sagittal Vertical Axis (C7SVA), on maintaining sagittal balance and adapting to spinal pathologies [4–8]. A high PI with high LL leads to a higher load on posterior parts of the spine and was observed to be associated with an increased risk for the development of spondylarthrosis and degenerative spondylolisthesis [2,3,7,9,10]. In comparison, the occurrence of DDD or OC is conditioned by higher load on anterior parts of the spine at low PI and minor LL [5,11,12]. In a previous study, it could be shown that PI plays an important role concerning the level of occurrence and severity of isthmic spondylolisthesis (iSPL) and that spinopelvic anatomy determines the pathogenesis of iSPL [13].

While PI is a static anatomical parameter, the spinopelvic alignment is dynamic and capable of compensatory mechanisms to preserve sagittal balance in degenerative spinal pathologies, including LDH [14]. A recent longitudinal study showed that the magnitude of PI significantly influences compensatory mechanisms associated with age-related deterioration of spinopelvic alignment [15].

The pathogenesis of LDH is multifactorial and the most affected segments are L4/5 and L5/S1 in 90% [16–19]. An association between a low PI and the occurrence of LDH in younger individuals and a trend towards the affection of upper lumbar segments in older patients have been observed [5,12,20–22]. Knowledge of the influence of individual spinopelvic interactions on the manifestation of LDH would give a more detailed understanding of the complex interactions between individual anatomy and resulting biomechanics of the spine in a common spinal pathology.

The aim of this study was therefore to investigate the association between spinopelvic anatomy with a comprehensive set of spinopelvic parameters including PI, PT, and Sacral Slope and the distinct level of LDH.

2. Materials and Methods

2.1. Study Design

The study was designed as a single-center, retrospective clinical study. Patients with symptomatic LDH at L4/5 and L5/S1 who underwent minimal invasive sequestrectomy at our institution (university hospital) between 2016 and 2023 were included in this study. Surgical intervention was specifically indicated for patients experiencing radicular symptoms due to LDH that had proven unresponsive to a period of comprehensive conservative treatment. To maintain a clear focus on isolated LDH and its relationship with spinopelvic anatomy, patients with any evidence of further degenerative changes at the index level were excluded from the study. Additional exclusion criteria encompassed a history of previous spinal surgery, multilevel LDH, spinal tumor, spinal trauma, or presentation as an emergency surgery, all of which could confound the analysis of inherent spinopelvic influences on LDH pathogenesis.

The study was approved by the local ethics committee prior to the initiation (EA 1/342/21).

2.2. Radiographic Analysis

Diagnosis of LDH was routinely confirmed by preoperative magnetic resonance imaging (MRI). To assess global spinal alignment and individual spinopelvic morphology, all patients underwent standardized preoperative standing lateral full spine radiographs.

Patients were positioned with their knees extended and arms supported during radiography. This protocol was specifically designed to ensure that radiographs were reproducible across examinations and to allow for pre- and postoperative comparative measurements. Spinopelvic parameters as defined by PT, PI, Sacral Slope, LL, C7SVA, PI-LL were manually measured preoperatively using the SurgiMap Spine software version 2.3.2.1 (Nemaris Inc.; New York, NY, USA). To maintain measurement consistency and reliability, all parameters were assessed by a single experienced spine surgeon. The methodologies for measuring each parameter were precisely defined: PI was assessed with the angle formed by a perpendicular line from the midpoint of the sacral endplate and a line connecting this midpoint to the center of the femoral head. PI represents a fixed anatomical pelvic shape. PT was defined as the angle between the plumb line and the line connecting the center of the femoral heads to the midpoint of the sacral endplate. The Sacral Slope was measured as the angle between superior endplate of the S1 vertebra and a horizontal reference line. LL was measured using the Cobb angle method. LL was measured as the angle between the superior endplate of L1 and the superior endplate of S1. C7SVA was established as the horizontal distance between a plumb line dropped from the center of the C7 vertebral body and the posterosuperior corner of the S1 vertebral body, serving as an indicator of global sagittal balance. Additionally, PI-LL mismatch was calculated as the absolute difference between PI and LL (PI-LL). Relative Lumbar Lordosis (RLL) was calculated by subtracting measured LL from ideal LL, with ideal LL being dependent on the patient's pelvic incidence (PI). RLL was calculated using this formula: RLL = L1 − S1 Lordosis minus ($0.62 \times$ PI + 29) [23,24]. All measurements were performed by an experienced spine surgeon using the SurgiMap Spine software (Nemaris Inc.; New York, NY, USA). Previously high intra- and interobserver reliability was shown for all sagittal parameters [3,6,7,25].

2.3. Statistical Analysis

Descriptive statistics were calculated for all demographic and radiographic parameters. Continuous variables were presented as mean standard deviation, while categorical variables were expressed as frequencies and percentages. We used the Kolmogorov–Smirnov test to assess normal distribution of data. To facilitate comparisons between the two groups (L4/5 LDH vs. L5/S1 LDH) independent samples t-tests were employed for normally distributed continuous variables. For non-parametric data we used the Mann–Whitney U test. For evaluation of associations between continuous variables Spearman's rank correlation coefficient was utilized. A p-value of < 0.05 was considered statistically significant. All statistical analyses were performed using SPSS Statistics software, Version 27 (IBM Corp., Armonk, NY, USA).

3. Results

A total of 57 patients was included in the study, of whom 32 were women and 25 were men. The overall mean age of the study population was 48.0 ± 12.8 years. Group allocation was performed according to the level of the affected spinal segment.

The L5/S1 group consisted of 23 patients, accounting for 40.4% of the total patient collective. The mean age of individuals in this group was 43.4 ± 12.4 years. Sex distribution within the L5/S1 group was composed of 13 women (56.5%) and 10 men (43.5%). The L4/L5 group included 34 patients (59.6%). Patients in this group had a mean age of 50.0 ± 11.7 years including 19 (55.9%) women and 15 (44%) men.

The mean age for men in the L4/L5 group was 53.4 years, compared to 43.9 years in the L5/S1 group. For women, the mean age was 47.3 years in the L4/L5 group and 42.9 years in the L5/S1 group. The distribution of sex did not differ significantly between the L4/L5 and L5/S1 groups ($\chi^2 = 0.002$, $p = 0.962$). There were no significant differences between age distribution in both groups calculated with the Mann–Whitney U Test (L5/S1—$p = 0.56$, U = 74.5; L4/L5—$p = 0.07$, U = 193.5) The detailed analysis of preoperative radiographic parameters according to group is listed in Table 1.

Table 1. Preoperative spinopelvic parameters. PI: pelvic incidence; PT: pelvic tilt; SS: sacral slope; LL: lumbar lordosis; RLL: relative lumbar lordosis; SVA: sagittal vertical axis. Standard deviation (±); *p*: statistical significance of values.

Segment		PI [°]	PT [°]	SS [°]	LL [°]	RLL [°]	SVA [mm]
L4/5 (*n* = 34)	Mean	52.1 (±9.2)	17.8 (±8.3)	34 (±9.7)	45.8 (±14.4)	−15.5 (±12.4)	49 (±41.5)
L5/S1 (*n* = 23)	Mean	44.7 (±8)	11 (±7.2)	33.6 (±7.3)	48.2 (±9.3)	−8.6 (±9.6)	22.7 (±41.7)
	p	0.004	0.006	0.67	0.33	0.026	0.019
Mean overall (*n* = 57)	Mean	50 (±9.8)	15.9 (±9.6)	34.3 (±8.8)	46.8 (±13)	−13.3 (±12.6)	39.4 (±43)

Diagram 1. Box plot diagram for L4/5 and L5/S1 group comparison for selected spinopelvic parameters given in degrees.

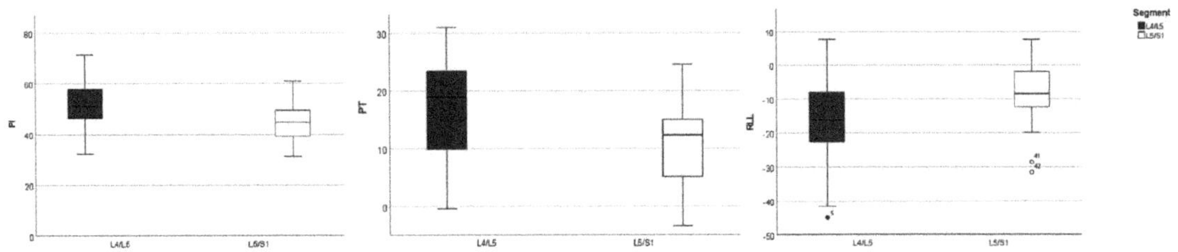

The mean PI of patients with LDH at L5/S1 was 44.7 ± 8.0° and was statistically significantly lower compared to the PI of 52.1 ± 9.2° for patients with LDH at L4/5 ($p = 0.004$). Similarly, the mean PI of patients with LDH at L5/S1 was significantly lower with 11 ± 8° compared to patients of the L4/5 group with 17.8 ± 8.3° ($p = 0.006$). The RLL showed significantly lower values for the L4/5 group (−15.5 ± 12.4°) compared to patients of the L5/S1 group (−8.6 ± 9.6°; $p = 0.026$). The PI-LL mismatch was significantly lower in the L5/S1 group, compared to patients with LDH at L4/5 (Spearman Rho Correlation Index= −0.380, $p = 0.004$). C7SVA was significantly lower for patients with LDH at L5/S1 (22.7 ± 41.7 mm) compared to patients with LDH at L4/5 (49 ± 41.5 mm; $p = 0.019$), whereas both groups had no sagittal imbalance with C7SVA greater than 100 mm. Patient age correlated positively with PT (Spearman Rho Correlation Index= −0.298, with $p = 0.026$). We observed no significant differences for Sacral Slope and LL between the two groups. A Mann–Whitney U Test was performed for subgroup analysis. The subgroup analysis by sex showed no significant differences for all spinopelvic parameters for the L4/5 group. In the L5/S1 group female patients showed significantly higher Sacral Slope compared to men ($p = 0.022$, effect size 0.51). There were no further significant sex-associated differences in this group. Figures 1 and 2 illustrate the routinely performed preoperative diagnostics and workup.

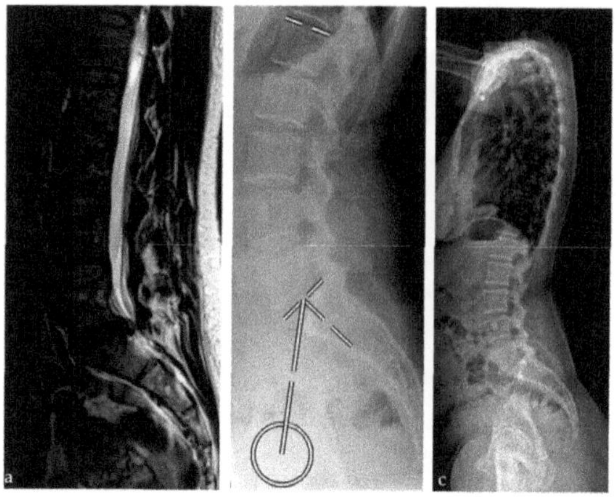

Figure 1. Patient example illustrating lumbar disc herniation at L5/S1 and mean spinopelvic parameters: pelvic incidence: 44.7°, lumbar lordosis: 48.2°, and relative lumbar lordosis: −8.6° displayed in an (**a**) MRI (**b**) lateral lumbar spine radiograph (**c**) standing lateral full spine radiograph.

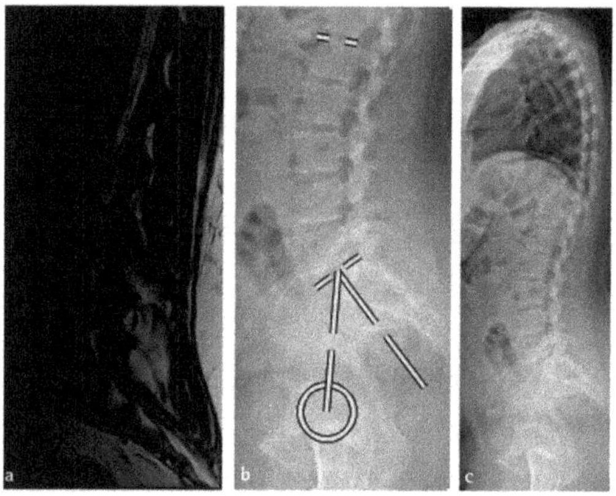

Figure 2. Patient example illustrating lumbar disc herniation at L4/L5 and mean spinopelvic parameters: pelvic incidence 52.1°, lumbar lordosis 45.8° and a relative lumbar lordosis of −15.5° displayed in an (**a**) MRI (**b**) lateral lumbar spine radiograph (**c**) standing lateral full spine radiograph.

4. Discussion

In this study, we investigated the association between the distinct level of LDH and spinopelvic anatomy. For the first time, it could be shown that the distinct level of LDH is associated with the extent of PI.

The influence of spinopelvic anatomy and biomechanics on the genesis of advanced patterns of spinal degeneration has been shown previously [2,26,27]. To our knowledge, there is no previously published study investigating the association between the distinct level of LDH as a pathology of initial segment degeneration and sagittal profile.

The etiology of LDH is multifactorial. Genetic predisposition, advancing age, sex, obesity, and specific lifestyle elements have been identified as risk factors for the development of LDH [1,4,16,18,19,27]. However, our findings emphasize the clinically relevant impact of spinopelvic anatomy and biomechanics on the pathogenesis of LDH.

In previous studies, a low PI and the sagittal profile types I and II of the Roussouly classification were identified as a risk factor for the development of LDH in younger patients [20,22].

In our study, LDH was likewise observed in patients with a low PI and spinopelvic anatomy according to Roussouly types I and II [1,4,5]. Types I and II of the Roussouly classification are characterized by Sacral Slopes values < 35 degrees associated with low to moderate PI. These types showed reduced load-bearing capacity and higher intradiscal pressures in movement in a finite element model [28]. The results of our clinical study support these findings. Just as LDH, symptomatic iSPL most commonly occurs in L5/S1 and L4/5 [29]. A previous study revealed the biomechanical impact of spinopelvic anatomy and, in particular, PI on the pathogenesis of iSPL. Patients with iSPL at L5/S1 had significantly higher PI values than patients with iSPL at L4/5. A higher PI could lead to increased shear stress in the predisposed segment and condition the observed higher grade of slippage of iSPL in L5/S1 [13].

These findings all correspond to the model of sagittal profile with the relevant impact of spinopelvic biomechanics in the development of spinal pathologies.

A low PI associated with low LL may predispose individuals to LDH due to increased anterior loading and elevated pressure on the intervertebral discs [5,30]. A high PI with increased LL may lead to iSPL as a result of elevated shear forces, which can promote facet joint degeneration and vertebral slippage [2,9,10,31,32]. These results and the results of previous studies indicate that specific mono- or multisegmental spinal pathologies are influenced and conditioned by distinct spinopelvic biomechanical forces.

A high PI and corresponding high LL allow for greater spinopelvic compensation mechanisms [3,33]. Despite significant differences in PI between the two groups in this study, there was no statistically significant difference in the preoperative mean values of LL between the two groups. This finding suggests the existence of a different capacity for compensation mechanisms. Patients with higher PI and LDH at L4/5 presented increased compensatory adaptation. These findings highlight the importance of RLL when assessing LL in the context of sagittal balance in patients with degenerative lumbar pathologies. Compensatory mechanisms have been observed in other degenerative spinal pathologies before and are suggested to exist in patients with LDH. The reduction in lordosis relative to the individual spinopelvic anatomy proves the existence of compensatory mechanisms in patients with LDH [34]. One of the key mechanisms of compensation is an increased PT. This is concordant with other studies investigating compensatory mechanisms in patients with chronic lower back pain [5]. Interestingly, patients with LDH at L4/5 were significantly older than patients with LDH at L5/S1. This supports previously published results, showing a trend towards more cranial segments for LDH occurrence with increasing age [21].

The findings of our study demonstrate the existence of compensatory mechanisms in older patients with LDH. The age difference between the two groups emphasizes the importance of biomechanics of the spine on the pathogenesis and compensatory adaptations after LDH.

Adaptation to degenerative spinal disorders and compensatory mechanisms are in focus of actual musculoskeletal research. A recently published study demonstrated that individuals with high PI compensate by decreasing thoracic kyphosis, while those with low PI show an increase in LL. This aligns with our observation that patients with LDH at L5/S1, which are associated with lower PI, exhibited reduced compensatory reduction in LL compared to those with L4/5 LDH with higher PI. Compensatory mechanisms are shaped by individual spinopelvic parameters. These findings proof that even in early degenerative conditions like LDH, dynamic compensatory mechanisms exist, allowing patients to maintain global sagittal balance [15]. All patients of the study had a SVA < 10 mm and therefore no sagittal imbalance, which can even be observed in LDH patients [35].

The comparison between female and male patients showed no statistically significant differences for the analyzed spinopelvic parameters except for a higher Sacral Slope for women in the L5/S1 group. Compensatory changes in sagittal profile in LDH influence the variable parameters of sagittal profile masking underlying sex differences. These compensatory mechanisms do not include PI, as a fixed anatomical parameter.

Limitations of this study can be seen in the retrospective design and the relatively small study population. The small study population resulted from the strict exclusion of patients with further degeneration of the affected segment and the exclusive consideration of surgically treated patients. The homogeneous cohort selection allowed focused analysis without the confounder of segment degeneration. Future longitudinal studies are needed to evaluate the long-term biomechanical influence of spinopelvic parameters on the genesis of specific spinal pathologies and their clinical implications.

5. Conclusions

This study provides novel insights into the pathogenesis of LDH, demonstrating for the first time a significant association between the distinct level of LDH and individual spinopelvic anatomy, particularly PI. Our findings indicate that specific inherent sagittal profiles may predispose individuals to LDH at the L4/5 versus the L5/S1 segment. PI differs between individuals with LDH at L4/5 compared to L5/S1. Compensatory mechanisms in patients with LDH exist independently of age. These findings underscore the relevant biomechanical influence of the sagittal profile on the development of LDH and highlight the clinical relevance of understanding individual spinopelvic anatomy. While sequestrectomy remains a reliable and commonly practiced treatment for symptomatic LDH without further segment degeneration, our knowledge of the underlying biomechanical predispositions could pave the way for establishing more targeted preventive measures for LDH in the future.

Author Contributions: Conceptualization, R.Z. and J.L.; methodology, R.Z. and J.L.; writing—original draft preparation, J.L., R.Z., K.L. and M.H.; writing—review and editing, M.H. and R.Z.; visualization, M.H. and J.B.; supervision, R.Z. and K.L. All authors have read and agreed to the published version of the manuscript.

Funding: We acknowledge support from the German Research Foundation (DFG) and the Open Access Publication Fund of Charité–Universitätsmedizin Berlin.

Informed Consent Statement: The study was approved by the local ethics committee (EA 1/342/21)—due to the retrospective nature of this study, informed consent was not deemed necessary.

Data Availability Statement: The data are not publicly available due to privacy.

Conflicts of Interest: The authors declare no conflicts of interest.

Abbreviations

The following abbreviations are used in this manuscript:

PI	Pelvic Incidence
PT	Pelvic Tilt
LL	Lumbar Lordosis
SS	Sacral Slope
RLL	Relative Lumbar Lordosis
SVA	Sagittal Vertical Axis
iPSL	Isthmic spondylolisthesis
C7SVA	C7 Sagittal Vertical Axis
LDH	Lumbar disc herniation

References

1. Kirnaz, S.; Capadona, C.; Wong, T.; Goldberg, J.L.; Medary, B.; Sommer, F.; McGrath, L.B., Jr.; Härtl, R. Fundamentals of Intervertebral Disc Degeneration. *World Neurosurg.* **2022**, *157*, 264–273. [CrossRef]
2. Fasser, M.R.; Furrer, P.R.; Fisler, L.; Urbanschitz, L.; Snedeker, J.G.; Farshad, M.; Widmer, J. The triadic relationship between spinal posture, loading, and degeneration. *Front. Bioeng. Biotechnol.* **2025**, *13*, 1444540. [CrossRef]
3. Roussouly, P.; Pinheiro-Franco, J.L. Biomechanical analysis of the spino-pelvic organization and adaptation in pathology. *Eur. Spine J.* **2011**, *20*, 609. [CrossRef]
4. Menezes-Reis, R.; Bonugli, G.P.; Dalto, V.F.; da Silva Herrero, C.F.P.; Defino, H.L.A.; Nogueira-Barbosa, M.H. Association Between Lumbar Spine Sagittal Alignment and L4–L5 Disc Degeneration Among Asymptomatic Young Adults. *Spine* **2016**, *41*, E1081–E1087. [CrossRef]
5. Barrey, C.; Jund, J.; Noseda, O.; Roussouly, P. Sagittal balance of the pelvis-spine complex and lumbar degenerative diseases. A comparative study about 85 cases. *Eur. Spine J.* **2007**, *16*, 1459–1467. [CrossRef] [PubMed]
6. Roussouly, P.; Gollogly, S.; Berthonnaud, E.; Dimnet, J. Classification of the normal variation in the sagittal alignment of the human lumbar spine and pelvis in the standing position. *Spine* **2005**, *30*, 346–353. [CrossRef]
7. Strube, P.; Pumberger, M.; Sonnow, L.; Zippelius, T.; Nowack, D.; Zahn, R.K.; Putzier, M. Association Between Lumbar Spinal Degeneration and Anatomic Pelvic Parameters. *Clin. Spine Surg.* **2018**, *31*, 263–267. [CrossRef] [PubMed]
8. Legaye, J.; Duval-Beaupère, G.; Hecquet, J.; Marty, C. Pelvic incidence: A fundamental pelvic parameter for three-dimensional regulation of spinal sagittal curves. *Eur. Spine J.* **1998**, *7*, 99–103. [CrossRef]
9. Sahin, M.S.; Ergün, A.; Aslan, A. The Relationship Between Osteoarthritis of the Lumbar Facet Joints and Lumbosacropelvic Morphology. *Spine* **2015**, *40*, E1058–E1062. [CrossRef]
10. Soydan, Z.; Bayramoglu, E.; Altas, O. The Impact of Spinopelvic Alignment on the Facet Joint Degeneration. *Glob. Spine J.* **2025**, *15*, 306–313. [CrossRef] [PubMed]
11. Bae, J.; Lee, S.H.; Shin, S.H.; Seo, J.S.; Kim, K.H.; Jang, J.S. Radiological analysis of upper lumbar disc herniation and spinopelvic sagittal alignment. *Eur. Spine J.* **2016**, *25*, 1382–1388. [CrossRef]
12. Wei, X.; Gengwu, L.; Chao, C.; Yifan, L.; Shang, S.; Ruixi, H.; Yunhan, J.; Xiaodong, Z.; Zhikun, L. Correlations between the sagittal plane parameters of the spine and pelvis and lumbar disc degeneration. *J. Orthop. Surg. Res.* **2018**, *13*, 137. [CrossRef] [PubMed]
13. Labbus, K.; Bürger, J.; Löchel, J.; Schäfer, F.M.; Putzier, M.; Zahn, R.K. Impact of Individual Spinopelvic Anatomy on the Localization and Severity of Symptomatic Isthmic Spondylolisthesis. *Glob. Spine J.* **2024**, *14*, 2311–2316. [CrossRef] [PubMed]
14. Yokoyama, K.; Tanaka, H.; Ito, Y.; Yamada, M.; Sugie, A.; Wanibuchi, M.; Kawanishi, M. Analgesic Posture and Pelvic Morphology in Patients with Lumbar Disc Herniation. *World Neurosurg.* **2021**, *147*, e411–e415. [CrossRef]
15. Yamato, Y.; Hasegawa, T.; Yoshida, G.; Banno, T.; Arima, H.; Oe, S.; Ide, K.; Yamada, T.; Murakami, Y.; Matsuyama, Y. The effect of pelvic incidence on age-related changes in spinopelvic sagittal alignment in older individuals: A longitudinal study for 10 years. *Eur. Spine J.* **2025**. [CrossRef] [PubMed]
16. Hadjipavlou, A.G.; Tzermiadianos, M.N.; Bogduk, N.; Zindrick, M.R. The pathophysiology of disc degeneration: A critical review. *J. Bone Jt. Surg. Br.* **2008**, *90*, 1261–1270. [CrossRef]
17. Zukowski, L.A.; Falsetti, A.B.; Tillman, M.D. The influence of sex, age and BMI on the degeneration of the lumbar spine. *J. Anat.* **2012**, *220*, 57–66. [CrossRef]
18. Farshad-Amacker, N.A.; Hughes, A.P.; Aichmair, A.; Herzog, R.J.; Farshad, M. Determinants of evolution of endplate and disc degeneration in the lumbar spine: A multifactorial perspective. *Eur. Spine J.* **2014**, *23*, 1863–1868. [CrossRef]
19. Hedman, T.; Rogers, A. Pathomechanics of Early-Stage Lumbar Intervertebral Disc Degradation Leading to Discogenic Pain—A Narrative Review. *Bioengineering* **2025**, *12*, 389. [CrossRef]
20. Beck, J.; Brisby, H.; Baranto, A.; Westin, O. Low lordosis is a common finding in young lumbar disc herniation patients. *J. Exp. Orthop.* **2020**, *7*, 38. [CrossRef]
21. Dammers, R.; Koehler, P.J. Lumbar disc herniation: Level increases with age. *Surg Neurol.* **2002**, *58*, 209–212; discussion 212–213. [CrossRef]
22. Ould-Slimane, M.; Ferracci, F.X.; Gillibert, A.; Szadkowski, M.; Lesage, C.; Vieira, T.D.; Sacco, R.; d'Astorg, H. Sagittal alignment of the spine and lumbar disc herniation in young adults: A historical, case-control study. *Orthop. Traumatol. Surg. Res.* **2025**, *111*, 104219. [CrossRef] [PubMed]
23. Yilgor, C.; Sogunmez, N.; Yavuz, Y.; Abul, K.; Boissiére, L.; Haddad, S.; Obeid, I.; Kleinstück, F.; Sánchez Pérez-Grueso, F.J.; Acaroğlu, E.; et al. European Spine Study Group. Relative lumbar lordosis and lordosis distribution index: Individualized pelvic incidence-based proportional parameters that quantify lumbar lordosis more precisely than the concept of pelvic incidence minus lumbar lordosis. *Neurosurg. Focus.* **2017**, *43*, E5. [CrossRef]
24. Yilgor, C.; Sogunmez, N.; Boissiere, L.; Yavuz, Y.; Obeid, I.; Kleinstück, F.; Pérez-Grueso, F.J.S.; Acaroglu, E.; Haddad, S.; Mannion, A.F.; et al. European Spine Study Group (ESSG). Global Alignment and Proportion (GAP) Score: Development and Validation of

a New Method of Analyzing Spinopelvic Alignment to Predict Mechanical Complications After Adult Spinal Deformity Surgery. *J. Bone Joint Surg. Am.* **2017**, *99*, 1661–1672. [CrossRef]
25. Löchel, J.; Putzier, M.; Dreischarf, M.; Grover, P.; Urinbayev, K.; Abbas, F.; Labbus, K.; Zahn, R. Deep learning algorithm for fully automated measurement of sagittal balance in adult spinal deformity. *Eur. Spine J.* **2024**, *33*, 4119–4124. [CrossRef]
26. Fei, H.; Li, W.S.; Sun, Z.R.; Ma, Q.W.; Chen, Z.Q. Analysis of Spino-pelvic Sagittal Alignment in Young Chinese Patients with Lumbar Disc Herniation. *Orthop. Surg.* **2017**, *9*, 271–276. [CrossRef] [PubMed]
27. Iatridis, J.C.; Nicoll, S.B.; Michalek, A.J.; Walter, B.A.; Gupta, M.S. Role of biomechanics in intervertebral disc degeneration and regenerative therapies: What needs repairing in the disc and what are promising biomaterials for its repair? *Spine J.* **2013**, *13*, 243–262. [CrossRef]
28. Wang, W.; Pei, B.; Wu, S.; Lu, D.; He, P.; Ma, C.; Wu, X. Biomechanical responses of human lumbar spine and pelvis according to the Roussouly classification. *PLoS ONE* **2022**, *17*, e0266954. [CrossRef]
29. Sonne-Holm, S.; Jacobsen, S.; Rovsing, H.C.; Monrad, H.; Gebuhr, P. Lumbar spondylolysis: A life long dynamic condition? A cross sectional survey of 4.151 adults. *Eur. Spine J.* **2007**, *16*, 821–828. [CrossRef] [PubMed]
30. Galbusera, F.; Brayda-Bruno, M.; Costa, F.; Wilke, H.J. Numerical evaluation of the correlation between the normal variation in the sagittal alignment of the lumbar spine and the spinal loads. *J. Orthop. Res.* **2014**, *32*, 537–544. [CrossRef]
31. Jentzsch, T.; Geiger, J.; Bouaicha, S.; Slankamenac, K.; Nguyen-Kim, T.D.; Werner, C.M. Increased pelvic incidence may lead to arthritis and sagittal orientation of the facet joints at the lower lumbar spine. *BMC Med. Imaging* **2013**, *13*, 34. [CrossRef] [PubMed]
32. Filardi, V.; Simona, P.; Cacciola, G.; Bertino, S.; Soliera, L.; Barbanera, A.; Pisani, A.; Milardi, D.; Alessia, B. Finite element analysis of sagittal balance in different morphotype: Forces and resulting strain in pelvis and spine. *J. Orthop.* **2017**, *14*, 268–275. [CrossRef]
33. Vaz, G.; Roussouly, P.; Berthonnaud, E.; Dimnet, J. Sagittal morphology and equilibrium of pelvis and spine. *Eur. Spine J.* **2002**, *11*, 80–87. [CrossRef]
34. Endo, K.; Suzuki, H.; Tanaka, H.; Kang, Y.; Yamamoto, K. Sagittal spinal alignment in patients with lumbar disc herniation. *Eur. Spine J.* **2010**, *19*, 435–438. [CrossRef] [PubMed]
35. Liang, C.; Sun, J.; Cui, X.; Jiang, Z.; Zhang, W.; Li, T. Spinal sagittal imbalance in patients with lumbar disc herniation: Its spinopelvic characteristics, strength changes of the spinal musculature and natural history after lumbar discectomy. *BMC Musculoskelet. Disord.* **2016**, *17*, 305. [CrossRef] [PubMed]

bioengineering

MDPI

Article

Using Two X-Ray Images to Create a Parameterized Scoliotic Spine Model and Analyze Disk Stress Adjacent to Spinal Fixation—A Finite Element Analysis

Te-Han Wang [1], Po-Hsing Chou [2,3] and Chen-Sheng Chen [1,*]

[1] Department of Physical Therapy and Assistive Technology, National Yang Ming Chiao Tung University, Taipei 112304, Taiwan
[2] Department of Orthopaedics and Traumatology, Taipei Veterans General Hospital, Taipei 112201, Taiwan
[3] Department of Orthopedics, School of Medicine, National Yang Ming Chiao Tung University, Taipei 112304, Taiwan
* Correspondence: cschen0623@nycu.edu.tw; Tel.: +886-2-2826-7353

Abstract

Posterior instrumentation is used to treat severe adolescent idiopathic scoliosis (AIS) with a Cobb angle greater than 40 degrees. Clinical studies indicate that AIS patients may develop adjacent segment degeneration (ASD) post-surgery. However, there is limited research on the biomechanical effects on adjacent segments after surgery, and straightforward methods for creating finite element (FE) models that reflect vertebral deformation are lacking. Therefore, this study aims to use biplanar X-ray images to establish a case-specific, parameterized FE model reflecting coronal plane vertebral deformation and employ FE analysis to compare pre- and postoperative changes in the range of motion (ROM), endplate stress, and intervertebral disk stress of adjacent segments. We developed an FE model from biplanar X-ray images of a patient with AIS, using ANSYS software to establish pre- and postoperative models. The shape of the preoperative model was validated using computed tomography (CT) reconstruction. A flexion moment was applied to C7 of the spine model to achieve the same forward bending angle in the pre- and postoperative models. This study successfully developed a case-specific parameterized FE model based on X-ray images. The differences between Cobb angle and thoracolumbar kyphosis angle measurements in X-ray images and CT reconstructions were 6.5 and 5.4 mm. This FE model was used to analyze biomechanical effects on motion segments adjacent to the fixation site, revealing a decrease in maximum endplate and disk stress in the cranial segment and an increase in stress in the caudal segment.

Keywords: scoliosis; finite element model; X-ray image; adjacent disk; biomechanics

1. Introduction

Adolescent idiopathic scoliosis (AIS) is a common spinal deformity with unknown etiology, characterized by spinal curvature along the coronal plane and associated spinal rotation. AIS typically affects adolescents aged 10 to 16 years, with an incidence rate of about 2–3% and a higher prevalence in female patients [1]. The severity of AIS is generally measured using the Cobb angle; when the angle exceeds 40 degrees, surgery is often recommended. Posterior instrumentation is the most common surgical method, using implanted metal devices to correct the abnormal curvature [2].

Despite the effectiveness of surgery in correcting scoliosis and improving patients' quality of life, clinical research indicates that adjacent segment degeneration (ASD) may occur post-operatively [3–6]. Potential risk factors for ASD include longer fusion segments [3], lower fixation levels [4,5], greater residual curvature, and larger post-operative thoracolumbar kyphosis (TLK) angles [6,7]. ASD often presents as compensatory increased ROM in unfused adjacent segments, leading to increased stress on intervertebral disks and endplates [8,9], which results in disk degeneration, pain, and functional impairments. An uneven stress distribution may also cause Modic changes in endplates [5,6], as well as degenerative lesions affecting the endplates. Although clinical research suggests that patients with AIS are at risk of ASD post-surgery, it remains unclear whether the surgery itself increases the ASD risk and how biomechanical effects on adjacent segments change before and after surgery, necessitating further research.

Finite element (FE) analysis is crucial for studying spinal biomechanics, enabling simulations based on a patient-specific parameterized FE model. Currently, the application of FE analysis in the field of AIS primarily includes examining the stress imbalance caused by scoliosis, the effects of the brace, and surgical treatment [10–15]. To assess stress imbalance, Zhang et al. [10] used CT-based lumbar spine models to show that flexion stress accumulates on the concave side of the scoliotic curve; Li et al. [12] used CT-based spine models to assess stress distribution and deformation under load in scoliotic spines. In brace treatment, FE analysis can evaluate how different force configurations and magnitudes affect scoliosis correction and brace performance, leading to optimized designs that make braces lighter and more effective. Chou et al. [13] found that rotating the thoracic pad by 20 degrees dorsally in a Boston brace provides optimal corrective effects; Ali et al. [14] demonstrated that using a soft brace instead of a traditional rigid brace allows for spinal mobility while correcting, helping to reduce common side effects of rigid braces, such as muscle atrophy and skin issues. FE analysis can assess the therapeutic outcomes and potential complications of surgical treatment. For instance, Somtua et al. [15] studied the stress on pedicle screws and strain on vertebrae in patients with different Cobb angles undergoing surgical correction, finding that, when the Cobb angle exceeds 40 degrees, the vertebral strain may increase the risk of fractures. Currently, however, there is limited FE analysis research focused on the biomechanical effects on adjacent segments before and after surgery for AIS.

Previous FE analysis studies have typically used computed tomography (CT) or X-ray imaging to reconstruct the spinal morphology. CT-based models can accurately reconstruct the three-dimensional (3D) spine but have limitations due to high radiation exposure and the inability to reconstruct soft tissues like intervertebral disks. X-ray imaging, conversely, involves lower radiation exposure and allows for 3D reconstruction through parameterization. However, parameterized models often struggle to replicate the vertebral deformities seen in severe scoliosis cases.

To investigate the biomechanical effects in adjacent joints before and after surgical treatment in AIS patients, and given that current FE models based on X-ray images cannot accurately simulate vertebral deformities in severe scoliosis, this study aims to develop a patient-specific parameterized FE model for AIS patients using X-ray imaging, incorporate coronal plane vertebral deformation to simulate pre- and postoperative biomechanics, and analyze the ROM and stress distribution changes in adjacent segments post-operatively.

2. Materials and Methods

2.1. Study Subject

This study recruited a patient with Lenke type 5 AIS and a Cobb angle of 52 degrees, who was eligible for posterior spinal instrumentation surgery. The study was approved by the Institutional Review Board of National Yang-Ming University (#YM111055F), and the participant provided written informed consent before enrollment. Based on the patient's biplanar X-ray images, a case-specific, parameterized FE model was constructed to simulate the changes in ROM and stress distribution pre- and post-surgery.

2.2. Parameterized Finite Element Model

The parameterized FE model was based on a lumbar spine model that has been validated and used in previous studies [16–19] and incorporated coronal plane vertebral deformation. The parameterization method followed the model developed by Chou et al. [13] and was implemented using ANSYS 14.5 with the parametric design language (APDL) (ANSYS, Inc., Houston, TX, USA). The model consists of the spinal section and implanted instrumentation. The spinal section includes 18 vertebrae from L5 to C7, 17 intervertebral disks, and associated ligaments. Each vertebra comprises cortical bone, cancellous bone, and posterior elements. The intervertebral disks include the nucleus pulposus, annulus fibrosus, disk fibers, and endplates. At the same time, the ligaments consist of anterior and posterior longitudinal ligaments, the interspinous ligament, the supraspinous ligament, the ligamentum flavum, the intertransverse ligament, and the facet joint capsules. Vertebral bodies were modeled using Solid45 elements, the nucleus pulposus with Fluid80 elements, and the annulus fibrosus and endplates with Solid45 elements, while the Link10 elements represented disk fibers and ligaments. The spinal model comprises 17,439 elements and 17,043 nodes. The material properties of the scoliotic FE model were listed in Table 1. The lumbar spine model has been validated in previous studies [17,20], showing stiffer behavior during flexion compared with in vitro cadaveric testing, with ROM values being 4° lower than those described in Rohlmann's in vitro study. Additionally, the model exhibited greater stiffness in extension and rotation, although the differences remained within 2°. The discrepancies between in vitro testing and FE simulations were within one standard deviation.

Table 1. Material properties of the scoliotic FE model [17].

Material	Young's Modulus (MPa)	Poisson's Ratio	Area (mm^2)
Bone			
Cortical	12,000	0.3	-
Cancellous	100	0.2	-
Posterior element	3500	0.25	-
Disc			
Nucleus pulposus	1	0.499	-
Ground Substance	4.2	0.46	-
Annulus Fibers	175	0.4	
Endplate	24	0.4	-
Ligament			
ALL	7.8		24
PLL	10		14.4
TL	10		3.6
LF	15		40
ISL	10		26
SSL	8		23
CL	7.5		30

ALL, anterior longitudinal ligament; CL, capsular ligament; ISL, interspinous ligament; LF, ligamentum flavum; PLL, posterior longitudinal ligament; SSL, supraspinous ligament; TL, transverse ligament.

The parameterized FE model of the spine was constructed using biplanar X-ray images, including posterior–anterior (PA) and lateral views, obtained with 3D Slicer version 4.10.2 [21]. The entire spine was reconstructed using a lumbar vertebral body L5. The relative location and angle of the whole spine were determined from two X-ray images. Sampling coordinates X0, Y0, and Z0 were extracted from the images (Figure 1) and transformed into FE model coordinates X, Y, and Z using Euler angle rotations in the Y-X-Z order (Equation (1)). The vertebral rotation was defined using the Nash–Moe method and judged by the location of the pedicle [22]. Adjustable parameters in the FE model include the following:

1. Coordinates of the superior posterior points of each vertebral body and vertebral body length along the three axes (Figure 2a);
2. Sagittal (θ) and transverse (φ) tilt angles of each vertebral body (Figure 2b);
3. Coronal tilt angles (β) of the superior and inferior endplates and their FE model (Figure 2c).

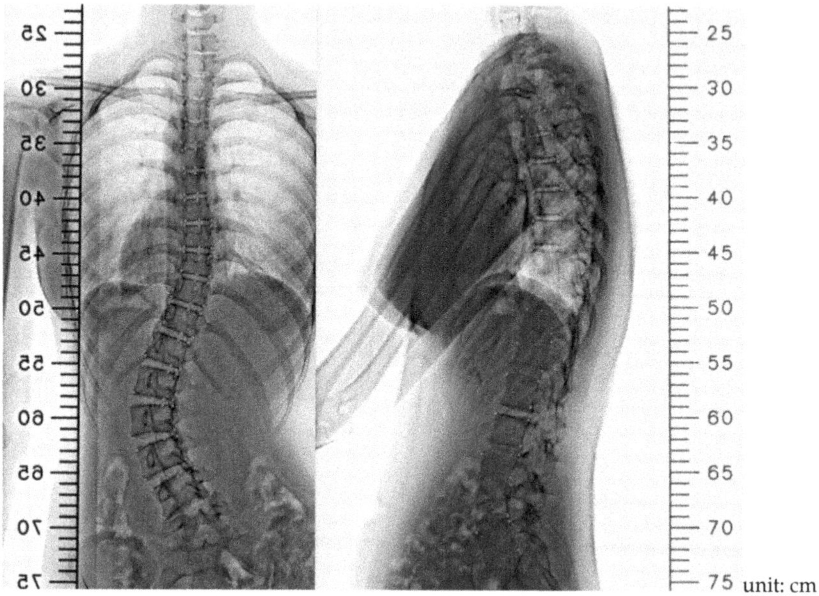

Figure 1. X-ray image parameters were obtained with a 3D Slicer in PA view and lateral view. Note: Pink markers indicate sampling points.

The user could input the above parameters to create the FE model of the entire spine automatically. The pre- (Figure 3a) and postoperative (Figure 3b) FE models, along with the instrumentation model (Figure 3c), were then established. Instrumentation includes pedicle screws and connecting rods. In the post-operative model, pedicle screws were placed bilaterally from T10 to T11 and L1 to L4, and unilaterally at T12. Two rods were used to secure screws on each side, with transverse connectors linking rods at T11, L2, and L3. The lumbar screws were 6.6 mm in diameter, the thoracic screws and rods were 5.5 mm, and the transverse connectors were 2 mm. Pedicle screws and rods were modeled using Beam188 elements, with 99 elements and 98 nodes. All materials were assumed to be linear and isotropic, with parameters based on previous studies [20,23].

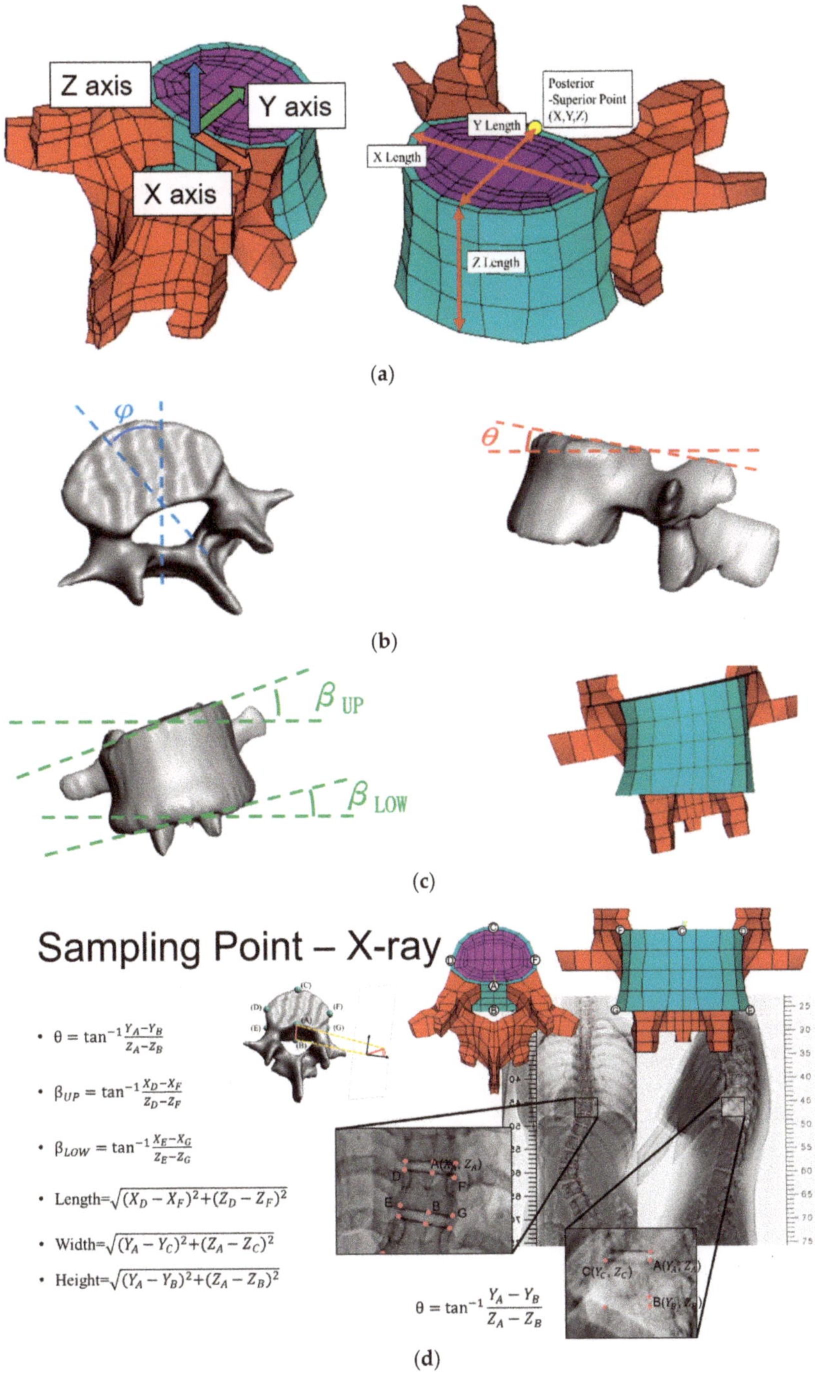

Figure 2. Adjustable spine parameters: (**a**) Coordinates of the superior posterior point and the vertebral body lengths along the three axes. (**b**) Sagittal tilt angle of the vertebral body and transverse tilt angle of the vertebral body. (**c**) Coronal tilt angles of superior and inferior endplates. (**d**) Calculation of angle and length from nodal locations. Note: 1. Sampling points included A, B, C, D, E, F, and G; 2. X, Y, Z indicate the coordinates.

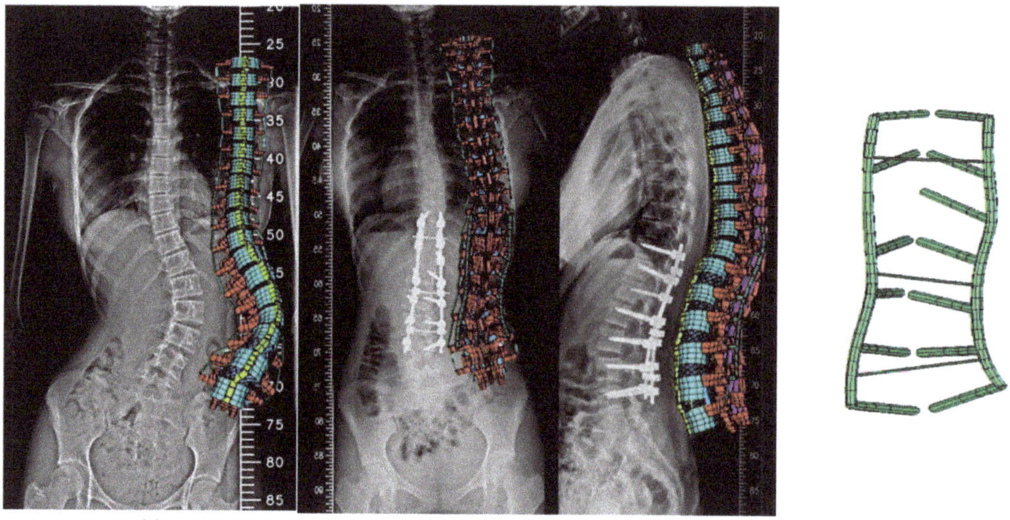

(**a**) (**b**) (c)

Figure 3. Posterior views of the case-specific FE model: (**a**) pre-op, (**b**) post-op, (**c**) instrumentation.

$$\begin{Bmatrix} \begin{bmatrix} X_1 \\ Y_1 \\ Z_1 \\ 1 \end{bmatrix} = \begin{bmatrix} \cos\beta & 0 & \sin\beta & 0 \\ 0 & 1 & 0 & 0 \\ -\sin\beta & 0 & \cos\beta & 0 \\ 0 & 0 & 0 & 1 \end{bmatrix} \begin{bmatrix} X_0 \\ Y_0 \\ Z_0 \\ 1 \end{bmatrix} \\ \begin{bmatrix} X_2 \\ Y_2 \\ Z_2 \\ 1 \end{bmatrix} = \begin{bmatrix} 1 & 0 & 0 & 0 \\ 0 & \cos\theta & -\sin\theta & 0 \\ 0 & \sin\theta & \cos\theta & 0 \\ 0 & 0 & 0 & 1 \end{bmatrix} \begin{bmatrix} X_1 \\ Y_1 \\ Z_1 \\ 1 \end{bmatrix} \\ \begin{bmatrix} X \\ Y \\ Z \\ 1 \end{bmatrix} = \begin{bmatrix} \cos\varphi & -\sin\varphi & 0 & 0 \\ \sin\varphi & \cos\varphi & 0 & 0 \\ 0 & 0 & 1 & 0 \\ 0 & 0 & 0 & 1 \end{bmatrix} \begin{bmatrix} X_2 \\ Y_2 \\ Z_2 \\ 1 \end{bmatrix} \end{Bmatrix} \quad (1)$$

Note:

1. X0, Y0, and Z0: original nodal coordinate in the vertebral body L5
2. X1, Y1, and Z1: new nodal coordinate in a new vertebral body after rotating an angle of β (frontal plane)
3. X2, Y2, and Z2: new nodal coordinate in a new vertebral body after rotating an angle of θ (sagittal plane)
4. X, Y, and Z: new nodal coordinate in a new vertebral body after rotating an angle of φ (transverse plane)
5. A total of 782 nodes were in the vertebral body L5

2.3. Model Validation

To ensure the geometric accuracy of the model, the FE model constructed from X-ray images was compared with the patient's CT images. CT imaging provides detailed 3D spinal data, allowing for model accuracy assessment. Using a software 3D Slicer 4.10.2, a 3D reconstruction of the CT images was performed, and the geometries were compared with the X-ray-derived model (Figure 4). The spatial coordinate error of each vertebra was calculated and presented as the mean absolute error (MAE) (Equation (2)).

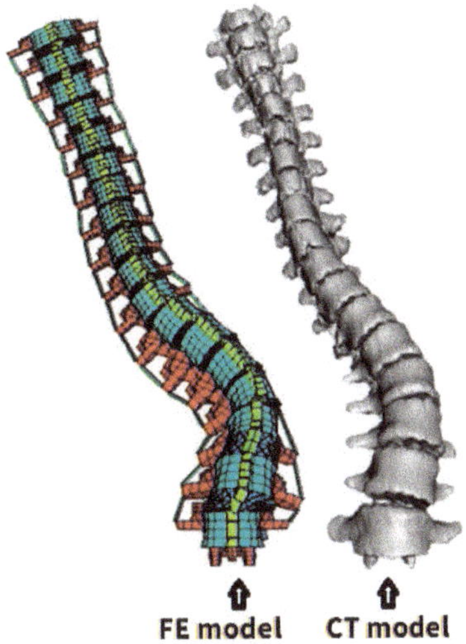

Figure 4. Comparison between FE- and CT-based reconstructed 3D model.

To verify the model's angular accuracy, the Cobb (Figure 5a) and thoracolumbar kyphosis (TLK) angles (Figure 5b), measured from the FE model, were compared with values obtained from imaging.

$$\mathrm{MAE} = \sum_{i=1}^{18} |\mathrm{FE\ model}_i - \mathrm{CT}_i| \quad (2)$$

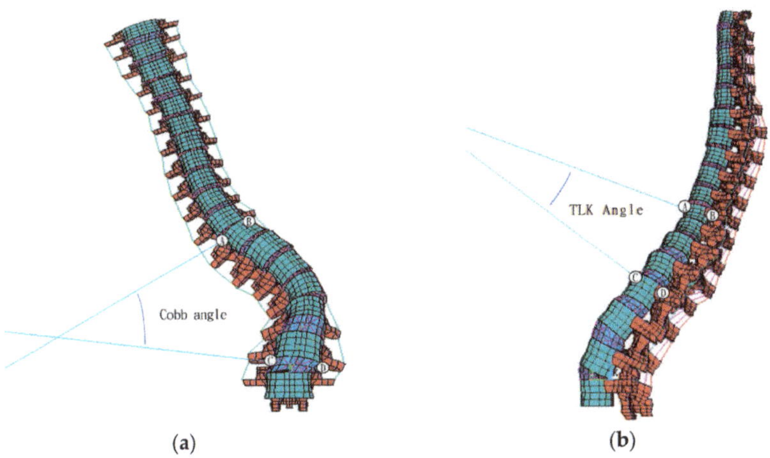

Figure 5. Calculation of FE model's (**a**) Cobb angle and (**b**) TLK angle. **Note:** A, B, C and D indicate the sampling point to measure the Cobb's and TLK angles.

Note:

1. i = 1~18 indicates each vertebral body from L5 to C7.
2. MAE indicates the mean absolute error between the FE model and the CT images.

2.4. Boundary and Loading Conditions

The inferior end of the L5 segment was fixed to simulate the patient's actual movements, representing lower body support. A flexion moment was applied to the C7 segment to replicate the loading conditions during forward bending. A flexion moment of 10 Nm was used in the preoperative spine model [17]. Incremental force was applied to the C7 vertebra for the postoperative model to control the overall spinal angle, ensuring that it matched the preoperative model during flexion [24]. FE analysis was performed on the pre- and postoperative models with the same flexion moment. Analysis parameters included the overall stress distribution, range of motion (ROM), annulus fibrosus stress, and endplate stress changes before and after surgery.

3. Results

3.1. Model Validation

The preoperative parameterized FE model closely resembled the CT images in overall shape. The MAE of the X-coordinate of vertebral bodies (median lateral direction) was 28.8 mm, with the most significant discrepancy being found at C7 (Figure 6a); for the Y-coordinate (anterior–posterior direction), the MAE was 33.7 mm, with the most significant error occurring at C7 (Figure 6b); for the Z-coordinate (vertical direction), the MAE was 7.4 mm, with the most considerable discrepancy being observed at T7 (Figure 6c).

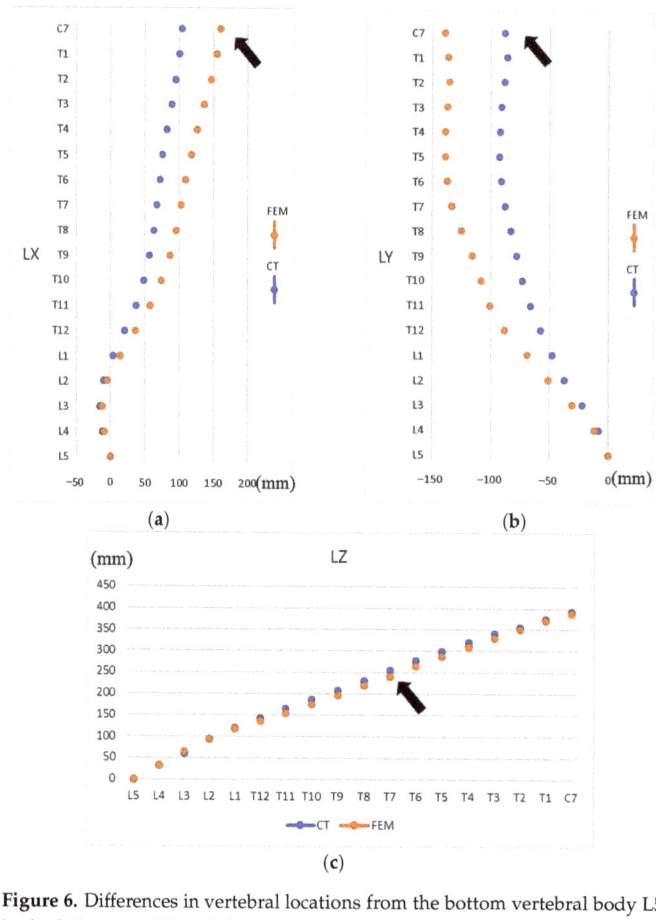

Figure 6. Differences in vertebral locations from the bottom vertebral body L5 to upper vertebral body C7 between FE model and CT images. (**a**) X-axis coordinates (ML direction), (**b**) Y-axis coordinates (AP direction), (**c**) Z-axis coordinates (vertical). Note: Arrows indicate the vertebrae with the most significant discrepancies. LX, LY, and LZ represent the X-, Y-, and Z-coordinates of the vertebral bodies.

The Cobb angle measured by the FE model was 45.5° preoperatively and 13.9° postoperatively, while X-ray measurements yielded 52° and 11°, respectively, indicating a discrepancy of less than 6.5° between the two methods.

The TLK angle measured using the FE model was 20.2° preoperatively and 10.0° postoperatively, while X-ray measurements showed 15.3° and 4.6°, respectively, leading to a discrepancy of less than 5.4° between the two methods.

3.2. Biomechanical Analysis

3.2.1. Segmental ROM

During forward bending, the ROM of the surgically treated segments (T10-L4) post-surgery decreased to 21.1% of the preoperative ROM (Table 2). The ROM of the cranial adjacent segment (T9-T10) increased by approximately 1.35 times compared with pre-surgery, while the caudal adjacent segment (L4-L5) decreased to 39.2% of the preoperative ROM. The ROM of other cranial segments (C7-T10) increased to about 1.7 times the preoperative value.

Table 2. Comparison of ROM and the maximum von Mise stress in adjacent segments.

	Pre-Op	Post-Op
ROM (°)		
C7-T10	21.52	36.31
T9-T10	1.71	2.31
T10-L4	14.53	3.06
L4-L5	5.21	2.04
Total	41.26	41.41
	Max Endplate Stress (MPa)	
T10	3.31	5.77
L4	6.43	1.71
	Max Annulus Fibrosus Stress (MPa)	
T9-T10	0.63	1.46
L4-L5	1.67	0.42

Note: The surgical segments are from T10 to L4; the adjacent segments are T9-T10 and L4-L5.

3.2.2. Maximum Stress in Endplate and Annulus Fibrosus

For the endplate, the maximum stress at the superior endplate of T10 increased from 3310 kPa preoperatively to 5770 kPa postoperatively during flexion. The stress concentration shifted from the concave side preoperatively to the ventral side postoperatively (Figure 7). For the inferior endplate of L4, the maximum stress decreased from 6430 kPa preoperatively to 1710 kPa postoperatively, about 26.6% of the preoperative level, with stress being concentrated on the dorsal side in both cases.

For the annulus fibrosus, the maximum stress in the cranial adjacent segment (T9-T10) increased from 634 kPa preoperatively to 1460 kPa postoperatively, approximately 2.3 times the preoperative level, with the stress concentrating on the ventral side. In the caudal adjacent segment (L4-L5), the maximum stress decreased from 1670 kPa preoperatively to 416 kPa postoperatively, about 24.9% of the preoperative level, with the stress being concentrated on the dorsal side.

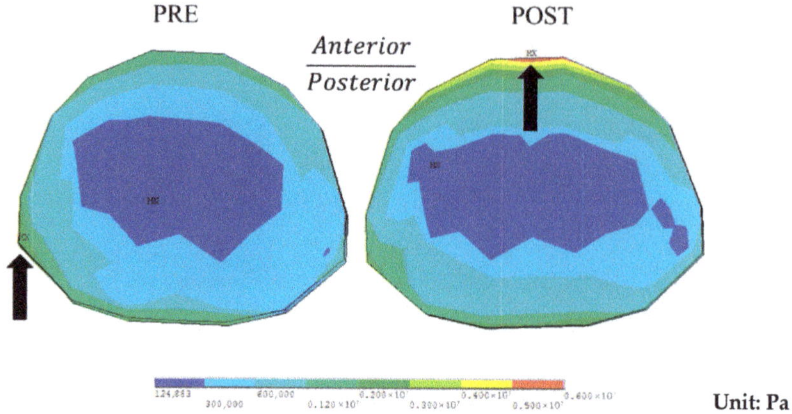

Figure 7. Stress distribution at the superior endplate of T10 during flexion preoperatively (PRE) and postoperatively (POST). Note: Black arrows indicate areas of high stress.

3.2.3. Overall Stress Distribution

The movement and stress distribution in the complete spinal model during flexion are illustrated in Figure 8. In the preoperative model, stress is concentrated on the concave side of the scoliotic curve, as indicated by the red arrows. In the post-operative model, most surgically treated segments (T10-L4) exhibit significantly reduced stress, except T10, which experiences higher stress levels.

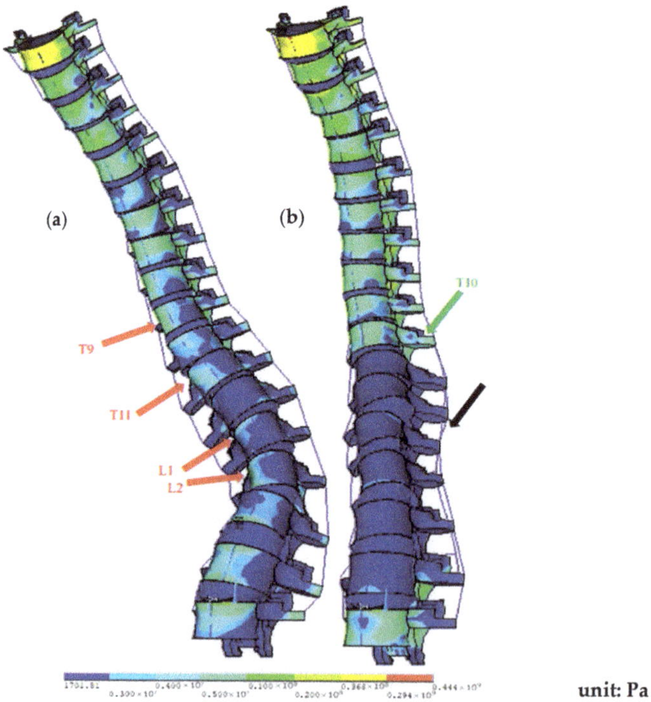

Figure 8. Stress distribution of the spine model during flexion: (**a**) pre-op; (**b**) post-op (fixed at T10-L4). Note: Red arrows show the pre-op stress concentration on the concave side of the curve; black arrows indicate reduced stress in the surgical segments; and green arrows indicate higher stress at T10.

4. Discussion

4.1. Model Reconstruction

In previous studies, FE models have been constructed based on CT and X-ray images. Using CT images to build models presents several difficulties, such as radiation exposure and accurately modeling the intervertebral disk. For example, Wei et al. [24] used reverse engineering software to construct the disk, and Somtua et al. [16] utilized CAD software like SolidWorks to build the disk. In contrast, models constructed from X-ray images struggle to capture vertebral shapes accurately. For instance, Ali et al. [15] used X-ray images and beam elements to build an FE model, representing each vertebral body with a single beam element, which cannot depict the deformations of the vertebral body that are often seen in severe scoliosis cases.

This study developed a parameterized FE model based on X-ray images to accurately represent vertebral body deformation, thus avoiding the high radiation dose associated with CT imaging while overcoming the limitations of X-ray in capturing vertebral deformities. This model benefits patients with AIS requiring surgical intervention and those with milder scoliosis (with Cobb angles below 40 degrees), who may only need bracing or physical therapy. Since CT scans are not routine for these patients, constructing high-precision models from CT images is challenging. Using low-radiation X-ray images, the parameterized model can reconstruct an FE model that captures vertebral body deformation, supporting biomechanical analysis of scoliosis and the design of corrective devices such as the Boston brace.

4.2. Model Validation

The Cobb angle calculated by the FE model and the X-ray measurement differed by 6.5 degrees preoperative and 2.9 degrees postoperative. A study [25] found that the Cobb angle measured by different observers may differ by 2.8 to 10 degrees within a 95% confidence interval. Therefore, the error in this study is acceptable. As for the reason why the Cobb angle measured by X-ray pre-operative is larger than the result calculated by the FE model, but the result measured by X-ray postoperative is smaller, there may be two reasons: the projection of the 3D angle and the vector skew. When the angle in 3D space is projected onto the 2D plane, its size may change, and it may become larger or smaller depending on the projection angle; while when using the vector angle to calculate the angle in space, if the vector is not coplanar but skewed, it may be affected by the angle between different planes. Therefore, the vector angle is affected not only by the coronal plane angle but also by the transverse plane angle, resulting in an angle larger than expected.

The TLK angle measured from the FE model also deviated by only 5.4 degrees from the X-ray measurements. Biomechanically, FE analysis revealed that post-operative flexion increased the ROM and maximum stress in the cranial adjacent segments, while the ROM and stress in the caudal adjacent segments decreased.

In terms of vertebral location, the MAE values of 28 mm (ML direction) and 34 mm (AP direction) were slightly higher, despite the subject having a spine length of 390 mm (C7-L5). Two reasons were attributed to these errors. First, the discrepancies between the FE model and CT images are primarily due to differences in the subject's posture during the imaging process. The FE model parameters were derived from standing, weight-bearing X-ray images, whereas the CT images were taken with the subject in a supine position [26]. Second, the error may be due to the measurement of the vertebral body L5. All vertebral bodies were generated based on the vertebral body L5. We found that inaccurately digitizing the location of the L5 vertebral body in X-ray images would result in more errors in the FE scoliotic model. Therefore, the user should carefully digitize the location of the vertebral body L5 in X-ray images.

Additionally, the errors in the Cobb and TLK angle calculations from the model compared with X-ray measurements arose mainly from the 3D-to-2D projection process, where the skew in vectors may also introduce measurement errors. Previous studies have reported discrepancies between Cobb angle measurements from 3D and 2D images [27,28], with several studies also finding that inter-observer variation in Cobb angle measurements can differ by 2.8 to 10 degrees within a 95% confidence interval [29], making the errors in this study acceptable.

Although there are some errors in the FE scoliotic model, an experienced technician could create a scoliotic FE model within 3 h using X-ray images, and the FE model could be rapidly constructed for clinical application.

4.3. Biomechanical Analysis

This study found that the ROM of cranial unfused segments increased to approximately 1.7 times the preoperative level for the segmental ROM. In contrast, the ROM of caudal unfused segments decreased to around 39.2% of the preoperative level. This study found that the mobility of the unfused cranial segment accounted for 52.2% of the total mobility preoperatively and 87.7% postoperatively, an increase of 35.5%. Ensberg et al.'s study [27] found that the mobility of the unfused cranial segment accounted for 38.5% of the total mobility preoperatively and 69.0% postoperatively, an increase of 30.5%. It is evident that in both studies, the unfused cranial segment compensated for most of the reduced ROM in the fused segment.

Regarding endplate stress, the maximum stress of caudal adjacent segments decreased post-surgery. In contrast, cranial adjacent segments experienced an increase in maximum stress, with a more balanced stress distribution. This outcome may be related to the increased ROM in the cranial segments and the decreased ROM in the caudal segments after surgery; furthermore, Akazawa et al. [5–7] suggest that Modic changes observed in patients with AIS may result from the concentration of stress on the concave side of the scoliotic curve. Therefore, the even stress distribution post-surgery might help reduce the likelihood of Modic changes.

For annulus fibrosus stress, the maximum stress of cranial adjacent segments increased post-operatively, while that in caudal adjacent segments decreased. This finding is consistent with the study by Nohara et al. [28], which, after a ten-year follow-up, found that AIS patients who had been treated with corrective surgery had significantly lower rates of disk degeneration at the L3-4 and L4-5 segments than untreated patients, with reductions of 35.1% and 33.4%, respectively. This may suggest that surgical treatment for AIS reduces the maximum stress on the intervertebral disks in caudal adjacent segments, lowering the likelihood of disk degeneration.

4.4. Study Limitations and Assumptions

This study has several limitations: (1) the quality of the X-ray images may affect the accuracy of the FE model; (2) the shape of the FE model is based on scaling and deforming adult L5 vertebrae in various directions, which may not accurately reflect all anatomical features of the human spine, such as the facet joint orientation; (3) since contact element was not assigned to the facet joints, only flexion movements—which impact the facet joints minimally—were analyzed, excluding extension, lateral bending, and axial rotation; (4) this study only measured the ROM in the sagittal plane, without accounting for the coupling effect in scoliosis; (5) material properties of the vertebral body and disc in the FE model were assumed to be linear and isotropic, which the isotropy of the vertebral body and disc would result in a stiffer behavior of the lumbar spine than anisotropy of the vertebral body and disc [13,20]; the more realistic material properties should be added to the FE model in future studies [30,31]; (6) the study was based on data from a single Lenke type 5 AIS case, limiting the generalizability to other AIS types or broader patient populations; (7) Since the study aimed to generate an FE model using APDL codes rapidly, the number of nodes in the vertebral body L5 was fixed. As a result, the mesh convergence was not conducted in the study because it would alter the nodal numbers of the original vertebral body. Therefore, the coarse mesh might lower the accuracy of the FE calculation.

5. Conclusions

This study successfully used biplanar X-ray images to obtain vertebral coordinates, size, and tilt angles to construct a patient-specific parameterized FE model. The model also incorporated coronal plane vertebral deformation to represent severe scoliosis accurately. The differences in Cobb and TLK angles between the model and X-ray measurements were within 6.5 and 5.4 degrees, respectively. Using this model, flexion—a common spinal movement—was simulated, and the ROM, endplate stress, and disk stress in adjacent segments were analyzed. The findings suggest that surgical treatment for patients with Lenke type 5 AIS may reduce the ROM of caudal adjacent segments to about 40% of the preoperative level, with maximum endplate stress decreasing to approximately 27% and maximum disk stress to about 25%.

Author Contributions: Conceptualization, C.-S.C.; Software, T.-H.W.; Formal analysis, T.-H.W.; Investigation, P.-H.C.; Data curation, T.-H.W.; Writing—original draft, T.-H.W.; Visualization, P.-H.C.; Supervision, C.-S.C.; Project administration, P.-H.C.; Funding acquisition, C.-S.C. All authors have read and agreed to the published version of the manuscript.

Funding: This research was supported by the National Science and Technology Council (NSTC 112-2221-E-A49-017-MY3).

Institutional Review Board Statement: The study was conducted in accordance with the Declaration of Helsinki, and approved by the Institutional Review Board (#YM111055F) of National Yang-Ming University (24 June 2022).

Informed Consent Statement: Informed consent was obtained from all subjects involved in the study.

Data Availability Statement: The original contributions presented in the study are included in the article, further inquiries can be directed to the corresponding author.

Acknowledgments: We thank the reviewers for their valuable feedback which helped improve the manuscript.

Conflicts of Interest: The authors declare no conflict of interest.

References

1. Weinstein, S.L.; Dolan, L.A.; Cheng, J.C.; Danielsson, A.; Morcuende, J.A. Adolescent idiopathic scoliosis. *Lancet* **2008**, *371*, 1527–1537. [CrossRef]
2. Goldberg, C.J.; Moore, D.P.; Fogarty, E.E.; Dowling, F.E. Scoliosis: A review. *Pediatr. Surg. Int.* **2008**, *24*, 129–144. [CrossRef] [PubMed]
3. Liu, F.; Liu, F.; Wang, H. Half of the adolescent idiopathic scoliosis patients may have lumbar adjacent segment degeneration following spinal fusion: A systemic review and meta-analysis. *J. Orthop. Surg.* **2024**, *32*, 1–9. [CrossRef] [PubMed]
4. Lonner, B.S.; Ren, Y.; Upasani, V.V.; Marks, M.M.; Newton, P.O.; Samdani, A.F.; Chen, K.; Shufflebarger, H.L.; Shah, S.A.; Lefton, D.R.; et al. Disc degeneration in unfused caudal motion segments ten years following surgery for adolescent idiopathic scoliosis. *Spine Deform.* **2018**, *6*, 684–690. [CrossRef]
5. Akazawa, T.; Kotani, T.; Sakuma, T.; Minami, S.; Orita, S.; Fujimoto, K.; Shiga, Y.; Takaso, M.; Inoue, G.; Miyagi, M.; et al. Spinal fusion on adolescent idiopathic scoliosis patients with the level of L4 or lower can increase lumbar disc degeneration with sagittal imbalance 35 years after surgery. *Spine Surg. Relat. Res.* **2017**, *1*, 72–77. [CrossRef]

6. Akazawa, T.; Watanabe, K.; Matsumoto, M.; Tsuji, T.; Kawakami, N.; Kotani, T.; Sakuma, T.; Yamamoto, T.; Demura, S.; Orita, S.; et al. Modic changes and disc degeneration in adolescent idiopathic scoliosis patients who reach middle age without surgery: Can residual deformity cause lumbar spine degeneration? *J. Orthop. Sci.* **2018**, *23*, 884–888. [CrossRef] [PubMed]
7. Akazawa, T.; Kotani, T.; Sakuma, T.; Minami, S.; Orita, S.; Inage, K.; Fujimoto, K.; Shiga, Y.; Torii, Y.; Umehara, T.; et al. Modic changes and disc degeneration of non-fused segments 27 to 45 years after Harrington instrumentation for adolescent idiopathic scoliosis: Comparison to healthy controls. *Spine* **2018**, *43*, 556–561. [CrossRef]
8. Hashimoto, K.; Aizawa, T.; Kanno, H.; Itoi, E. Adjacent segment degeneration after fusion spinal surgery—A systematic review. *Int. Orthop.* **2019**, *43*, 987–993. [CrossRef]
9. Debnath, U.K.; Scammell, B.E.; Freeman, B.J.C.; McConnell, J.R. Predictive factors for the outcome of surgical treatment of lumbar spondylolysis in young sporting individuals. *Glob. Spine J.* **2018**, *8*, 121–128. [CrossRef]
10. Zhang, Q.; Chon, T.E.; Zhang, Y.; Baker, J.S.; Gu, Y. Finite element analysis of the lumbar spine in adolescent idiopathic scoliosis subjected to different loads. *Comput. Biol. Med.* **2021**, *136*, 104745. [CrossRef]
11. Jie, Y.; Li, M.; Dong, A.; Luo, Y.-Y.; Luo, C.-L.; Li, J.; Zheng, P.; Zhang, X.; Wong, M.S.; Ma, C.Z.-H.; et al. Digitalized 3D Spinal decompression and correction device improved initial brace corrections and patients' comfort among adolescents with Idiopathic scoliosis: A single-centre, single-blinded randomized controlled trial. *Bioengineering* **2024**, *11*, 1246. [CrossRef]
12. Wei, W.; Zhang, T.; Huang, Z.; Yang, J. Finite element analysis in brace treatment on adolescent idiopathic scoliosis. *Med. Biol. Eng. Comput.* **2022**, *60*, 907–920. [CrossRef] [PubMed]
13. Chou, W.K.; Liu, C.L.; Liao, Y.C.; Cheng, F.H.; Zhong, Z.C.; Chen, C.S. Using finite element method to determine pad positions in a Boston brace for enhancing corrective effect on scoliotic spine: A preliminary analysis. *J. Med. Biol. Eng.* **2012**, *32*, 29–35. [CrossRef]
14. Ali, A.; Fontanari, V.; Schmölz, W.; Agrawal, S.K. Active soft brace for scoliotic spine: A finite element study to evaluate in-brace correction. *Robotics* **2022**, *11*, 37. [CrossRef]
15. Somtua, C.; Aroonjarattham, P.; Saengpetch, N.; Rattanapan, N.; Aroonjarattham, K. Biomechanical analysis of scoliosis adjusted by screw fixation system with finite element analysis. *Eng. J.* **2024**, *28*, 25–34. [CrossRef]
16. Chen, C.S.; Cheng, C.K.; Liu, C.L. A Biomechanical Comparison of posterolateral fusion and posterior fusion in the lumbar spine. *J. Spinal Disord. Tech.* **2002**, *15*, 53–63. [CrossRef]
17. Chen, C.S.; Cheng, C.K.; Liu, C.L.; Lo, W.H. Stress analysis of the disc adjacent to interbody fusion in lumbar spine. *Med. Eng. Phys.* **2001**, *23*, 485–493. [CrossRef]
18. Chen, C.S.; Feng, C.K.; Cheng, C.K.; Tzeng, M.J.; Liu, C.L.; Chen, W.J. Biomechanical analysis of the disc adjacent to posterolateral fusion with laminectomy in lumbar spine. *J. Spinal Disord. Tech.* **2005**, *18*, 58–65. [CrossRef]
19. Pan, J.H.; Chen, C.S.; Liu, C.L.; Chou, P.H. Biomechanical effects of a novel pedicle screw W-Type rod fixation for lumbar spondylolysis: A finite element analysis. *Bioengineering* **2023**, *10*, 451. [CrossRef]
20. Zhong, Z.C.; Chen, S.H.; Hung, C.H. Load- and displacement controlled finite element analyses on fusion and non-fusion spinal implants. *Proc. Inst. Mech. Eng. Part H J. Eng. Med.* **2009**, *223*, 143–157. [CrossRef]
21. Fedorov, A.; Beichel, R.; Kalpathy-Cramer, J.; Finet, J.; Fillion-Robin, J.-C.; Pujol, S.; Bauer, C.; Jennings, D.; Fennessy, F.M.; Sonka, M.; et al. 3D Slicer as an Image Computing Platform for the Quantitative Imaging Network. *Magn. Reson. Imaging* **2012**, *30*, 1323–1341. [CrossRef]
22. Ng, S.Y.; Bettany-Saltikov, J. Imaging in the diagnosis and monitoring of children with idiopathic scoliosis. *Open Orthop. J.* **2017**, *11*, 1500–1520. [CrossRef]
23. Goel, V.K.; Monroe, B.T.; Brinkmann, P. Interlaminar shear stresses and laminae separation in a disc. Finite element analysis of the L3-L4 motion segment subjected to axial compressive loads. *Spine* **1995**, *20*, 689–698. [CrossRef] [PubMed]
24. Wei, W.; Zhang, T.; Yang, J.; Qian, Y.; Dong, Y. Material sensitivity of patient-specific finite element models in the brace treatment of scoliosis. *Front. Bioeng. Biotechnol.* **2023**, *16*, 1111449. [CrossRef]
25. Elfiky, T.; Patil, N.; Shawky, M.; Siam, A.; Ragab, R.; Allam, Y. Oxford Cobbometer versus computer assisted-software for measurement of Cobb Angle in adolescent idiopathic scoliosis. *Neurospine* **2020**, *17*, 304–311. [CrossRef]
26. Liang, Z.; Wang, Q.; Xia, C.; Chen, Z.; Xu, M.; Liang, G.; Zhang, Y.; Ye, C.; Zhang, Y.; Yu, X.; et al. From 2D to 3D: Automatic measurement of the Cobb angle in adolescent idiopathic scoliosis with the weight-bearing 3D imaging. *Spine J.* **2024**, *24*, 1282–1292. [CrossRef]
27. Engsberg, J.R.; Lenke, L.G.; Reitenbach, A.K.; Hollander, K.W.; Bridwell, K.H.; Blanke, K. Prospective evaluation of trunk range of motion in adolescents with idiopathic scoliosis undergoing spinal fusion surgery. *Spine* **2002**, *27*, 1346–1354. [CrossRef] [PubMed]
28. Nohara, A.; Kawakami, N.; Tsuji, T.; Ohara, T.; Saito, T.; Kawakami, K. Intervertebral disc degeneration during postoperative follow-up more than 10 years after corrective surgery in idiopathic scoliosis: Comparison between patients with and without surgery. *Spine* **2018**, *43*, 255–261. [CrossRef] [PubMed]
29. d'Astorg, H.; Bourret, S.; Ramos-Pascual, S.; Szadkowski, M.; Le Huec, J.C. Comparison of Cobb angle measurements for scoliosis assessment using different imaging modalities: A systematic review. *EFORT Open Rev.* **2023**, *8*, 489–498. [CrossRef]

30. Kim, J.H.; Hyeok, J.J.; Woo, J.H.; Kim, S.M. Correlation analysis of suture anchor pull-out strength with cortical bone thickness and cancellous bone density on a finite element model. *Bioengineering* **2025**, *12*, 863. [CrossRef]
31. Fada, R.; Shahgholi, M.; Azimi, R.; Babadi, N.F. Estimation of porosity effect on mechanical properties in calcium phosphate cement reinforced by strontium nitrate nanoparticles: Fabrication and FEM analysis. *Arab. J. Sci. Eng.* **2024**, *49*, 1815–1825. [CrossRef]

bioengineering

MDPI

Article

Does Resistance Indicate Malposition? A Standardized Comparison of Pedicle Screw Placement

Sascha Kurz [1,2,*], Benjamin Fischer [1,2], Janine Schultze [1,2], Florian Metzner [1,2], Toni Wendler [1,2], Christoph-Eckhard Heyde [1,2] and Stefan Schleifenbaum [1,2]

1 ZESBO—Center for Research on Musculoskeletal Systems, Faculty of Medicine, Leipzig University, 04103 Leipzig, Germany

2 Department of Orthopaedic, Trauma and Plastic Surgery, University of Leipzig Medical Center, 04103 Leipzig, Germany

* Correspondence: sascha.kurz@medizin.uni-leipzig.de

Abstract

Pedicle screw malpositioning remains a frequent complication, with reported rates from 2% to 15%, often leading to revision surgeries. Analyzing mechanical resistance and torque encountered during screw insertion has been implicated as a promising approach for real-time detection. Five fresh-frozen human thoracolumbar spine specimens were utilized in this study. Using 3D-printed templates, correct trajectories were systematically compared against four defined malpositions (medial, lateral, superior, superolateral), with offsets ranging from 2.0 mm to 3.5 mm. Drilling, tapping, and insertion phases were conducted at a constant speed and defined feed force. Contrary to the anticipated behavior, malpositioned trajectories showed no statistically significant difference in peak torque compared to correct trajectories across all phases (e.g., tapping $p = 0.944$, $r = 0.01$; insertion $p = 0.693$, $r = 0.05$). Regional stratification between thoracic and lumbar spine also failed to yield significant differences. The only statistically significant difference was observed between the correct trajectory and the superolateral malposition during drilling ($p = 0.038$). Under the tested standardized conditions, torque-based mechanical resistance during pedicle screw placement is generally not a reliable and consistent real-time indicator of malposition.

Keywords: biomechanics; spine; experiment; pedicle screw; mechanical resistance; screw misplacement; malposition; torque

1. Introduction

Pedicle screw fixation is a widely accepted technique for posterior spinal stabilization in the treatment of degenerative diseases, traumatic injuries, and spinal deformities [1,2]. Despite its routine use and the availability of advanced intraoperative navigation systems, pedicle screw malpositioning remains a frequent and clinically relevant complication. Reported misplacement rates vary between 2% and 15%, depending on the definition and detection method used [3,4]. These misplacements are among the leading causes of reoperation within the first 30 days after spinal surgery [5], and in pediatric deformity cases, they represent the most common instrumentation-related complication [6].

Malpositioning can occur in various directions—such as medial, lateral, caudal, superior or superolateral—and can result in cortical breaches of varying severity. While some breaches remain asymptomatic, others pose significant risks to neural and vascular structures, particularly in cases involving the medial or inferior pedicle wall [7,8]. Notably, there is considerable variability among surgeons regarding which malpositions warrant revision, with neurological symptoms being the most decisive factor [8].

A systematic framework for evaluating pedicle screw placement was introduced by Gertzbein and Robbins [9], which grades cortical breaches based on their extent and potential neurological impact. Although originally developed for postoperative assessment, this system can be used as a clinical reference point for defining screw misplacement. Current intraoperative strategies to detect malpositioning include fluoroscopy, CT-based navigation, and neuromonitoring. While these methods can reduce the risk of screw misplacement, they are associated with increased radiation exposure, cost, and limited real-time responsiveness [3]. Consequently, there is growing interest in alternative feedback mechanisms that can provide immediate intraoperative information about screw trajectory.

One promising approach involves analyzing the mechanical resistance and torque encountered during screw insertion. These parameters reflect the interaction between screw geometry, bone quality, and trajectory. In ideal insertions, the resistance profile follows a predictable curve; deviations from this pattern may indicate cortical breaches or trajectory errors. Preliminary in vitro studies have implicated that both peak torque and resistance fluctuations differ significantly between correctly placed and malpositioned screws [10–13].

Other studies postulate a correlation between peak torque and stability, which depends on correct positioning [14], suggesting their potential as real-time indicators of screw alignment. Tai et al. [15] found that screws with a cortical bone trajectory demonstrated significantly higher insertion torque than traditional or modified trajectory screws. This indicates that there are altered resistance curves based on bone engagement patterns. These results suggest that malpositioned screws traversing different bone structures would exhibit characteristic resistance signatures. Peak torque typically occurs when the screw shank passes through the pedicle, approximately at the middle of screw length. Malpositioned screws would theoretically alter this peak timing and magnitude due to Cortical wall breach [16] (sudden resistance loss when violating pedicle walls), bone density variations [17] (different resistance patterns when engaging cortical versus cancellous bone) and angular trajectory changes [18] (modified force vectors affecting torque requirements).

In this study, we systematically investigate the resistance and torque profiles recorded during the drilling, tapping, and insertion phases of pedicle screw placement under controlled laboratory conditions. Using thoracic and lumbar vertebrae from human cadavers, we compare correct trajectories with defined malpositions—medial, lateral, superior, and superolateral—based on parallel displacement from the optimal path. Our aim is to identify mechanical markers that reliably distinguish between accurate and faulty screw placements.

Therefore, the primary research question of this study is whether pedicle screw malpositions could be reliably detected by differences in mechanical resistance during insertion. We hypothesize that malpositioned screws exhibit distinct resistance curves and altered peak torque values compared to correctly aligned insertions. Validating this hypothesis could enable the integration of torque-based feedback into surgical workflows, improving intraoperative decision-making and reducing the incidence of pedicle screw misplacement.

2. Materials and Methods

2.1. Specimen Acquisition

All body donors provided informed written consent for the donation of their bodies for teaching and research purposes while they were alive. As part of the body donor program regulated by the Saxonian Death and Funeral Act of 1994 (third section, paragraph 18, item 8), the Ethics Committee of the University of Leipzig Medical Center obtained institutional approval for the use of postmortem tissues from the Institute of Anatomy at the University of Leipzig (ethical approval no. 129/21-ck, approval date 1 March 2023).

The authors declare that all experiments were conducted according to the principles of the Declaration of Helsinki (as revised in 2013).

Five fresh-frozen human cadaveric thoracolumbar spine specimens (T8–L5) were harvested from donors enrolled in the institutional body-donor program (Table 1). Each specimen comprised five thoracic (T8–T12) and five lumbar vertebrae (L1–L5), yielding ten vertebrae per cadaver. Soft tissue was removed to expose the bony anatomy of the pedicles and vertebral bodies while preserving osseous integrity. After preparation of the spinal segments by the Institute of Anatomy of Leipzig University, the specimens were stored at −80 °C. Prior to the start of the experimental phase, all samples were scanned in a frozen state using quantitative computed tomography (QCT) (Ingenuity CT/Brilliance Big Bore | Koninklijke Philips N.V., Amsterdam, NL-NH) with 1 mm slice thickness and a tube voltage of 120 kV, utilizing a calcium-hydroxyapatite phantom (Bone Density Calibration Phantom 6 H200—QRM-50124 | QRM GmbH, Möhrendorf, DE-BY). Mean cancellous Hounsfield units (HU) were measured in Mimics Innovation Suite (Version 24 | Materialise NV, Leuven, BE-VBR) and converted to volumetric bone mineral density (vBMD) for each vertebra according to Metzner et al. [19] using linear regression based on the procedure of Brett and Brown [20].

Table 1. Cadaver data.

ID	Sex	Age in yr	Body Mass in kg	Height in m	BMI in kg m^{-2}
1	Male	87	99	1.72	33.5
2	Male	51	61	1.74	20.1
3	Male	83	81	1.72	27.4
4	Male	84	66	1.74	21.8
5	Female	81	82	1.89	23.0

BMI = body mass index.

2.2. Preliminary Planning

The previously performed CT scans were used to plan pedicle screw trajectories and insertion depths for each vertebral level. Based on Gertzbein and Robbins [9], five trajectory types were defined: correct (c), lateral (l), medial (m), superior (s) and superolateral (sl) (Table 2).

Table 2. Planned trajectory offsets.

Trajectory Misalignment	Deviation Value [9]	Planned Offset
correct (c)	—	0.0 mm
lateral (l)	1–5 mm	3.0 mm
medial (m)	2–5 mm	3.5 mm
superior (s)	1–3 mm	2.0 mm
superolateral (sl)	1–5 mm	3.0 mm

To ensure a systematic and balanced distribution of screw misalignments across vertebral levels and cadavers, an initial planning table was created. The assignment of malpositions followed a structured rotational principle: for each vertebral level, the type of trajectory deviation—such as medial, lateral, superior, or superolateral—was shifted downward by one cadaver in a cyclic manner. This approach ensured that each cadaver received a distinct combination of malpositions throughout the thoracolumbar spine, avoiding clustering or repetition of specific trajectory types.

One vertebra per spinal region (thoracic and lumbar) of each cadaver was instrumented bilaterally with correct trajectories to serve as a reference level. The side on which each malposition was applied—left or right—was determined by randomization using

Microsoft Excel (Microsoft 365 MSO, Version 2502, Build 16.0.18526.20416, 64 Bit | Microsoft Corporation, Redmond, US-WA). This introduced variability while maintaining control over the overall distribution of trajectory deviations (Table 3).

Table 3. Initially planned screw malpositions.

Vertebra	Cadaver				
	1	2	3	4	5
L5	l – R	c – B	s – R	sl – L	m – R
L4	m – L	l – L	c – B	s – R	sl – R
L3	sl – L	m – R	l – L	c – B	s – R
L2	s – R	sl – R	m – L	l – L	c – B
L1	c – B	s – L	sl – R	m – R	l – L
T12	l – R	c – B	s – L	sl – R	m – R
T11	m – R	l – R	c – B	s – R	sl – L
T10	sl – R	m – L	l – R	c – B	s – L
T9	s – R	sl – R	m – L	l – L	c – B
T8	c – B	s – R	sl – R	m – L	l – L

Trajectories: c = correct, l = lateral, m = medial, s = superior, sl = superolateral. Side: L = left, R = right, B = bilateral. Format: [trajectory] – [side].

Following segmentation of the vertebrae from CT data, the initial trajectory plan was reviewed and adjusted based on individual anatomical characteristics, including structural anomalies and fractures. In cases where the originally assigned malposition was deemed biomechanically unfeasible or anatomically incompatible, the trajectory was modified accordingly. If anatomical exclusion criteria—such as severe fractures or compromised pedicle integrity—prevented safe screw placement altogether, the affected vertebrae were excluded from planning and subsequent testing. This applied to two vertebrae (L1 in Cadaver 1 and L5 in Cadaver 4), resulting in four omitted screw placements prior to template design.

Specimen-specific drilling templates were created to enable precise and reproducible screw placement during the experiment. The templates were designed using CAD software (Rhino 7 SR37 | Robert McNeel & Associates, Seattle, US-WA), based on segmented vertebral models derived from CT data. For each vertebra, the entry point and transpedicular trajectory were defined in three-dimensional space, and the corresponding screw diameter was represented as a cylindrical volume within the software environment. Once the trajectory was established, the drill path was integrated into the template design to ensure alignment with the pedicle axis. Stable and anatomically conforming fit to the bone surface was guaranteed by equipping each template with pre-shaped contact surfaces and structural support elements tailored to the individual vertebral geometry. The finalized templates were manufactured using PolyJet (Stratasys J850 DAP | Stratasys Inc., Minnetonka, US-MN) 3D printing from VeroPureWhite (VeroPureWhite | Stratasys Inc., Minnetonka, US-MN) material, resulting in high-resolution, rigid guides suitable for the intended experimental application. The preliminary planning was based on the usage of M.U.S.T. pedicle screws (Medacta International, Castel San Pietro, CH-TI) and associated instruments; screw dimensions were planned per manufacturer specifications and recorded in the dataset. The final screw lengths and diameters listed in Table 4 were assigned to the respective vertebrae.

Table 4. Planned screw trajectories and dimensions per vertebra and cadaver.

Vertebra	Cadaver 1		Cadaver 2		Cadaver 3		Cadaver 4		Cadaver 5	
	L	R	L	R	L	R	L	R	L	R
L5	sl —	c —	c 6/50	s 6/50	c 7/55	c 7/55	c 7/50	l 7/50	c 6/50	m 6/50
L4	c 7/50	s 7/50	c 6/50	c 6/50	l 7/50	c 7/50	m 7/50	c 7/50	c 6/50	sl 6/50
L3	c 7/50	c 7/50	l 6/50	c 6/50	c 7/50	m 7/50	sl 6/50	c 6/50	c 6/50	s 6/50
L2	l 7/50	c 7/50	m 6/50	c 6/50	c 7/45	sl 7/50	c 6/50	s 6/50	c 6/50	c 6/50
L1	c 6/50	m 6/50	c 6/50	sl 6/50	s 6/50	c 6/50	c —	c —	l 6/50	c 6/50
T12	c 5/50	sl 6/50	s 6/50	c 6/50	c 6/50	c 6/50	c 6/50	l 6/50	c 6/50	m 6/50
T11	c 6/50	s 6/50	c 6/50	c 6/50	c 6/50	l 6/50	c 6/50	m 6/50	sl 6/50	c 6/50
T10	c 6/45	c 6/50	c 6/40	l 6/50	m 6/50	c 6/50	c 6/50	sl 6/50	s 5/50	c 5/50
T9	l 5/45	c 5/50	m 5/45	c 5/50	c 6/50	sl 6/50	c 5/50	s 5/50	c 5/45	c 5/50
T8	m 5/40	c 5/50	c 5/40	sl 5/50	s 5/45	c 5/50	c 5/40	c 5/50	l 5/50	c 5/50

Trajectories: c = correct; l = lateral; m = medial; s = superior; sl = superolateral; "—" = not plannable. Side: L = left; R = right. Format: $\frac{\text{[trajectory]}}{\text{[diameter]/[length] in mm}}$.

2.3. Specimen Preparation

Before the tests began, the samples were removed from the ultra-low temperature freezer and gently thawed to 20 °C for 24 h. On the day of testing, all vertebrae were separated and the remaining soft tissue was removed. The vertebrae were then embedded in aluminum sleeves using two-component polyurethane rapid casting resin (RenCast FC 52/53 isocyanate, FC 52 polyol | Huntsman Advanced Materials, East Lansing, US-MI), which was filled with aluminum hydroxide powder ($Al(OH)_3$) (Gössl + Pfaff GmbH, Karlskron, DE-BY) (Figure 1). The mass-related mixing ratio of the individual components isocyanate:polyol:aluminum hydroxide was 1:1:3.

(a) (b) (c)

Figure 1. Specimen preparation. (**a**) Separated dissected lumbar vertebra. (**b**) Vertebra in sleeve, held by embedding device. (**c**) Vertebra embedded in rapid casting resin.

2.4. Experimental Setup

A dedicated test rig was developed to enable reproducible biomechanical evaluation of pedicle screw insertion, as shown in Figure 2. The specimen embedding sleeve was securely mounted on a unit for multi-axis rotational compensation, which is coupled to a precision linear slide to ensure reproducible colinear alignment of the insertion axis of all tools with the pre-planned trajectory. To guide the linear slide, a deflected feeder weight was utilized, which applied 27.6 N directly in the slide's traversing axis.

The average rotational speed of a typical surgical screwdriver was chosen based on the OrthoDrive MBQ-700 motor system (De Soutter Medical Limited, Aston Clinton, GB-BKM). For actuation, a servomotor (Lexium BMH type 0703T01A2A | Schneider Electric GmbH, Düsseldorf, DE-NW) driven by a electric frequency converter (CN6 I/O | Schneider Electric GmbH, Düsseldorf, DE-NW) was coupled to an economy planetary gearbox (PLE060-008-SSSA3AD-R14 | Neugart GmbH, Kippenheim, DE-BW). Torque was measured using a pre-amplified slip-ring torque transducer (4501A20HA, 2 N m to 1000 N m—valid manufacturer test certificate for measurement uncertainty of 0.1% | Kistler, Winterthur, CH-ZH) and connected to a chuck. A linear variable differential transformer (LVDT) (RACC-50—calibrated in advance of experiments via two-point calibration | MEGATRON Elektronik GmbH & Co. KG, Munich, DE-BY) was used to measure the travel distance resulting from the insertion depth. Signals were amplified by a measurement amplifier (GSV-1A8USB K6D/M16 | ME Systeme GmbH, Hennigsdorf, DE-BB) and logged at 1 kHz via the graphical programming environment LabVIEW (Version 2017 | National Instruments, Austin, US-TX).

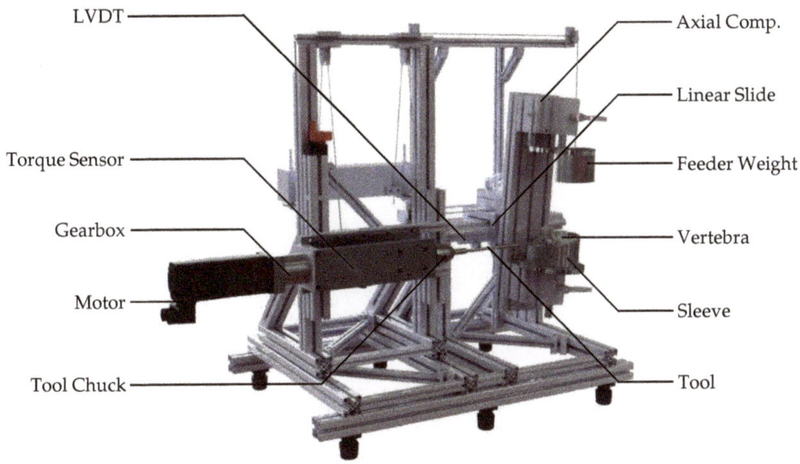

Figure 2. Custom rig for pedicle screw insertion. The specimen and drill-guide template are mounted on a linear carriage; torque and displacement are measured in real time.

2.5. Screw Insertion

Each vertebra, embedded in its aluminum sleeve, was mounted in the test rig. With the feeder weight unloaded, the pedicle axis was precisely aligned using the integrated alignment unit and multi-axis compensation mechanism (see Figure 2), starting with the left pedicle. The pre-manufactured drill-guide template was positioned over the spinous process and aligned with both pedicles using anatomically contoured support elements.

Once alignment was completed, the surgical drill was mounted in the tool chuck. The travel distance of the linear slide was set according to the planned screw length, measured via the integrated caliper system. This value was entered into the test software to calibrate the automatic depth stop, after which the feeder weight was armed and pilot hole drilling initiated. Following tool exchange, the tap was used to cut the thread to full depth. After a final tool change, the pedicle screw was implanted, which was advanced into the prepared tract to the predefined insertion depth.

All three procedural steps—drilling, tapping, and screw insertion—were performed at a constant speed of $25\,\text{min}^{-1}$, using original instrumentation (Medacta International, Castel San Pietro, CH-TI). The entire insertion process is illustrated exemplarily in Figure 3 on a right pedicle screw. After completion of the left pedicle insertion, the vertebra was repositioned within the test rig to align the right pedicle. The drill-guide template was trimmed around the support shaft of the inserted left screw and the procedure was repeated under identical conditions.

Events that could lead to data corruption were marked with the corresponding error note and excluded from the evaluation. Any other deviations from the planned test procedure were also noted. For example, if the insertion depth was not reached to avoid contact between the screw head and bone, this was noted. Later, the data series was evaluated using the recorded actual measured insertion depth.

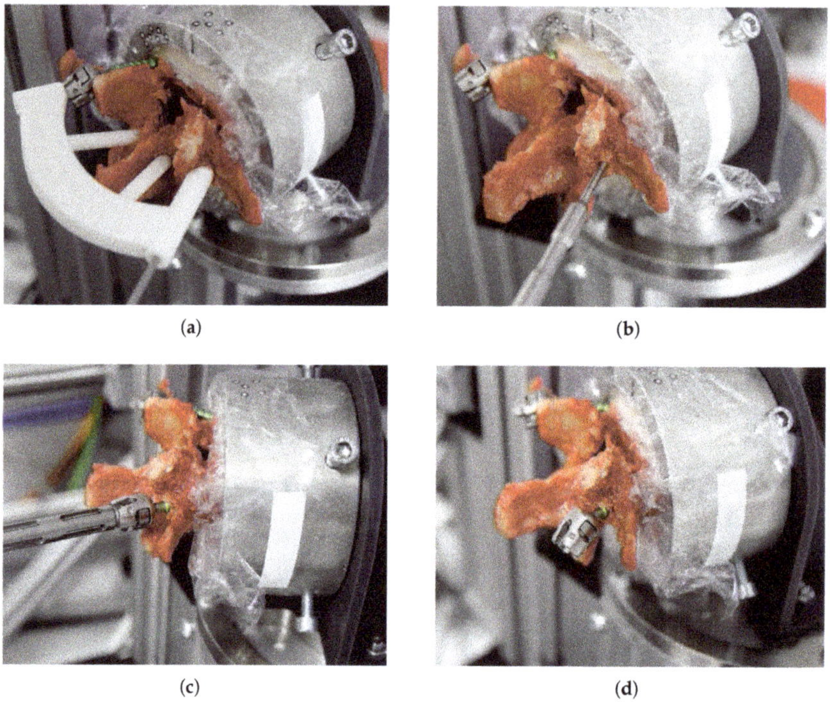

Figure 3. Screw insertion process. (**a**) Drilling the pilot hole using the pre-manufactured drill-guide template. (**b**) Tapping. (**c**) Screw insertion. (**d**) Bileaterally implanted vertebra.

2.6. Data Acquisition

Torque, insertion depth, and time were continuously recorded during all three procedural phases: drilling, tapping, and screw insertion. Data acquisition was controlled via a custom LabVIEW interface, which also governed motor actuation via the electric frequency converter and sensor synchronization. Upon initiation, the software prompted the user to enter the target rotational speed, the planned travel distance—previously determined via the test rig's integrated caliper system—and the current procedural step (i.e., drilling, tapping, or screw insertion).

Once confirmed, all connected sensors were automatically zeroed, the offset values were recorded, and real-time recording of torque, time, and displacement was initiated. From these primary signals, feed rate was derived. All four parameters were visualized in synchronized control graphs throughout the procedure. Upon reaching the predefined insertion depth, the measurement was automatically terminated, and the data were exported in `.xlsx` format using Microsoft Excel.

To ensure traceability and automation during post-processing, a standardized file naming convention was implemented. Each filename encoded the cadaver number, vertebral level and laterality, target revolutions, as well as screw dimensions. This structure ensured that all procedural metadata were embedded directly in the file name, allowing for efficient identification and cross-referencing of measurements across specimens and conditions (see Supplementary Material File S1).

All measurements from the same cadaver were stored within a single file, with separate worksheets created for each procedural step and side. This structure enabled specimen-wise accumulation of data and facilitated subsequent analysis. All data were checked for conclusiveness. Datasets with errors that appeared to result from the data export were excluded from the analysis.

Given the employment of specimen-specific drill-guide templates, experimenters were invariably aware of the trajectory type during screw placement. However, data acquisition was fully automated, so the recorded signals were not influenced by operator knowledge.

Peak torque was determined for each trial by first identifying the three highest individual torque measurements recorded during the respective procedural phase. The three peak values were averaged to yield the top-3-average (top-3-avg), which provides a single, representative measure of maximal torque output for that phase. This metric reduces the influence of transient fluctuations, measurement noise, and isolated artifacts that can occur in a single peak value while capturing the upper performance range of each trial. Prior to calculating the top-3-avg, the torque data were filtered to remove artifactual or mechanically biased segments. For drilling, the distance-referenced data were truncated to exclude the initial and final 2 mm of the trajectory. This eliminated start-up effects when the drill bit first contacted the bone surface and end-stop effects at the completion of the bore. For tapping and screw insertion, only the final 2 mm were excluded to remove end-stop effects. The resulting peak torque (top-3-avg, filtered) values were used as the dependent variable in all subsequent statistical analyses. To ensure objective grouping, the datasets were processed using standardized file names and metadata.

2.7. Statistical Methods

Initially, descriptive analyses were performed using box plots in Excel based on the raw data. All subsequent statistical analyses were performed using IBM SPSS Statistics (Version 29 | IBM Corporation, Armonk, US-NY). Given the very limited sample size ($N = 5$ cadavers), formal normality testing was not performed due to insufficient statistical power to reliably detect departures from normality. Consequently, non-parametric paired comparisons were conducted using the Wilcoxon signed-rank test for left- and right-sided measurements for each procedural phase (drilling, tapping and screw insertion). Only cadaver-vertebra combinations with complete bilateral data across all three phases were included to ensure direct comparability. All tests were two-sided and exact p-values were calculated to account for the small sample size. Effect sizes were expressed as rank-biserial correlations to quantify the magnitude and direction of differences.

To examine whether peak torque differed between correctly positioned (c) and malpositioned (m, l, s, or sl) trajectories, side-specific observations were pooled across all vertebrae (independent of laterality) and compared for each procedural phase (drilling, tapping, and screw insertion). The dependent variable was peak torque (top-3-avg, filtered). Group differences were tested using the two-tailed Mann–Whitney U test. Effect sizes r were calculated to quantify the magnitude of differences. The enabling of a more detailed assessment of potential differences in peak torque was accomplished by the stratification of the dataset into thoracic (T8–T12) and lumbar (L1–L5) subsets according to vertebral level. All measurements were included within each subset, irrespective of side. Peak torque (top-3-avg, filtered) values were then compared between correctly positioned (c) and malpositioned (m, l, s, sl) trajectories, separately for each procedural phase (drilling, tapping, screw insertion). Group comparisons were performed using the two-tailed Mann–Whitney U test, and effect sizes r were derived from the standardized test statistic Z to quantify the magnitude of the differences.

Furthermore, trajectory-based comparisons were performed to assess whether peak torque differed between correct and malpositioned trajectories. For each procedural phase all available observations were pooled and grouped according to trajectory type: correct (c) versus malpositioned (m, l, s or sl). Due to the distribution and unequal group sizes, group differences were tested using the two-tailed Mann–Whitney U test.

In addition, Spearman's rank correlation was used to determine whether the top 3-avg values are contingent on the vBMD values that were determined for the respective vertebra side.

3. Results

In total, 100 screws were initially planned across all cadavers and vertebral levels, corresponding to 300 possible datasets from drilling, tapping, and insertion phases. After applying exclusion criteria, 226 datasets and 67 placed screws remained for analysis. The distribution of planned versus analyzed screw trajectories is summarized in Table 5, which provides detailed counts per trajectory type.

Table 5. Planned versus analyzed screw trajectories (based on Table 3).

Trajectory Type	Planned	Excluded	Analyzed
Correct (c)	60	18	42
Medial (m)	10	4	6
Lateral (l)	10	4	6
Superior (s)	10	2	8
Superolateral (sl)	10	5	5
Total	100	33	67

Four cadaver-vertebra combinations with complete bilateral measurements across all three procedural phases were included in the analysis of side-dependence. The box plots in Figure 4 illustrate the three phases of drilling, tapping, and screw insertion for each donor and side. As expected, the tapping and screw insertion phases exhibit higher resistance due to the significantly larger tool and implant diameters.

The results of the Wilcoxon signed-rank test showed that no statistically significant difference was observed in the Drilling phase ($Z = -0.552$, exact $p = 0.750$). Similarly, in the tapping phase, no significant difference was found ($Z = -0.365$, exact $p = 0.875$). In the screw insertion phase, values tended to be higher on the left; however, the difference did not reach statistical significance at the 0.05 level ($Z = -1.826$, exact $p = 0.125$) as shown in Table 6.

Table 6. Wilcoxon signed-rank test: Side dependence for correct trajectories.

Parameter	Drilling	Tapping	Screw Insertion
Z	−0.552 *	−0.365 **	−1.826 **
Asymp. sig. ***	0.581	0.715	0.068
Exact sig. ***	0.750	0.875	0.125

* Based on negative ranks, ** based on positive ranks, *** 2-sided.

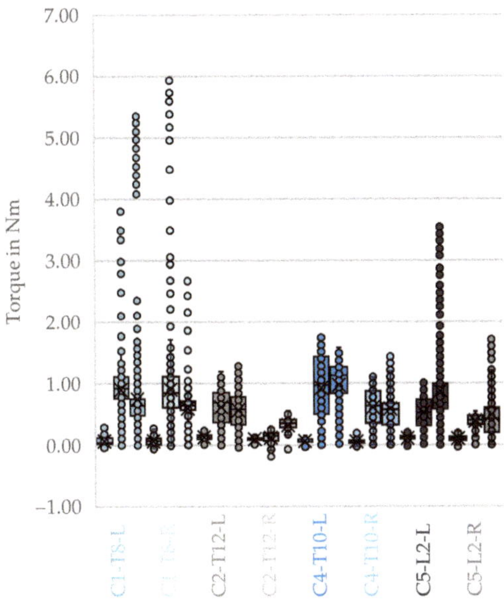

Figure 4. Side-by-side comparison of fully correctly implanted vertebrae containing all phases. From left to right per color: drilling, tapping, and screw insertion with [cadaver]-[vertebral segment]-[side]. X: mean value of the dataset, circles: outliers located more than 1.5 times the interquartile range.

The derived box plots for top-3-avg comparing correct trajectories and malpositions did not show any visible differences Figure 5.

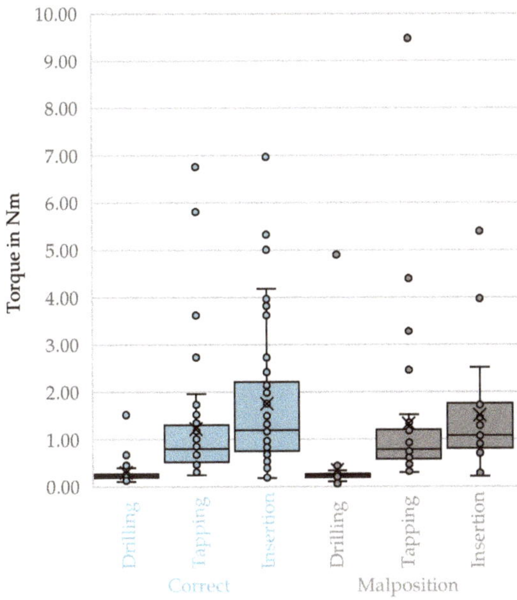

Figure 5. Top-3-avg of the phases of drilling, tapping, and screw insertion for correct trajectories compared to malpositions. X: mean value of the dataset, circles: outliers located more than 1.5 times the interquartile range.

No statistically significant differences in peak torque (top-3-avg, filtered) were observed between correctly and malpositioned trajectories across all vertebrae for any procedural phase (Table 7). For drilling, the Mann–Whitney U test yielded $U = 852.000$, $Z = -0.714$, $p = 0.475$, $r = 0.08$. For tapping, $U = 609.000$, $Z = -0.070$, $p = 0.944$, $r = 0.01$. For screw insertion, $U = 494.500$, $Z = -0.395$, $p = 0.693$, $r = 0.05$. All effect sizes were negligible, indicating that the observed differences in rank distributions were not practically relevant.

Table 7. Mann–Whitney U test; correctly vs. malpositioned trajectories.

Parameter	Drilling	Tapping	Screw Insertion
U	852.000	609.000	494.500
Z	−0.714	−0.070	−0.395
Asymp. sig. *	0.475	0.944	0.693

* two-sided.

No statistically significant differences in peak torque for the thoracic subset (T8–T12 upper part of Table 8) between correct and malpositioned trajectories during drilling ($U = 260.000$, $Z = -0.022$, $p = 0.983$, $r = 0.00$), tapping ($U = 174.500$, $Z = -0.575$, $p = 0.566$, $r = 0.09$), or screw insertion ($U = 160.500$, $Z = -0.227$, $p = 0.820$, $r = 0.04$). All observed effect sizes were negligible.

Table 8. Mann–Whitney U test; correctly vs. malpositioned trajectories for thoracic and lumbar spine.

Region	Parameter	Drilling	Tapping	Screw Insertion
Thoracic	U	260.000	174.000	160.500
	Z	−0.022	−0.575	−0.227
	Asymp. sig. *	0.983	0.566	0.820
Lumbar	U	172.500	103.500	98.500
	Z	−0.908	−0.541	−0.022
	Asymp. sig. *	0.364	0.589	0.982

* two-sided.

For the lumbar subset (L1–L5, lower part of Table 8), Mann–Whitney U tests revealed no significant differences in peak torque for correct versus malpositioned trajectories during drilling ($U = 172.500$, $Z = -0.908$, $p = 0.364$, $r = 0.14$), tapping ($U = 103.5$, $Z = -0.541$, $p = 0.589$, $r = 0.10$), or screw insertion ($U = 98.500$, $Z = -0.022$, $p = 0.982$, $r = 0.00$). All observed effect sizes were negligible to small.

No visible differences could be discerned from the box plots of top-3-avg comparing the correct trajectories to the malpositioning during the drilling, thread cutting, and screw insertion phases in Figure 6.

A statistically significant difference in peak torque was observed between the correct trajectory and the superolateral malposition ($p = 0.038$) during the drilling phase. This indicates reduced torque performance in the presence of this deviation. However, all other comparisons within this phase—medial ($p = 0.605$), lateral ($p = 0.744$) and superior ($p = 0.521$)—showed no statistically significant differences ($p > 0.05$). In the tapping phase, none of the comparisons between the correct trajectory and the malpositioned trajectories yielded statistically significant differences in peak torque values (all $p > 0.05$). Similarly, in the screw insertion phase, no significant differences were found between correct placement and any of the tested malpositions (all $p > 0.05$). The detailed results can be found in Table 9 below.

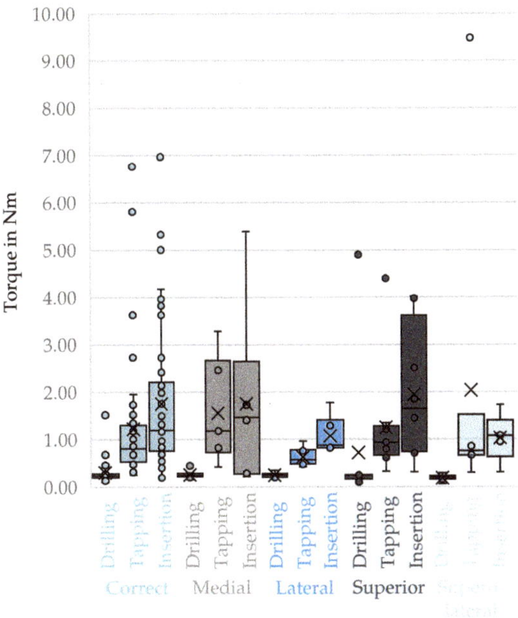

Figure 6. Top-3-avg of the phases drilling, Tapping, and screw insertion for correct trajectories compared to individual malpositions. X: mean value of the dataset, circles: outliers located more than 1.5 times the interquartile range.

Table 9. Mann–Whitney *U* test; correct versus malposition per phase.

Phase	Comparison	*N*	Asymp. Sig. *
Drilling	c vs. m	52/10	0.605
	c vs. l	52/9	0.744
	c vs. s	52/9	0.521
	c vs. sl	52/8	0.038
Tapping	c vs. m	41/6	0.232
	c vs. l	41/8	0.137
	c vs. s	41/9	0.649
	c vs. sl	41/7	0.872
Screw insertion	c vs. m	42/6	0.988
	c vs. l	42/6	0.427
	c vs. s	42/8	0.615
	c vs. sl	42/5	0.417

* two-sided. Trajectory classification: c = correct; l = lateral; m = medial; s = superior; sl = superolateral.

The correlation between top-3-avg and HU values of cortical bone (HU_{cb}) revealed a small but non-significant positive association ($\rho = 0.056$, $p = 0.200$, $N = 226$). Similarly, the correlation between top-3-avg and the HU values of spongy bone (HU_{sb}) showed a weak, non-significant association ($\rho = 0.050$, $p = 0.229$, $N = 226$) as shown in Table 10.

Table 10. Correlation tests for the dependence of the top-3-avg on derived bone density.

Spearman's ρ		**HU_{cb}**	**HU_{sb}**
Top-3-avg	Correlation coefficient	0.056	0.050
	Significance *	0.200	0.229
	N	226	226

* one-sided, $_{cb}$ cortical bone, $_{sb}$ spongy bone.

4. Discussion

4.1. Background and Objectives

Pedicle screw fixation is a widely used procedure, but malpositioning is common (with rates varying from 2% to 15% [3,4]. Such malpositions are one of the main causes of reoperations within 30 days of spinal surgery and are the most common instrumentation-related complication, particularly in pediatric deformities. According to experimental studies, higher insertion torque and better anchoring stability are achieved when the screws are placed correctly, while malpositioning is associated with reduced insertion torque and a higher risk of micro-movements and screw loosening [21]. This study aimed to identify mechanical markers that could reliably distinguish between correctly aligned and malpositioned screws. This was achieved by systematically examining the resistance and torque profiles during the drilling, thread cutting and screw insertion phases. It was hypothesized that malpositioned screws would exhibit distinct resistance curves and altered peak torque values compared to correctly inserted screws.

4.2. Interpretation of the Main Results

Contrary to the central hypothesis, the results showed that there was no statistically significant difference in the measured peak torque (top-3-avg, filtered) between correctly positioned and generally misaligned trajectories (pooled) across all vertebral segments and procedural phases (drilling, tapping and screw insertion). The effects were negligible.

Phase-related results: No significant differences in peak torque were found between correctly and malpositioned screws, either in the tapping phase ($p = 0.944$, $r = 0.01$) or in the screw-insertion phase itself ($p = 0.693$, $r = 0.05$). This suggests that, when measured as a top-3 average, peak torque may not be sufficiently sensitive to detect defined cortical breakthroughs (medial, lateral, superior and superolateral) in these later phases.

Regional analysis: The stratification of data into thoracic (T8–T12) and lumbar (L1–L5) spinal segments revealed no statistically significant differences in peak torque between correct and misaligned trajectories within individual phases.

Correlation with screw trajectories: Even though top-3-avg did not correlate with screw trajectories, this lack of correlation is most plausibly explained by the confounding effects of bone quality and engagement patterns. The nature of the defined malpositions often preserved cancellous bone contact, which also contributed to the lack of correlation. Additionally, while there may be characteristic signal features in the torque/resistance profile, they are likely too small to be detected by the current setup. Consequently, such features could not be reliably measured.

Correlation with vBMD and HU: Although the studies by Oh et al. [22] and Walter [21] suggest a correlation between torque and vBMD or HU, we were unable to confirm this in our investigation. This discrepancy may be explained by the fact that we used a controlled, standardized setup, whereas Oh et al. [22] used a surgical screw-in procedure via a conventional posterior approach and freehand technique. Walter [21] also used a machine-based setup but deliberately created screw channels that were 1 mm undersized.

4.3. Singular Significant Observation

The study identified only one statistically significant difference: A significant difference in peak torque ($p = 0.038$) was observed when comparing the correct trajectory with the superolateral malposition during drilling. This suggests that deviation in the superolateral direction can result in reduced torque performance. All other specific trajectory comparisons (correct vs. medial, correct vs. lateral, correct vs. superior) during drilling, thread cutting and screw insertion were not significant (all $p > 0.05$).

The fact that only the superolateral malposition showed a difference during drilling may suggest that loss of cortical resistance in the superior/lateral direction during initial pilot hole creation has a more immediate mechanical effect than strictly medial or lateral breakthroughs, which often maintain stronger cortical interactions until final positioning. In addition, the superolateral trajectory combines both angular deviation and cortical wall involvement, which alters the interaction between drill and bone compared to single-axis deviations. This dual deviation could lead to reduced bone stock along the drilling path and earlier cortical breach, resulting in a measurable change in resistance. Furthermore, the superolateral orientation frequently traverses thinner pedicle regions, amplifying local variations in bone density and cortical thickness. These anatomical and geometric factors together may explain why the superolateral malposition produced a distinct resistance signature, whereas other deviations did not. Future studies with larger sample sizes are required to determine whether this effect persists under clinical conditions.

4.4. Clinical Implications

The present findings have several important implications for the clinical management of spinal instrumentation during surgery, including error prevention and patient safety. While earlier studies [10–13] suggested that peak insertion torque may differ between correctly and malpositioned pedicle screws, our results confirm that relying solely on peak insertion torque is insufficient for detecting malpositioned pedicle screws. Since similar peak values can be produced by both correct and incorrect placements [22–25]. This insight is crucial for intraoperative decision-making as it prevents false reassurance and highlights the need for real-time monitoring of complex torque profiles and fluctuations in order to reliably capture subtle biomechanical changes [22–24].

Conversely, evaluating comprehensive torque profiles, including fluctuations, cyclic insertion torque (CIT) patterns and total resistance work (the integral of the torque curve), offers a more sensitive and specific approach to detecting subtle biomechanical changes during screw insertion [26–28]. Hee et al. [24] demonstrated that torque profiles differ significantly between cylindrical and conical screws during cortical wall breaches. Conical screws show a continuous increase in torque even after perforation, which can mask malposition events if only peak values are considered. Weidling et al. [28] mathematically demonstrated that the product of outer diameter and insertion torque correlates better with pull-out strength than peak torque alone. Meanwhile, Battula et al. [26] emphasized that optimizing pilot hole size requires consideration of the entire insertion process. Loggia et al. [27] recently described the CIT pattern during robot-assisted screw placement. This method provides both visual and haptic feedback, and may serve as a more reliable indicator of optimal screw trajectory and bone engagement.

The importance of resistance work and integral analysis is further supported by Martin et al. [29], who proposed a simulation method for determining intraoperative torque curves. This enables the development of standardized reference curves for screw insertion. These approaches could facilitate the development of intraoperative guidance systems that provide real-time feedback based on deviations from expected torque profiles [25,27,29]. Furthermore, Martin et al. [29] has proposed that continuous intraoperative torque monitoring could form the basis of machine learning–based predictive models to enhance intraoperative safety and training.

Our findings on side independence are corroborated by robust biomechanical evidence. Oh et al. [22] reported a strong correlation between left and right side torque values in correct insertions; Tai et al. [15] confirmed no significant stiffness differences between sides in controlled biomechanical tests; and Yang et al. [30] validated anatomical symmetry for bilateral pedicle screw fixation. This reproducibility strengthens the case for implementing torque profile analysis as a routine adjunct to navigation and robotic systems [25,27].

From a clinical perspective, these results support the integration of advanced intraoperative assistance systems. Sensor-based instruments, potentially enhanced by machine learning algorithms, could provide real-time feedback on insertion dynamics, including fluctuations, characteristic patterns, and cumulative resistance. This would alert surgeons to critical deviations from expected profiles [27,29]. These approaches align with the broader trend of data-driven, standardized surgical protocols, which improve intraoperative safety, enhance training, and improve quality assurance by providing objective, reproducible biomechanical parameters. Furthermore, integrating cumulative resistance and torque integral values enables personalized therapy planning, including tailoring implant design and pilot hole preparation to specific bone quality [26,28].

Looking ahead, larger, clinically oriented studies are required to validate these concepts under real surgical conditions. These trials should stratify patients by spinal region, bone quality, and deformity type to capture patient-specific variability. Prospective comparisons between conventional insertion and systems enhanced by dynamic profile monitoring would allow for an assessment of not only screw accuracy, but also clinical outcomes such as reoperation rates and neurological safety. Ultimately, establishing standardized reference curves across patient populations may provide the foundation for integration into intraoperative guidance systems and robotic platforms.

4.5. Limitations

The results and following conclusions of this study are subject to several limitations:

Sample size and statistical significance: Due to the very limited sample size of five human cadavers ($N = 5$), it was not possible to assume a normal distribution. Thus, non-parametric tests were necessary, such as the Wilcoxon signed-rank test and the Mann–Whitney U test. The small sample size severely limits the statistical power of the comparisons and increases the likelihood of a type II error, meaning subtle differences may have been overlooked. Although the highly standardized experimental setup and controlled trajectories strengthen internal validity, the generalizability of the findings to broader clinical populations is limited. Larger in vivo studies are required to validate whether the absence of discriminatory power persists under clinical conditions.

Standardised laboratory conditions: The experiments were conducted using a special test rig at a constant speed of $25\,\text{min}^{-1}$ and a defined feed force of 27.6 N, under highly standardized laboratory conditions. While this ensures repeatability, it does not fully reflect the variable speeds and forces applied during manual or navigation-guided insertion in vivo. In addition, variations in patient-specific anatomy, such as pedicle morphology, bone density, and spinal deformities, can substantially alter torque profiles. Intraoperative conditions, such as limited surgical exposure, bleeding, and soft tissue interference, can also obscure subtle resistance changes. These factors may reduce the detectability of malposition in clinical practice, highlighting the need for future studies under surgical conditions.

Use of drilling templates: Specific, 3D-printed drilling templates were used to ensure precise and consistent drilling trajectories. Although these templates could implement the planned positions systematically, they could not eliminate the variability and haptic feedback that would occur in an operating room.

Geometric definition of misalignment: Misalignments were defined as parallel displacements (offsets) of between 2.0 and 3.5 mm from the optimal path. Although the Gertzbein and Robbins [9] classification was used as a reference framework, the actual severity of the cortical breakthrough (e.g., the degree of perforation) was not quantified; instead, the mechanical response to the defined offset was examined.

Signal resolution and detection threshold: Due to their very small magnitude, potential signal features that could differentiate correct from incorrect trajectories could not be detected. We also examined the complete torque curves, but any characteristics indicating malposition appeared too subtle to be reliably captured with the current setup. Therefore, the statistical analysis was based on a robust, peak-related parameter (top-3-avg), which reduces the influence of transient fluctuations and ensures comparability across specimens.

5. Conclusions

This study shows that peak insertion torque alone is not a reliable indicator of pedicle screw malposition during surgery. No consistent or statistically significant differences were observed across all tested deviations and vertebral regions, with the sole exception of reduced torque during superolateral drilling. These findings contradict the initial hypothesis that malpositioning generally alters peak torque values, confirming that mechanical resistance alone lacks discriminatory power. Consistent with previous evidence on screw loosening, our results emphasize that intraoperative decision-making should not rely on peak torque feedback as the primary parameter.

Instead, a comprehensive analysis of the entire torque profile—including cyclic insertion torque patterns, resistance work and integral-based metrics—offers greater sensitivity in detecting subtle biomechanical changes and may provide a more robust foundation for intraoperative guidance systems. However, given the small sample size, the use of highly standardized laboratory conditions, and the geometric definition of misalignment, these conclusions should be regarded as proof of concept. Consequently, future research endeavors should pursue two primary avenues. Firstly, larger in vivo studies are needed to validate whether the absence of discriminatory power for peak torque persists under less predictable clinical conditions; secondly, and perhaps more promisingly, efforts should focus on developing standardized reference curves based on comprehensive torque metrics, which are likely essential for integration into navigation and robotic systems.

Supplementary Materials: The following supporting information can be downloaded at https://www.mdpi.com/article/10.3390/bioengineering12111254/S1: File S1: Experimental data.

Author Contributions: Conceptualization, S.K.; methodology, S.K. and F.M.; validation, S.K. and S.S.; formal analysis, S.K.; investigation, S.K., B.F. and J.S.; data curation, S.K.; writing—original draft, S.K.; writing—review and editing, S.K., B.F., J.S., F.M., T.W., C.-E.H. and S.S.; visualization, S.K.; project administration, S.K.; software, T.W.; resources, C.-E.H. and S.S.; supervision, S.S.; funding acquisition, S.K., C.-E.H. and S.S. All authors have read and agreed to the published version of the manuscript.

Funding: This study was funded by the Federal Ministry for Economic Affairs and Energy (grant number ZIM KK5575501SK3) and supported by the Open Access Publishing Fund of Leipzig University. The funders had no role in the study design, data collection and analysis, the decision to publish, or the preparation of the manuscript.

Institutional Review Board Statement: All tissues originated from the Institute of Anatomy of Leipzig University. Being part of the body donor program regulated by the Saxonian Death and Funeral Act of 1994 (third section, paragraph 18, item 8), the Ethics Committee of the University of Leipzig Medical Center obtained institutional approval for the use of postmortem tissues from the Institute of Anatomy at the University of Leipzig (ethical approval no. 129/21-ck, approval date 1 March 2023). The authors declare that all experiments were conducted according to the principles of the Declaration of Helsinki (as revised in 2013).

Informed Consent Statement: While alive, all body donors provided informed and written consent to donate their post-mortem tissues for education and research purposes.

Data Availability Statement: Body donor-related data were anonymized to prevent inferences about their identity. Additional data are available from the corresponding author upon reasonable request.

Acknowledgments: We extend our utmost gratitude to the body donors and their families who made this research possible.

Conflicts of Interest: The authors declare no conflicts of interest. The funders had no role in the design of the study; in the collection, analyses, or interpretation of data; in the writing of the manuscript; or in the decision to publish the results.

Abbreviations

The following abbreviations are used in this manuscript:

BMI	Body Mass Index
CIT	Cyclic Insertion Torque
CT	Computed Tomography
HU	Hounsfield unit
HU_{cb}	Hounsfield unit of cortical bone
HU_{sb}	Hounsfield unit of spongy bone
LVDT	Linear Variable Differential Transformer
QCT	Quantitative Computed Tomography
SPSS	Statistical Package for the Social Sciences
vBMD	Volumetric Bone Mineral Density

Symbols

The following symbols are used in this manuscript:

N	Number of observations (or number of pairs for hypothesis tests)
U	Mann–Whitney U statistic (rank-based test)
Z	Wilcoxon signed-rank statistic (standardized)

References

1. Burval, D.J.; McLain, R.F.; Milks, R.; Inceoglu, S. Primary pedicle screw augmentation in osteoporotic lumbar vertebrae: Biomechanical analysis of pedicle fixation strength. *Spine* **2007**, *32*, 1077–1083. [CrossRef]
2. Cho, W.; Cho, S.K.; Wu, C. The biomechanics of pedicle screw-based instrumentation. *J. Bone Jt. Surgery. Br. Vol.* **2010**, *92*, 1061–1065. [CrossRef]
3. Nevzati, E.; Marbacher, S.; Soleman, J.; Perrig, W.N.; Diepers, M.; Khamis, A.; Fandino, J. Accuracy of pedicle screw placement in the thoracic and lumbosacral spine using a conventional intraoperative fluoroscopy-guided technique: A national neurosurgical education and training center analysis of 1236 consecutive screws. *World Neurosurg.* **2014**, *82*, 866–871.e2. [CrossRef] [PubMed]
4. Woo, E.J.; DiCuccio, M.N. Clinically significant pedicle screw malposition is an underestimated cause of radiculopathy. *Spine J. Off. J. N. Am. Spine Soc.* **2018**, *18*, 1166–1171. [CrossRef] [PubMed]
5. Ouyang, H.; Hu, Y.; Hu, W.; Zhang, H.; Sun, Z.; Tang, Y.; Jiang, Y.; Chen, J.; Dong, S.; Li, W.; et al. Incidences, causes and risk factors of unplanned reoperations within 30 days of spine surgery: A single-center study based on 35,246 patients. *Spine J. Off. J. N. Am. Spine Soc.* **2022**, *22*, 1811–1819. [CrossRef] [PubMed]
6. Demura, S.; Ohara, T.; Tauchi, R.; Takimura, K.; Watanabe, K.; Suzuki, S.; Uno, K.; Suzuki, T.; Yanagida, H.; Yamaguchi, T.; et al. Incidence and causes of instrument-related complications after primary definitive fusion for pediatric spine deformity. *J. Neurosurg. Spine* **2023**, *38*, 192–198. [CrossRef]
7. Floccari, L.V.; Larson, A.N.; Crawford, C.H.; Ledonio, C.G.; Polly, D.W.; Carreon, L.Y.; Blakemore, L. Which Malpositioned Pedicle Screws Should Be Revised? *J. Pediatr. Orthop.* **2018**, *38*, 110–115. [CrossRef]
8. Aoude, A.; Ghadakzadeh, S.; Alhamzah, H.; Fortin, M.; Jarzem, P.; Ouellet, J.A.; Weber, M.H. Postoperative Assessment of Pedicle Screws and Management of Breaches: A Survey among Canadian Spine Surgeons and a New Scoring System. *Asian Spine J.* **2018**, *12*, 37–46. [CrossRef]
9. Gertzbein, S.D.; Robbins, S.E. Accuracy of pedicular screw placement in vivo. *Spine* **1990**, *15*, 11–14. [CrossRef]
10. Ge, D.W.; Chen, H.T.; Qian, Z.Y.; Zhang, S.; Zhuang, Y.; Yang, L.; Cao, X.J.; Sui, T. Biomechanical strength impact of lateral wall breach on spinal pedicle screw fixation. *Eur. Rev. Med. Pharmacol. Sci.* **2018**, *22*, 63–68. [CrossRef]
11. Goda, Y.; Higashino, K.; Toki, S.; Suzuki, D.; Kobayashi, T.; Matsuura, T.; Fujimiya, M.; Hutton, W.C.; Fukui, Y.; Sairyo, K. The Pullout Strength of Pedicle Screws Following Redirection After Lateral Wall Breach or End-plate Breach. *Spine* **2016**, *41*, 1218–1223. [CrossRef] [PubMed]

12. Maeda, T.; Higashino, K.; Manabe, H.; Yamashita, K.; Hayashi, F.; Goda, Y.; Tsuruo, Y.; Sairyo, K. Pullout Strength of Pedicle Screws Following Redirection After Lateral or Medial Wall Breach. *Spine* **2018**, *43*, E983–E989. [CrossRef]
13. Tsagkaris, C.; Fasser, M.R.; Farshad, M.; Passaplan, C.; Cornaz, F.; Widmer, J.; Spirig, J.M. Stability of medially and laterally malpositioned screws: A biomechanical study on cadavers. *Spine J. Off. J. N. Am. Spine Soc.* **2025**, *25*, 380–388. [CrossRef] [PubMed]
14. Nakashima, D.; Ishii, K.; Matsumoto, M.; Nakamura, M.; Nagura, T. A study on the use of the Osstell apparatus to evaluate pedicle screw stability: An in-vitro study using micro-CT. *PLoS ONE* **2018**, *13*, e0199362. [CrossRef] [PubMed]
15. Tai, C.L.; Chen, W.P.; Liu, M.Y.; Li, Y.D.; Tsai, T.T.; Lai, P.L.; Hsieh, M.K. Biomechanical comparison of pedicle screw fixation strength among three different screw trajectories using single vertebrae and one-level functional spinal unit. *Front. Bioeng. Biotechnol.* **2022**, *10*, 1054738. [CrossRef]
16. Isik, C.; Kose, K.C.; Inanmaz, M.E.; Tagil, S.M.; Sarman, H. The mechanisms of medial pedicle wall violation: Insertion method is as important as correct cannulation of the pedicle. *Adv. Orthop.* **2014**, *2014*, 283783. [CrossRef]
17. Oda, K.; Ohba, T.; Hiroshi, Y.; Fujita, K.; Tanaka, N.; Koyma, K.; Haro, H. Factors Affecting Pedicle Screw Insertional Torque in Spine Deformity Surgery. *Spine* **2021**, *46*, E932–E938. [CrossRef]
18. Sterba, W.; Kim, D.G.; Fyhrie, D.P.; Yeni, Y.N.; Vaidya, R. Biomechanical analysis of differing pedicle screw insertion angles. *Clin. Biomech.* **2007**, *22*, 385–391. [CrossRef]
19. Metzner, F.; Reise, R.; Heyde, C.E.; von der Höh, N.H.; Schleifenbaum, S. Side specific differences of Hounsfield-Units in the osteoporotic lumbar spine. *J. Spine Surg.* **2024**, *10*, 232–243. [CrossRef]
20. Brett, A.D.; Brown, J.K. Quantitative computed tomography and opportunistic bone density screening by dual use of computed tomography scans. *J. Orthop. Transl.* **2015**, *3*, 178–184. [CrossRef]
21. Walter, F. Testung der Primärstabilität Unzementierter und PMMA-Augmentierter Pedikelschrauben in Einem In-Vitro-Modell mit Osteoporotischen Knochen. Ph.D. Thesis, Universität Tübingen, Tübingen, Germany, 2018. [CrossRef]
22. Oh, B.H.; Kim, J.Y.; Lee, J.B.; Hong, J.T.; Sung, J.H.; Than, K.D.; Lee, H.J.; Kim, I.S. Screw Insertional Torque Measurement in Spine Surgery: Correlation With Bone Mineral Density and Hounsfield Unit. *Neurospine* **2023**, *20*, 1177–1185. [CrossRef]
23. Inceoglu, S.; Ferrara, L.; McLain, R.F. Pedicle screw fixation strength: Pullout versus insertional torque. *Spine J.* **2004**, *4*, 513–518. [CrossRef]
24. Hee, H.T.; Khan, M.S.; Goh, J.C.; Wong, H.K. Insertion torque profile during pedicle screw insertion of the thoracic spine with and without violation of the pedicle wall: Comparison between cylindrical and conical designs. *Spine* **2006**, *31*, E840–E846. [CrossRef] [PubMed]
25. Mulvaney, G.; Sowah, M.N.; DeCarlo, R.; Drossopoulos, P.; Holland, C.M.; Bohl, M.A. Biomechanical Analysis of Pedicle Screw Reinsertion Along the Same Trajectory in a Validated 3D-Printed Synthetic Bone Model. *World Neurosurg.* **2024**, *190*, e675–e683. [CrossRef] [PubMed]
26. Battula, S.; Schoenfeld, A.J.; Sahai, V.; Vrabec, G.A.; Tank, J.; Njus, G.O. The effect of pilot hole size on the insertion torque and pullout strength of self-tapping cortical bone screws in osteoporotic bone. *J. Trauma* **2008**, *64*, 990–995. [CrossRef] [PubMed]
27. Loggia, G.; Avrumova, F.; Lebl, D.R. Visuohaptic Feedback in Robotic-Assisted Spine Surgery for Pedicle Screw Placement. *J. Clin. Med.* **2025**, *14*, 3804. [CrossRef]
28. Weidling, M.; Heilemann, M.; Schoenfelder, S.; Heyde, C.E. Influence of thread design on anchorage of pedicle screws in cancellous bone: An experimental and analytical analysis. *Sci. Rep.* **2022**, *12*, 8051. [CrossRef]
29. Martin, F.L.; Seidler, A.; Paetzold-Byhain, K. Proposal for a simulation method to determine the intraoperative torque load for pedicle screws with cellular functional areas. In Proceedings of the DS 125: Proceedings of the 34th Symposium Design for X (DFX2023), Radebeul, Germany, 14–15 September 2023; pp. 215–224. [CrossRef]
30. Yang, S.; Sun, T.; Zhang, L.; Cong, M.; Guo, A.; Liu, D.; Song, M. Stress Distribution of Different Pedicle Screw Insertion Techniques Following Single-Segment TLIF: A Finite Element Analysis Study. *Orthop. Surg.* **2023**, *15*, 1153–1164. [CrossRef]

bioengineering

MDPI

Article

Parametric Finite Element Evaluation of Load Redistribution Under Progressive Lumbar Disc Degeneration

Oleg Ardatov [1,2,*], Sofia Rita Fernandes [3], Artūras Kilikevičius [2] and Vidmantas Alekna [1]

1 Faculty of Medicine, Vilnius University, LT-03101 Vilnius, Lithuania; vidmantas.alekna@mf.vu.lt
2 Faculty of Mechanics, Vilnius Gediminas Technical University, LT-10223 Vilnius, Lithuania; arturas.kilikevicius@vilniustech.lt
3 Instituto de Biofísica e Engenharia Biomédica, Faculdade de Ciências, Universidade de Lisboa, 1749-016 Lisboa, Portugal; srcfernandes@ciencias.ulisboa.pt
* Correspondence: oleg.ardatov@vilniustech.lt

Abstract

This study presents a finite element (FE) investigation of intervertebral disc (IVD) degeneration in the human lumbar spine (L1–L3 segment). The model, based on CT-derived geometry and isotropic hyperelastic representation of disc tissues, incorporates controlled simplifications, detailed in the limitations section. Degenerative changes were parametrically simulated across healthy, mild, moderate, and severe stages by reducing disc height (up to 60%), nucleus pulposus volume (up to 70%), and adjusting tissue stiffness to reflect dehydration and fibrosis. Displacement-controlled compressive loading was applied to assess von Mises stress distributions, reaction forces, and load transfer mechanisms. Results indicate significant load redistribution: annulus fibrosus stresses increased by up to 175% in severe degeneration, while nucleus pulposus stresses decreased by ~70%, indicating a diminished compressive load-bearing contribution of the nucleus. Model predictions were validated against cadaveric and in vivo data, confirming trends in intradiscal pressure (IDP) reductions (40–70%) and stress elevations. The parametric framework elucidates interactions between geometric and material changes, providing clinicians with insights into degeneration progression and guiding biomedical engineers in implant design and interventions.

Keywords: annulus fibrosus; degenerative spinal diseases; finite element method; intervertebral discs; material modeling; nucleus pulposus; numerical modeling; spinal biomechanics

1. Introduction

Degeneration of the intervertebral disc involves structural and biochemical alterations that impair its mechanical function. Key manifestations include reduced water and proteoglycan content, loss of disc height, annular tears, endplate sclerosis, and osteophyte formation, all of which influence spinal load distribution and mobility [1].

One of the earliest degenerative changes is the decreased hydration of the nucleus pulposus due to a decline in proteoglycan concentration. Water content, which is approximately 85% in young individuals, drops to about 75% in older age, accompanied by a reduction in osmotic pressure from ~0.19 MPa at age 37 to ~0.05 MPa at age 91 [2]. This diminishes the disc's ability to absorb loads and maintain uniform pressure distribution.

Progressive matrix degradation also alters tissue mechanical properties. Early degeneration may increase segmental flexibility due to reduced intradiscal pressure and annular weakening. At advanced stages, fibrosis and calcification lead to tissue stiffening and restricted mobility [2,3]. Accordingly, the elastic modulus decreases in the nucleus pulposus, while the annulus fibrosus may exhibit either reduced elasticity or pathological stiffening depending on the degeneration stage.

Geometric changes in the disc include a reduction in its height, which is particularly pronounced in the later stages of degeneration. Normally, even with aging, the height of the intervertebral disc remains relatively stable or may even slightly increase, but in pathological degeneration, a significant reduction in disc space height occurs, leading to decreased mobility of the spinal segment and alterations in the mechanics of adjacent vertebrae. The loss of height is associated with both dehydration of the nucleus pulposus and the destruction of structural elements in the annulus fibrosus. Additionally, thinning and weakening of the annulus fibrosus occur, increasing the risk of tears and fissures [4]. The studies [5–7] also provide specific data on the progressive reduction in intervertebral disc height due to degeneration. In cases of mild degeneration, the disc height decreases by 20%, while in moderate degeneration, the reduction reaches 40%. In severe degeneration, the height loss becomes even more pronounced, reaching 60%. These structural changes are accompanied by a reduction in the volume of the nucleus pulposus, leading to an increased load on the annulus fibrosus and alterations in the distribution of mechanical forces within the spinal segment.

Numerical modeling, particularly the finite element method, has proven to be an essential tool in studying the biomechanics of intervertebral discs, providing detailed insights into mechanical responses under various loading scenarios, tissue configurations, and degenerative conditions challenging to replicate experimentally [8,9]. Zhu et al. [10] utilized FE modeling to explore how disc degeneration affects the mechanical and electrical signals at the interface between the disc and vertebral bodies, highlighting its importance for understanding vertebral adaptation and remodeling processes. Holzapfel and Ogden [11] introduced an analytical model accounting for residual stresses and their effect on anisotropic disc behavior under different loading scenarios. Ghezelbash et al. [12] and Kandil et al. [13] adopted advanced microstructural-based modeling approaches to accurately represent the heterogeneous collagen and elastic fiber networks of the annulus fibrosus, demonstrating the necessity of detailed structural modeling to accurately reproduce mechanical disc responses. Xi et al. [14] examined the effects of fibrosis and disc-height reduction, concluding that these factors collectively alter load distributions, promoting vertebral osteoporosis and joint degeneration.

Despite extensive FE research on lumbar biomechanics, many degeneration models vary only one factor at a time (e.g., disc height, nucleus dehydration, or tissue stiffness) rather than combining these changes within a unified parametric framework. This limits assessment of how degenerative mechanisms interact to drive load redistribution.

To address this gap, we propose an integrated parametric degeneration model that simultaneously varies disc height (20–60%), nucleus pulposus volume (25–70%), and soft-tissue stiffness across four stages. The main contribution is a sensitivity-aware, mechanically consistent framework for comparative trends rather than identification of new biomechanical phenomena.

Applied to a two-level L1–L3 segment with L2–L3 kept intact as an internal reference, the approach enables systematic quantification of coupled geometric–material effects, including adjacent-segment interactions. Verification against in vivo and cadaveric trends, together with sensitivity analysis, supports robustness. Overall, the framework is reproducible and modular for comparative biomechanical analysis and engineering applications such as implant design.

https://doi.org/10.3390/bioengineering13020234

2. Materials and Methods

2.1. Initial Data and Numerical Model Development

The numerical model was developed using computed tomography scans of a sixty-year-old woman exhibiting typical age-related degenerative changes and involved several sequential steps. Initially, using the 3D Slicer program [15], the projections were aligned into three-dimensional geometry of the vertebral bodies (L1–L3), which was then converted to STL format. CT data were processed using threshold-based segmentation to isolate bony structures (global threshold range 200–2000 HU), followed by manual refinement around the endplates and pedicles. During the segmentation of the bony components, special attention was given to approximate the patient-specific sagittal curvature of the L1–L3 segment based on the midsagittal CT contours, while mesh smoothing was kept minimal to avoid altering the overall alignment. After extracting the L1–L3 vertebrae, surface meshes were decimated to reduce the initial number of elements (from approximately 60,000 to 39,000 per vertebra) while preserving anatomical fidelity. Subsequently, using the MeshLab 2022.22 software package [16], the geometry underwent noise reduction and smoothing. Mesh smoothing was performed using a Laplacian smoothing filter (2 iterations, relaxation factor 0.3) to remove noise artefacts without altering the overall geometry. Then, the processed STL file was exported to the SolidWorks 2024 software package [17], where using the 3D Scan module, the external geometry of the vertebrae was processed further. In SolidWorks, the model was refined by creating an internal continuous cavity representing cancellous bone, while assigning a thickness of 0.5 mm to the outer shell reflecting cortical bone [18]. The intervertebral discs (L1–L2 and L2–L3), consisting of the annulus fibrosus and the nucleus pulposus, were subsequently added using SolidWorks Features tool. The heights of the intervertebral discs were set to 10 mm, which corresponds to the average height of a healthy intervertebral disc [19]. At this stage of modeling, special attention was paid to accurately represent the cartilaginous endplates, which were assigned a uniform thickness of 0.5 mm [18]. The intact endplate assumption was adopted to isolate the mechanical effects of disc degeneration from endplate-related pathology. While endplate sclerosis commonly accompanies advanced degeneration and would locally increase stiffness and alter stress transfer to the trabecular bone, modeling it would introduce additional parametric uncertainty (e.g., sclerosis extent, mineralization degree). By keeping the endplate unchanged across all stages, we ensured that observed stress redistribution patterns are attributable solely to disc geometric and material changes. This simplification may underestimate local stress concentrations at the bone–disc interface in moderate and severe cases but does not affect the comparative trends central to this study.

Another notable simplification concerns the trabecular tissue representation. In this model, trabecular bone is depicted as a continuous solid body, visible in cross-section (Figure 1b). This discrepancy from actual anatomy is partially offset by reducing the assigned modulus of elasticity, as detailed further in Section 2.3. This approach captures the macroscopic load-bearing contribution of trabecular bone while avoiding the complexity and uncertainty associated with patient-specific trabecular microarchitecture. As the study focuses on comparative behavior between degeneration stages, this level of approximation is sufficient for the intended analysis. An additional significant simplification pertains to the perfectly flat surfaces at the external boundaries, specifically the lower endplate of vertebra L3 and the upper surface of vertebra L1, clearly illustrated in Figure 1b. This simplification was deliberately chosen to facilitate stable fixation and effective application of compressive loading, which constitutes the focus of this study. Conversely, the vertebral endplate surfaces adjacent to the intervertebral discs L1–L2 and L2–L3 were modeled with particular care, preserving their natural curvature due to the substantial impact this geometry has on biomechanical behavior.

https://doi.org/10.3390/bioengineering13020234

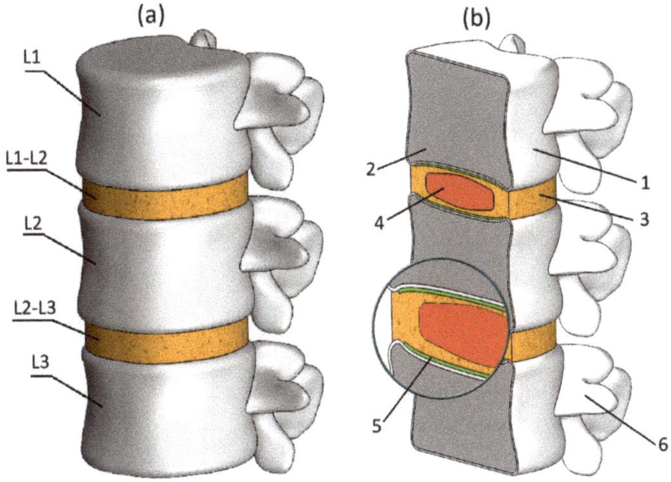

Figure 1. (**a**) Trimetric view of numerical lumbar spine model; (**b**) Section view of the model: 1—cortical shell, 2—cancellous bone, 3—annulus fibrosus, 4—nucleus pulposus, 5—bony endplate, 6—posterior elements.

2.2. Modelling the Geometrical Impact of Degenerative Changes on IVD

In the present study, the geometrical effects of intervertebral disc degeneration were modeled at the L1–L2 level by modifying two primary morphological parameters: disc height and nucleus pulposus volume. Specifically, mild, moderate, and severe degenerative stages correspond to reductions in disc height by 20%, 40%, and 60%, respectively, compared to a healthy intervertebral disc [4]. Both anterior and posterior disc heights were decreased proportionally during the modelling process. The reduction in disc height was applied uniformly across the anterior–posterior direction to represent the predominant pattern of symmetric disc space narrowing commonly observed in early-to-moderate degenerative disc disease without significant sagittal imbalance or endplate deformity. While advanced degeneration may sometimes produce anterior wedge-shaped collapse (particularly in cases with kyphotic deformity or asymmetric endplate damage), uniform reduction was chosen here to maintain comparability across degeneration stages and to focus on the direct biomechanical consequences of height and volume loss rather than postural adaptations.

The percentage reductions in disc height and nucleus pulposus volume were selected within the empirical ranges reported in radiological and MRI-based studies of lumbar degeneration, where advanced stages commonly exhibit marked height loss and nucleus shrinkage [5–7]. These values were therefore selected to reflect typical structural progression rather than patient-specific anatomy. The use of four sequential degeneration stages also provides an inherent sensitivity-type assessment of how progressively increasing geometric and material alterations affect mechanical response.

The L2–L3 disc was intentionally kept unchanged and served as an internal reference, allowing us to clearly identify the mechanical effects produced solely by the degenerative alterations at L1–L2.

Furthermore, with the progression of degeneration, the cross-sectional area (on the horizontal plane) of nucleus pulposus is reduced by 25% (mild degeneration), 50% (moderate degeneration), and ultimately to 70% (severe degeneration). The cross-sectional views of the degenerative models are presented in Figure 2.

https://doi.org/10.3390/bioengineering13020234

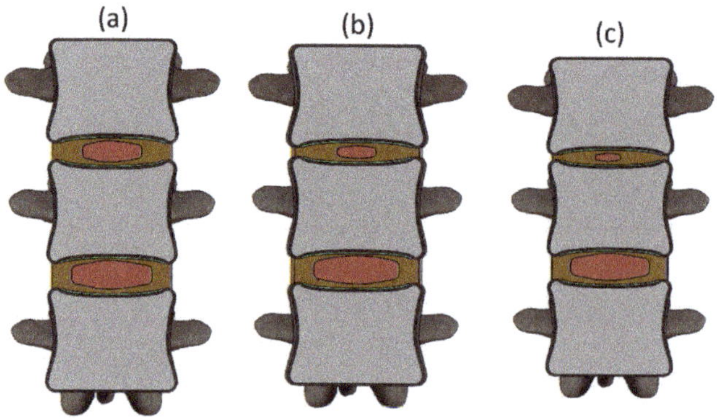

Figure 2. Different stages of the L1–L2 IVD degeneration (sectioned frontal view): (**a**) Mild degeneration; (**b**) Moderate degeneration; (**c**) Severe degeneration.

As seen in Figure 2, each stage of degeneration is characterized by progressive reductions in disc height and nucleus pulposus volume. In model 2a (mild degeneration), the disc height is slightly reduced, and the nucleus pulposus retains most of its volume, occupying a substantial portion of the disc space. In model 2b (moderate degeneration), the reduction in disc height becomes more pronounced, and the nucleus pulposus is visibly smaller in cross-sectional area. Finally, in model 2c (severe degeneration), there is a substantial collapse of the intervertebral disc space, and the nucleus pulposus is significantly diminished, appearing as a narrow horizontal structure.

Anterior and posterior height reductions were applied proportionally as a modeling simplification consistent with the percentage-based parametrization of disc height loss. This approach ensured controlled geometric modification without introducing additional asymmetry-related effects.

2.3. Mechanical Properties of Model Components

Bone structures as well as the cartilaginous endplates were modeled as perfectly elastic, homogeneous continuum materials. The mechanical properties assigned to these components, such as Young's modulus and Poisson's ratio, were selected based on consistently reported values in previously published studies [4,18,20] and are summarized in Table 1.

Table 1. Mechanical properties of bone and endplates.

Model Component	Young's Modulus, MPa	Poisson's Ratio	Yield Strength, MPa
Cortical bone	8000	0.3	64
Cancellous Bone	100	0.3	
Posterior elements	3500	0.3	
Endplates	24	0.4	

The soft tissues (annulus fibrosus and nucleus pulposus) were modelled as hyperelastic isotropic bodies. The Mooney–Rivlin (M-R) material model was applied, where M-R strain energy density function is defined as a two-constant formulation [21]:

$$w = C_1(I_1 - 3) + C_2(I_2 - 3) + \frac{1}{2}K(I_3 - 1)^2, \quad (1)$$

where C_1 and C_2 are the first and second material constants, respectively, associated with the distortion response, while K is the material constant governing the volumetric response.

https://doi.org/10.3390/bioengineering13020234

The terms I_1, I_2 and I_3 represent the reduced invariants of the Cauchy–Green deformation tensor, defined through the principal stretch ratios.

The volumetric material constant K can be expressed as follows:

$$K = \frac{6(C_1 + C_2)}{3(1 - 2v)}, \tag{2}$$

where v represents the Poisson's ratio.

The annulus fibrosus was modeled as an isotropic hyperelastic material (Mooney–Rivlin), which is a major simplification compared to its real anisotropic fiber-reinforced structure. Comparative studies indicate that isotropic or simplified embedded-fiber annulus formulations can be less accurate than fully anisotropic models for predicting absolute range of motion (R^2 = 0.69–0.89 vs. 0.95) and intradiscal pressure (R^2 = 0.80–0.86 vs. 0.87) [12,22]. However, the largest reported discrepancies in local stress–strain fields (up to 16-fold) are largely associated with differences in specimen geometry and boundary conditions, rather than isotropy alone [12,22]. Accordingly, when the same isotropic constitutive model and loading setup are applied consistently across all degeneration stages—as in the present study—this simplification is expected to preserve comparative degeneration-driven trends, while absolute peak stresses remain sensitive to constitutive choices; for example, peak AF von Mises stresses may vary by >100% between hyperelastic formulations [23].

The degree of degeneration of the annulus fibrosus and nucleus pulposus was simulated by altering the elastic constants, assuming that higher effective stiffness corresponds to a more deteriorated tissue state [24,25]. The range of elastic constants for cartilage was derived from previously reported studies [19,26–29]. The values of the constants C_1 and C_2 are summarized in Table 2. The Poisson's ratio was considered constant and set at 0.45 for the annulus fibrosus [30] and 0.4995 for the nucleus pulposus [30].

Table 2. Elastic constants of IVD components accordingly to degeneration [19,26–29].

Degeneration Stage	C_1, MPa (AF)	C_2, MPa (AF)	C_1, MPa (NP)	C_2, MPa (NP)
Healthy	0.18	0.045	0.12	0.030
Mild	0.40	0.100	0.14	0.035
Moderate	0.60	0.150	0.17	0.041
Severe	0.90	0.230	0.19	0.045

It should be noted that material properties for model components were taken from experimentally established studies commonly used in FE modeling of the lumbar spine. Because the literature lacks a unified, age- and sex-specific dataset, values were selected from sources reporting physiologically consistent ranges rather than matched patient-specific properties.

2.4. Problem Formulation

To accurately capture the mechanical response of the lumbar spine—especially under conditions involving large deformations and the nonlinear material behavior characteristic of intervertebral disc tissues—the framework of nonlinear elasticity was employed in this study. This approach enables a more realistic simulation of the complex stress–strain relationships occurring within hyperelastic soft tissues, such as the annulus fibrosus and nucleus pulposus, when subjected to physiological loading.

In this study a nonlinear static analysis was performed. Although the solver internally uses the standard dynamic equilibrium formulation (including mass and damping matrices), inertial and damping terms were suppressed, and the solution proceeded through incremental load steps combined with the Newton–Raphson iterative method:

https://doi.org/10.3390/bioengineering13020234

$$^{t+\Delta t}[K]^{(i)}\{\Delta U\}^{(i)} = {}^{t+\Delta t}\{R\}^{(i)}, \quad (3)$$

where $^{t+\Delta t}\{R\}^{(i)}$ denotes the effective load vector, $[\Delta U]^{(i)}$ is the vector of incremental displacements at iteration (*i*) and $^{t+\Delta t}[K]^{(i)}$ represents the effective stiffness matrix. The three-dimensional nonlinear analysis was performed using Intel Direct Sparse solver.

It should be noted that the time increments reported by the software [17] correspond to load-increment steps, not to physical time integration. At each increment, the stiffness matrix was updated, and the equilibrium was solved in a fully nonlinear manner.

To assess the stress distribution within the intervertebral disc and surrounding bony structures, the von Mises stress criterion was employed. Although originally developed for isotropic, ductile metals, this scalar metric has gained widespread acceptance in biomechanical simulations [18–20,31–36] due to several advantages. First, von Mises stress provides a single-value representation of the multi-axial stress state, facilitating comparative analysis across complex anatomical regions. Second, despite the anisotropic and heterogeneous nature of biological tissues, studies [31,35,36] have demonstrated that von Mises stress correlates well with regions prone to mechanical failure or tissue remodeling, particularly in cortical and trabecular bone, as well as in the annulus fibrosus under compression and bending.

In the context of intervertebral discs, where large deformation gradients occur, von Mises stress remains a practical indicator for identifying peak stress zones and assessing the structural integrity of hyperelastic materials [31–34]. It is also widely used for evaluating load-bearing performance in bone tissue, highlighting regions susceptible to micro-damage accumulation or structural failure [34]. At the same time, von Mises stress does not capture direction-dependent fiber mechanics of the annulus fibrosus and therefore cannot fully describe anisotropic tissue behavior. For this reason, it is interpreted here as a relative indicator rather than a physiologically complete measure.

The von Mises criterion is expressed in Equation (4):

$$\sigma_y = \sqrt{\frac{(\sigma_1 - \sigma_2)^2 + (\sigma_2 - \sigma_3)^2 + (\sigma_3 - \sigma_1)^2}{2}}, \quad (4)$$

where σ_1, σ_2 and σ_3 are the maximum, intermediate, and minimum principal stresses respectively, and σ_y is a von Mises stress.

2.5. Boundary Conditions and Finite Element Mesh

The loading configuration is illustrated in Figure 3a. The inferior surface of the L3 vertebra was fully constrained to prevent any motion, while a uniform prescribed displacement of 2.5 mm was applied directly to the superior endplate of L1 across all models developed in this study. The loaded surface was defined as a rigid kinematic boundary (equivalent to a rigid plate), ensuring all nodes moved uniformly. This combination of fixed inferior constraint and rigid displacement loading is known to influence apparent segment stiffness and reaction forces by restricting global deformation modes and enforcing uniform axial compression.

Alternative boundary condition formulations, such as force-controlled loading, compliant endplate constraints, or follower load application, would allow different load-sharing mechanisms and deformation patterns to develop, potentially affecting absolute stiffness and reaction force magnitudes. However, since the present study applies identical boundary conditions across all degeneration stages, the observed comparative trends—driven by progressive disc height reduction and material property changes—remain internally consistent and suitable for relative analysis.

https://doi.org/10.3390/bioengineering13020234

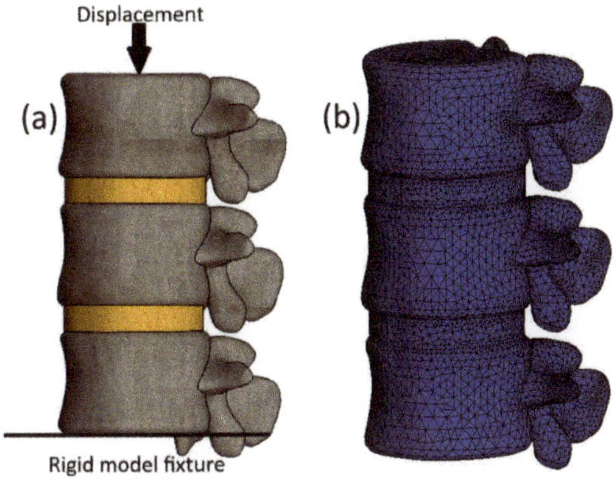

Figure 3. 3D model of a healthy lumbar spine: (**a**) Loading scheme; (**b**) Finite element mesh.

The prescribed 2.5 mm axial displacement was selected to represent a moderate, non-extreme compression range for a two-disc lumbar segment (L1–L3). This corresponds to approximately 1–1.5 mm of disc compression per level [37], which is consistent with expected deformation under moderate compressive loading in neutral posture.

Displacement-controlled loading was used to impose identical kinematic conditions across all models, enabling direct comparison between healthy and degenerated discs and isolating the effects of geometric and material changes on stress distribution. This approach improves reproducibility and numerical stability by avoiding stiffness-dependent deformation variability typical of force-controlled loading. Follower loads and muscle forces were omitted to maintain a controlled setup; although this affects absolute stiffness and load sharing, comparative trends across degeneration stages remain valid.

Facet articulation in the present study was represented using two formulations: (1) no facet contact in the healthy and mild stages; and (2) facet contact in the moderate and severe stages. Under pure axial compression in neutral posture, facet joints are expected to contribute minimally to load sharing in healthy lumbar segments [38]. Thus, the no-contact assumption (the facet contact solver was inactive) provides a reasonable approximation of low facet involvement during early degeneration. Conversely, disc height loss in advanced degeneration can reduce joint space and promote posterior approximation; the contact formulation therefore serves as an upper-bound scenario that intentionally exaggerates posterior load transfer [39] to explore its mechanical consequences. Therefore, the observed load shift to posterior elements in moderate and severe degeneration stages also results from this degeneration-driven modeling approach. The coefficient of friction was set to 0.5 for the moderate degeneration model (corresponding to Grade 2 per Weishaupt classification) and 0.9 for the severe degeneration model (Grade 4, simulates near-complete cartilage loss and high frictional resistance typical of advanced osteoarthritis) [39].

The finite element mesh consisted of quadratic 10-node tetrahedral elements (Figure 3b), chosen to accurately represent complex vertebral geometry, particularly in the posterior elements and endplate interfaces. This formulation enables efficient local refinement, minimizes distortion, and ensures stable and accurate stress computation. Mesh convergence was verified through successive refinements, with changes in maximum von Mises stress and displacement remaining below 5% (Table 3). All models exhibited sufficient spatial resolution, with element sizes ranging from 0.0029 to 3 mm. Although the severe model contains fewer elements due to geometric volume loss, the element size in the L1–L2 disc region was controlled to maintain sufficiently fine local mesh density in the

https://doi.org/10.3390/bioengineering13020234

remaining thin disc, and the reported peak AF stress (6.3 MPa) remained within the <5% convergence threshold under mesh refinement.

Table 3. Finite element mesh parameters for calculation cases reflecting different stages of IVD degeneration.

Model	Number of Elements	Number of Nodes	Number of Degrees of Freedom	Element Size, mm	Skewness; Aspect Ratio	Number of Jacobian Points
Healthy	576,345	704,576	2,111,946			
Mild	559,054	683,457	2,009,363	0.0029–3	<0.8; <3	16
Moderate	536,692	656,108	1,922,396			
Severe	509,857	623,296	1,832,490			

Each simulation required approximately 345 min of computation time on a workstation equipped with a 3.2 GHz quad-core AMD Ryzen 5 processor and 16 GB of RAM.

2.6. Modeling Assumptions and Their Implications for Interpretation

The finite element framework employed in this study relies on a set of deliberate modeling assumptions introduced to enable a controlled and systematic investigation of degeneration-related load redistribution in the lumbar spine. These assumptions ensure numerical robustness, minimize geometric and material uncertainty, and allow consistent comparison across degeneration stages under identical loading conditions. Table 4 summarizes the key simplifications, their rationale, and implications.

Table 4. Summary of key modeling simplifications, their rationale, and implications for interpretation.

Simplification	Justification	Implications for Interpretation
Absence of follower load and muscle forces	Avoids additional assumptions related to curvature-dependent load paths and subject-specific muscle activation, preserving a controlled parametric framework.	Load transmission occurs through simplified axial compression; deformation modes and reaction forces may differ from physiological loading scenarios, while comparative degeneration-driven trends remain interpretable.
No ligament structures	Avoids adding nonlinear, tension-only ligaments with subject-specific properties and attachment uncertainty, ensuring a controlled, reproducible comparison across degeneration stages under compression load.	The segmental mechanical response under compression primarily reflects disc- and bone-driven behavior. Ligament-mediated stabilization is not included, so results characterize internal load redistribution in the absence of soft-tissue tension contributions.
Flat endplates	Controls geometric variability and isolates the mechanical contribution of disc degeneration without additional curvature-related effects.	Flattened boundaries may over constrain the segment, restricting small, coupled motions under compression and thereby increasing apparent stiffness and potentially inflating absolute reaction forces.
Homogeneous cancellous bone representation	CT-based heterogeneity was not the focus; uniform cancellous properties improve numerical robustness and reduce computational cost.	Local stress heterogeneity within vertebral bodies is not resolved; the vertebral body response represents an effective, averaged mechanical behavior.

https://doi.org/10.3390/bioengineering13020234

Table 4. *Cont.*

Simplification	Justification	Implications for Interpretation
Isotropic hyperelastic annulus fibrosus	Prioritizes numerical stability and enables a consistent parametric degeneration framework; explicit fiber-level anisotropy was beyond the intended scope.	The annulus response represents bulk mechanical behavior; direction-dependent fiber effects are not explicitly captured, which may influence absolute stress magnitudes.
Fixed inferior constraint and rigid displacement loading	Ensures numerical stability and reproducible kinematic conditions across all degeneration stages, allowing controlled comparison of disc-driven mechanical effects.	The flattened external boundaries may increase apparent stiffness. However, relative mechanical trends across degeneration stages remain valid due to consistent application.
Single-sample geometry	Reduces inter-subject anatomical variability and isolates the mechanical effects of progressive disc degeneration, enabling a controlled comparison across degeneration stages without confounding geometric differences.	The results reflect degeneration-driven mechanical trends within a fixed anatomical configuration and may not capture population-level variability associated with subject-specific geometry; therefore, findings should be interpreted as mechanistic rather than statistical generalizations.

Although the assumptions and lack of direct experimental validation prevent the model from being a fully comprehensive physiological representation, it provides a robust numerical basis for identifying characteristic stress-transfer patterns in progressive disc degeneration. Results should be interpreted as relative computational indicators rather than definitive physiological predictions.

Within this framework, the model elucidates how coupled geometric and material changes (disc height reduction, nucleus pulposus volume loss, and tissue stiffening) affect load sharing between the disc, vertebral bodies, and posterior elements. The consistent parametric structure enables transparent examination of these relative influences across degeneration stages.

3. Results and Discussion

3.1. Stress Analysis Across Degeneration Stages in the L1–L3 Segment

The von Mises stress plots for models with different degeneration types in case of 2.5 mm displacement are presented in Figure 4.

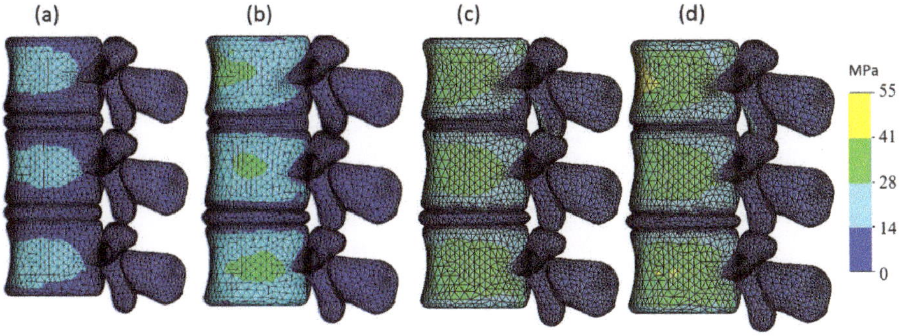

Figure 4. Von Mises stress plots of the three-dimensional model with different stages of IVD degeneration: (**a**) Healthy IVD model; (**b**) Model with mild IVD degeneration; (**c**) Model with moderate IVD degeneration; (**d**) Model with severe IVD degeneration.

https://doi.org/10.3390/bioengineering13020234

In the healthy model (Figure 4a), most stress is concentrated in the cortical shell of the vertebral bodies with stress magnitudes predominantly ranging between 14 and 28 MPa (cyan zones). The distribution appears uniform, while stress within the intervertebral discs and posterior elements remains minimal, typically below 14 MPa (blue zones).

In the mild degeneration model (Figure 4b), stress increases slightly in the vertebral bodies, with more prominent green areas (peak value reaches ~40 MPa), especially in the central areas of the vertebral bodies. With moderate degeneration (Figure 4c), stress further intensifies and green areas (~40 MPa) become more widespread in the cortical regions; however, higher levels (above 41 MPa) on the cortical shell are still not observed. In the severe degeneration model (Figure 4d), the stress becomes even more pronounced: the cortical bone exhibits larger green regions approaching the 41 MPa threshold; however, even in this case, the peak cortical stress does not exceed 55 MPa.

In the healthy and mildly degenerated models, the posterior elements remained largely unloaded, whereas in the moderate and severe stages these structures began to carry noticeable stress, reaching up to 55 MPa in the severe case. This behavior results from the axial displacement-controlled loading used in the present study: under pure compression, the facet joints do not engage in the healthy configuration, and their involvement appears only when disc height loss reduces the joint space. The simplified representation of facet articulation employed here does not reproduce physiological facet mechanics, as no contact formulation was solved and neither contact area nor frictional behavior was modeled. Therefore, the absolute stress magnitudes should be interpreted cautiously. Nevertheless, the increase in posterior loading with progressive disc height reduction reflects a mechanically plausible consequence of reduced joint spacing under degeneration.

3.2. Analysis of Reaction Forces Under Different Degrees of Disc Degeneration

Figure 5 illustrates the reaction forces measured on the upper surface of the L1 vertebra in response to applied displacement across different stages of intervertebral disc (IVD) degeneration. As observed, the magnitude of reaction force progressively increases with the severity of degeneration.

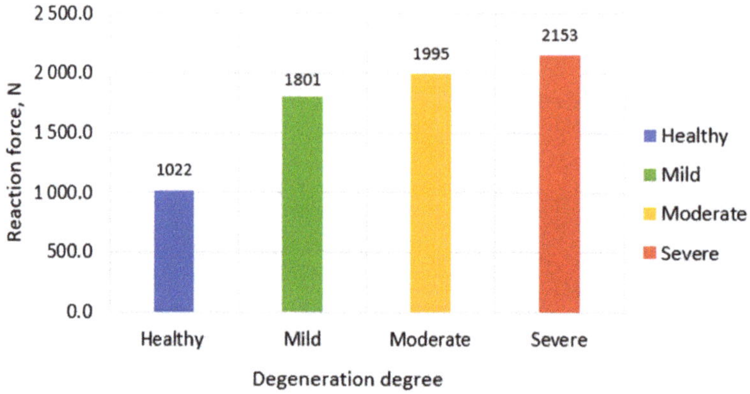

Figure 5. Reaction forces on upper L1 vertebral surface in case of different IVD degeneration degrees.

It should be noted that in this study, we consider the reaction forces as comparative indicators of relative segmental stiffness, rather than as absolute biomechanical measures requiring additional normalization. In the healthy disc model, the total reaction force remains at approximately 1022 N, which aligns with expected biomechanical responses under physiological loading. For mild degeneration, this value increases to around 1801 N,

https://doi.org/10.3390/bioengineering13020234

reflecting the initial rise in disc stiffness and reduction in energy absorption capacity. In the case of moderate degeneration, the reaction force further elevates to approximately 1995 N, while in the severe degeneration scenario, the maximum reaction force reaches nearly 2153 N.

The most significant jump in reaction force occurs between the healthy and mildly degenerated models, increasing almost twofold—from 1022 N to 1801 N. This sharp rise is primarily associated with the stiffening of the annulus fibrosus and the reduction in disc height, which together impair the nucleus pulposus's ability to effectively dissipate compressive loads.

In contrast, the transition from mild to moderate degeneration (1801 N to 1995 N), and further to severe degeneration (2153 N), results in a more gradual increase. This plateauing effect suggests that beyond a certain threshold of structural degradation, additional increases in stiffness do not proportionally raise the reaction force. This phenomenon can be explained by the redistribution of mechanical load: as the disc height—particularly in the L1–L2 region—diminishes, the posterior elements begin to bear a greater share of the compressive load. This compensatory mechanism reduces the rate at which reaction force increases with degeneration, as part of the load is offloaded onto these secondary load-bearing structures.

To account for geometric changes across degeneration stages, reaction forces were normalized by the total vertical height of the L1–L3 segment, measured from the inferior surface of L3 to the superior surface of L1. The normalized values were healthy (97 mm) → 10.5 N/mm; mild degeneration (95 mm) → 19.0 N/mm; moderate degeneration (93 mm) → 21.5 N/mm; severe degeneration (91 mm) → 23.7 N/mm. This normalization reveals that effective segmental stiffness increases by approximately 126% from healthy to severe degeneration, reflecting the combined effects of material stiffening and geometric compaction (Figure 6).

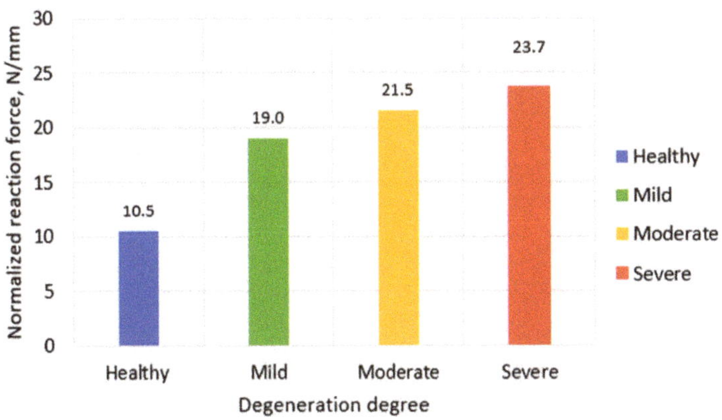

Figure 6. Normalized reaction forces on upper L1 vertebral surface in case of different IVD degeneration degrees.

These findings highlight the biomechanical complexity of degeneration progression and emphasize the need to consider the entire spinal segment, rather than the intervertebral disc alone, in finite element analyses.

3.3. Stress Patterns in the Intervertebral Disc Across Degeneration Stages

Figure 7 presents von Mises stress distributions within the L1–L2 and L2–L3 intervertebral discs under 2.5 mm compression for all degeneration stages.

https://doi.org/10.3390/bioengineering13020234

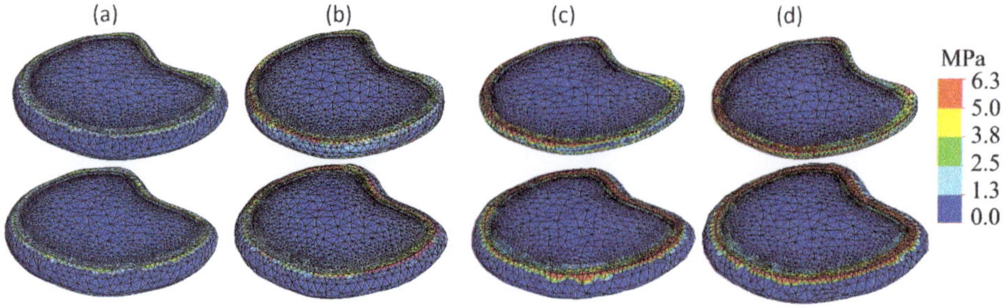

Figure 7. Von Mises stress plots: (**a**) Healthy IVD model; (**b**) Model with mild IVD degeneration; (**c**) Model with moderate IVD degeneration; (**d**) Model with severe IVD degeneration. Upper row presents L1–L2 IVD; bottom row presents L2–L3 IVD.

In the healthy model (Figure 7a), stress remains evenly distributed, with peak values localized at the periphery of the annulus fibrosus, not exceeding 1.5 MPa. The nucleus pulposus exhibits minimal stress, consistent with its fluid-like biomechanical behavior and effective load distribution. It should be noted that these simplifications of flat and uniform endplates may introduce localized artifacts in von Mises stress patterns near the bone–disc interface, and these values should therefore be interpreted with caution. However, since the same simplification was applied consistently across all degeneration stages, the comparative trends in stress redistribution remain reliable.

As degeneration progresses, clear shifts in stress distribution patterns are observed. In the mild degeneration model (Figure 7b), stress concentrations begin to form near the posterior-lateral regions of the annulus, indicating altered load transmission pathways. Moderate degeneration (Figure 7c) further intensifies this trend: peak stress zones extend deeper into the annulus and appear bilaterally. In the severe degeneration model (Figure 7d), maximum stress zones become highly localized at the outer edge of the annulus fibrosus, particularly in the posterior region, with von Mises values exceeding 6 MPa.

This shift indicates that progressive degeneration not only reduces disc height but also leads to mechanical overloading of annular fibers, increasing the risk of structural failure, fissures, and herniation.

The bar chart in Figure 8 quantifies peak von Mises stress values within the annulus fibrosus for all degeneration stages. The stress rises from approximately 2.3 MPa in the healthy disc to 4.2 MPa in mild degeneration, followed by 4.7 MPa in moderate degeneration, and peaking at nearly 6.3 MPa in severe degeneration.

This progressive increase reflects both the compromised energy dissipation ability of the disc and the redistribution of loads from the central nucleus pulposus to peripheral annular structures. The elevated stress values in advanced stages also correlate with increased risk of annular tears and further degeneration of the disc [10].

It is observed that, regardless of degeneration severity, the difference in stress levels between the two discs within the same model remains relatively small. However, the L1–L2 disc consistently exhibits slightly higher maximum stress values compared to L2–L3. This trend is most prominent in the healthy degeneration scenario, where peak stress in L1–L2 reaches 2.3 MPa, while in L2–L3 it is 2 MPa.

The consistently higher stress in L1–L2 is primarily due to the proximity of the applied load, as explained by Saint-Venant's principle. This proximity effect may be amplified by the displacement-controlled setup with a rigid kinematic boundary (uniform endplate motion) and would likely be reduced under force-controlled or follower-load conditions that allow more natural coupled motions and load sharing. Because the compressive displacement is applied to the superior surface of L1, the adjacent L1–L2 disc undergoes

https://doi.org/10.3390/bioengineering13020234

greater localized deformation, while the L2–L3 disc is partially shielded by load dissipation through the upper vertebra. This behavior highlights the influence of boundary conditions in finite element analyses.

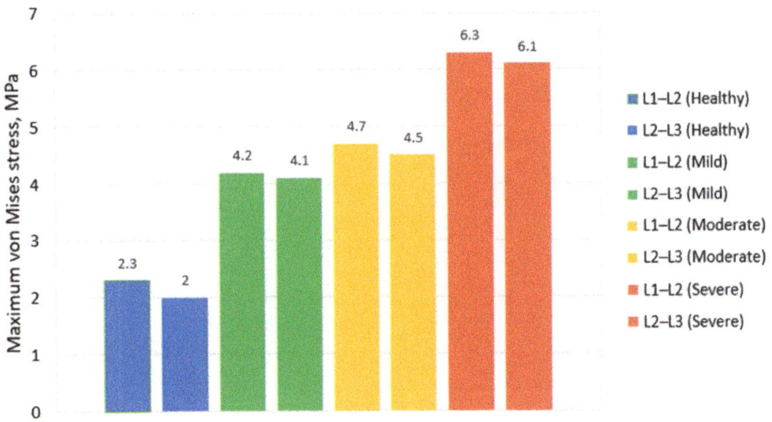

Figure 8. Peak von Mises stress values in the annulus fibrosus under different degeneration scenarios.

Figure 9 illustrates the maximum von Mises stress values in the nucleus pulposus of the L1–L2 and L2–L3 discs for different stages of IVD degeneration. In the healthy models, the peak stresses in the nucleus pulposus of both discs are nearly identical (0.42 MPa for L1–L2 and 0.40 MPa for L2–L3). In mild degeneration at L1–L2, the stress in the degenerated L1–L2 nucleus slightly decreases to 0.35 MPa, consistent with early loss of hydrostatic capacity. Interestingly, the adjacent intact L2–L3 disc experiences a marked increase in nucleus stress (up to 0.78 MPa), likely due to increased load transmission through the mildly stiffened L1–L2 segment under the fixed displacement boundary condition. This effect diminishes in moderate and severe degeneration as the L1–L2 disc loses most of its load-bearing capacity, shifting stresses primarily to the annulus and posterior elements.

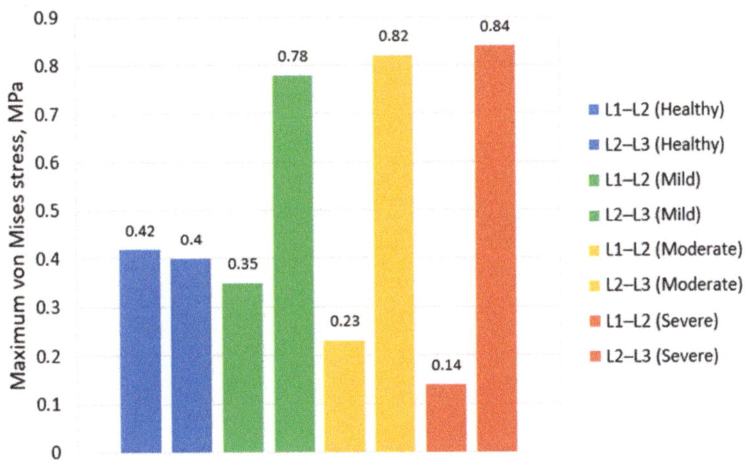

Figure 9. Peak von Mises stress values in the nucleus pulposus under different degeneration scenarios.

However, in the moderate and severe degeneration stages, a sharp decline is observed—stress drops to just 0.23 MPa for L1–L2 (moderate degeneration) and 0.14 MPa for L1–L2 in case of severe degeneration degree. This significant reduction is associated

https://doi.org/10.3390/bioengineering13020234

with the collapse of the nucleus pulposus structure and its replacement by fibrotic or calcified tissue. As the nucleus loses its hydrostatic pressure-retaining properties, it no longer effectively transmits compressive loads, resulting in a stress shift toward the annulus fibrosus and posterior elements. Thus, the stress dynamics in the nucleus pulposus across degeneration stages reflect the progressive loss of its biomechanical function, transitioning from a primary load-bearing component to a largely inactive remnant in the degenerated disc.

3.4. Results Verification and Discussion

Direct quantitative validation of the present FE model is constrained by the absence of reference data that reproduce the same geometry, degeneration pattern, and boundary conditions used in this study. Existing investigations—whether computational or experimental—differ substantially in anatomical detail, modelling assumptions, loading strategies, and degeneration definitions, which prevents a strict one-to-one comparison with the current subject-specific two-level model. In such situations, verification is appropriately based on demonstrating that the predicted stresses fall within biomechanically reasonable ranges and that the overall mechanical trends align with established understanding of lumbar load transfer, rather than on exact numerical matching with any single dataset.

While direct experimental deformation–load validation was not possible, the model was validated through multi-metric consistency with published FE and experimental trends.

The reaction forces obtained for the healthy disc model (~1022 N at 2.5 mm displacement) closely match experimental lumbar stiffness measurements reported by Park et al. [29]. Similarly, stress distributions observed in the cortical shell and annulus fibrosus are consistent with the finite element studies by [27,28]. Cortical shell stresses (14–55 MPa) are consistent with previously published ranges (8–61 MPa [18] and 7.4–64 MPa [40]), indicating that the model does not produce unrealistically high stress concentrations under axial compression.

For healthy discs, the model's IDP estimates (approximately 0.5–1.0 MPa under displacement-controlled compression) align closely with in vivo measurements reported by [41], where standing postures yielded 0.5 MPa at L4–L5. Similarly, [42] documented baseline IDP values of 0.3–0.5 MPa in healthy individuals during standing, supporting the model's representation of uniform load distribution in non-degenerated states. In degenerated scenarios, the observed 70% reduction in nucleus pulposus stress across mild to severe stages corresponds to IDP declines of 40–70% in cadaveric studies by [43], where Pfirrmann-graded discs showed IDP drops from 0.4–1.3 MPa (healthy) to 0.1–0.8 MPa (degenerated), alongside height losses of 0.2–1.0 mm. This is further corroborated by in vivo data from [42], indicating 40–60% IDP reductions (to 0.1–0.3 MPa) in degenerative disc disease patients with MRI grading. The model's 160% increase in annulus fibrosus stress mirrors elevated stresses (3–10 MPa) and failure loads in degenerated annuli [44], with stiffness increases of 10–20%. Additionally, trends in range of motion (ROM)—initial increases followed by decreases—align with [45], reporting 20–30% ROM elevations at moderate degeneration (grade 4) and 15–25% reductions at severe stages (grade 5). Posterior load shifts in advanced degeneration are consistent with [46], where telemetry data showed IDP in sitting (0.2–0.6 MPa) and standing (0.3–0.4 MPa) for degenerative disc diseases, indicating no exacerbated risk from sitting compared to standing. A further decline in nucleus pressure has been reported during early degeneration, corresponding to 25–30% reductions in IDP from 0.25–0.45 MPa to 0.19–0.34 MPa [47].

Overall, these comparisons affirm the model's qualitative and quantitative trends in load redistribution and place the predicted responses within biomechanically plausible ranges reported in vivo and in cadaveric studies. Absolute magnitudes should still be in-

https://doi.org/10.3390/bioengineering13020234

terpreted cautiously, as simplifications such as isotropy and idealized boundary conditions can affect peak values, as discussed in the limitations.

The variation of peak stress values in the nucleus pulposus requires separate consideration and detailed explanation. According to different sources, disc degeneration may lead either to an increase [27] or a decrease [10] in intradiscal stresses, while some studies report opposite trends. In our work, both patterns were observed, and each requires careful interpretation. In the healthy model, the peak stresses in both discs are nearly identical (0.42 and 0.40 MPa). Under mild degeneration, however, the stress in L2–L3 nearly doubles, whereas the stress in L1–L2 slightly decreases, which aligns with the trends described in [10]. As degeneration progresses, stresses in the deteriorated disc continue to decrease, while the intact L2–L3 disc exhibits only a minor increase.

The non-monotonic changes in NP stress across degeneration stages can be explained by the shift in its mechanical role. In the healthy state, the NP behaves as a hydrated, hydrostatic structure that carries a large portion of the axial load, which leads to relatively high internal stress. During mild degeneration, the NP still retains its ability to sustain pressure, while the reduction in disc height and the early stiffening of the annulus can increase nucleus compression, resulting in a modest stress elevation at the adjacent level.

With progression to moderate and severe degeneration, the NP loses volume and hydration and becomes mechanically less distinct from the surrounding annulus. As a result, it can no longer sustain hydrostatic pressure and contributes less to load bearing, so NP stress decreases despite the overall increase in segmental stiffness. The load path shifts toward the stiffer annulus fibrosus and endplates, leading to a more deviatoric NP response. These findings underscore the value of parametric studies, since stress responses depend on the coupled effects of tissue properties and geometry, which can generate markedly different mechanical patterns.

3.5. Sensitivity Analysis

To assess the robustness of the model and the influence of input parameter variations on the observed trends, a limited sensitivity analysis was performed. Key parameters were varied around their baseline values, and the impact on peak von Mises stresses in the annulus fibrosus (AF) and nucleus pulposus (NP), as well as on the total reaction force at the superior surface of L1, was evaluated. The results are presented in Table 5.

The following parameters were tested in the healthy and severe degeneration configurations:

- Disc height: ±10% from baseline (healthy: 10 mm → 9 mm/11 mm; severe: 4 mm → 3.6 mm/4.4 mm).
- Nucleus pulposus volume (cross-sectional area): ±15% from baseline.
- Annulus fibrosus stiffness (Mooney–Rivlin constants C_1 and C_2): ±20% from baseline values.

All other parameters (boundary conditions, mesh setup, material properties of bone/endplates) remained unchanged. Simulations were performed using the same displacement-controlled loading (2.5 mm axial compression) as in the main study.

The conducted sensitivity analysis helped identify the following key observations:

- The model is most sensitive to disc height variations: a 10% reduction increases peak AF stress by 13–18% and total reaction force by 15–23%, while a 10% increase reduces these values accordingly. This confirms disc height as one of the dominant geometric factors influencing segmental stiffness and annulus loading.
- Nucleus pulposus volume has a moderate effect, primarily on NP stress (changes up to ±13–15%), with minimal impact on AF stress (±3–4%) and reaction force (±8–11%).

https://doi.org/10.3390/bioengineering13020234

- Annulus fibrosus stiffness (C_1 and C_2) exerts the strongest influence on peak AF stress (up to ±26%) and reaction force (±13–19%), but has negligible effect on NP stress (±2–6%).
- Importantly, all qualitative trends remain consistent across the tested ranges: progressive degeneration still leads to increased AF stress, decreased NP stress, and a shift of load toward posterior elements. No reversal of trends was observed.

Table 5. Results of the sensitivity analysis (percentage change relative to baseline values).

Parameter Varied	Variation	Configuration	% Change in Peak AF Stress (L1–L2)	% Change in Peak NP Stress (L1–L2)	% Change in Total Reaction Force
Disc height	+10%	Healthy	−14%	+9%	−16%
Disc height	−10%		+18%	−11%	+23%
Disc height	+10%	Severe	−13%	+6%	−12%
Disc height	−10%		+16%	−8%	+15%
Nucleus pulposus volume	+15%	Healthy	−3%	+13%	−8%
Nucleus pulposus volume	−15%		+4%	−14%	+11%
AF stiffness (C_1 and C_2)	+20%	Healthy	+21%	−6%	+15%
AF stiffness (C_1 and C_2)	−20%		−17%	+5%	−13%
AF stiffness (C_1 and C_2)	+20%	Severe	+26%	−4%	+19%
AF stiffness (C_1 and C_2)	−20%		−20%	+2%	−16%

Stress evaluation in the present study is primarily based on peak von Mises values. This choice was motivated by the highly localized nature of stress concentrations observed in degenerated discs, particularly within the annulus fibrosus following nucleus pulposus volumetric collapse. Such local stress maxima are mechanically relevant, as they are more closely associated with the initiation of annular tears and structural failure than spatially averaged measures.

Despite the deliberate modeling simplifications adopted in this study, several key findings can be considered robust. In particular, the qualitative trends of load redistribution with progressive disc degeneration—namely the increase in annulus fibrosus stress, the marked reduction of nucleus pulposus stress, and the posterior shift of load sharing in advanced stages—remain consistent across all tested parameter variations and degeneration grades. Sensitivity analysis confirmed that these trends persist under ±10–20% variations in disc height, nucleus volume, and annulus stiffness, indicating that the observed mechanisms are primarily driven by geometric collapse and relative tissue stiffening rather than by secondary stabilizing structures.

At the same time, results related to absolute segment stiffness, reaction force magnitude, and the quantitative contribution of posterior elements are likely sensitive to the omission of ligaments and follower load. Ligamentous tension and muscle-induced follower loads are known to influence spinal stability and load-sharing, particularly under physiological posture and motion. Their exclusion may therefore affect the absolute values of reaction forces and posterior element stresses. However, because these simplifications were applied consistently across all degeneration stages, the comparative interpretation of degeneration-driven mechanical trends remains valid.

4. Conclusions

This study presents a robust parametric finite element framework for evaluating load redistribution in progressive lumbar disc degeneration, demonstrating significant shifts in biomechanical behavior across healthy, mild, moderate, and severe stages. Key findings include a 175% increase in annulus fibrosus von Mises stresses, elevating risks of tears and herniation; a ~70% decrease in nucleus pulposus stresses, indicating loss of hydrostatic function; and a progressive load transfer to posterior elements, suggesting mechanically unfavorable load redistribution patterns that may be associated with conditions such as facet joint degeneration or vertebral bone deconditioning. The non-monotonic nucleus stress patterns—initial rise in adjacent intact discs due to early stiffening, followed by sharp declines—highlight the complex interactions of geometric and material changes. The observed reduction in nucleus pulposus stress in advanced degeneration should not be interpreted as stress relief. Instead, it reflects a loss of the nucleus load-bearing function due to volumetric collapse, whereby the nucleus becomes mechanically embedded within the annulus fibrosus and no longer participates effectively in hydrostatic load sharing.

Verified against in vivo intradiscal pressure reductions (40–70%) and cadaveric data, and confirmed robust through sensitivity analysis (±10–20% parameter variations yielding <26% quantitative changes), these trends provide reliable insights into degeneration mechanics. Unlike prior single-factor models, presented approach offers a comprehensive, reproducible pipeline that elucidates coupled mechanisms and adjacent-segment effects, empowering clinicians to better assess progression risks and tailor interventions, while guiding biomedical engineers in designing more effective implants and therapies.

While simplifications (e.g., isotropic materials, no poroelasticity) limit absolute physiological accuracy, their consistent application ensures valid comparative analysis. These interpretations should be understood within the scope of the present simplified model, which does not include ligamentous structures, active muscle forces, or population-level anatomical variability, and therefore aims to describe relative mechanical trends rather than direct clinical outcomes. Future extensions could incorporate anisotropic fiber models, follower loads, or multi-subject geometries to further enhance predictive power and clinical applicability.

Author Contributions: Conceptualization, O.A., S.R.F., A.K. and V.A.; methodology, O.A., S.R.F., A.K. and V.A.; software, O.A.; formal analysis, O.A., S.R.F., A.K. and V.A.; investigation, O.A., S.R.F., A.K. and V.A.; resources, O.A., A.K. and V.A.; data curation, V.A.; writing—original draft preparation, O.A., S.R.F., A.K. and V.A.; writing—review and editing, V.A.; visualization, O.A., S.R.F., A.K. and V.A.; supervision, V.A.; project administration, O.A., A.K. and V.A.; funding acquisition, O.A., A.K. and V.A. All authors have read and agreed to the published version of the manuscript.

Funding: This project has received funding from the Research Council of Lithuania (LMTLT), agreement No. S-PD-24-141, and from Fundação para a Ciência e a Tecnologia, I.P. (FCT, https://ror.org/00snfqn58) under Grant UID/00645/2025. (https://doi.org/10.54499/UID/00645/2025) and/or UID/PRR/00645/2025. (https://doi.org/10.54499/UID/PRR/00645/2025). For the purpose of Open Access, the author has applied a CC BY public copyright licence to any Author Accepted Manuscript (AAM) version arising from this submission.

Institutional Review Board Statement: Not applicable.

Informed Consent Statement: Not applicable.

Data Availability Statement: Data presented in this study are available on request from the corresponding author.

Conflicts of Interest: The authors declare no conflicts of interest.

References

1. Oktay, A.B.; Albayrak, N.B.; Akgul, Y.S. Computer Aided Diagnosis of Degenerative Intervertebral Disc Diseases from Lumbar MR Images. *Comput. Med. Imaging Graph.* **2014**, *38*, 613–619. [CrossRef]
2. Galbusera, F.; Van Rijsbergen, M.; Ito, K.; Huyghe, J.M.; Brayda-Bruno, M.; Wilke, H.J. Ageing and Degenerative Changes of the Intervertebral Disc and Their Impact on Spinal Flexibility Ageing and Degenerative Changes of the Intervertebral Disc and Their Impact on Spinal Flexibility. *Eur. Spine J.* **2014**, *23*, 324–332. [CrossRef]

https://doi.org/10.3390/bioengineering13020234

3. Liu, Y.; Dou, Y.; Sun, X.; Yang, Q. Mechanisms and Therapeutic Strategies for Senescence-Associated Secretory Phenotype in the Intervertebral Disc Degeneration Microenvironment. *J. Orthop. Transl.* **2024**, *45*, 56–65. [CrossRef] [PubMed]
4. Zhang, X.Y.; Han, Y. Comparison of the Biomechanical Effects of Lumbar Disc Degeneration on Normal Patients and Osteoporotic Patients: A Finite Element Analysis. *Med. Eng. Phys.* **2023**, *112*, 103952. [CrossRef] [PubMed]
5. Griffith, J.F.; Wang, Y.-X.J.; Antonio, G.E.; Choi, K.C.; Yu, A.; Ahuja, A.T.; Leung, P.C. Modified Pfirrmann Grading System for Lumbar Intervertebral Disc Degeneration. *Spine* **2007**, *32*, E708–E712. [CrossRef] [PubMed]
6. Salamat, S.; Hutchings, J.; Kwong, C.; Magnussen, J.; Hancock, M.J. The Relationship between Quantitative Measures of Disc Height and Disc Signal Intensity with Pfirrmann Score of Disc Degeneration. *Springerplus* **2016**, *5*, 829. [CrossRef]
7. Teichtahl, A.J.; Urquhart, D.M.; Wang, Y.; Wluka, A.E.; Heritier, S.; Cicuttini, F.M. A Dose-Response Relationship between Severity of Disc Degeneration and Intervertebral Disc Height in the Lumbosacral Spine. *Arthritis Res. Ther.* **2015**, *17*, 297. [CrossRef]
8. Jo, M.; Chae, S.W. Stress Analysis of Intervertebral Disc during Occupational Activities. *Comput. Methods Programs Biomed.* **2021**, *208*, 106298. [CrossRef]
9. Ruiz, C.; Noailly, J.; Lacroix, D. Material Property Discontinuities in Intervertebral Disc Porohyperelastic Finite Element Models Generate Numerical Instabilities Due to Volumetric Strain Variations. *J. Mech. Behav. Biomed. Mater.* **2013**, *26*, 1–10. [CrossRef]
10. Zhu, Q.; Gao, X.; Chen, S.; Gu, W.; Brown, M.D. Effect of Intervertebral Disc Degeneration on Mechanical and Electric Signals at the Interface between Disc and Vertebra. *J. Biomech.* **2020**, *104*, 109756. [CrossRef]
11. Holzapfel, G.A.; Ogden, R.W. Biomechanical Stresses in a Residually Stressed Idealized Intervertebral Disc. *Int. J. Non-Linear Mech.* **2024**, *161*, 104687. [CrossRef]
12. Ghezelbash, F.; Eskandari, A.H.; Shirazi-Adl, A.; Kazempour, M.; Tavakoli, J.; Baghani, M.; Costi, J.J. Modeling of Human Intervertebral Disc Annulus Fibrosus with Complex Multi-Fiber Networks. *Acta Biomater.* **2021**, *123*, 208–221. [CrossRef] [PubMed]
13. Kandil, K.; Zaïri, F.; Messager, T.; Zaïri, F. A Microstructure-Based Model for a Full Lamellar-Interlamellar Displacement and Shear Strain Mapping inside Human Intervertebral Disc Core. *Comput. Biol. Med.* **2021**, *135*, 104629. [CrossRef]
14. Xi, Z.; Xie, Y.; Sun, S.; Wang, N.; Chen, S.; Wang, G.; Li, J. IVD Fibrosis and Disc Collapse Comprehensively Aggravate Vertebral Body Disuse Osteoporosis and Zygapophyseal Joint Osteoarthritis by Posteriorly Shifting the Load Transmission Pattern. *Comput. Biol. Med.* **2024**, *170*, 108019. [CrossRef] [PubMed]
15. 3D Slicer Image Computing Platform. Available online: https://www.slicer.org/ (accessed on 23 March 2023).
16. MESHLAB. Available online: https://www.meshlab.net/ (accessed on 23 March 2023).
17. SOLIDWORKS Systemes. Available online: https://help.solidworks.com/ (accessed on 23 March 2023).
18. Kim, Y.H.; Wu, M.; Kim, K. Stress Analysis of Osteoporotic Lumbar Vertebra Using Finite Element Model with Microscaled Beam-Shell Trabecular-Cortical Structure. *J. Appl. Math.* **2013**, *2013*, 285165. [CrossRef]
19. Cai, X.; Sun, M.; Huang, Y.; Liu, Z.; Liu, C.; Du, C.; Yang, Q. Biomechanical Effect of L4–L5 Intervertebral Disc Degeneration on the Lower Lumbar Spine: A Finite Element Study. *Orthop. Surg.* **2020**, *12*, 917–930. [CrossRef]
20. McDonald, K.; Little, J.; Pearcy, M.; Adam, C. Development of a Multi-Scale Finite Element Model of the Osteoporotic Lumbar Vertebral Body for the Investigation of Apparent Level Vertebra Mechanics and Micro-Level Trabecular Mechanics. *Med. Eng. Phys.* **2010**, *32*, 653–661. [CrossRef]
21. Jaramillo, H.E. Evaluation of the Use of the Yeoh and Mooney-Rivlin Functions as Strain Energy Density Functions for the Ground Substance Material of the Annulus Fibrosus. *Math. Probl. Eng.* **2018**, *2018*, 1570142. [CrossRef]
22. Gruber, G.; Nicolini, L.F.; Ribeiro, M.; Lerchl, T.; Wilke, H.; Jaramillo, H.E.; Senner, V.; Kirschke, J.S.; Nispel, K. Comparative FEM Study on Intervertebral Disc Modeling: Holzapfel-Gasser-Ogden vs. Structural Rebars. *Front. Bioeng. Biotechnol.* **2024**, *12*, 1391957. [CrossRef]
23. Xie, F.; Zhou, H.; Zhao, W.; Huang, L. A Comparative Study on the Mechanical Behavior of Intervertebral Disc Using Hyperelastic Finite Element Model. *Technol. Health Care* **2017**, *25*, 177–187. [CrossRef]
24. Mimura, M.; Panjabi, M.; Oxland, T.; Crisco, J.; Yamamoto, I.; Vasavada, A. Disc Degeneration Affects the Multidirectional Flexibility of the Lumbar Spine. *Spine* **1994**, *19*, 1371–1380. [CrossRef] [PubMed]
25. Zirbel, S.; Stolworthy, D.; Howell, L.; Bowden, A. Intervertebral Disc Degeneration Alters Lumbar Spine Segmental Stiffness in All Modes of Loading under a Compressive Follower Load. *Spine J.* **2013**, *13*, 1134–1147. [CrossRef] [PubMed]
26. Ruberté, L.M.; Natarajan, R.N.; Andersson, G.B. Influence of Single-Level Lumbar Degenerative Disc Disease on the Behavior of the Adjacent Segments—A Finite Element Model Study. *J. Biomech.* **2009**, *42*, 341–348. [CrossRef] [PubMed]
27. Galbusera, F.; Schmidt, H.; Neidlinger-Wilke, C.; Gottschalk, A.; Wilke, H.J. The Mechanical Response of the Lumbar Spine to Different Combinations of Disc Degenerative Changes Investigated Using Randomized Poroelastic Finite Element Models. *Eur. Spine J.* **2011**, *20*, 563–571. [CrossRef]
28. Schmidt, H.; Kettler, A.; Rohlmann, A.; Claes, L.; Wilke, H.J. The Risk of Disc Prolapses with Complex Loading in Different Degrees of Disc Degeneration—A Finite Element Analysis. *Clin. Biomech.* **2007**, *22*, 988–998. [CrossRef]

https://doi.org/10.3390/bioengineering13020234

29. Poojara, V.; Thaker, M.; Trivedi, R.; Modi, B.; Patel, R. Three Dimensional Analysis of Mechanical Behavior of Lumbar Vertebra under Impact Load and Its Biomechanical Importance. *Mater. Today Proc.* **2020**, *38*, 402–406. [CrossRef]
30. Kurutz, M. Finite Element Modelling of Human Lumbar Spine. In *Finite Element Analysis*; IntechOpen: London, UK, 2010; ISBN 978-953-307-123-7.
31. Ma, J.; Ding, Y.; Rozi, R.; Han, J.; Jiang, Q. From Single Ligament to Multi-Ligament Injury: A Finite Element Study on the Contribution of the Posterior Ligamentous Complex to Segmental Stability and Intervertebral Disc Stress Distribution. *BMC Musculoskelet. Disord.* **2025**, *26*, 820. [CrossRef]
32. Zhang, Q.; Jin, Y.; He, L.; Zhang, K.; Chen, L.; Chen, W.; Feng, H. Biomechanical Effect of Cage Size in Single—Level Anterior Cervical Discectomy and Fusion: A Finite Element Analysis. *BMC Musculoskelet. Disord.* **2025**, *26*, 594. [CrossRef]
33. Wang, B.; Liu, Y.; Wang, Z.; Niu, C.; Tang, J.; Sun, L. Osteoporosis in Adjacent Cervical Segments Exacerbates Disc Herniation. *Sci. Rep.* **2025**, *15*, 22901. [CrossRef]
34. Wan, C.; Shen, X.; Wu, X.; Yu, C.; Shao, Y.; Zhang, R.; Shang, J.; Li, J.; Zhang, Y.; Li, Y. Assessing the Biomechanics of Scheuermann' s Kyphosis Affected Thoracolumbar Spine in Forward Flexion at the Tissue-Level Using a Finite Element Model. *Sci. Rep.* **2025**, *15*, 27408. [CrossRef]
35. Flor, O.; Rubio, J.E.; Tong, J.; Sundaramurthy, A.; Pant, A.; Nagaraja, S.; Owen, M.K.; Samaan, M.A.; Noehren, B.; Reifman, J. Effect of Walking with an Active Ankle Exoskeleton on the Biomechanical Responses of the Lumbar Spine. *Front. Bioeng. Biotechnol.* **2025**, *13*, 1654585. [CrossRef] [PubMed]
36. Mu, X.; Wei, X.; Nong, J.; Ye, H.; Li, Z.; Wei, M.; Wei, J. Clinical Evaluation and Fi Nite Element Analysis of Bone Cement-Augmented Anterolateral Screw Fi Xation versus Percutaneous Bilateral Pedicle Screw Fi Xation Co-Applied with Oblique Lumbar Interbody Fusion for Single-Level Lumbar Degenerative Diseases with Osteoporosis. *Front. Bioeng. Biotechnol.* **2025**, *13*, 1571849. [CrossRef]
37. Park, W.M.; Kim, Y.H.; Lee, S. Effect of Intervertebral Disc Degeneration on Biomechanical Behaviors of a Lumbar Motion Segment under Physiological Loading Conditions. *J. Mech. Sci. Technol.* **2013**, *27*, 483–489. [CrossRef]
38. Inoue, N.; Orías, A.A.E.; Segami, K. Biomechanics of the Lumbar Facet Joint. *Spine Surg. Relat. Res.* **2020**, *4*, 1–7. [CrossRef]
39. Park, J.S.; Goh, T.S.; Lee, J.S.; Lee, C. Analyzing Isolated Degeneration of Lumbar Facet Joints: Implications for Degenerative Instability and Lumbar Biomechanics Using Fi Nite Element Analysis. *Front. Bioeng. Biotechnol.* **2024**, *12*, 1294658. [CrossRef]
40. Chabarova, O.; Selivonec, J. Virtual Modelling the Impact of Torsional Loading on Osteoporotic Vertebrae Buckling. *Acta Bioeng. Biomech.* **2024**, *26*, 13–22. [CrossRef]
41. Wilke, H.J.; Neef, P.; Caimi, M.; Hoogland, T.; Claes, L.E. New in Vivo Measurements of Pressures in the Intervertebral Disc in Daily Life. *Spine* **1999**, *24*, 755–762. [CrossRef]
42. Sato, K.; Kikuchi, S.; Yonezawa, T. In Vivo Intradiscal Pressure Measurement in Healthy Individuals and in Patients with Ongoing Back Problems. *Spine* **1999**, *24*, 2468–2474. [CrossRef]
43. Schmidt, H.; Heuer, F.; Drumm, J.; Klezl, Z.; Claes, L.; Wilke, H.-J. Application of a Calibration Method Provides More Realistic Results for a Finite Element Model of a Lumbar Spinal Segment. *Clin. Biomech.* **2007**, *22*, 377–384. [CrossRef]
44. Newell, N.; Little, J.P.; Christou, A.; Adams, M.A.; Adam, C.J.; Masouros, S.D. Biomechanics of the Human Intervertebral Disc: A Review of Testing Techniques and Results. *J. Mech. Behav. Biomed. Mater.* **2017**, *69*, 420–434. [CrossRef]
45. Fujiwara, A.; Tamai, K.; An, H.S.; Kurihashi, T.; Lim, T.H.; Yoshida, H.; Saotome, K. The Relationship between Disc Degeneration, Facet Joint Osteoarthritis, and Stability of the Degenerative Lumbar Spine. *J. Spinal Disord.* **2000**, *13*, 444–450. [CrossRef]
46. Dreischarf, M.; Shirazi-Adl, A.; Arjmand, N.; Rohlmann, A.; Schmidt, H. Estimation of Loads on Human Lumbar Spine: A Review of in Vivo and Computational Model Studies. *J. Biomech.* **2016**, *49*, 833–845. [CrossRef]
47. Heuer, F.; Schmidt, H.; Klezl, Z.; Claes, L.; Wilke, H.-J. Stepwise Reduction of Functional Spinal Structures Increase Range of Motion and Change Lordosis Angle. *J. Biomech.* **2007**, *40*, 271–280. [CrossRef]

https://doi.org/10.3390/bioengineering13020234

MDPI AG
Grosspeteranlage 5
4052 Basel
Switzerland
Tel.: +41 61 683 77 34

Bioengineering Editorial Office
E-mail: bioengineering@mdpi.com
www.mdpi.com/journal/bioengineering

www.ingramcontent.com/pod-product-compliance
Lightning Source LLC
Chambersburg PA
CBHW061759210926
51850CB00006B/593
* 9 7 8 3 7 2 5 8 7 7 2 9 4 *